Motion Optimization and Control of Single and Multiple Autonomous Aerial, Land, and Marine Robots

Motion Optimization and Control of Single and Multiple Autonomous Aerial, Land, and Marine Robots

Editors

Reza Ghabcheloo

Antonio M. Pascoal

MDPI • Basel • Beijing • Wuhan • Barcelona • Belgrade • Manchester • Tokyo • Cluj • Tianjin

Editors
Reza Ghabcheloo
Tampere University
Finland

Antonio M. Pascoal
Instituto Superior Tecnico (IST)
Portugal

Editorial Office
MDPI
St. Alban-Anlage 66
4052 Basel, Switzerland

This is a reprint of articles from the Special Issue published online in the open access journal *Sensors* (ISSN 1424-8220) (available at: https://www.mdpi.com/journal/sensors/special_issues/Robot-Motion-Control).

For citation purposes, cite each article independently as indicated on the article page online and as indicated below:

LastName, A.A.; LastName, B.B.; LastName, C.C. Article Title. *Journal Name* **Year**, *Volume Number*, Page Range.

ISBN 978-3-0365-6328-2 (Hbk)
ISBN 978-3-0365-6329-9 (PDF)

Contents

About the Editors

Reza Ghabcheloo

Reza Ghabcheloo received his BSc degree in Electronics from the Iran University of Science and Technology (IUST) and his MSc degree in Control Engineering from the K.N Toosi University of Technology, Tehran, Iran in 1997 and 1999, respectively. He then received a Ph.D. in Electrical Engineering from the Technical University of Lisbon, Portugal in 2007 with his thesis titled "Coordinated Path Following of Multiple Autonomous Vehicles". Important visits during his doctoral studies included to the Dept. of Engineering Cybernetics of the Norwegian University of Science and Technology (NTNU), Trondheim, Norway and the Naval Postgraduate School (NPS), California, USA.

From 2007–2008, he was a postdoc researcher with the Instituto Superior Técnico, Institute for Systems and Robotics, Lisbon, Portugal, where he conducted research on networked control systems. He then moved to Finland as a senior researcher from 2008–2014 with the Department of Intelligent Hydraulics and Automation of Tampere University of Technology, Finland working within the Academy of Finland Center of Excellence on Generic Intelligent Machines. In 2015, he started an Associate Professor position with Tampere University and, since then, has been teaching several courses including the Fundamentals of Mobile Robots, Advanced Robotics, Electrohydraulic Servo Systems, and Autonomous Mobile Machines. Since 2018 he has led the curriculum planning team for Robotics in Tampere University.

He is the founder of the autonomous mobile machines research group at the Faculty of Engineering and Natural Sciences of Tampere University. His research interests include the intersection of robotics and working machines, optimization and machine learning, motion control, and perception. He has published over 80 journal and conference articles in these areas. His motto is to conduct practically relevant research rooted in solid theory. He has a strong position of trust among the leading working machine industry, with eight of his doctoral students co-supervised by industry partners in the field of heavy duty mobile machinery.

Antonio M. Pascoal

MSc degree in Electrical Engineering and PhD degree in Control Science from the University of Minnesota, Minneapolis, Minnesota, USA in 1983 and 1987, respectively. He is currently a senior researcher and Associate Professor of Control and Robotics at the Institute for Systems and Robotics (ISR) of IST. He is the coordinator of the Thematic Line Ocean Exploration and Exploitation of the Laboratory for Robotics and Engineering Systems (LARSyS). He was an Adjunct Scientist with the National Institute of Oceanography, Goa, India from 2012–2014 and a Visiting Faculty with the Department of Ocean vii Engineering, IIT Madras, under the Indian Sparc Programme, from 2018–2022. He has coordinated and participated in a large number of international projects that have led to the design, development, and field-testing of single and multiple autonomous marine and air vehicles and systems in cooperation with partners in India, USA, Korea, Brazil, Peru, and Europe.

He has been a member of the International Program Committees of numerous conferences on dynamical systems and control and marine and aerial robotics. He has supervised or co-supervised 11 postdoctoral and 15 PhD students. He has published a total of 113 books, book chapters, and peer-reviewed journal papers, and more than 260 conference papers (12,423 citations, h-index 59, i10-index 241/ Google Scholar). He received the IEEE OES AUV Distinguished Lifetime Technical Achievement Award in 2020.

His expertise includes Dynamical Systems Theory, Marine Robotics, Navigation, Guidance, and Control of Autonomous Vehicles, and Networked Control and Estimation with applications to air and underwater robots.

sensors

MDPI

Editorial

Motion Optimization and Control of Single and Multiple Autonomous Aerial, Land, and Marine Robots

Reza Ghabcheloo [1,*] and António Pascoal [2]

1 Faculty of Engineering and Natural Sciences, Tampere University, P.O. Box 1001, 33014 Tampere, Finland
2 Institute for Systems and Robotics (ISR), Instituto Superior Tecnico (IST), Torre Norte, Piso 8, Av. Rovisco Pais, 1, 1049-001 Lisbon, Portugal
* Correspondence: reza.ghabcheloo@tuni.fi

You always admire what you really don't understand (Blaise Pascal)

Fast-paced developments in the fields of aerial, land, and marine robotics are steadily paving the way for a wide spectrum of scientific and commercial applications of autonomous vehicles with far-reaching societal implications. Autonomous robots equipped with advanced sensors and manipulators afford humans the capability to operate seamlessly in remote and hazardous environments, as if they were extensions of our eyes and hands. Robots have become the tools par excellence for scientists and commercial operators to explore and monitor the state of heterogenous environments on Earth, inspect offshore wave and energy infrastructures, monitor the growth of crops, and transport goods, among a myriad of other activities. Groups of robots acting in cooperation have started to impact the development of multiple system platforms for adaptive environmental sampling, search and rescue operations in hard-to-access regions, and even coordinated image-taking in the movie and sports industries. The types of robots used are highly heterogeneous and cater to specific user-defined requirements for operations in the air, on land, and at sea. Notwithstanding this diversity, they have in common a number of attributes that are key to their capability to explore or act upon the environment with great agility while exhibiting high levels of performance, resilience, adaptability, and safety.

It is against this backdrop of ideas that this reprint addresses fundamental problems that are at the root of the development of a new breed of heterogenous robots that can act in isolation or cooperatively towards the execution of a wide spectrum of mission scenarios. Representative examples include the study of single and cooperative motion-planning methods with a view to meeting temporal and energy constraints in the presence of robot dynamic constraints, while taking explicitly into account the topology of the underlying communication networks and inter-vehicle and vehicle–obstacle avoidance requisites; the creation of safe and emergent behaviors in a distributed manner, at both the motion planning and control levels; the incorporation of event-driven communication strategies to try and reduce the amount of information exchanged among the different agents; the study of new methods to solve constrained optimal control problems efficiently in a receding horizon fashion; the development of effective techniques for adaptive and robust control in the presence of plant model uncertainty of partially known models, especially for safety critical systems; and the study of how advanced perception can be brought to bear on the reformulation of the above problems in a sensor-based context, yielding challenging questions in the area of visual and acoustic-based servoing, object tracking, and obstacle detection and avoidance. The reprint includes a number of chapters that focus on theoretical and practical issues pertaining to motion optimization and control, with special emphasis on, but not limited to, single and multiple autonomous aerial, land, and marine robots.

Citation: Ghabcheloo, R.; Pascoal, A. Motion Optimization and Control of Single and Multiple Autonomous Aerial, Land, and Marine Robots. *Sensors* **2023**, *23*, 87. https://doi.org/10.3390/s23010087

Received: 14 December 2022
Accepted: 19 December 2022
Published: 22 December 2022

Kielas-Jensen et al. [1] present a method for the generation of trajectories for autonomous vehicles that exploits the use of Bernstein polynomial approximations to transcribe infinite-dimensional optimization problems into nonlinear programming problems. These, in turn, can be solved efficiently using off-the-shelf optimization solvers. The main motivation for this approach stems from the fact that Bernstein polynomials possess favorable geometric properties and yield computationally efficient algorithms that enable a trajectory planner to efficiently evaluate and enforce constraints along the vehicles' trajectories, including maximum speed and angular rates as well as the minimum distance between trajectories and between the vehicles and obstacles. To support the application of the method described, an open-source toolbox called BeBOT (Bernstein/Bézier Optimal Trajectories) is introduced that implements key operations and algorithms involving Bernstein polynomials. The toolbox can be used to efficiently generate feasible and collision-free trajectories for single and multiple vehicles.

Sabetghadam et al. [2] introduce a distributed algorithm to generate collision-free trajectories for a group of quadrotors flying through a common workspace. In the setup adopted, each vehicle replans its trajectory, in a receding horizon manner, by solving a small-scale optimization problem that only involves its own individual variables. A Voronoi partitioning of space is adopted to derive local constraints that guarantee collision avoidance with all neighbors for a certain time horizon. The obtained set of collision avoidance constraints explicitly take into account the vehicle's orientation to avoid infeasibility issues caused by ignoring the quadrotor's rotational motion. Moreover, the resulting constraints can be expressed as Bézier curves and thus can be evaluated efficiently, without discretization, to ensure that collision avoidance requirements are satisfied at any time instant, even for an extended planning horizon. The proposed approach is validated through extensive simulations with up to 100 drones. The results show that the proposed method has a higher success rate at finding collision-free trajectories for large groups of drones compared to other Voronoi diagram-based methods.

Andreasson et al. [3] describe a local planning approach for pseudo-omnidirectional vehicles, that is, vehicles that can drive sideways and rotate in place. The local planner, named MSD, is rooted in optimal control theory and relies on the formulation of a nonlinear optimization problem formulation that exploits the omni-motion capabilities of the vehicle to drive the vehicle to the goal in a smooth and efficient manner while avoiding obstacles and singularities. MSDU is designed for a real platform for mobile manipulation, where one key function is the capability to drive in narrow and confined areas. Real-world evaluations show that MSDU planned paths are smoother and more accurate than those obtained with a comparable local path planner, Timed Elastic Band (TEB), with a mean (translational, angular) error for MSDU of (0.0028 m, 0.0010 rad) compared to (0.0033 m, 0.0038 rad) for TEB. MSDU also generated paths that were consistently shorter than TEB, with a mean (translational, angular) distance traveled of (0.6026 m, 1.6130 rad) for MSDU compared to (0.7346 m, 3.7598 rad) for TEB.

Srikanth et al. [4] address the problem of the fast adaptation of manipulator trajectories for task perturbations. The main objective is to deal with the fact that manipulator joint space trajectory optimization under end-effector task constraints leads to a challenging non-convex problem. Thus, a real-time adaptation of prior computed trajectories to perturbations in task constraints often becomes intractable. Existing works use the so-called warm-starting of trajectory optimization to improve computational performance. A fundamentally different approach that relies on deriving analytical gradients of the optimal solution with respect to the task constraint parameters is introduced. The proposed algorithm provides near real-time adaptation of joint trajectories for a diverse class of task perturbations, such as (i) changes in initial and final joint configurations of end-effector orientation-constrained trajectories and (ii) changes in the end-effector goal or waypoints under end-effector orientation constraints. These two examples are related to real-world applications ranging from learning from demonstration to obstacle avoidance.

Tian et al. [5] offer a solution to problems that arise in path-planning strategies that exploit the use of rapidly exploring random trees, namely long path planning time and a large number of redundant points. To this end, an improved algorithm based on a parent point priority determination strategy and a real-time optimization strategy is derived to optimize rapidly exploring random tree algorithms. First, in order to shorten the path-planning time, the parent point is determined before generating a new point, which eliminates the complicated process of traversing the random tree to search the parent point when generating a new point. Second, a real-time optimization strategy is combined, whose core idea is to compare the distance of a new point, its parent point, and two ancestor points to the target point when a new point is generated, choosing a new point that is helpful for the growth of the random tree to reduce the number of redundant points. Simulation results of a three-dimensional path planning showed that the success rate of the proposed algorithm was close to 100%. Compared with the rapidly exploring random trees algorithm, the number of points was reduced by more than 93.25%, the path planning time was reduced by more than 91.49%, and the path length was reduced by more than 7.88%. The IRB1410 manipulator was used as a test platform in a laboratory environment to assess the efficacy of the new algorithm.

Papadimitrakis et al. [6] is a contribution to the field of automatic collision avoidance for surface vessels, which has been the subject of intensive research in recent years, aiming for the development of decision support systems to aid officers in conventional vessels, or for the creation of autonomous vessel controllers. A multi-ship control problem is addressed using a model predictive controller (MPC) that makes use of obstacle ship-trajectory-prediction models that build upon the radial basis function (RBF) framework and are trained on real AIS data sourced from an open-source database. The usage of such sophisticated trajectory-prediction models enables the controller to correctly infer the existence of a collision risk and apply evasive control actions in a timely manner, thus accounting for the slow dynamics of a large vessel, such as container ships, and enhancing the cooperation between controlled vessels. The proposed method is evaluated on a real-life case from the Miami port area, and the generated trajectories are assessed in terms of safety, economy, and COLREG compliance by comparison with an identical MPC controller utilizing straight-line predictions for the obstacle vessel.

Oftadeh et al. [7] present a nonlinear and universal path-following controller for Wheeled Mobile Robots (WMRs). In contrast to the previous algorithm, the new controller solves the path-following problem for all common categories of holonomic and nonholonomic WMRs, such as omnidirectional, unicycle, car-like, and all steerable wheels. This generality is the consequence of a two-stage solution that separately tackles the platform path-following constraints and the wheels' kinematic constraints. During the first stage, for a virtual mobile platform free from the wheels' constraints, a strategy is developed to drive the WMR asymptotically to the desired path. The second stage accounts for the kinematic constraints imposed by the wheels. This is accomplished by casting the otherwise intractable wheels' kinematic and nonholonomic constraints in the form of explicit proportional functions between the velocity of the platform and those of the wheels. This result leads to a closed-form trajectory generation scheme for the asymptotic path that constantly keeps the wheels' steering and driving velocities within their pre-specified bounds. Extensive experimental results on several types of WMRs, along with simulation results for the other types, are provided to demonstrate the performance and efficacy of the method developed.

Maurya et al. [8] tackle the problem of path following of marine vehicles along straight lines in the presence of currents by resorting to an inner–outer control loop strategy, with due account for the presence of currents. The inner–outer loop control structures exhibit a fast–slow temporal scale separation that yields simple "rules of thumb" for controller tuning. Stated intuitively, the inner-loop dynamics should be much faster than those of the outer loop. Conceptually, the procedure described has three key advantages: (i) it decouples the design of the inner and outer control loops, (ii) the structure of the outer-loop controller does not require exact knowledge of the vehicle dynamics, and (iii) it affords practitioners a very convenient method to effectively implement path-following controllers on a wide range of vehicles. The path-following controller is designed at the kinematic outer loop level and issues heading commands to the inner loop. The key underlying idea is to provide a seamless implementation of path-following control algorithms on heterogeneous vehicles, which are often equipped with heading autopilots. To this end, it is assumed that the heading control system is characterized in terms of an input–output stability (IOS)-like relationship without detailed knowledge of the vehicle dynamics parameters. The stability of the combined inner–outer loops is shown formally by resorting to nonlinear control theory, wherein the cascade and feedback systems of interest are characterized in terms of their IOS properties. Tests with AUVs and one ASV in real-life conditions have shown the efficacy of the path-following control structure developed.

Jacinto et al. [9] address the problem of formation control of a quadrotor and one (or more) marine vehicles operating at the surface of the water with the end goal of encircling the boundary of a chemical spill, enabling such vehicles to carry and release chemical dispersants used during ocean cleanup missions to break up oil molecules. Firstly, the mathematical models of the Medusa class of marine robots and quadrotor aircrafts are introduced, followed by the design of single-vehicle motion controllers that allow these vehicles to follow a parameterized path individually using Lyapunov-based techniques. At the second stage, a distributed controller using event-triggered communications is introduced, enabling the vehicles to perform a cooperative path following missions according to a pre-defined geometric formation. In the next step, a real-time path-planning algorithm is developed that makes use of a camera sensor, installed onboard the quadrotor. This sensor enables the detection in the image of which pixels encode parts of a chemical spill boundary and their use to generate and update, in real time, a set of smooth B-spline-based paths for all the vehicles to follow cooperatively. The performance of the complete system is evaluated by resorting to 3-D simulation software, making it possible to visually simulate a chemical spill. Results from real water trials are also provided for parts of the system, where two Medusa vehicles are required to perform a static lawn-mowing path following the mission cooperatively at the surface of the water.

Delbene et al. [10] introduce a procedure for the autonomous landing of a quadrotor on an unmanned surface vehicle in a marine environment. The relative pose and velocity of the vehicle with respect to the quadrotor are estimated using a combination of data coming from a vision system, which recognizes a set of fiducial markers (AprilTags) located on the vehicle itself, and an ultrasonic sensor, to achieve further robustness during the final landing phase. Details on the landing strategy and on the hardware and software architectures used to implement it are provided. Software-in-the-loop tests were performed as a validation step for the proposed algorithms; to recreate realistic conditions, the movements of the landing platform have been replicated using data from a test in a real marine environment. In order to provide further proof of the reliability of the vision system, a video sequence from a manual landing of a quadrotor on the surface vehicle in a real marine environment has been processed, and the results are presented.

Walton et al. [11] address the problem of optimal defense of a high-value unit (HVU) against a large-scale swarm attack. Multiple models for intra-swarm cooperation strategies are discussed, and a framework is proposed to combine the cooperative models with HVU tracking and adversarial interaction forces. Using this setup, the problem of defending against a swarm attack is cast in the framework of optimal control under uncertain parameters. Numerical solution methods to the latter are discussed, and a consistent result for the dual problem of this framework is derived, providing a tool for verifying computational results. It is further shown that the dual conditions can be computed numerically, providing further computational utility. Finally, the numerical results are applied to derive optimal defender strategies against a 100-agent swarm attack.

Dang et al. [12] discuss important topics in the areas of over-actuated underwater robots whose actuators are propeller thrusters. In general, the positions and orientations of the latter follow classical configurations. This poses limitations on the capability of the robots and does not optimize their performance in terms of energy efficiency, reactivity, and versatility, especially when the robots operate in confined environments. In order to optimize the thruster configuration designs for underwater over-actuated systems, performance indices (manipulability, energetic, reactive, and robustness indices) are introduced. A multi-objective optimization problem is formulated and analyzed. To deal with different objectives with different units, the goal-attainment method, which can avoid the difficulty of choosing a weighting vector to obtain a good balance among these objectives, was selected to solve the problem. A solution design procedure was proposed and discussed. The efficacy efficiency of the proposed method was proven by simulations and experimental results.

Finally, Zhao et al. [13] present a tracking controller for nonlinear systems with matched uncertainties based on contraction metrics and disturbance estimation that provides exponential convergence guarantees. Within the proposed approach, a disturbance estimator is derived to estimate the pointwise value of the uncertainties, with a pre-computable estimation error bound (EEB). The estimated disturbance and the EEB are then incorporated in a robust Riemannian energy condition to compute the control law that guarantees exponential convergence of actual state trajectories to the desired ones. Simulation results on aircraft and planar quadrotor systems demonstrate the efficacy of the proposed controller, which yields better tracking performance than existing controllers for both systems.

The editors and the authors express their sincere gratitude to the publisher and members of the staff for their unwavering commitment and invaluable advice and encouragement that contributed in a very decisive manner to enriching the quality of this reprint.

To all the parents who want a future with peace for their children.
Reza Ghabcheloo

To Stephanie, Ricardo, Ana, and Madalena, the gentle pillars of my life.
António Pascoal

Conflicts of Interest: The authors declare no conflict of interest.

References

1. Kielas-Jensen, C.; Cichella, V.; Berry, T.; Kaminer, I.; Walton, C.; Pascoal, A. Bernstein Polynomial-Based Method for Solving Optimal Trajectory Generation Problems. *Sensors* **2022**, *22*, 1869. [CrossRef] [PubMed]
2. Sabetghadam, B.; Cunha, R.; Pascoal, A. A Distributed Algorithm for Real-Time Multi-Drone Collision-Free Trajectory Replanning. *Sensors* **2022**, *22*, 1855. [CrossRef] [PubMed]
3. Andreasson, H.; Larsson, J.; Lowry, S. A Local Planner for Accurate Positioning for a Multiple Steer-and-Drive Unit Vehicle Using Non-Linear Optimization. *Sensors* **2022**, *22*, 2588. [CrossRef] [PubMed]
4. Srikanth, S.; Babu, M.; Masnavi, H.; Kumar Singh, A.; Kruusamäe, K.; Krishna, K.M. Fast Adaptation of Manipulator Trajectories to Task Perturbation by Differentiating through the Optimal Solution. *Sensors* **2022**, *22*, 2995. [CrossRef] [PubMed]
5. Tian, L.; Zhang, Z.; Zheng, C.; Tian, Y.; Zhao, Y.; Wang, Z.; Qin, Y. An Improved Rapidly-Exploring Random Trees Algorithm Combining Parent Point Priority Determination Strategy and Real-Time Optimization Strategy for Path Planning. *Sensors* **2021**, *21*, 6907. [CrossRef] [PubMed]
6. Papadimitrakis, M.; Stogiannos, M.; Sarimveis, H.; Alexandridis, A. Multi-Ship Control and Collision Avoidance Using MPC and RBF-Based Trajectory Predictions. *Sensors* **2021**, *21*, 6959. [CrossRef] [PubMed]
7. Oftadeh, R.; Ghabcheloo, R.; Mattila, J. Universal Path-Following of Wheeled Mobile Robots: A Closed-Form Bounded Velocity Solution. *Sensors* **2021**, *21*, 7642. [CrossRef] [PubMed]
8. Maurya, P.; Morishita, H.M.; Pascoal, A.; Aguiar, A.P. A Path-Following Controller for Marine Vehicles Using a Two-Scale Inner-Outer Loop Approach. *Sensors* **2022**, *22*, 4293. [CrossRef] [PubMed]
9. Jacinto, M.; Cunha, R.; Pascoal, A. Chemical Spill Encircling Using a Quadrotor and Autonomous Surface Vehicles: A Distributed Cooperative Approach. *Sensors* **2022**, *22*, 2178. [CrossRef] [PubMed]
10. Delbene, A.; Baglietto, M.; Simetti, E. Visual Servoed Autonomous Landing of an UAV on a Catamaran in a Marine Environment. *Sensors* **2022**, *22*, 3544. [CrossRef] [PubMed]
11. Walton, C.; Kaminer, I.; Gong, Q.; Clark, A.H.; Tsatsanifos, T. Defense against Adversarial Swarms with Parameter Uncertainty. *Sensors* **2022**, *22*, 4773. [CrossRef] [PubMed]

12. Dang, T.; Lapierre, L.; Zapata, R.; Ropars, B.; Lepinay, P. Over-Actuated Underwater Robots: Configuration Matrix Design and Perspectives. *Sensors* **2021**, *21*, 7729. [CrossRef] [PubMed]
13. Zhao, P.; Guo, Z.; Hovakimyan, N. Robust Nonlinear Tracking Control with Exponential Convergence Using Contraction Metrics and Disturbance Estimation. *Sensors* **2022**, *22*, 4743. [CrossRef] [PubMed]

sensors

Article

Bernstein Polynomial-Based Method for Solving Optimal Trajectory Generation Problems

Calvin Kielas-Jensen [1,*], Venanzio Cichella [1], Thomas Berry [2], Isaac Kaminer [3], Claire Walton [4] and Antonio Pascoal [2]

1 Cooperative Autonomous Systems (CAS) Lab, Department of Mechanical Engineering, University of Iowa, Iowa City, IA 52242, USA; venanzio-cichella@uiowa.edu
2 Laboratory of Robotics and Engineering Systems (LARSyS), ISR/IST, University of Lisbon, 1049-001 Lisbon, Portugal; thomasdpberry@tecnico.ulisboa.pt (T.B.); antonio@isr.tecnico.ulisboa.pt (A.P.)
3 Department of Mechanical and Aerospace Engineering, Naval Postgraduate School, Monterey, CA 93943, USA; kaminer@nps.edu
4 Department of Electrical Engineering, University of Texas at San Antonio, San Antonio, TX 78249, USA; claire.walton@utsa.edu
* Correspondence: calvin-kielas-jensen@uiowa.edu

Abstract: This paper presents a method for the generation of trajectories for autonomous system operations. The proposed method is based on the use of Bernstein polynomial approximations to transcribe infinite dimensional optimization problems into nonlinear programming problems. These, in turn, can be solved using off-the-shelf optimization solvers. The main motivation for this approach is that Bernstein polynomials possess favorable geometric properties and yield computationally efficient algorithms that enable a trajectory planner to efficiently evaluate and enforce constraints along the vehicles' trajectories, including maximum speed and angular rates as well as minimum distance between trajectories and between the vehicles and obstacles. By virtue of these properties and algorithms, feasibility and safety constraints typically imposed on autonomous vehicle operations can be enforced and guaranteed independently of the order of the polynomials. To support the use of the proposed method we introduce BeBOT (Bernstein/Bézier Optimal Trajectories), an open-source toolbox that implements the operations and algorithms for Bernstein polynomials. We show that BeBOT can be used to efficiently generate feasible and collision-free trajectories for single and multiple vehicles, and can be deployed for real-time safety critical applications in complex environments.

Keywords: optimal trajectory generation; Bernstein polynomials; Bézier curves; optimal control

Citation: Kielas-Jensen, C.; Cichella, V.; Berry, T.; Kaminer, I.; Walton, C.; Pascoal, A. Bernstein Polynomial-Based Method for Solving Optimal Trajectory Generation Problems. *Sensors* **2022**, *22*, 1869. https://doi.org/10.3390/s22051869

Academic Editor: Roberto Teti

Received: 6 December 2021
Accepted: 2 February 2022
Published: 27 February 2022

1. Introduction

The field of autonomous guidance has exploded in the past decade. Significant progress has been made in self driving vehicles, bringing them one step closer to reality [1]. Precision agriculture utilizes autonomous aerial vehicles to monitor crops and spray pesticides [2], and the development in autonomous weed pulling robots may reduce or eliminate the need for potentially harmful pesticides [3]. Underactuated marine surface vehicles can be controlled using a flatness-based approach [4]. Companies such as Amazon, Starship, and Zipline have already begun making autonomous deliveries [5–7]. In fact the first autonomous aerial vehicle has already been flown on a different planet [8]. This progress has led to high demand for computationally efficient algorithms that may yield safe and optimal trajectories to be planned for groups of autonomous vehicles. Our proposed method aims to accomplish these tasks by formulating the optimal trajectory generation problem as a nonlinear programming problem and exploiting the useful features of Bernstein polynomials.

Most techniques for planning and control of autonomous systems fall into one of two categories: closed-loop methods or open-loop methods. Closed-loop methods, sometimes referred to as feedback or reactive methods, use the current state knowledge to determine, in real time, what the next control input should be. On the other hand, open-loop methods determine control values or compute motion trajectories out to a specified time horizon with the use of the system's model.

One common closed-loop technique that originally stemmed from maze solving algorithms is the bug algorithm. The bug algorithm, e.g., [9,10], uses local knowledge of the environment and a global goal to either follow a wall or move in a straight line towards the goal. This algorithm can be implemented on very simple devices due to typically requiring only two tactile sensors. However, it does not account for a vehicle's dynamics and constraints. Moreover, bug algorithms are non-optimal methods and cannot be used for the execution of complex missions that require the optimization of some cost. For a review and comparison of bug algorithms, the reader is referred to [11].

Rather than working on an agent's positions, the velocity obstacle (VO) algorithm uses relative velocities between the agent and obstacles to determine trajectories which will avoid collisions. The original term velocity obstacle was presented in [12]. Variations on the VO method include Common Velocity Obstacle [13], Nonlinear Velocity Obstacles [14], and Generalized Velocity Obstacles [15]. Other relevant closed-loop methods use artificial potential fields, which leverage a potential function providing attractive forces towards the goal and repulsive forces away from obstacles [16–18].

Among the advantages of closed-loop methods are fast computation and the ability to react to changing environments and unforeseen events. Furthermore, theoretical tools aimed at deriving safety guarantees of closed-loop methods are fairly well developed, and are mostly rooted in nonlinear systems analysis and robust and adaptive control. Despite these benefits, closed-loop methods are difficult to employ for multiple vehicle teams. They also generally lack the capability of presenting a human operator with a predicted trajectory and act rather like a black box, which can result in a lack of trust between the operator and the autonomous system.

In contrast to closed-loop methods, open-loop methods can generate solutions in one-shot for the whole mission time, and are therefore able to present an operator with an intuitive representation of the future trajectory. This representation is typically shown as a 2D or 3D path and may also include speed, acceleration, and higher derivatives of the vehicle's motion. Randomized algorithms such as probabilistic roadmaps (PRMs) [19] and rapidly exploring random trees (RRT, RRT*) [20,21] randomly sample the work space to reduce computational complexity. PRMs randomly sample feasible regions within the work space to construct a dense graph. A graph-based solver can then be used to determine the optimal route. RRTs compute trajectories by using directed sampling to build trees. This approach can find feasible solutions in situations involving both a high number of constraints and high dimensional search spaces. Unfortunately, random sampling algorithms may be difficult to use in real-time applications due to computational complexity and may end up exploring regions that will not lead to a solution.

Similar to PRMs, other graph-based approaches aim to efficiently build and then search a graph. Cell decomposition methods, e.g., [22,23], build a graph of their environment by recursively increasing the resolution of areas of interest resulting in a few large nodes of open space and many small nodes near obstacles. Once a graph has been built, a graph solver can be used. A popular graph solver is the A* algorithm [24], which is an extension of Dijkstra's algorithm that uses a heuristic function to improve the search speed. Many modifications to the A* algorithm also exist, such as Lifelong Planning A* [25], which replans a path anytime an obstacle appears on the existing path and utilizes Dijkstra's algorithm to transition the robot from its current pose to the new path. Similarly, Anytime Dynamic A* (ADA*) [26] iteratively decreases a suboptimality bound to improve the plan's optimality within a specified maximum computation time limit. Iterative approaches such as ADA*, sometimes called "anytime" methods, compute a coarse solution and then refine it until a computation timeout is reached. For example, Ref. [27] investigates the addition of committed trajectories and branch-and-bound tree adaptation to the RRT* algorithm to produce an online anytime method.

In addition to graph-based representations of trajectories, polynomial approximation methods can be used as well. In [28], trajectories are represented as piecewise polynomial functions and are generated in a manner that minimizes their snap. In TrajOpt [29], a sequential quadratic program is solved to generate optimal polynomial trajectories while performing continuous time collision checking.

Other open-loop methods include CHOMP [30,31], STOMP [32], and HOOP [33]. In CHOMP, infeasible trajectories are pulled out of collisions while simultaneously smoothing the trajectories using covariant gradient descent. STOMP adopts a similar cost function to that found in CHOMP, but generalizes to cost functions whose gradients are not available. This is done by stochastically sampling noisy trajectories. HOOP utilizes a problem formulation which computes vehicle trajectories in two steps. In the first step, a path is planned quickly by considering only the vehicle's kinematics. The second step then refines this trajectory into a higher order piecewise polynomial using a quadratic program. Other methods, such as WASPAS-mGqNS [34], balance the optimality of motion plans with respect to the mission objectives against exploring unknown environments.

Open-loop methods provide useful tools for dealing with high dimensional problems such as multiple vehicles and several constraints. They are also capable of producing trajectories that accomplish multiple goals. However, due to the curse of dimensionality, the computational complexity of open-loop methods grows significantly with the number of vehicles, constraints, and goals. For the most part, motion planning methods trade optimality and/or safety for computational speed. Our goal is to introduce a method that mitigates this trade-off, and that provides provably safe solutions for high dimensional problems while retaining the computational efficiency of low-order trajectory planning algorithms. This is achieved by exploiting the useful features of Bernstein polynomials.

The Bernstein basis was originally introduced by Sergei Natanovich Bernstein (1880–1968) in order to provide a constructive proof of Weierstrass's theorem. Bernstein polynomials were not widely used until the advent of digital computers due to their slow convergence as function approximants. Widespread adoption eventually occurred when it was realized that the coefficients of Bernstein polynomials could be intuitively manipulated to change the shape of curves described by these polynomials. In the 1960s, two French automotive engineers became interested in using this idea: Paul de Faget de Casteljau and Pierre Étienne Bézier.

Designing complex shapes for automobile bodies by sculpting clay models proved to be a time consuming and expensive process. To combat this, de Casteljau and Bézier sought to develop mathematical tools that would allow designers to intuitively construct and manipulate complex shapes. Due to de Casteljau publishing most of his research internally at his place of employment, Bézier's name became more widely associated with Bernstein polynomials, frequently referred to as Bézier curves. Building on existing research and modern technology, Bernstein polynomials provide several useful properties for many fields.

The Bernstein basis provides numerical stability [35], as well as useful geometric properties and computationally efficient algorithms that can be used to derive and implement efficient algorithms for the computation of trajectory bounds, trajectory extrema, minimum temporal and spatial separation between two trajectories and between trajectories and obstacles, and collision detection. Bernstein polynomials also allow for the representation of continuous time trajectories using low-order approximations.

Our method for trajectory generation builds upon [36–38], where Bernstein polynomials were introduced as a tool to approximate the solutions of nonlinear optimal control problems with provable convergence guarantees. While the results concerning the convergence of the Bernstein approximation method are out of the scope of the present paper, here we focus on the design of algorithms and functions for Bernstein polynomials. These include evaluating bounds, minimum spatial distance, collision detection, and penetration distance. Additionally, we show how these properties can be used for trajectory generation in realistic mission scenarios such as trajectory generation for swarms, navigating cluttered environments, and motion planning for vehicles operating in a Traveling Salesman mission. For the interested reader, an open-source implementation is provided. This paper extends the results initially presented in [39]. In particular, in [39] we focused on autonomous vehicle trajectories representation by Bernstein polynomials, and proposed a preliminary implementation of BeBOT for minimum distance computation and collision detection for safe autonomous operation. In the present paper, we extend previous work by exploiting properties and proposing algorithms for both Bernstein polynomials and rational Bernstein polynomials. BeBOT includes an open-source Python implementation of these algorithms, which enables the user to exploit the properties of (rational) Bernstein polynomials for trajectory generation. Furthermore, we address several applications for multiple autonomous systems and show the efficacy of BeBOT in enabling safe autonomous operations. We added a new algorithm, the penetration algorithm, and several additional examples including air traffic control, navigation of a cluttered environment, vehicle overtaking, 1000 vehicle swarming, a marine vehicle example, and two examples of a vehicle routing problem. An implementation of the examples presented in this paper is available at our GitHub website [40] and can be customized to facilitate the toolbox's usability.

The goal of this manuscript is to provide a general framework which can be applied to a plethora of different systems ranging from mobile robots to manipulators. However, we do provide several numerical examples for mobile robots and include the governing motion equations for ease of implementation, e.g., see examples in Sections 5.1 and 5.6.

In brief, the main contributions of this article are:

1. Novel algorithms which exploit the useful properties of (rational) Bernstein polynomials for use in trajectory generation.
2. Several examples implementing the aforementioned algorithms in realistic mission scenarios.

The paper is structured as follows. In Section 2 we introduce Bernstein polynomials and their properties. Section 3 introduces the use of Bernstein polynomials to parameterize 2D and 3D trajectories. In Section 4 we present computationally efficient algorithms for the computation of state and input constraints typical of trajectory generation applications. In Section 5 we demonstrate the efficacy of these algorithms through several numerical examples. The paper ends with Section 6, which draws some conclusions. A Python implementation of the properties and algorithms presented, as well as the scripts used to generate the plots and examples found throughout this paper, can be found on our GitHub webpage [40].

In what follows, vectors are denoted by bold letters, e.g., $\mathbf{p} = [p_x, p_y]^\top$ and $||\cdot||$ denotes the Euclidean norm (or magnitude), e.g., $||\mathbf{p}|| = \sqrt{p_x^2 + p_y^2}$.

2. Mathematical Preliminaries

The motion planning problems addressed in this work can be in general formulated as optimal control problems. Letting the states and control inputs of the vehicles be denoted by $\boldsymbol{x}(t)$ and $\boldsymbol{u}(t)$, respectively, the optimal motion planning problem can formally be stated as follows:

$$\min_{\boldsymbol{x}(t),\boldsymbol{u}(t)} I(\boldsymbol{x}(t),\boldsymbol{u}(t)) = E(\boldsymbol{x}(0),\boldsymbol{x}(t_f)) + \int_0^{t_f} F(\boldsymbol{x}(t),\boldsymbol{u}(t))dt \tag{1}$$

subject to

$$\dot{\boldsymbol{x}}(t) = \boldsymbol{f}(\boldsymbol{x}(t),\boldsymbol{u}(t)), \quad \forall t \in [0,t_f], \tag{2}$$

$$\boldsymbol{e}(\boldsymbol{x}(0),\boldsymbol{x}(t_f)) = \mathbf{0}, \tag{3}$$

$$\boldsymbol{h}(\boldsymbol{x}(t),\boldsymbol{u}(t)) \le \mathbf{0}, \quad \forall t \in [0,t_f], \tag{4}$$

where $I : \mathbb{R}^{n_x} \times \mathbb{R}^{n_u} \to \mathbb{R}$, $E : \mathbb{R}^{n_x} \times \mathbb{R}^{n_x} \to \mathbb{R}$, $F : \mathbb{R}^{n_x} \times \mathbb{R}^{n_u} \to \mathbb{R}$, $\boldsymbol{f} : \mathbb{R}^{n_x} \times \mathbb{R}^{n_u} \to \mathbb{R}^{n_x}$, $\boldsymbol{e} : \mathbb{R}^{n_x} \times \mathbb{R}^{n_x} \to \mathbb{R}^{n_e}$, and $\boldsymbol{h} : \mathbb{R}^{n_x} \times \mathbb{R}^{n_u} \to \mathbb{R}^{n_h}$.

Here, I defined in Equation (1) is a Bolza-type cost functional, with end point cost E and running cost F. The constraint in Equation (2) enforces the dynamics of the vehicles considered, Equation (3) enforces the boundary conditions, e.g., initial and final position, speed, heading angles of the vehicles, and Equation (4) describes feasibility and mission specific constraints, e.g., minimum and maximum speed, acceleration, collision avoidance constraints, etc.

In previous work [36–38] we presented a discretization method to approximate state and input by nth order Bernstein polynomials. This approximation allows us to transcribe the optimal control problem into a non-linear programming problem, which can then be solved by off-the-shelf optimization solvers. In particular, we show that the solution to the non-linear programming problem converges to the solution of the original optimal control problem as n increases. The present paper focuses on the geometric properties of Bernstein polynomials and their implementation for computationally efficient and safe trajectory generation. In the following, we report the properties of Bernstein polynomials and rational Bernstein polynomials which are relevant to this paper.

An nth order Bernstein polynomial defined over an arbitrary interval $[t_0, t_f]$ is given by

$$\mathbf{C}_n(t) = \sum_{i=0}^{n} \mathbf{P}_{i,n} B_{i,n}(t), \quad t \in [t_0, t_f], \tag{5}$$

where $\mathbf{P}_{i,n} \in \mathbb{R}^D$ is the ith Bernstein coefficient, D is the number of dimensions, and $B_{i,n}(t)$ is the Bernstein polynomial basis defined as

$$B_{i,n}(t) = \binom{n}{i} \frac{(t - t_0)^i (t_f - t)^{n-i}}{(t_f - t_0)^n}, \qquad \binom{n}{i} = \frac{n!}{i!(n-i)!},$$

for all $i = 0, \ldots, n$. Typically the dimensionality of a Bernstein polynomial, D, is either 2 or 3 for 2D or 3D spatial curves, respectively. In this case, Bernstein polynomials are often referred to as Bézier curves and their Bernstein coefficients are known as *control points*. While Bézier's original work did not explicitly use the Bernstein basis [41,42], it was later shown that the original formulation is equivalent to the Bernstein form polynomial [43].

An nth order *rational* Bernstein polynomial, $R_n(t)$, is defined as

$$R_n(t) = \frac{\sum_{i=0}^{n} P_{i,n} w_{i,n} B_{i,n}(t)}{\sum_{i=0}^{n} w_{i,n} B_{i,n}(t)}, \quad t \in [t_0, t_f], \tag{6}$$

where $w_{i,n} \in \mathbb{R}$, $i = 0, \ldots, n$, are referred to as weights. A list of relevant properties of Bernstein polynomials used throughout this article can be found in Appendix A.

3. Generation of 2D and 3D Trajectories Using (Rational) Bernstein Polynomials

3.1. 2D Trajectories

Here we will examine several illustrative examples of the properties of Bernstein polynomials and rational Bernstein polynomials in 2D. All the plots presented can be generated using the example code available at [40].

Figures 1–9 contain several examples of 2D trajectories in the spatial domain. Two trajectories are plotted in Figure 1 along with an obstacle. The trajectories $\mathbf{C}^{[1]}(t)$ and $\mathbf{C}^{[2]}(t)$ are defined as in Equation (5) with $t_0 = 10$ s and $t_f = 20$ s where the Bernstein coefficients are temporally equidistant. The vector of Bernstein coefficients of trajectory $\mathbf{C}^{[1]}(t)$ is

$$\mathbf{P}_5^{[1]} = \begin{bmatrix} 0 & 2 & 4 & 6 & 8 & 10 \\ 5 & 0 & 2 & 3 & 10 & 3 \end{bmatrix}.$$

The vector of Bernstein coefficients of trajectory $\mathbf{C}^{[2]}(t)$ is

$$\mathbf{P}_5^{[2]} = \begin{bmatrix} 1 & 3 & 6 & 8 & 10 & 12 \\ 6 & 9 & 10 & 11 & 8 & 8 \end{bmatrix}.$$

The circular obstacle has a radius of 1 and is centered at point $[3,4]^\top$. Figure 2 highlights the endpoints property (Property A2). Note that the trajectory $\mathbf{C}^{[1]}(t)$ passes through its first and last Bernstein coefficients $\mathbf{P}_{0,5}^{[1]} = [0,5]^\top$ and $[10,3]^\top$, respectively. Likewise, the trajectory $\mathbf{C}^{[2]}(t)$ passes through its first and last Bernstein coefficients $[1,6]^\top$ and $[12,8]^\top$, respectively. The convex hull property (Property A1) is illustrated in Figure 3.

Useful operations can be efficiently performed on Bernstein polynomials by manipulating only their coefficients. The de Casteljau algorithm (Property A5) allows one to split a Bernstein polynomial into two separate polynomials. This is shown in Figure 4 where trajectories $\mathbf{C}^{[1]}(t)$ and $\mathbf{C}^{[2]}(t)$ are split at $t_{div} = 15$ s. Degree elevation (Property A6) is performed on both trajectories in Figure 5. Note that in both Figures 4 and 5, the convex hulls are more accurate than the conservative convex hulls in Figure 3. This idea will be expanded upon in the next section.

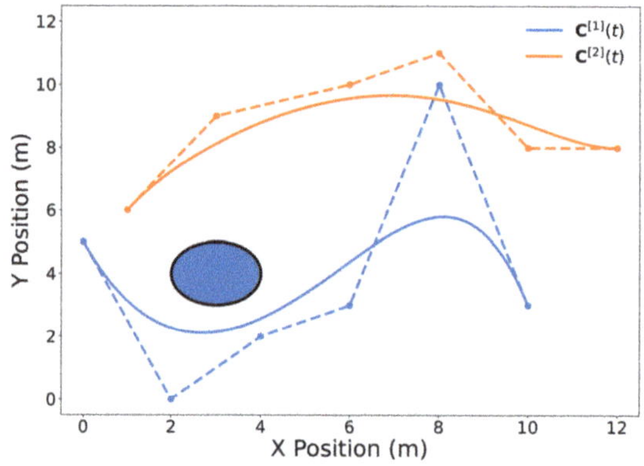

Figure 1. Spatial representation of two Bernstein polynomial trajectories in 2D near a circular obstacle. Trajectory $\mathbf{C}^{[1]}(t)$ is drawn in blue and trajectory $\mathbf{C}^{[2]}(t)$ is drawn in orange. The solid lines are the polynomials and the dashed lines connect the Bernstein coefficients for convenience. Note that the temporal aspect of the trajectories above has been omitted for clarity of presenting the geometric properties of Bernstein polynomials.

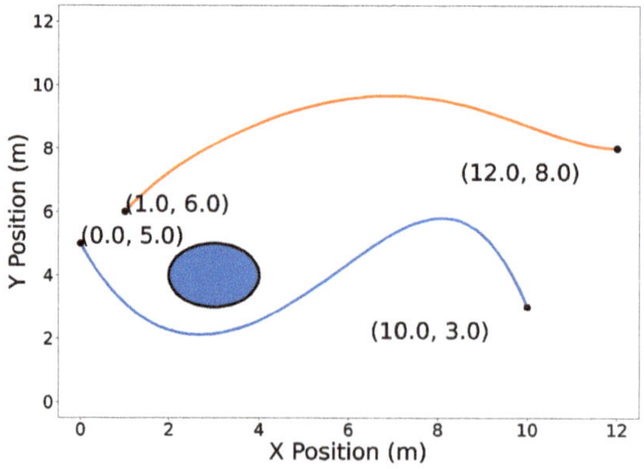

Figure 2. Trajectories $\mathbf{C}^{[1]}(t)$ (blue) and $\mathbf{C}^{[2]}(t)$ (orange) with their endpoints highlighted in 2D.

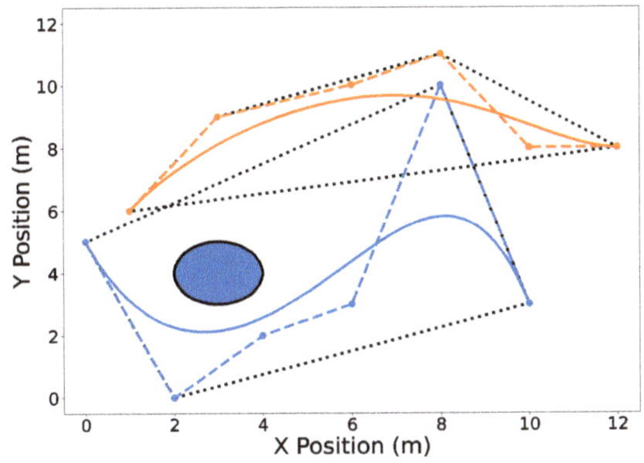

Figure 3. Convex hulls drawn as black dotted lines around the Bernstein coefficients of trajectories $\mathbf{C}^{[1]}(t)$ (blue) and $\mathbf{C}^{[2]}(t)$ (orange). The dashed lines connect the control points of their corresponding trajectories.

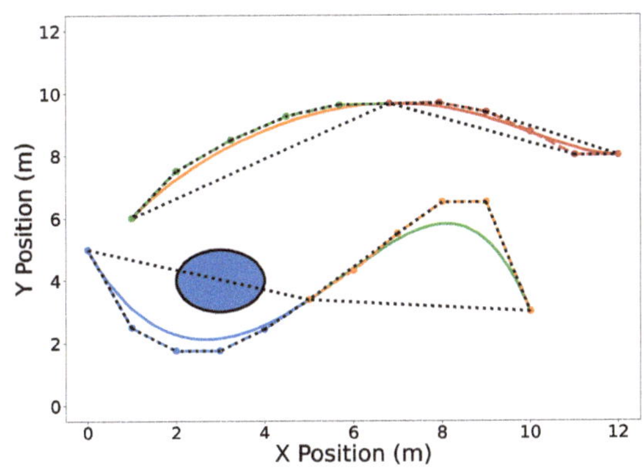

Figure 4. Trajectories $\mathbf{C}^{[1]}(t)$ (split into blue and green) and $\mathbf{C}^{[2]}(t)$ (split into orange and red) split at $t_{div} = 15$ s. Convex hulls are drawn around the Bernstein coefficients of the new split trajectories.

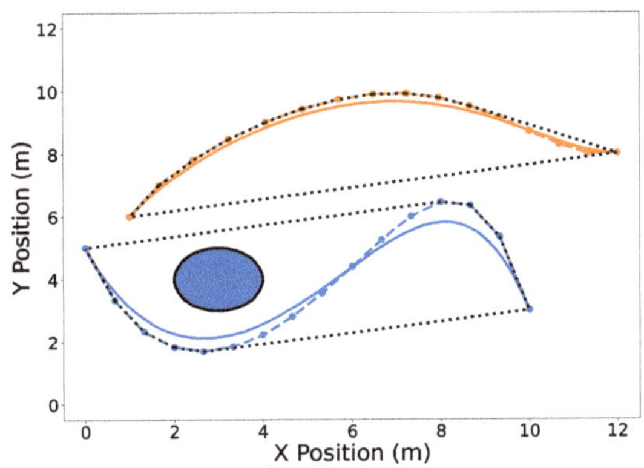

Figure 5. Convex hull drawn around the elevated Bernstein coefficients of trajectories $\mathbf{C}^{[1]}(t)$ (blue) and $\mathbf{C}^{[2]}(t)$ (orange).

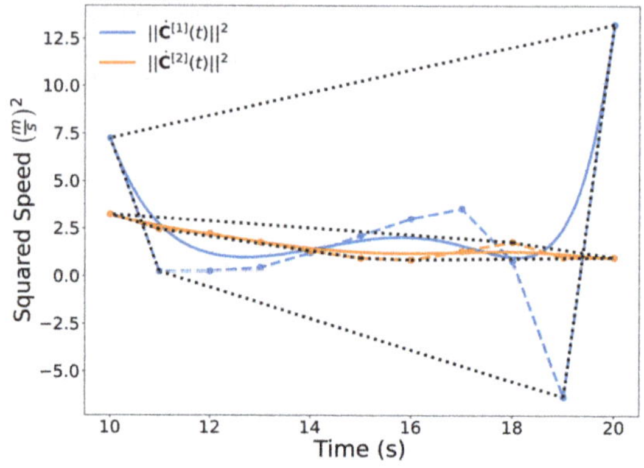

Figure 6. Squared speed of the trajectories $\mathbf{C}^{[1]}(t)$ and $\mathbf{C}^{[2]}(t)$. A convex hull is drawn around the Bernstein coefficients. Note that even though the coefficients may be negative, the actual curve is not.

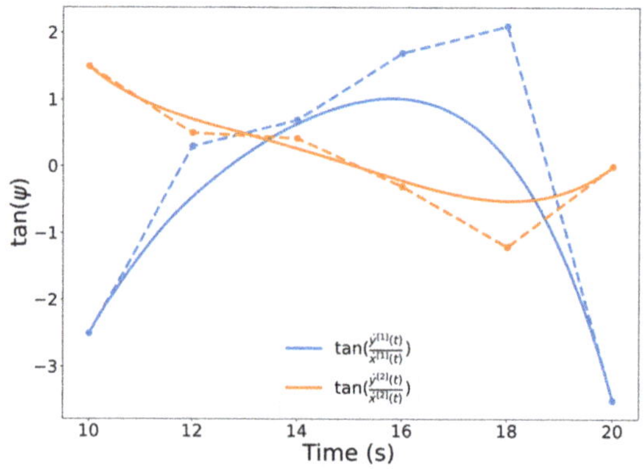

Figure 7. Illustration of the quantity expressed in Equation (8) for trajectories $\mathbf{C}^{[1]}(t)$ and $\mathbf{C}^{[2]}(t)$.

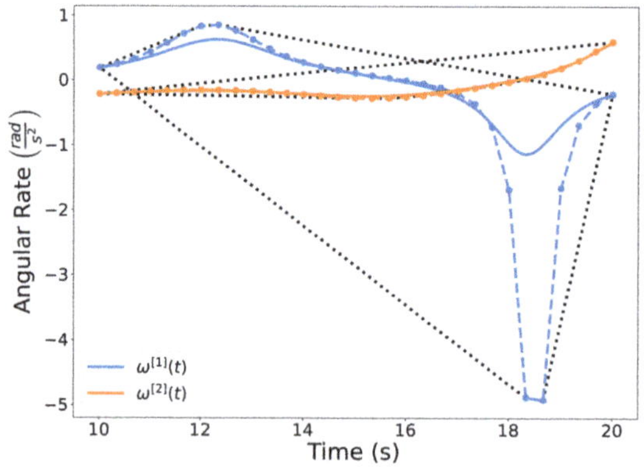

Figure 8. Angular rates of trajectories $\mathbf{C}^{[1]}(t)$ and $\mathbf{C}^{[2]}(t)$. Note that the angular rates are rational Bernstein polynomials.

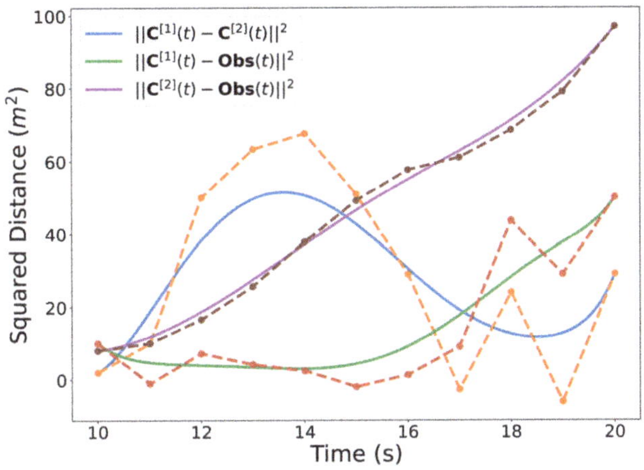

Figure 9. Squared distance between trajectories and the center of the circular obstacle.

Bernstein polynomials can also be used to extract useful information about the dynamics of the trajectories. In Figure 6, Bernstein polynomials representing the squared speed of trajectories $\mathbf{C}^{[1]}(t) = [x^{[1]}(t), y^{[1]}(t)]^\top$ and $\mathbf{C}^{[2]}(t) = [x^{[2]}(t), y^{[2]}(t)]^\top$ are shown along with their corresponding coefficients and convex hulls. The squared speed is computed using the derivative and arithmetic operation properties (Properties A3 and A7) as follows

$$(v^{[1]}(t))^2 = (\dot{x}^{[1]}(t))^2 + (\dot{y}^{[1]}(t))^2$$

Note that the squared speed of a trajectory described by a Bernstein polynomial is also a Bernstein polynomial.

Letting $\mathbf{C}^{[1]}(t) = [x^{[1]}(t), y^{[1]}(t)]^\top$, the heading angle $\psi(t)$ of a trajectory can be computed as

$$\psi^{[1]}(t) = \tan^{-1}\left(\frac{\dot{y}^{[1]}(t)}{\dot{x}^{[1]}(t)}\right). \tag{7}$$

Since the inverse tangent of a Bernstein polynomial is not a Bernstein polynomial, we take the tangent of both sides of the equation, yielding

$$\tan(\psi^{[1]}(t)) = \left(\frac{\dot{y}^{[1]}(t)}{\dot{x}^{[1]}(t)}\right), \tag{8}$$

which is a rational Bernstein polynomial. Figure 7 illustrates the quantity expressed in Equation (8) computed for trajectories $\mathbf{C}^{[1]}(t)$ and $\mathbf{C}^{[2]}(t)$.

To determine the angular rate along the trajectory at any point in $t \in [t_0, t_f]$, we can take the derivative of the heading angle, yielding

$$\omega^{[1]}(t) = \dot{\psi}^{[1]}(t) = \frac{\dot{x}^{[1]}(t)\ddot{y}^{[1]}(t) - \ddot{x}^{[1]}(t)\dot{y}^{[1]}(t)}{(\dot{x}^{[1]}(t))^2 + (\dot{y}^{[1]}(t))^2}. \tag{9}$$

Since the angular rate can be determined using Properties A3 and A7, it can be represented as a rational Bernstein polynomial. The angular rates of trajectories $\mathbf{C}^{[1]}(t)$ and $\mathbf{C}^{[2]}(t)$ are shown in Figure 8.

Finally, the Bernstein polynomial representing the squared distance between two trajectories at every point in time can be computed from

$$d^2(t) = (x^{[2]}(t) - x^{[1]}(t))^2 + (y^{[2]}(t) - y^{[1]}(t))^2, \\ \forall t \in [t_0, t_f] \tag{10}$$

The center point of a circular, static obstacle $\mathbf{Obs}(t)$, can be represented as a Bernstein polynomial whose coefficients are all identical and set to the position of the obstacle, i.e.,

$$\mathbf{P}^{[Obs]} = \begin{bmatrix} x^{[Obs]} & \dots & x^{[Obs]} \\ y^{[Obs]} & \dots & y^{[Obs]} \end{bmatrix}.$$

The degree of the Bernstein polynomial representing the center point of the circular obstacle is equal to that of the order of the Bernstein polynomials representing the trajectories.

3.2. 3D Trajectories

We now introduce two 3D Bernstein polynomials with $t_0 = 10$ s and $t_f = 20$ s, where the coefficients are equidistant in time, and illustrate their properties in Figures 10–17. The Bernstein coefficients of trajectory $\mathbf{C}^{[3]}(t)$ are

$$\mathbf{P}_5^{[3]} = \begin{bmatrix} 7 & 3 & 1 & 1 & 3 & 7 \\ 1 & 2 & 3 & 8 & 3 & 5 \\ 0 & 2 & 1 & 9 & 8 & 10 \end{bmatrix},$$

and the Bernstein coefficients of trajectory $\mathbf{C}^{[4]}(t)$ are

$$\mathbf{P}_5^{[4]} = \begin{bmatrix} 1 & 1 & 4 & 4 & 8 & 8 \\ 5 & 6 & 9 & 10 & 8 & 6 \\ 1 & 1 & 3 & 5 & 11 & 6 \end{bmatrix}.$$

These polynomials are drawn in Figure 10.

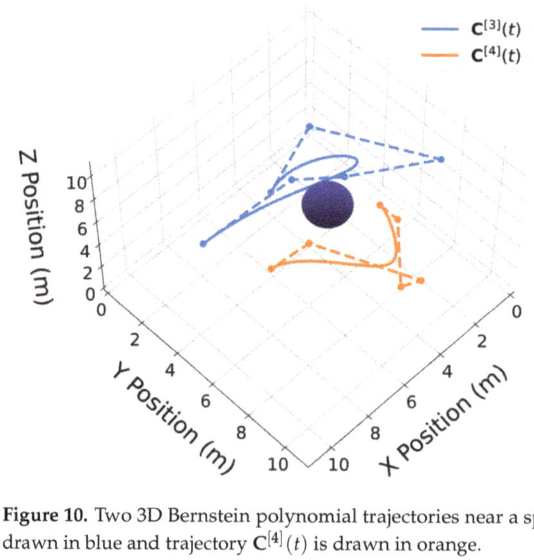

Figure 10. Two 3D Bernstein polynomial trajectories near a spherical obstacle. Trajectory $\mathbf{C}^{[3]}(t)$ is drawn in blue and trajectory $\mathbf{C}^{[4]}(t)$ is drawn in orange.

Similar to the 2D examples, Figures 11–14 illustrate the end points, convex hull, de Casteljau, and elevation properties, respectively. Figures 15 and 16 show the squared speed and squared acceleration of trajectories $\mathbf{C}^{[3]}(t)$ and $\mathbf{C}^{[4]}(t)$, respectively. These values were computed using the derivative and arithmetic properties. Finally, Figure 17 shows the squared Euclidean distance between the trajectories and the center of the spherical obstacle at every point in time. The distance was found using Property A7.

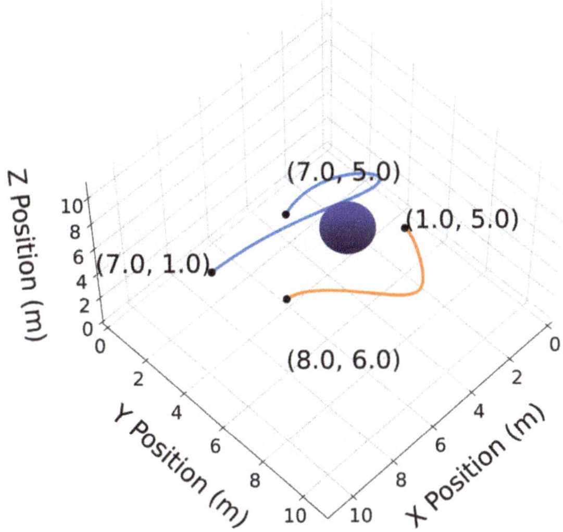

Figure 11. 3D trajectories $\mathbf{C}^{[3]}(t)$ and $\mathbf{C}^{[4]}(t)$ with their endpoints highlighted.

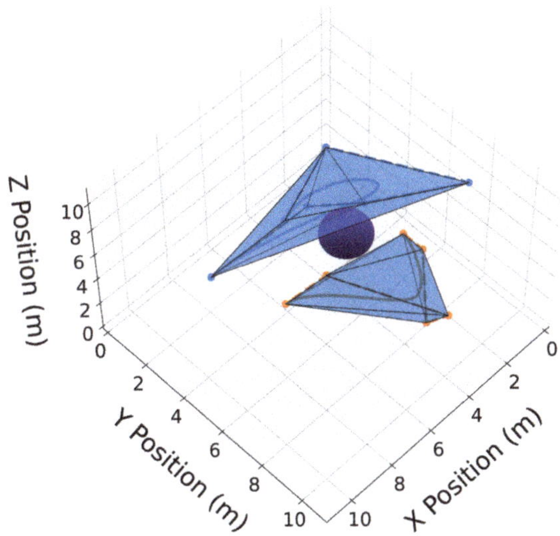

Figure 12. 3D convex hulls drawn as transparent blue surfaces around the Bernstein coefficients of trajectories $\mathbf{C}^{[3]}(t)$ and $\mathbf{C}^{[4]}(t)$.

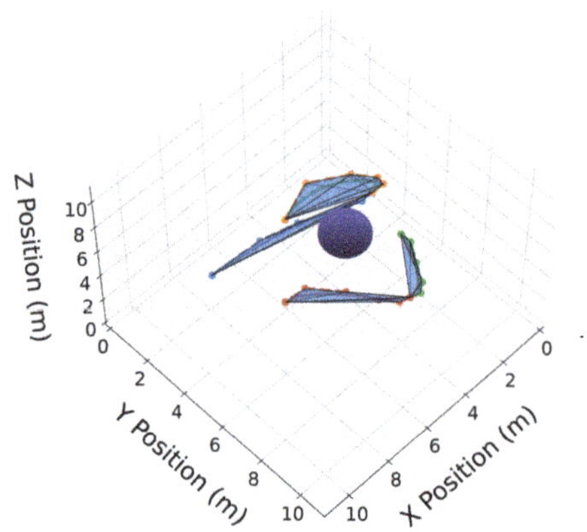

Figure 13. Trajectories $\mathbf{C}^{[3]}(t)$ and $\mathbf{C}^{[4]}(t)$ split at $t_{div} = 15$ s. Convex hulls are drawn around the Bernstein coefficients of the new split trajectories.

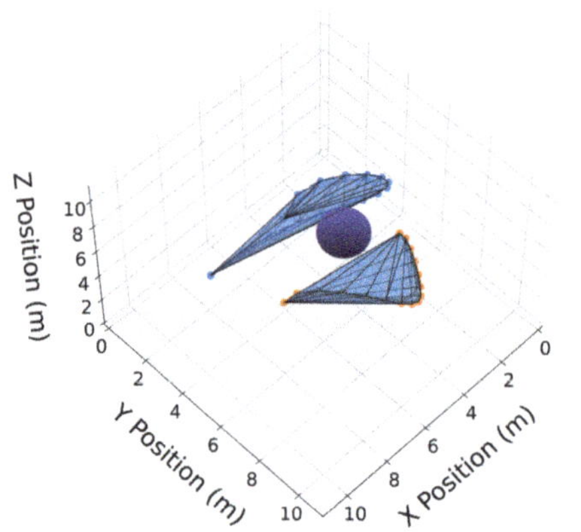

Figure 14. Convex hull drawn around the elevated Bernstein coefficients of trajectories $\mathbf{C}^{[3]}(t)$ and $\mathbf{C}^{[4]}(t)$.

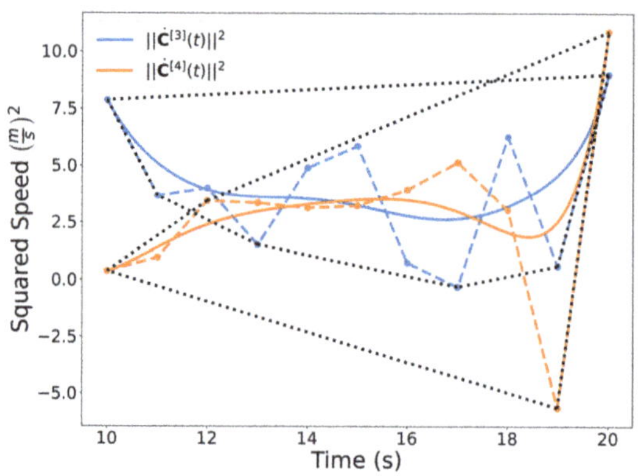

Figure 15. Squared speed of the trajectories $\mathbf{C}^{[3]}(t)$ and $\mathbf{C}^{[4]}(t)$. A convex hull is drawn around the Bernstein coefficients. Note that even though the coefficients may be negative, the actual curve is not.

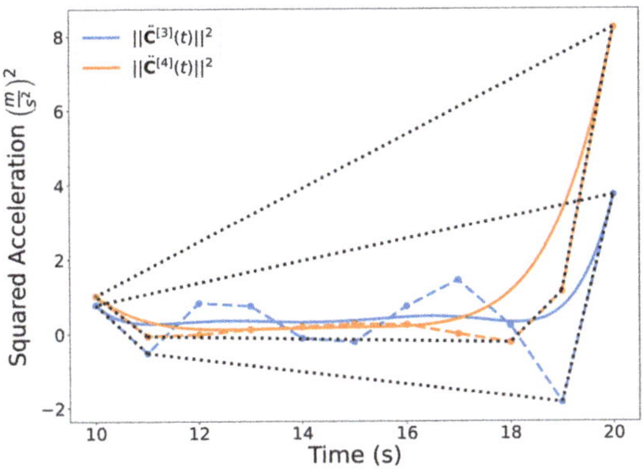

Figure 16. Squared acceleration of the trajectories $\mathbf{C}^{[3]}(t)$ and $\mathbf{C}^{[4]}(t)$ with corresponding convex hulls.

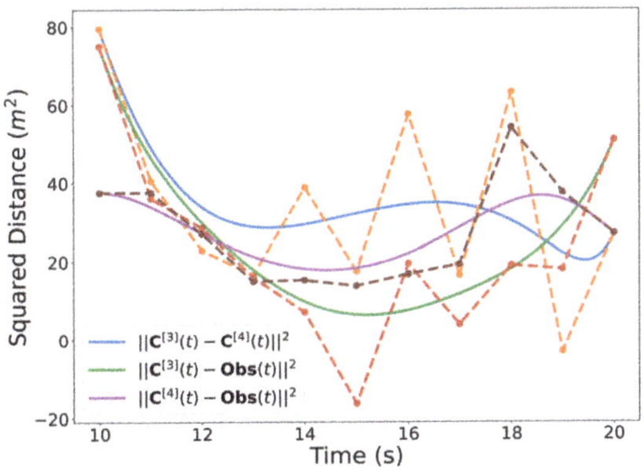

Figure 17. Squared distances between the trajectories and then center of the spherical obstacle.

4. Algorithms for (Rational) Bernstein Polynomials

This section contains algorithms and procedures for Bernstein polynomials that make use of the properties presented in Section 2. These functions include: evaluating bounds, using the convex hull property to quickly find conservative bounds; evaluating extrema, through an iterative procedure that computes a solution within a desired tolerance; minimum spatial distance, applying a similar iterative procedure to find the minimum spatial distance between two Bernstein polynomials; and collision detection, which quickly determines whether a collision may exist or does not exist.

4.1. Evaluating Bounds

Property A1 allows one to quickly establish conservative bounds on the Bernstein polynomial. For example, given the 2D Bernstein polynomial, see Equation (5), with coefficients given by

$$\mathbf{P}_5 = \begin{bmatrix} 0 & 1 & 2 & 3 & 4 & 5 \\ 5 & 0 & 2 & 5 & 7 & 5 \end{bmatrix},$$

lower and upper bounds C_{min} and C_{max} satisfying $C_{\text{min}} \le C(t) \le C_{\text{max}}, \forall t \in [t_0, t_f]$ can be derived using Equation (A1). Figure 18 exhibits the Bernstein polynomial (solid blue line) given the above coefficients (orange dots connected with dashes). The most conservative estimate of the minimum and maximum Y values of the Bernstein polynomial is given by the coefficients with the lowest and highest Y values, respectively. The lower bound is 0 and the upper bound is 7. As mentioned in Property A6 and Equation (A8), the Bernstein coefficients converge to the curve as the polynomial is degree elevated. This fact can be used to derive tighter bounds. A degree elevation of 20 results in a lower bound of 1.93 and an upper bound of 5.89. This is a closer estimate of the actual minimum and maximum, 2.26 and 5.70, respectively (see red dotted lines and Section 4.2). Figure 18 also illustrates degree elevations of 5, 10, and 15. Since the degree elevation matrix, see Equation (A7), is independent of the Bernstein coefficients, a database of elevation matrices can be computed ahead of time to produce tight estimates of the bounds at a low computational cost.

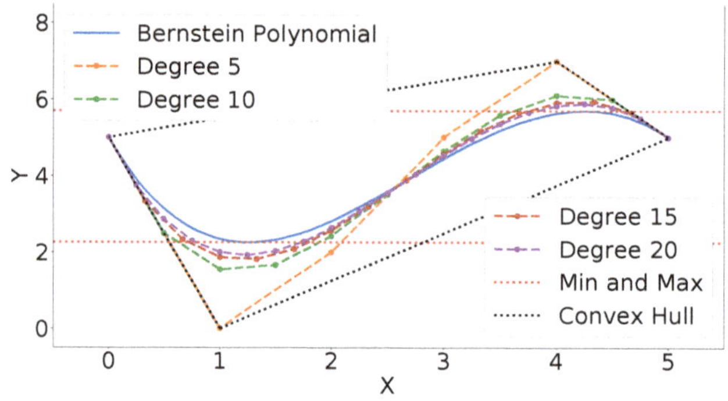

Figure 18. Bounds for Bernstein polynomials. The solid blue line is the Bernstein polynomial, the dashed lines connect the coefficients of each different order, the black dotted line represents the convex hull of the 5th degree Bernstein polynomial, and the red dotted lines represent the actual extrema.

4.2. Evaluating Extrema

The extrema of a Bernstein polynomial are calculated using an iterative procedure similar to the one proposed in [44]. This is done by recursively splitting the curve and using the Convex Hull (Property A1) to obtain an estimate within some desired tolerance. Algorithm 1 outlines the process for determining the maximum of a Bernstein polynomial and can easily be modified to determine the minimum of a Bernstein polynomial.

The inputs to Algorithm 1 are the Bernstein polynomial's coefficients, $\mathcal{P} = \{\mathbf{P}_n\}$, $\mathbf{P}_n = [P_{0,n}, \ldots, P_{n,n}]$, an arbitrarily large *negative* global lower bound, α, and a desired tolerance, ϵ. Note that in order to compute a reliable maximum, $\alpha \leq \min(P)$. In practice, α is set to the lowest possible value that can be reliably represented in the computer.

Line 1 finds the maximum of the two endpoints of the Bernstein polynomial, where n is the degree of the polynomial. This makes use of the End Points (Property A2). Next, we determine the upper bound by simply finding the maximum of $\mathcal{P}$. The *if* statement on line 3 determines whether the global lower bound should be replaced with the current lower bound. The next *if* statement, line 6, will prune the current set of Bernstein coefficients. This is valid because α always provides a lower bound on the global maximum. If the upper bound of any subset is below α, then we know that it is impossible for any point on that subset to be the global maximum. The final *if* statement, line 9, determines whether the difference between the upper and lower bounds is within the desired tolerance and returns the global minimum bound α if the tolerance is met.

The *else* statement, starting on line 11, splits the curve and then recursively calls Algorithm 1 again. The function split() utilizes the de Casteljau algorithm (Property A5). One of two different splitting points, t_{div}, can be employed. The first option simply splits the curve in half, i.e., $t_{div} = t_0 + \frac{t_f - t_0}{2}$. The second option uses the index of the largest valued coefficient, $i_{ub} = \text{argmax}(\mathcal{P})$, to determine the splitting point, i.e., $t_{div} = t_0 + (t_f - t_0)\frac{i_{ub}}{n}$.

Algorithm 1 (and its converse) is employed to find the minimum and maximum of the 5th degree Bernstein polynomial depicted in Figure 18 (red lines). The execution time to compute the minimum is 320 μs on a Lenovo ThinkPad laptop using an Intel Core i7-8550U, 1.80 GHz CPU. The implementation can be found in [40].

Algorithm 1: Evaluating Maximum Value of a Bernstein Polynomial

```
   Input: P, α, ε
1  P_lb = max{P[0], P[n]}
2  P_ub = max(P)
3  if P_lb > α then
4  |  α = P_lb
5  end
6  if α > P_ub then
7  |  return α
8  end
9  if P_ub − P_lb < ε then
10 |  return α
11 else
12 |  P^A, P^B = split(P)
13 |  α_A = Algorithm 1(P^A, α, ε)
14 |  α_B = Algorithm 1(P^B, α, ε)
15 |  α = max(α_A, α_B)
16 end
17 return α
```

4.3. *Minimum Spatial Distance*

The minimum spatial distance between two Bernstein polynomials can be computed using the method outlined in [44]. This is done by exploiting the Convex Hull property (Property A1), the End Point Values property (Property A2), the de Casteljau Algorithm (Property A5), and the Gilbert-Johnson-Keerthi (GJK) algorithm [45]. The latter is widely used in computer graphics to compute the minimum distance between convex shapes.

The algorithm for the minimum spatial distance between two Bernstein polynomials is shown in Algorithm 2. The first two inputs to the function are the sets of Bernstein coefficients, $\mathcal{P} = \{\mathbf{P}_m\}$ and $\mathcal{Q} = \{\mathbf{Q}_n\}$, which define the two Bernstein polynomials in question. The last two inputs are the global upper bound on the minimum distance, α, and a desired tolerance ϵ.

The upper_bound() function on line 1 finds the furthest distance between the end points of the two Bernstein polynomials, i.e., lower_bound$(\mathcal{P}, \mathcal{Q}) = \max\{\mathcal{P}[0] - \mathcal{Q}[0], \mathcal{P}[0] - \mathcal{Q}[n], \mathcal{P}[m] - \mathcal{Q}[0], \mathcal{P}[m] - \mathcal{Q}[n]\}$ where m and n are the degrees of the polynomials represented by $\mathcal{P}$ and $\mathcal{Q}$, respectively. This is a valid upper bound on the minimum distance between the two polynomials due to End Point Values (Property A2).

The lower_bound() function on line 2 finds the lower bound on the distance between the two polynomials by using the GJK algorithm. This is a valid lower bound because of Property A1, Convex Hull. The next three *if* statements on lines 3, 6, and 9 are very similar to those seen in Algorithm 1. Line 3 updates the global upper bound α if the current upper bound is smaller. Line 6 prunes the current iteration since it is impossible the current lower bound, *lower*, to be the minimum distance if it is larger than the global upper bound. Line 9 returns α if the desired tolerance is met.

The lines within the *else* statement split the Bernstein polynomials defined by $\mathcal{P}$ and $\mathcal{Q}$ and recursively call Algorithm 2. Like in Algorithm 1, the first option for splitting would be to simply split at the halfway point. The second option for splitting the curves is outlined in [44] and uses the location at which the minimum distance occurs to choose the splitting point. Figure 19a illustrates the minimum distance between several different Bernstein polynomials. The code to generate this plot can be found in [40]. The execution time to compute the minimum spatial distance is 3.29 ms on a Lenovo ThinkPad laptop using an Intel Core i7-8550U, 1.80 GHz CPU.

Algorithm 2: Minimum Distance Between Two Bernstein Polynomials

Input: $\mathcal{P}, \mathcal{Q}, \alpha, \epsilon$
$upper = \text{upper\_bound}(\mathcal{P}, \mathcal{Q})$
$lower = \text{lower\_bound}(\mathcal{P}, \mathcal{Q})$
if $upper < \alpha$ **then**
  $\alpha = upper$
end
if $\alpha < lower$ **then**
  return α
end
if $upper - lower < \epsilon$ **then**
  return α
else
  $\mathcal{P}^A, \mathcal{P}^B = \text{split}(\mathcal{P})$
  $\mathcal{Q}^A, \mathcal{Q}^B = \text{split}(\mathcal{Q})$
  $\alpha = \min(\alpha, \text{Algorithm 2 } (\mathcal{P}^A, \mathcal{Q}^A, \alpha))$
  $\alpha = \min(\alpha, \text{Algorithm 2 } (\mathcal{P}^A, \mathcal{Q}^B, \alpha))$
  $\alpha = \min(\alpha, \text{Algorithm 2 } (\mathcal{P}^B, \mathcal{Q}^A, \alpha))$
  $\alpha = \min(\alpha, \text{Algorithm 2 } (\mathcal{P}^B, \mathcal{Q}^B, \alpha))$
end
return α

Remark 1. *Note that Algorithm 2 can also be employed to compute the minimum distance between a Bernstein polynomial and a point or a convex shape. This is shown in Figure 19b.*

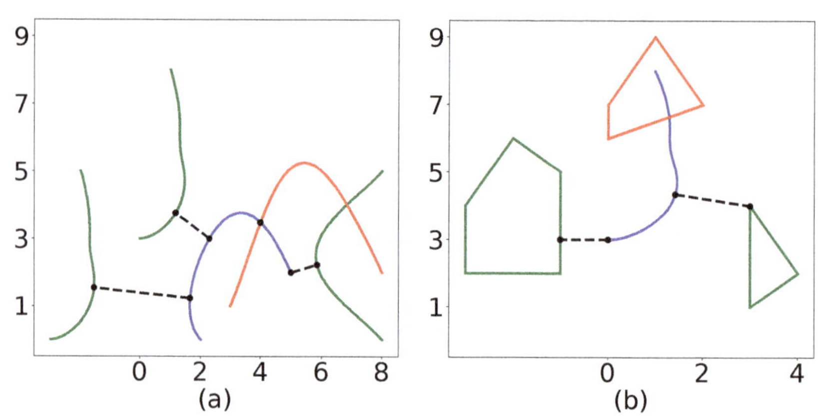

Figure 19. (**a**) Minimum distance between curves. (**b**) Minimum distance between a curve and a polygon. All distances are measured to the blue curve. A red curve or polygon indicates that a collision exists.

4.4. Collision Detection

In some cases it may be desirable to quickly check the feasibility of a trajectory rather than finding a minimum distance. The collision detection algorithm can be used in these cases. The two major differences between the Collision Detection Algorithm and the Minimum Distance Algorithm described previously are a modification of the GJK algorithm and a change in the stopping criteria. Rather than having the GJK algorithm return a minimum distance, it simply returns whether a collision has been detected (i.e., convex hulls intersecting). The stopping criteria is set to return the moment no collisions are found rather than continuing iterations to meet a desired tolerance. For example, if the original convex hulls of two Bernstein polynomials do not intersect, the collision detection algorithm will return *no collision* after the first iteration while the minimum distance algorithm will continue to iterate until the desired tolerance is met. Therefore, this algorithm is computationally inexpensive compared to the minimum distance algorithm, with the drawback that it only returns a binary value (no collision or collision possible) rather than a minimum distance.

The collision detection algorithm is shown in Algorithm 3. The inputs are the coefficients of the Bernstein polynomials being compared, $\mathcal{P}$ and $\mathcal{Q}$, and the maximum number of iterations *max_iter*. The *while* loop beginning on line 2 runs until it is determined that a collision does not exist or until the maximum number of iterations is met. The find_collisions() function on line 3 uses the modified GJK algorithm to determine which convex hulls from the set $\mathcal{P}$ collide with those from the set $\mathcal{Q}$. The *if* statement on line 4 checks to see whether collisions were found. If both $\mathcal{P}_{col}$ and $\mathcal{Q}_{col}$ are empty, then no collisions exist. If collisions do exist then the *for* loops starting on lines 7 and 11 split all the convex hulls that were found to collide and add them to the set to be checked. Note that the parent set that is split is removed from the set of convex hulls to check. If the maximum number of iterations is met, then the algorithm returns that a collision is possible. The execution time when a collision is possible is 1.10 ms on a Lenovo ThinkPad laptop using an Intel Core i7-8550U, 1.80 GHz CPU. However, when a collision does not exist, the execution time is only 7.25 μs.

Algorithm 3: Collision Detection

Input: $\mathcal{P}, \mathcal{Q}, max\_iter$

1 $k = 0$
2 **while** $k < max\_iter$ **do**
3     $\mathcal{P}_{col}, \mathcal{Q}_{col} = \text{find\_collisions}(\mathcal{P}, \mathcal{Q})$
4   **if** $\mathcal{P}_{col} \cup \mathcal{Q}_{col} = \{\}$ **then**
5   **return** No Collision
6   **end**
7   **for** $\mathcal{P}_i \in \mathcal{P}_{col}$ **do**
8         $\mathcal{P}^A, \mathcal{P}^B = \text{split}(\mathcal{P}_i)$
9         $\mathcal{P} = \mathcal{P} \cup \{\mathcal{P}^A, \mathcal{P}^B\} \setminus \mathcal{P}_i$
10   **end**
11   **for** $\mathcal{Q}_i \in \mathcal{Q}_{col}$ **do**
12         $\mathcal{Q}^A, \mathcal{Q}^B = \text{split}(\mathcal{Q}_i)$
13         $\mathcal{Q} = \mathcal{Q} \cup \{\mathcal{Q}^A, \mathcal{Q}^B\} \setminus \mathcal{Q}_i$
14   **end**
15     $k{+}{+}$
16 **end**
17 **return** Collision Possible

4.5. Penetration Algorithm

If two convex shapes intersect, in order to derive information such as the penetration depth and vector, the EPA [46] can be used. A slight modification of the EPA algorithm is proposed here, which is referred to as the DEPA, whose objective is to find the penetration of one convex shape relative to another along a specific direction $\vec{d}$. The top left plot of Figure 20 shows two shapes intersecting each other, and the remaining plots show examples of penetration vectors, i.e., the vector $\vec{d}$ needed to move the second shape so that it no longer intersects the first shape. The DEPA algorithm finds the shortest possible penetration vector. The pseudocode is reported below (see Algorithm 4).

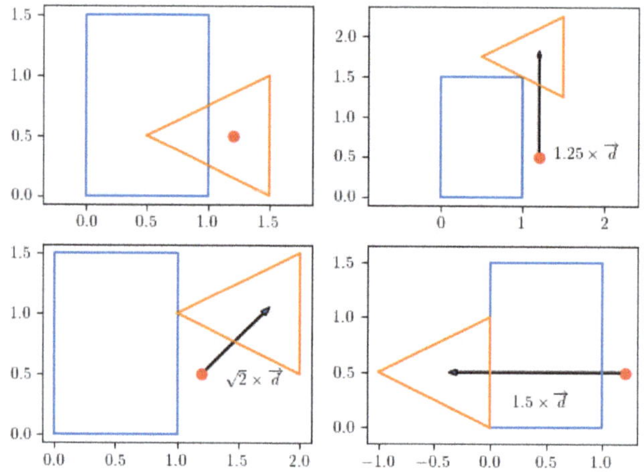

Figure 20. Illustration of penetration.

Algorithm 4: Directed Extended Polytope Algorithm

Data: $\vec{d}$
Data: $MinkowskiDifference$
Data: $simplex$
ContainsOrigin($simplex$);
$A, B \leftarrow$ InterectingEdge($simplex$, Ray($\vec{0}$, $\vec{d}$));
Loop
    $\vec{n} \leftarrow$ TripleProduct($\vec{AB}$, $\vec{A}$, $\vec{AB}$);
    $C \leftarrow$ Support($MinkowskiDifference$, $\vec{n}$);
    if *Parallel*($\vec{C}$, $\vec{d}$) **then**
        $\vec{vec} \leftarrow \vec{C}$;
        return;
    else if *Equals(C,B) OR Equals(C,A)* **then**
        break;
    else if *Intersection(Ray(0, $\vec{d}$), $\overline{CA}$)* **then**
        $B \leftarrow C$;
    else if *Intersection(Ray(0, $\vec{d}$), $\overline{CB}$)* **then**
        $A \leftarrow C$;
    $\vec{vec} \leftarrow$Intersection(Ray(0, $\vec{d}$), $\overline{AB}$);
EndLoop

Figure 21 demonstrates the algorithm in 4 steps. The first plot shows the Minkowski Difference of the shapes depicted in Figure 22, which contains the origin and with a triangle simplex that contains the origin. This is the desired direction to move a polytope A (blue polytope) of Figure 22 such that it no longer contains polytope B (beige polytope). Once the norm of the point along the edge of the Minkowski Difference parallel to $\vec{d}$ is found, A can then move in the same direction with the same length to no longer intersect B.

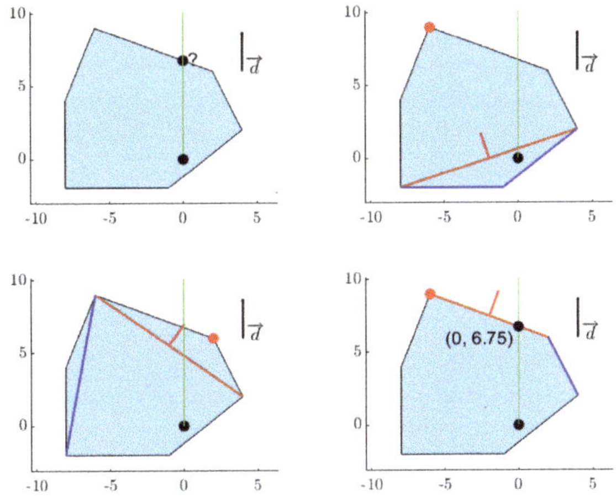

Figure 21. Iteration of the DEPA algorithm.

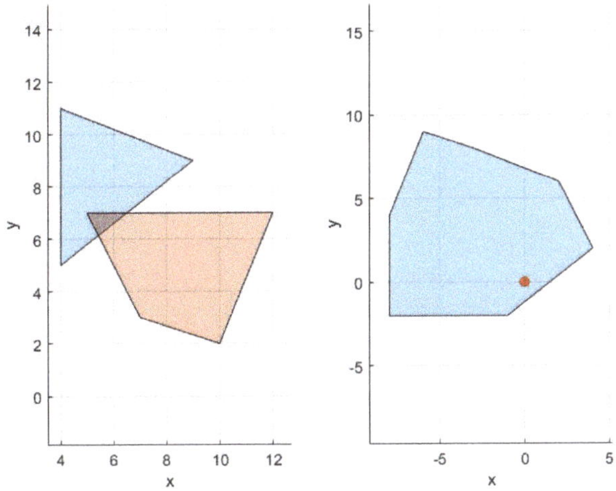

Figure 22. Two intersecting polygons and resulting Minkowski Difference.

5. Numerical Examples

In this section, numerical examples using the BeBOT toolkit and Python's Scipy Optimization package are examined (flight tests are available at [47]). The implementation of the following examples can be found in [40].

5.1. Dubins Car—Time Optimal

In this simple example, several trajectories for a vehicle with Dubins car dynamics are generated to illustrate the properties of Bernstein polynomials. We let the desired trajectory assigned to the vehicle be given by the Bernstein polynomial

$$\begin{bmatrix} C_n^{[x]}(t) \\ C_n^{[y]}(t) \end{bmatrix} = \mathbf{C}_n(t) = \sum_{i=0}^{n} \mathbf{P}_{i,n} B_{i,n}(t), \quad t \in [t_0, t_f]. \tag{11}$$

The square of the speed of the vehicle is a 1D Bernstein polynomial given by

$$v^2(t) = ||\dot{\mathbf{C}}_n(t)||^2.$$

The heading angle is

$$\psi(t) = \tan^{-1}\frac{\dot{C}_n^{[y]}(t)}{\dot{C}_n^{[x]}(t)}, \tag{12}$$

and the angular rate is a 1D rational Bernstein polynomial given by

$$\omega(t) = \frac{\ddot{C}_n^{[y]}(t)\dot{C}_n^{[x]}(t) - \dot{C}_n^{[y]}(t)\ddot{C}_n^{[x]}(t)}{||\dot{\mathbf{C}}_n(t)||^2}. \tag{13}$$

The objective at hand is to find a trajectory that arrives at a desired destination in the minimum possible time while adhering to feasibility and safety constraints. In particular, the trajectory generation problem is as follows:

$$\min_{\mathbf{P}_n, t_f} t_f$$

subject to

$$\begin{aligned}
&\mathbf{C}_n(t_0) = \mathbf{C}_0, \quad \mathbf{C}_n(t_f) = \mathbf{C}_f, \\
&\psi(t_0) = \psi_0, \quad \psi(t_f) = \psi_f \\
&||\dot{\mathbf{C}}_n(t_0)|| = v_0, \quad ||\dot{\mathbf{C}}_n(t_f)|| = v_f, \\
&||\dot{\mathbf{C}}_n(t)||^2 \le v_{\max}^2, \quad \forall t \in [t_0, t_f] \\
&||\dot{\psi}(t)|| \le \omega_{\max}, \quad \forall t \in [t_0, t_f] \\
&||\mathbf{C}_n(t) - \mathbf{O}_i||^2 \ge d_s^2, \quad \forall t \in [t_0, t_f], \quad i = 1, 2.
\end{aligned}$$

We set the initial and final position, heading, and speed to $\mathbf{C}_0 = [3, 0]^\top$ m, $\mathbf{C}_f = [7, 10]^\top$ m, $\psi_0 = \psi_f = \frac{\pi}{2}$ rad, and $v_0 = v_f = 1\,\frac{\text{m}}{\text{s}}$. The maximum speed, maximum angular rate, and minimum safe distance constraints are $v_{\max} = 5\,\frac{\text{m}}{\text{s}}$, $\omega_{\max} = 1\,\frac{\text{rad}}{\text{s}}$, and $d_s^2 = 1$ m, respectively. The positions of the obstacles are $\mathbf{O}_1 = [3, 2]^\top$ m and $\mathbf{O}_2 = [6, 7]^\top$ m.

In the problem above, the initial and final constraints for position, heading, and speed are enforced using the End Point Values property (Property A2) together with Equations (A2), (12) and (13). Similarly, the same property is used to enforce the initial and final speeds and headings (see (A2)). Note that the norm squared of the speed and of the distance between the trajectory and the obstacles can be expressed as Bernstein polynomials (the sum, the difference, and the product between Bernstein polynomials are also Bernstein polynomials). A similar argument can be made for the norm square of the angular rate, which can be expressed as a rational Bernstein polynomial (see Property A7). Thus, the maximum speed and angular rate, and collision avoidance constraints can be enforced using the Evaluating Bounds or Evaluating Extrema procedures described in Sections 4.1 and 4.2.

Figure 23 shows the results with $n = 10$. The blue curve is obtained by enforcing the constraints using the Evaluating Bounds procedure. The optimal time of arrival at the final destination is $t_f = 9.14$ s. Next, we solve the same problem by enforcing the constraints using the Evaluating Bound procedure together with Degree Elevation. Recall that by degree elevating a Bernstein polynomial, the Bernstein coefficients converge towards the curve. Thus, degree elevation can be used to enforce constraints with tighter bounds. The orange and green lines show the trajectories obtained using degree elevations of 30 and 100, respectively. Degree elevation to degree 30 results in an optimal final time $t_f = 7.64$ s. The elevation to degree 100 provides an optimal value $t_f = 7.12$ s. Finally, the trajectory with smallest optimal final time, $t_f = 6.45$ s, depicted as the red curve in

Figure 23, is obtained by enforcing the constraints using the Evaluating Extrema algorithm (Section 4.2). While higher degree elevations or evaluating the exact extrema can produce more optimal trajectories, that optimality comes at the cost of additional computation time. Using a Lenovo Thinkpad P52s with an Intel Core i7-8550U CPU with a 1.8 GHz clock and 8 GB of memory, the computation time required for no degree elevation, a degree elevation of 30, a degree elevation of 100, and the exact extrema algorithm was 0.105 s, 0.146 s, 0.201 s, and 0.573 s, respectively.

Figure 24 illustrates the squared speed of each example. Figure 25 shows the angular rate of each trial. It can be seen that the vehicle correctly adheres to the speed and angular rate constraints for each trial with the only differences being the final time and proximity to the obstacles.

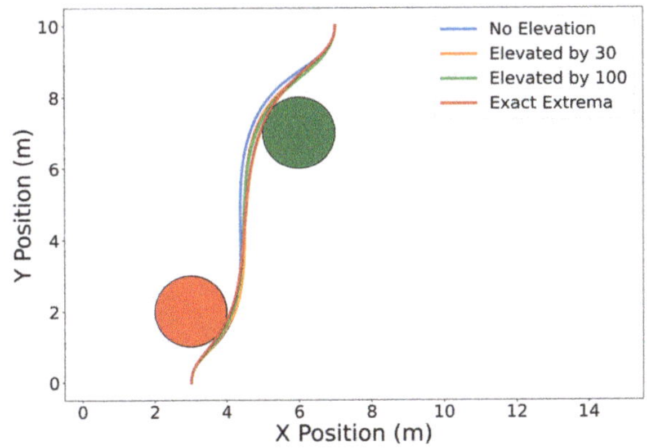

Figure 23. Time optimal trajectory for vehicle with initial and final speeds and headings, maximum speed, maximum angular rate, and maximum safe distance constraints ranging from least to most conservative distance estimates.

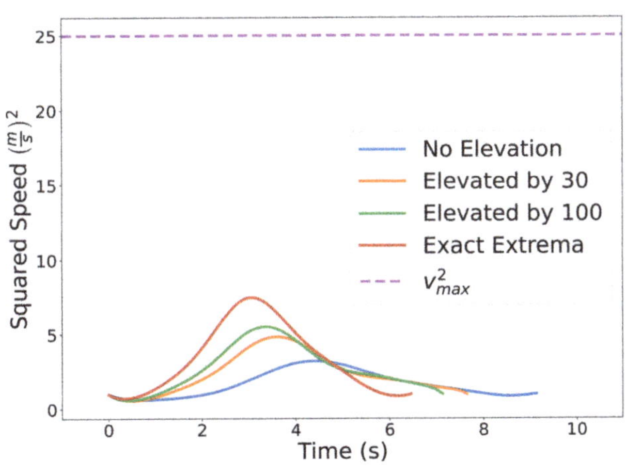

Figure 24. Plot of the squared speed constraints for each separate trial.

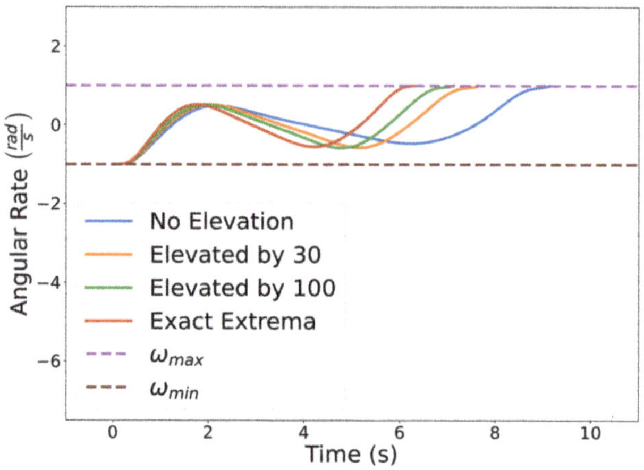

Figure 25. Plot of the angular rate constraints for each separate trial.

Remark 2. *The Exact Extrema function is a complex non-linear and non-smooth function. When it is used to enforce constraints, gradient-based optimization solvers such as the one used in this work can fail to converge to a feasible solution, especially if the initial guess is not feasible. One option is to use an iterative procedure where (i) a feasible sub-optimal solution is obtained by enforcing the collision avoidance constraint using the Evaluating Bounds function, and (ii) this solution is then used as an initial guess to solve the (more accurate) problem with the Exact Extrema constraint.*

5.2. Air Traffic Control—Time Optimal

In this example, we consider the problem of routing several commercial flights between major US cities in two dimensions (i.e., constant altitude). Assuming that each flight departs at the same time, the goal is to minimize the combined flight time of all the vehicles. Let the position, speed, heading, and angular rate of each vehicle under consideration be parameterized as in Section 5.1. We shall also make the assumption that the trajectories are on a 2D plane rather than on the surface of the Earth.

The goal is to compute cumulatively time optimal trajectories subject to maximum speed and angular velocity bounds, initial and final position, angle, and speeds. The vehicles must also maintain a minimum safe distance between each other. This problem can be formulated as follows:

$$\min_{\mathbf{P}_n, \mathbf{t}_f} \sum_{k=1}^{m} t_f^{[k]}$$

subject to

$$\begin{aligned}
&\mathbf{C}_n^{[k]}(0) = \mathbf{C}_0^{[k]}, \quad \mathbf{C}_n^{[k]}\left(t_f^{[k]}\right) = \mathbf{C}_f^{[k]}, \\
&\psi^{[k]}(0) = \psi_0^{[k]}, \quad \psi^{[k]}\left(t_f^{[k]}\right) = \psi_f^{[k]}, \\
&||\dot{\mathbf{C}}_n^{[k]}(0)|| = v_0^{[k]}, \quad ||\dot{\mathbf{C}}_n^{[k]}\left(t_f^{[k]}\right)|| = v_f^{[k]}, \\
&v_{\min}^2 \leq ||\dot{\mathbf{C}}_n^{[k]}(t)||^2 \leq v_{\max}^2, \quad \forall t \in [0, t_f^{[k]}], \\
&|\dot{\psi}^{[k]}(t)| \leq \omega_{\max}, \quad \forall t \in [0, t_f^{[k]}], \\
&||\mathbf{C}_n^i(t) - \mathbf{C}_n^j(t)||^2 \geq d_s^2, \quad \forall i,j \in \{1, \ldots, m\}, i \neq j.
\end{aligned}$$

where the superscript $[k]$ corresponds to the kth vehicle out of m vehicles, $\mathbf{C}_0^{[k]}$ and $\mathbf{C}_f^{[k]}$ are the initial and final positions, $\psi_0^{[k]}$ and $\psi_f^{[k]}$ are the initial and final headings, $v_0^{[k]}$ and $v_f^{[k]}$ are the initial and final speeds, v_{min} and v_{max} are the minimum and maximum speeds, ω_{max} is the maximum angular velocity, d_s is the minimum safe distance, and $t_f^{[k]}$ is the final time of the kth vehicle.

The departure cities, in vehicle order, are: San Diego, New York, Minneapolis, and Seattle. The arrival cities, in vehicle order, are: Minneapolis, Seattle, Miami, and Denver. The initial and final speeds are all $v_0^{[k]} = v_f^{[k]} = 205 \frac{\mathrm{m}}{\mathrm{s}}\ \forall k \in \{1, \ldots, m\}$, the initial headings are $\boldsymbol{\psi}_0 = [0, \pi, 0, 0]^\top$ rad, the final headings are $\boldsymbol{\psi}_f = [0, \pi, -\frac{\pi}{2}, 0]$ rad, the minimum speed is $v_{\mathrm{min}} = 200\ \frac{\mathrm{m}}{\mathrm{s}}$, the maximum speed is $v_{\mathrm{max}} = 260\ \frac{\mathrm{m}}{\mathrm{s}}$, the maximum angular velocity is $\omega_{\mathrm{max}} = 3 \frac{\mathrm{deg}}{\mathrm{s}} = 0.0524 \frac{\mathrm{rad}}{\mathrm{s}}$, the minimum safe distance is $d_s = 5$ km, and the degree of the Bernstein polynomials being used is 5.

The initial and final position constraints are enforced using the End Point Values property (Property A2). Similarly, the same property is used to enforce the initial and final speeds and headings (see (A2)). Note that the norm square of the speed and the norm square of the distance between vehicles can be expressed as 1D Bernstein polynomials (the sum, difference, and product between Bernstein polynomials are also Bernstein polynomials). A similar argument can be made for the norm square of the angular rate, which can be expressed as a rational Bernstein polynomial (see Property A7). Thus, the maximum speed and angular rate, and collision avoidance constraints can be enforced using the Evaluating Bounds or Evaluating Extrema procedures described in Sections 4.1 and 4.2.

The optimized flight plans can be seen in Figure 26. The squared speed of each vehicle is shown in Figure 27. Note that each vehicle begins and ends with the same speed. The vehicles never slow down less than their initial speeds which means they never reach the minimum speed constrain, nor do the vehicles go faster than the maximum speed. In Figure 28, the angular velocity of each vehicle is shown. The minimum and maximum angular rate constraints are shown by the dotted lines. The vehicles' angular rates never approach the minimum or maximum angular rate constraints due to the large area being covered. Finally, the squared euclidean distance between vehicles is shown in Figure 29. As expected, the squared Euclidean distance between two vehicles never falls below the minimum safe distance. Note that curves within the constraint plots end at different times. This is expected since each vehicle has a different final time. The furthest time reached in Figure 29 is less than that of the other plots because the other vehicles have already reached their final time before the longest flight reaches its final time.

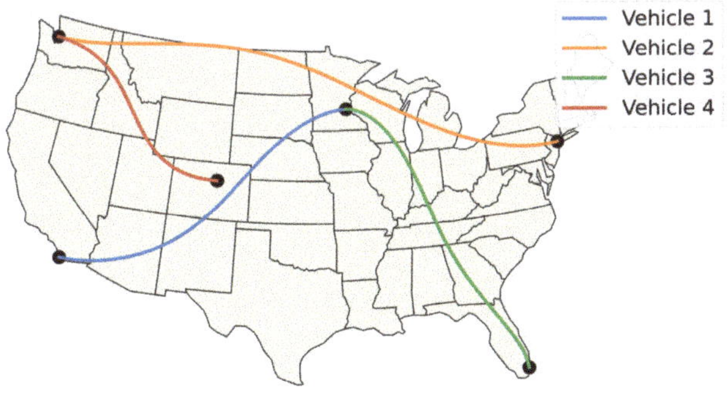

Figure 26. Commercial flight trajectories between major US cities.

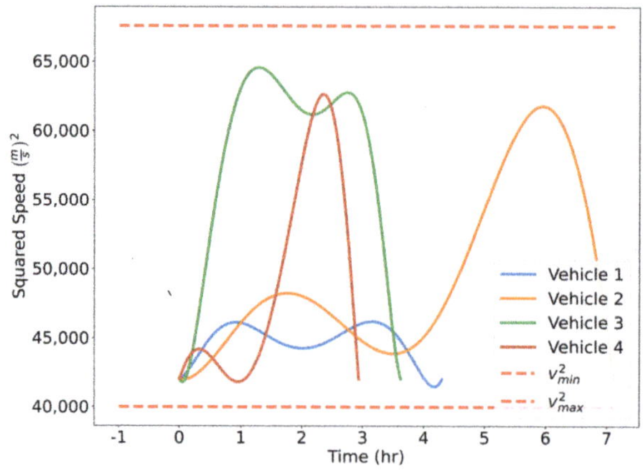

Figure 27. Verifying speed constraints for the Air Traffic Control example.

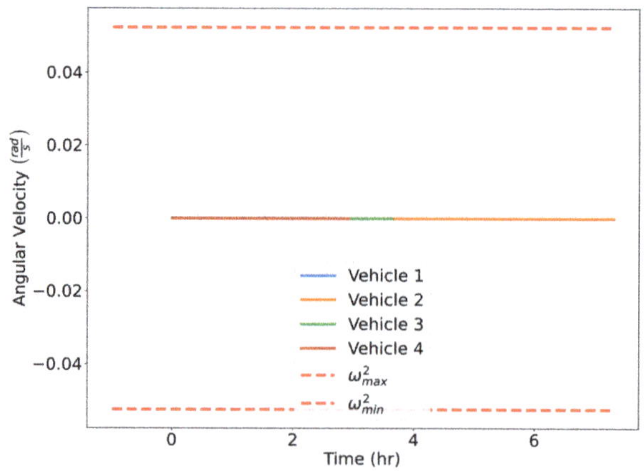

Figure 28. Verifying angular rate constraints for the Air Traffic Control example.

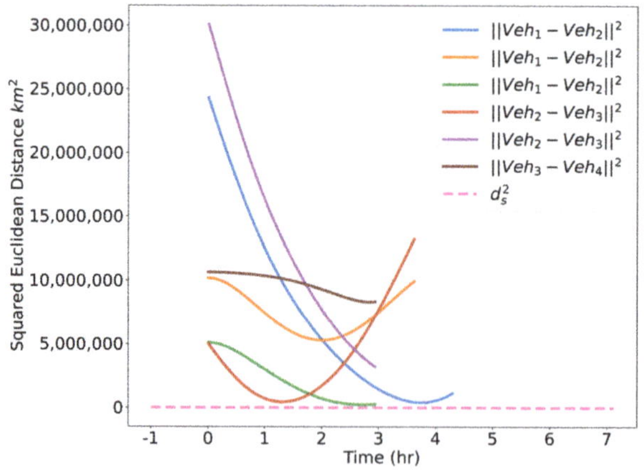

Figure 29. Verifying minimum safe distance constraints for the Air Traffic Control example.

5.3. Cluttered Environment

In many real world scenarios, robots must safely traverse cluttered environments. In this example, three aerial vehicles traveling at a constant altitude must navigate around several obstacles while also adhering to dynamic and minimum safe distance constraints. Let the position, speed, heading angle, and angular rate of each vehicle be defined as in Section 5.1. The goal of this example is to compute trajectories whose arc length is minimized subject to maximum speed constraints along with initial and final positions, heading angles, and speeds. The vehicles should also adhere to a minimum safe distance between each other and between obstacles. We formulate the problem as follows:

$$\min_{\mathbf{P}_n} \sum_{i=1}^{m} \sum_{k=0}^{n-1} ||\mathbf{P}_{k+1,n}^{[i]} - \mathbf{P}_{k,n}^{[i]}|| \tag{14}$$

subject to

$$\begin{aligned}
&\mathbf{C}_n^{[k]}(0) = \mathbf{C}_0^{[k]}, \quad \mathbf{C}_n^{[k]}(t_f) = \mathbf{C}_f^{[k]}, \\
&\psi^{[k]}(0) = \psi_0^{[k]}, \quad \psi^{[k]}(t_f) = \psi_f^{[k]}, \\
&||\dot{\mathbf{C}}_n^{[k]}(0)|| = v_0^{[k]}, \quad ||\dot{\mathbf{C}}_n^{[k]}(t_f)|| = v_f^{[k]}, \\
&||\dot{\mathbf{C}}_n^{[k]}(t)||^2 \leq v_{\max}^2, \quad \forall t \in [0, t_f], \\
&||\mathbf{C}_n^{[i]}(t) - \mathbf{C}_n^{[j]}(t)||^2 \geq d_s^2, \quad \forall i, j \in \{1, \ldots, m\}, i \neq j, \\
&||\mathbf{C}_n^{[i]}(t) - \mathbf{O}_j||^2 \geq d_{obs}^2, \quad \forall t \in [0, t_f], \quad i \in \{1, \ldots, m\}, \\
&\quad j \in \{1, \ldots, b\},
\end{aligned}$$

where $\mathbf{O}_j$ is the position of the jth obstacle out of b obstacles.

The initial positions for each vehicle, in order, are $[0,0]^\top$ m, $[10,0]^\top$ m, and $[20,0]^\top$ m. The initial speeds are all 1 $\frac{\text{m}}{\text{s}}$ and the initial heading angles are all $\frac{\pi}{2}$ rad. The final positions for each vehicle are, in order, $[20,30]^\top$ m, $[0,30]^\top$ m, and $[10,30]^\top$ m. The final speeds and final heading angles are the same as the initial speeds and heading angles. The order of the Bernstein polynomials being used is 7, the final time is $t_f = 30$ s, the minimum safe distance between vehicles is $d_s = 1$ m, the minimum safe distance between vehicles and obstacles is $d_{obs} = 2$ m, and the maximum speed is $v_{\max} = 10$ $\frac{\text{m}}{\text{s}}$. The vehicles traversing the cluttered environment can be seen in Figure 30. This experiment has been repeated in the Cooperative Autonomous Systems (CAS) lab using three AR Drones 2.0. The flight tests can be viewed at [47].

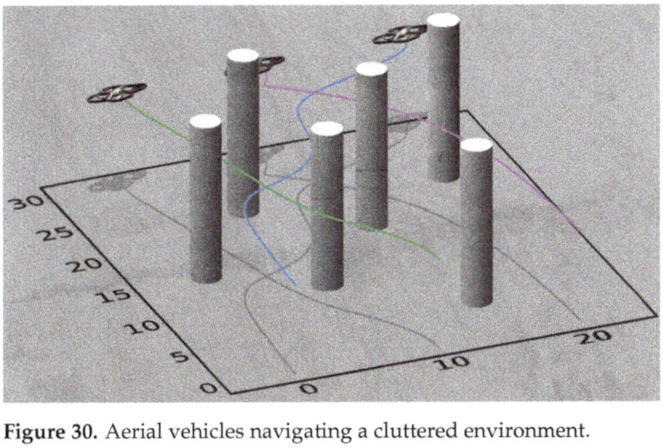

Figure 30. Aerial vehicles navigating a cluttered environment.

5.4. Vehicle Overtake

Here we consider an autonomous driving example in which one vehicle attempts to overtake another vehicle while driving around a 90° corner. The corner is defined by two arcs with a center point located at $[140, 0]^\top$ m. The inner track has a radius of $r_{inner} = 125$ m and the outer track has a radius of $r_{outer} = 140$ m. To clearly distinguish the vehicle being overtaken, it will be referred to as the opponent.

For simplicity, we consider the objective of minimizing the arc-length of the trajectory, which can be done by minimizing the sum of the squared Euclidean norm of consecutive control points, i.e.,

$$E(\mathbf{P}_n) = \sum_{i=1}^{n} ||\mathbf{P}_{i,n} - \mathbf{P}_{i-1,n}||^2. \tag{15}$$

The desired endpoint of the vehicle is at the end of the corner. This is computed by measuring the angle between the vehicle's position and the end of the curve,

$$A(\mathbf{P}_n) = \left(\arctan 2\left(\mathbf{P}_{n,n}^{[y]} - \mathbf{q}^{[y]}, \mathbf{P}_{n,n}^{[x]} - \mathbf{q}^{[x]}\right) - \frac{\pi}{2}\right)^2, \tag{16}$$

where the function arctan 2 returns an angle in the correct quadrant [48]. Given Equations (15) and (16), we formulate the problem as

$$\min_{\mathbf{P}_n}((1-\alpha)E(\mathbf{P}_n) + \alpha A(\mathbf{P}_n))\beta \tag{17}$$

subject to

$$\begin{aligned}
&\mathbf{C}_n(0) = \mathbf{C}_0, \quad \psi(0) = \psi_0, \quad ||\dot{\mathbf{C}}_n(0)|| = v_0, \\
&||\dot{\mathbf{C}}_n(t)||^2 \leq v_{\max}^2, \quad \forall t \in [0, t_f], \\
&|\dot{\psi}(t)| \leq \omega_{\max}, \quad \forall t \in [0, t_f], \\
&r_{inner}^2 \leq ||\mathbf{C}_n(t) - \mathbf{q}||^2 \leq r_{outer}^2, \\
&||\mathbf{C}_n(t) - \mathbf{O}_n(t)||^2 \geq d_s^2,
\end{aligned}$$

where α and β are tuning parameters. $\mathbf{C}_0$, ψ_0, and v_0 are the initial position, heading, and speed of the vehicle, respectively. $v_{\max}$ and $\omega_{\max}$ are the maximum speed and angular rate, respectively. The predicted trajectory of the opponent is represented as the Bernstein polynomial $\mathbf{O}_n(t)$ and the minimum safe distance to the opponent is d_s. Using a sensor such as a camera or LiDAR, one could measure the state of the opponent and then predict its future position using a method such as the one presented in [49].

At time $t = t_0$, when planning occurs, the position of the vehicle is $[5, 0]^\top$ m, its speed is 50 $\frac{\text{m}}{\text{s}}$, and its initial heading angle is $\frac{\pi}{2}$ rad. The control points of the Bernstein polynomial representing the opponent's trajectory are

$$\begin{bmatrix} 10 & 25 & 50 & 70 \\ 10 & 75 & 90 & 105 \end{bmatrix} \text{m}.$$

The maximum speed is 65 $\frac{\text{m}}{\text{s}}$, the maximum angular rate is $\frac{\pi}{5}\,\frac{\text{rad}}{\text{s}}$, and the minimum safe distance is 3 m. The tuning parameters are $\alpha = 1 - 10^{-6}$ and $\beta = 100$.

Figure 31 illustrates the optimized vehicle trajectory overtaking the opponent's trajectory. Figure 32 shows the squared speed of the vehicle along with the maximum speed constraint. As expected, the vehicle's speed approaches the maximum speed in order to successfully overtake the opponent. Figure 33 shows the vehicle's angular rate and its upper and lower constraints. It is clear that the vehicle remains within the desired bounds. Figure 34 shows the squared distance between the vehicle and the opponent. While the vehicle does come close to the opponent, it is never closer than the minimum safe distance.

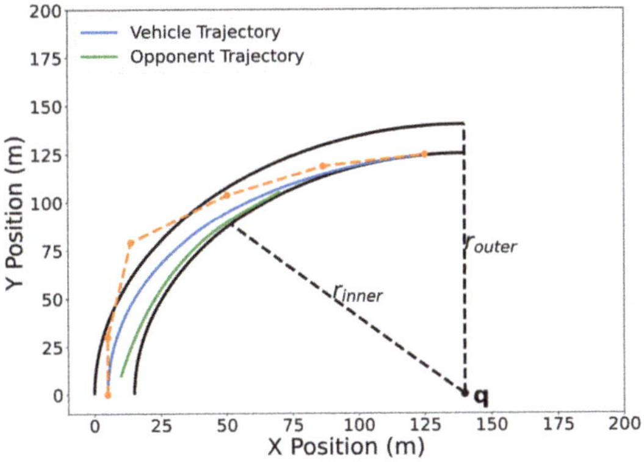

Figure 31. Overtaking trajectory. The track follows a quarter circle whose center point is at **q**. The inner track line has a radius of r_{inner} and the outer track line has a radius of r_{outer}.

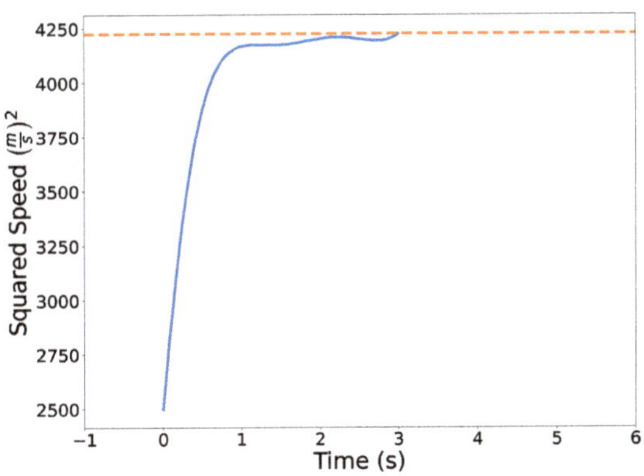

Figure 32. Squared speed profile of the vehicle.

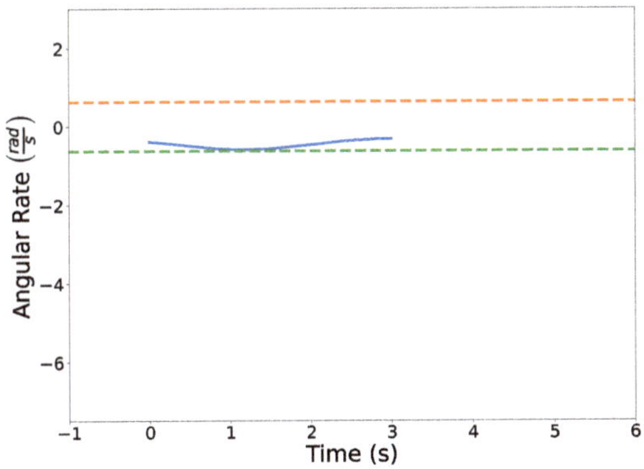

Figure 33. Angular rate profile of the vehicle.

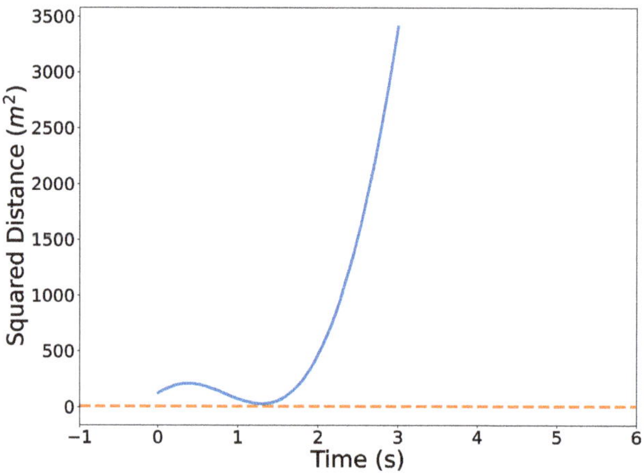

Figure 34. Squared distance between vehicle and opponent.

5.5. Swarming

This section examines two methods for generating trajectories for large groups of autonomous aerial vehicles. The centralized method optimizes every trajectory at once. On the other hand, the decentralized method generates trajectories one at a time and compares them to previously generated trajectories.

The position of each vehicle in a swarm of m vehicles for the following examples is parameterized as a 3D Bernstein polynomial, i.e.,

$$\sum_{i=0}^{n} \mathbf{P}_{i,n}^{[j]} B_{i,n}(t) = \mathbf{C}_n^{[j]}(t), \quad \forall j \in \{1,\ldots,m\}, \quad \mathbf{P}_n^{[j]} \in \mathbb{R}^{3\times n}.$$

5.5.1. 101 Vehicle—Centralized

The centralized method optimizes the trajectories for each vehicle simultaneously. The goal is to minimize the arc length of each trajectory. There are m vehicles with 3rd order Bernstein polynomials representing their trajectories which are constrained to a minimum safe distance between each other and initial and final positions. This is formulated as follows:

$$\min_{\mathbf{P}_n} \sum_{i=1}^{m} \sum_{k=0}^{n-1} ||\mathbf{P}_{k+1}^{[i]} - \mathbf{P}_k^{[i]}||,$$

subject to

$$\mathbf{C}_n^{[i]}(0) = \mathbf{C}_0^i, \quad \mathbf{C}_n^{[i]}(t_f) = \mathbf{C}_f^{[i]}, \quad \forall i \in \{1,\ldots,m\},$$
$$||\mathbf{C}_n^{[i]}(t) - \mathbf{C}_n^{[j]}(t)||^2 \geq d_s^2, \quad \forall i,j \in \{1,\ldots,m\}, i \neq j.$$

The initial positions for each vehicle were chosen randomly from a 25 m × 25 m grid at an altitude of $z = 0$ m. The final positions were chosen to spell out "CAS", as seen in Figure 35, at an altitude of $z = 100$ m. In the next section we significantly reduce the number of dimensions in the optimization vector by using the decentralized approach.

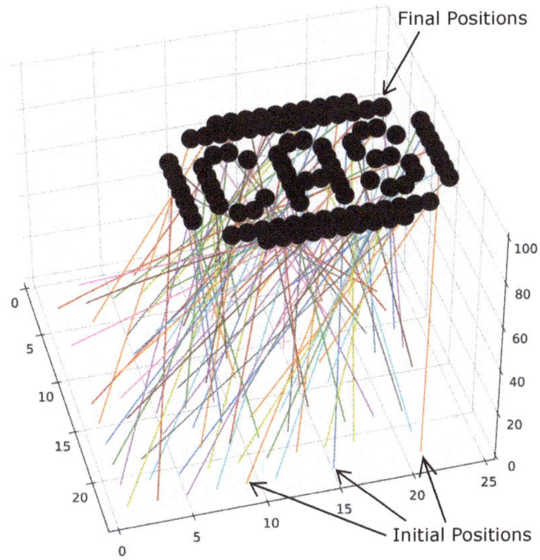

Figure 35. 101 vehicles spelling out CAS using the centralized method.

5.5.2. 101 Vehicle—Decentralized

The decentralized method iteratively computes trajectories for the ith vehicle. Each new iteration is compared to the previously computed trajectories so that the minimum safety distance constraint is met. The problem that is solved at each iteration is written as

$$\min_{\mathbf{P}_n^{[i]}} \sum_{i=1}^{m} \sum_{k=0}^{n-1} ||\mathbf{P}_{k+1}^{[i]} - \mathbf{P}_k^{[i]}||$$

subject to

$$\mathbf{C}_n^{[i]}(0) = \mathbf{C}_0^{[i]}, \quad \mathbf{C}_n^{[i]}(t_f) = \mathbf{C}_f^{[i]},$$
$$||\mathbf{C}_n^{[i]}(t) - \mathbf{C}_n^{[j]}(t)||^2 \geq D_s^2, \quad \forall j \in \{1,\dots,i-1\}, \quad i > 1.$$

Note that the first vehicle does not need to satisfy the minimum safe distance constraint since no trajectories have been computed before it.

The parameters used in this example were identical to that of the previous subsection. The resulting figure has been omitted due to its similarity to Figure 35.

5.5.3. 1000 Vehicle—Decentralized

The decentralized method can be used to compute 1000 trajectories. In this example, it is employed to generate the paths seen in Figure 36 to display the University of Iowa Hawkeye logo. The initial points are equally dispersed at an altitude of $z = 0$ m on a 100 m × 100 m grid. The final points are the pattern shown at an altitude of $z = 100$. The cost function aims to maximize the temporal distance between the current ith trajectory and the previously generated jth trajectories by taking the reciprocal of the sum of the Bernstein coefficients of the norm squared difference, i.e.

$$\min_{\mathbf{P}_n^{[i]}} \frac{1}{\sum_{j=1}^{i-1} \mathbf{P}^{[norm,j]}}, \quad i > 1,$$

subject to

$$\mathbf{C}_n^{[i]}(0) = \mathbf{C}_0^{[i]}, \quad \mathbf{C}_n^{[i]}(t_f) = \mathbf{C}_f^{[i]},$$

where $\mathbf{P}^{[norm,j]}$ are the Bernstein coefficients of the Bernstein polynomial representing the squared temporal distance between the ith and jth trajectories, i.e.,

$$||\mathbf{C}^{[i]}(t) - \mathbf{C}^{[j]}(t)||^2 = \sum_{i=0}^{n} P^{[norm,j]} B_{i,n}(t).$$

It should be noted that this formulation of cost function and constraints is used as a proof of concept. For other possible cost function and constraint formulations, the reader is referred to [50,51].

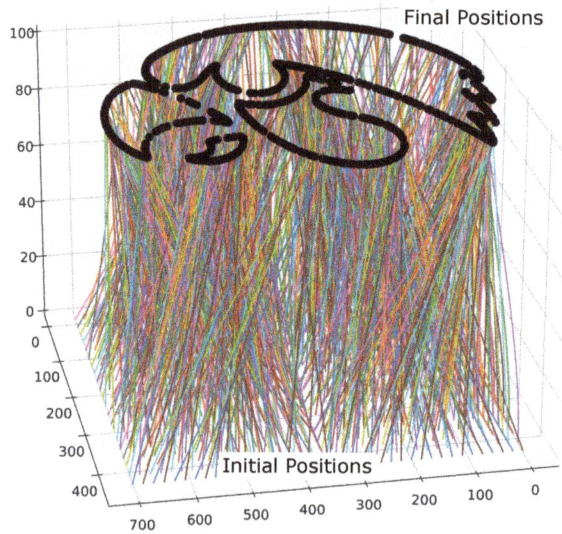

Figure 36. Trajectories for 1000 aerial vehicles with initial and final position and minimum safety distance constraints.

5.6. Marine Vehicle Model

In this example, we consider a marine vehicle model known as the medusa. The equations of motion of the medusa are as follows

$$\begin{aligned} \dot{x} &= u\cos\psi - v\sin\psi, \\ \dot{y} &= u\sin\psi + v\cos\psi, \\ \dot{\psi} &= r. \end{aligned} \tag{18}$$

$$\begin{aligned} m_u\dot{u} - m_v vr + d_u u &= \tau_u, \\ m_v\dot{v} + m_u ur + d_v v &= 0, \\ m_r\dot{r} - m_{uv} uv + d_r r &= \tau_r, \end{aligned} \tag{19}$$

where x and y represent the vehicle's position, ψ is the orientation, u (surge) and v (sway) are the linear velocities, r is the turning rate, and $\tau = [\tau_u, \tau_r]^T$ is the vector of forces and torques due to thrusters/surfaces (control input).

In this example, we let the state, $[x, y, \psi, u, v, r]^\top$, and input, $[\tau_u, \tau_r]^\top$, be approximated by Bernstein polynomials, and impose the vehicle's dynamics directly through Bernstein polynomial differentiation. Using Property A3, the dynamics constraints given by Equations (18) and (19) reduce to a set of algebraic constraints. Additional constraints imposed on this problem include collision avoidance and input saturation constraints. Figure 37 shows an example of motion planning for a medusa vehicle, which is required to reach a final destination in the minimum time. Ten markers are plotted along the trajectory (shown in blue) to represent the heading of the vehicle at that point in time. It can easily be seen that the vehicle's trajectory avoids the (inflated) unsafe region illustrated by the orange circle.

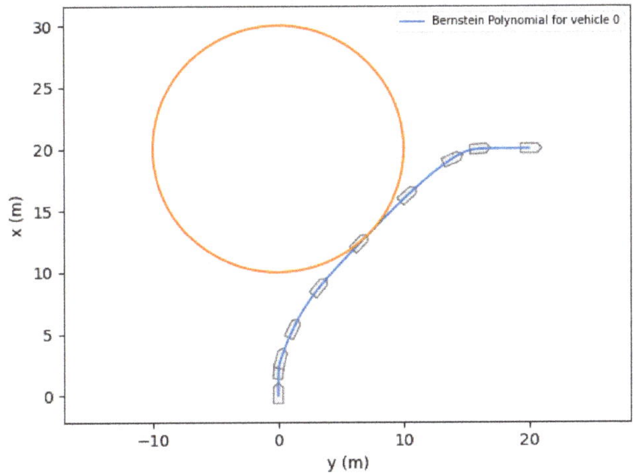

Figure 37. Medusa Example.

5.7. Dynamic Routing Problem

5.7.1. Single Vehicle Case

We consider a problem where a single vehicle is supposed to visit M neighborhoods $B_i = \{x \in R^2 : \|x - b_i\| \leq r\}$ in minimum time t_f. Here $r > 0$ and the vectors $b_i \in R^2$, $i \in [1, M]$ represent a sequence of points of interest that has been generated by a Traveling Salesman Problem (TSP) algorithm. Let t_i denote a time instance when the vehicle's position satisfies $\mathbf{C}_n(t_i) \in B_i$. Then, the dynamic routing problem for a single vehicle can be formulated as follows,

$\mathbf{DR}_1$

$$\min_{t_i, i \in [1,M],\ \mathbf{P}_n} t_f \tag{20}$$

subject to

$$\begin{aligned} \mathbf{C}_n(t_i) &\in B_i \\ t_{i-1} < t_i < t_{i+1},\ \forall i &= 2, M-1 \\ t_1 > 0,\ t_f &> t_M \\ \|\dot{\mathbf{C}}_n\|_\infty &\leq 1, \\ \|\ddot{\mathbf{C}}_n\|_\infty &\leq 1, \\ \mathbf{C}_n(0) = \mathbf{C}_n(t_f) &= \mathbf{C}_0 \end{aligned}$$

5.7.2. Numerical Solution: Single Vehicle Case

Let

$$\mathbf{C}_n^i(t) = \begin{cases} \sum_{k=0}^{n} \mathbf{P}_{i,k} B_{k,n}(t), t \in [t_{i-1}, t_i], t_0 = 0, t_{M+1} = t_f & , i \in [1, M+1] \\ 0, \text{ o.w.} \end{cases} \tag{21}$$

Define

$$\mathbf{C}_n(t) = \sum_{i=1}^{M+1} \mathbf{C}_n^i(t) \tag{22}$$

Then the numerical solution of the dynamic routing problem $\mathbf{DR}_1$ was obtained by solving the optimization problem

$$\min_{t_i, i \in [1,M], \ \mathbf{P}_n^i} t_f \tag{23}$$

subject to

$$\begin{aligned} \|\mathbf{C}_n^i(t_i) - b_i\| &\leq r \\ t_{i-1} < t_i < t_{i+1}, \ \forall i &= 2, M-1 \\ t_1 > 0, \ t_f &> t_M \\ \|\dot{\mathbf{C}}_n\|_\infty &\leq 1, \\ \|\ddot{\mathbf{C}}_n\|_\infty &\leq 1, \\ \mathbf{C}_n(0) = \mathbf{C}_n(t_f) &= \mathbf{C}_0 \\ \mathbf{C}_n^i(t_i) = \mathbf{C}_n^{i+1}(t_i), \ i &\in [1, M-1] \\ \dot{\mathbf{C}}_n^i(t_i) = \dot{\mathbf{C}}_n^{i+1}(t_i) \ i &\in [1, M-1] \end{aligned}$$

A simulation was performed illustrating a single agent visiting 30 neighborhoods. The resulting trajectory is shown in Figure 38. The agent is limited to velocities of arbitrary units ranging from -1 to 1 and is similarly limited to accelerations of arbitrary units also from -1 to 1. The velocities and accelerations of the vehicle can be seen in Figure 39 and Figure 40, respectively.

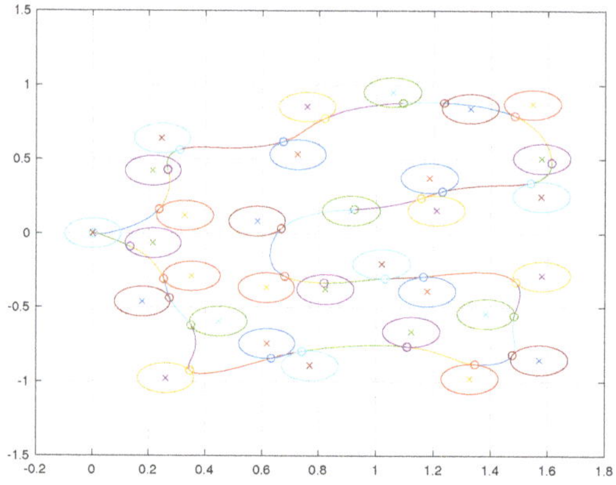

Figure 38. Trajectory of a single agent visiting 30 neighborhoods. The circles with an X in their center represent the neighborhoods and the circles along the vehicle's trajectory represent the points at which the vehicle makes its delivery.

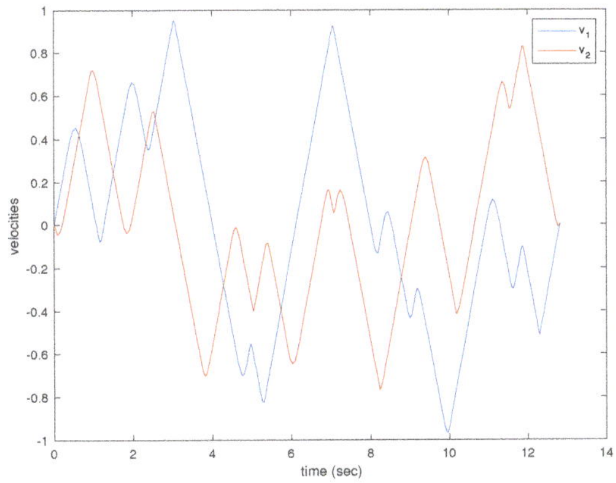

Figure 39. Velocity components of a single agent visiting 30 neighborhoods.

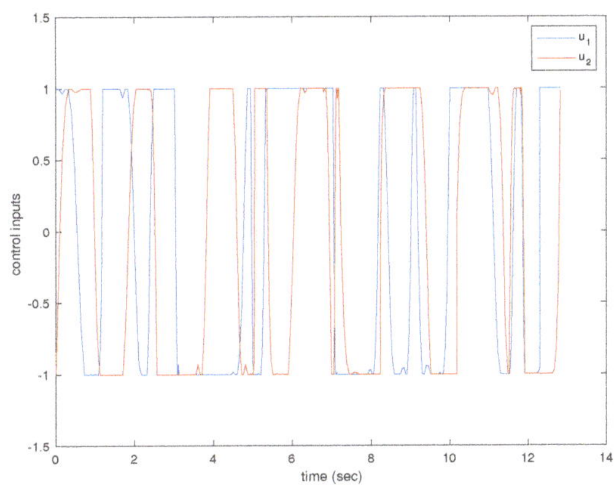

Figure 40. Acceleration components of a single agent visiting 30 neighborhoods.

5.7.3. Multiple Vehicle Case

In this case, K drones are assigned a total of K neighborhood sets to visit. Each neighborhood set, $\mathcal{P}_k$, consists of an equal number of neighborhoods B_{ij} which are defined by a set of points of interest $b_{ik} \in R^2$, i.e.,

$$B_{ik} = \{x \in R^2 : \|x - b_{ik}\| \leq r,\ i \in [1, M],\ k \in [1, K]\},$$

and

$$\mathcal{P}_k = \{B_{1k}, \ldots\ldots, B_{Mk},\ k \in [1, K]\}.$$

Let $t_{fk}, k \in [1, K]$ denote the total time it takes for the kth vehicle to visit every neighborhood in the set $\mathcal{P}_k$ once and let t_{ik} denote a time instance when the kth vehicle's position satisfies $\mathbf{C}_n^k(t_{ik}) \in B_{ik}$. Using this notation, we propose the following definition of the multi-vehicle dynamic routing problem for given positive number w_k and d

DR$_2$

$$\min_{t_{ik}, i \in [1,M], k \in [1,K],\ \mathbf{P}_n^k} \sum_{k=1}^{K} w_k t_{fk} \tag{24}$$

subject to

$$
\begin{aligned}
\mathbf{C}_n^k(t_{ik}) &\in B_{ik} \in \mathcal{P}_k, \forall k \in [1, K] \\
t_{i-1,k} < t_{ik} < t_{i+1,k}, &\ \forall i \in [2, M-1], k \in [1, K] \\
t_{1k} > 0, &\ t_{fk} > t_{Mk} \\
|\ddot{\mathbf{C}}_n^k(t)| &\leq u_{max}, \forall t \in [0, t_{fk}] \\
\mathbf{C}_n^k(0) = \mathbf{C}_n^k(t_f) &= \mathbf{C}_{k0}, \forall k \in [1, K] \\
|\mathbf{C}_n^k(t) - \mathbf{C}_n^l(t)| \geq d, k \neq l, \forall t \in [0, \max_{j \in [1,K]} t_{fj}], &\ k \in [1, K], l \in [1, K].
\end{aligned}
$$

Simulations were performed for a two agent and ten agent case each visiting 10 neighborhoods per agent. The resulting trajectories for the two agent case are shown in Figure 41 and for the ten agent case in Figure 42.

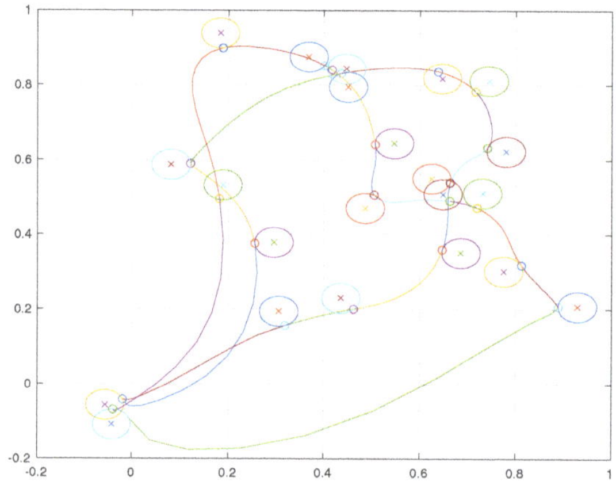

Figure 41. Two agents visiting ten neighborhoods each.

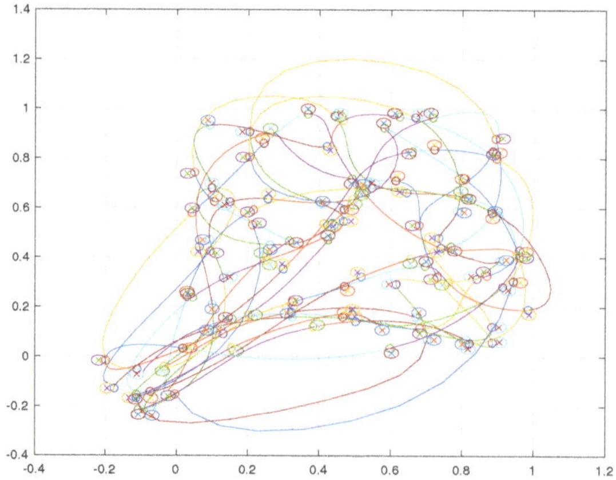

Figure 42. Ten agents visiting ten neighborhoods each.

6. Conclusions

We presented a method to generate optimal trajectories by using Bernstein polynomials to transcribe the problem into a nonlinear programming problem. By exploiting the useful properties of Bernstein polynomials, our method provides computationally efficient algorithms that can also guarantee safety in continuous time which are useful in optimization routines. These algorithms include evaluating bounds, evaluating extrema, minimum spatial distance between two Bernstein polynomials, minimum spatial distance between a Bernstein polynomial and a convex polygon, collision detection, and the penetration algorithm. We also developed an open source toolbox which makes these transcription methods readily available in the Python programming language.

Numerical examples were provided to demonstrate the efficacy of the method. Simple cost functions and constraints were implemented to generate trajectories for several realistic mission scenarios including air traffic control, navigating a cluttered environment, overtaking a vehicle, trajectory generation for a large swarm of vehicles, trajectory generation for a marine vehicle, and navigation for vehicles operating in a Traveling Salesman mission. Our formulation offers a powerful tool for users to generate optimal trajectories in real time scenarios for single or multiple robot teams. Future work includes developing new cost functions, exploring different optimization frameworks, and replanning trajectories to react to a changing environment.

Author Contributions: Conceptualization, C.K.-J., V.C., I.K., A.P., C.W.; methodology, C.K.-J., V.C., I.K., A.P., C.W.; software, C.K.-J., T.B., I.K., C.W.; validation, C.K.-J., T.B., I.K., C.W.; investigation, C.K.-J., V.C., I.K., A.P., C.W.; resources, V.C., I.K., A.P.; writing—original draft preparation, C.K.-J., V.C., I.K.; writing—review and editing, C.K.-J.; supervision, V.C., I.K., A.P., C.W.; project administration, V.C.; funding acquisition, V.C., I.K., A.P. All authors have read and agreed to the published version of the manuscript.

Funding: This research was supported by the Office of Naval Research, grants N000141912106, N000142112091 and N0001419WX00155. Antonio Pascoal was supported by H2020-EU.1.2.2—FET Proactive RAMONES, under Grant GA 101017808 and LARSyS-FCT under Grant UIDB/50009/2020. Isaac Kaminer was supported by the Office of Naval Research grant N0001421WX01974.

Institutional Review Board Statement: Not applicable.

Informed Consent Statement: Not applicable.

Data Availability Statement: Not applicable.

Conflicts of Interest: The authors declare no conflict of interest.

Appendix A

In this section we present and examine relevant properties of Bernstein polynomials and rational Bernstein polynomials, which are used throughout the article.

Property A1. *Convex Hull*

Both the Bernstein polynomial, Equation (5), and the rational Bernstein polynomial (6) (provided that $w_{i,n} > 0, i = 0, \dots, n$) satisfy

$$\min_{k \in \{0,\dots,n\}} P_k \leq C_n(t) \leq \max_{k \in \{0,\dots,n\}} P_k, \forall t \in [t_0, t_f]. \tag{A1}$$

It follows that 2D (or 3D) Bernstein polynomials and rational Bernstein polynomials lie within the convex hull defined by their Bernstein coefficients. An example of this property is depicted in Figure A1, which shows a 2D Bernstein polynomial contained within its convex hull.

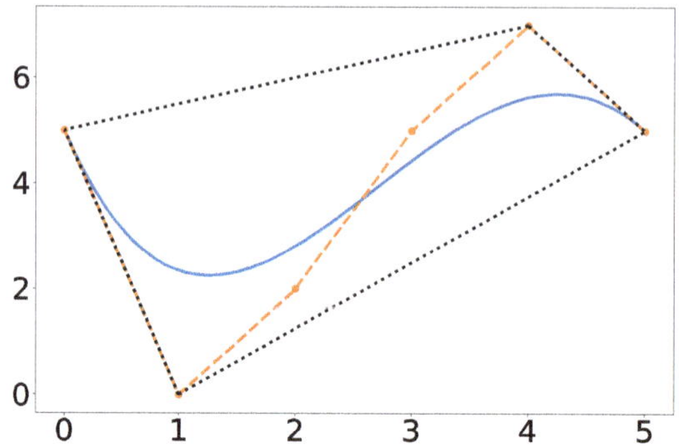

Figure A1. Convex hull of a Bernstein polynomial where the solid blue line is the curve, the black dotted line is the convex hull, and the orange points connected by dashed lines are the Bernstein coefficients.

Property A2. *End Point Values*

The first and last Bernstein coefficients of the Bernstein polynomial introduced in Equation (5), as well as the rational Bernstein polynomial in Equation (6), are their endpoints, i.e.,

$$C_n(t_0) = P_0 \quad \text{and} \quad C_n(t_f) = P_n.$$

Furthermore, the tangents to the curve at its first and last coefficients are the same as the lines connecting the first and second coefficients and the penultimate and ultimate coefficients, respectively. It follows that the first derivative of the Bernstein polynomial at the end points is as follows

$$\begin{aligned} \dot{C}_n(t_0) &= \frac{n}{t_f - t_0}(P_{1,n} - P_{0,n}), \\ \dot{C}_n(t_f) &= \frac{n}{t_f - t_0}(P_{n,n} - P_{n-1,n}). \end{aligned} \tag{A2}$$

Similarly, for a rational Bernstein polynomial we have

$$\begin{aligned} \dot{C}_n(t_0) &= \frac{nw_1}{(t_f - t_0)w_0}(P_{1,n} - P_{0,n}), \\ \dot{C}_n(t_f) &= \frac{nw_{n-1}}{(t_f - t_0)w_n}(P_{n,n} - P_{n-1,n}). \end{aligned} \tag{A3}$$

Property A3. *Derivatives*

The derivative of the Bernstein polynomial introduced in Equation (5) is an $(n-1)$th order Bernstein polynomial given by

$$\dot{C}_{n-1}(t) = \sum_{i=0}^{n-1} P'_{i,n-1} B_{i,n-1}(t), \tag{A4}$$

with the vector of Bernstein coefficients $P'_{n-1} = [P'_{0,n-1}, \ldots, P'_{n-1,n-1}]$ given by

$$P'_{n-1} = P_n \mathbf{D}_n.$$

In the equation above, $\mathbf{D}_n$ denotes the differentiation matrix given by

$$\mathbf{D}_n = \frac{n}{t_f - t_0}\begin{bmatrix} -1 & 0 & \cdots & 0 \\ 1 & \ddots & \ddots & \vdots \\ 0 & \ddots & \ddots & 0 \\ \vdots & \ddots & \ddots & -1 \\ 0 & \cdots & 0 & 1 \end{bmatrix} \in \mathbb{R}^{n+1\times n}. \tag{A5}$$

Remark A1. *Properties A2 and A3 can be easily extended to compute higher order derivatives of Bernstein polynomials.*

Property A4. *Integrals*

The definite integral of a Bernstein polynomial is given by

$$\int_{t_0}^{t_f} C_n(t)dt = \frac{t_f - t_0}{n+1}\sum_{i=0}^{n} P_{i,n}\,. \tag{A6}$$

Property A5. *The de Casteljau Algorithm*

The de Casteljau algorithm computes a Bernstein polynomial defined over an interval $[t_0, t_f]$ at any given $t_{div} \in [t_0, t_f]$. Moreover, it can be used to split a single Bernstein polynomial into two independent Bernstein polynomials. Given the Bernstein polynomial introduced in Equation (5) with the vector of coefficients $P_n = [P_{0,n}, \ldots, P_{n,n}]$, and a scalar $t_{div} \in [t_0, t_f]$, the Bernstein polynomial at t_{div} is computed using the following recursive relationship:

$$P^0_{i,n} = P_{i,n}\,, \quad i = 0, \ldots, n\,,$$

$$P^j_{i,n} = \frac{t_f - t_{div}}{t_f - t_0}P^{j-1}_{i,n} + \frac{t_{div} - t_0}{t_f - t_0}P^{j-1}_{i+1,n}\,,$$

with $i = 0, \ldots, n-j$, and $j = 1, \ldots, n$. Then, the Bernstein polynomial evaluated at t_{div} is $C_n(t_{div}) = P^n_{0,n}$. Note that the superscript here signifies an index and not an exponent. Moreover, the Bernstein polynomial can be subdivided at t_{div} into two nth order Bernstein polynomials with coefficients

$$P^0_{0,n}, P^1_{0,n}, \ldots, P^n_{0,n} \quad \text{and} \quad P^n_{0,n}, P^{n-1}_{1,n}, \ldots, P^0_{n,n}\,.$$

A geometric example of a 3rd order Bernstein polynomial being split at $t_{div} = 0.5$ using the de Casteljau is shown in Figure A2. Note that at the final iteration only a single point remains, $\mathbf{P}^3_{0,n}$, and lies on the original curve. The points $\{\mathbf{P}^0_{0,n}, \mathbf{P}^1_{0,n}, \mathbf{P}^2_{0,n}, \mathbf{P}^3_{0,n}\}$ become the Bernstein coefficients of the left curve and the points $\{\mathbf{P}^3_{0,n}, \mathbf{P}^2_{1,n}, \mathbf{P}^1_{2,n}, \mathbf{P}^0_{3,n}\}$ become the coefficients of the right curve.

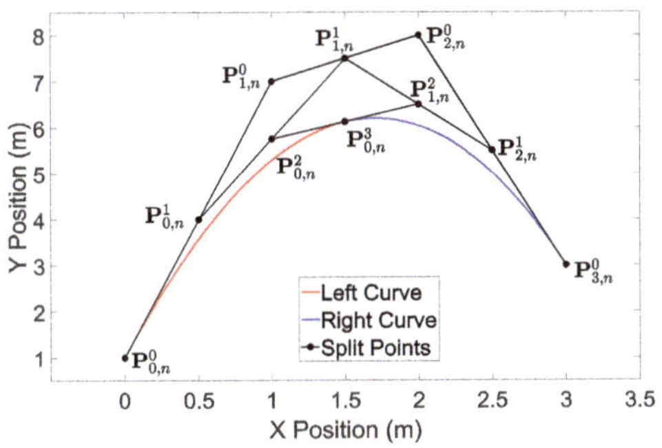

Figure A2. Geometric example of the de Casteljau algorithm splitting a 2D Bernstein polynomial at $t_{div} = 0.5$. The original curve is defined by the Bernstein coefficients $\{\mathbf{P}^0_{0,n}, \ldots, \mathbf{P}^0_{3,n}\}$, the left hand curve is shown in red, the right hand curve is shown in blue, and the superscript corresponds to the current iteration of the algorithm.

Remark A2. *For high order Bernstein polynomials, the exponential terms in Equation (5) can cause overflows. The de Casteljau algorithm can be used to overcome these issues on a computer when computing the value of a Bernstein polynomial, especially if lower bit floating point values are to be used such as 16 or 32 bits.*

Property A6. *Degree Elevation*

The degree of a Bernstein polynomial can be elevated by one using the following recursive relationship

$$P_{i,n+1} = \frac{i}{n+1} P_{i-1,n} + \left(1 - \frac{i}{n+1}\right) P_{i,n},$$

where $P_{0,n+1} = P_{0,n}$ *and* $P_{n+1,n+1} = P_{n,n}$. *Furthermore, any Bernstein polynomial of degree n can be expressed as a Bernstein polynomial of degree m,* $m > n$. *The vector of coefficients of the degree elevated Bernstein polynomial, namely* $\mathbf{P}_m = [P_{0,m}, \ldots, P_{m,m}]$, *can be calculated as*

$$\mathbf{P}_m = \mathbf{P}_n \mathbf{E},$$

where $\mathbf{E} = \{e_{j,k}\} \in \mathbb{R}^{(n+1)\times(m+1)}$ *is the degree elevation matrix with elements given by*

$$e_{i,i+j} = \frac{\binom{m-n}{j}\binom{n}{i}}{\binom{m}{i+j}}, \tag{A7}$$

where $i = 0, \ldots, n$ *and* $j = 0, \ldots, m-n$, *all other values in the matrix are zero, and* $\mathbf{P}_n = [P_{0,n}, \ldots, P_{n,n}]$ *is the vector of Bernstein coefficients of the curve being elevated (see [52]). Because* $\mathbf{E}$ *is conveniently independent of the coefficients, it can be computed ahead of time and stored for efficient computation of degree elevated Bernstein polynomials. By degree elevating a Bernstein polynomial, the coefficients converge to the polynomial, i.e.*

$$\max_i \left| P_{i,m} - C_m\left(\frac{i}{m}(t_f - t_0) + t_0\right) \right| \leq \frac{k}{m}, \tag{A8}$$

where k is a positive constant independent of m (see [53]).

Property A7. *Arithmetic Operations*

Arithmetic operations of Bernstein polynomials require that both curves be defined on the same time interval $[t_0, t_f]$. *In case they are not, the de Casteljau algorithm can be used to split them at an intersecting time interval.*

The sum and difference of two polynomials of the same order can be performed by simply adding and subtracting their Bernstein coefficients, respectively. For Bernstein polynomials of different order, the Degree Elevation (Property A6) may be used to increase the order of the lower order Bernstein polynomial.

Given two Bernstein polynomials, $F_m(t)$ *and* $G_n(t)$, *with degrees m and n, respectively, and having Bernstein coefficients* $X_{0,m}, \ldots, X_{m,m}$ *and* $Y_{0,n}, \ldots, Y_{n,n}$, *the product* $C_{m+n}(t) = F_m(t)G_n(t)$ *is a Bernstein polynomial of degree* $(m+n)$ *with coefficients* $P_{k,m+n}$, $k \in \{0, \ldots, m+n\}$ *given by*

$$P_{k,m+n} = \sum_{j=\max(0,k-n)}^{\min(m,k)} \frac{\binom{m}{j}\binom{n}{k-j}}{\binom{m+n}{k}} X_{j,m} Y_{k-j,n}. \tag{A9}$$

The ratio between two Bernstein polynomials, $F_n(t)$ *and* $G_n(t)$, *with coefficients* $X_{0,n}, \ldots, X_{n,n}$ *and* $Y_{0,n}, \ldots, Y_{n,n}$, *i.e.,* $R_n(t) = F_n(t)/G_n(t)$, *can be expressed as a rational Bernstein polynomial as defined in Equation (6), with coefficients and weights*

$$P_{i,n} = \frac{X_{i,n}}{Y_{i,n}}, \quad w_{i,n} = Y_{i,n},$$

respectively.

Property A8. *The de Casteljau Algorithm Extended to Rational Bernstein Polynomials*

The de Casteljau algorithm can be extended to rational Bernstein polynomials (see [54]). Letting

$$w_{i,n}^r(t) = \sum_{j=0}^{r} w_{i+j,n} B_{i,r}(t), \tag{A10}$$

we can determine a point on an nth order rational Bernstein polynomial using the recursive equation

$$\begin{aligned} P_{i,n}^r(t) = &\left(\frac{t_f - t}{t_f - t_0}\right) \frac{w_{i,n}^{r-1}(t)}{w_{i,n}^r(t)} P_{i,n}^{r-1}(t) \\ &+ \left(\frac{t - t_0}{t_f - t_0}\right) \frac{w_{i+1,n}^{r-1}(t)}{w_{i,n}^r(t)} P_{i+1,n}^{r-1}(t). \end{aligned} \tag{A11}$$

Moreover, the recursive relationship can be used to split the rational Bernstein polynomial into two nth order rational Bernstein polynomials with weights

$$w_{0,n}^0, w_{0,n}^1, \ldots, w_{0,n}^n \quad \text{and} \quad w_{0,n}^n, w_{1,n}^{n-1}, \ldots, w_{n,n}^0,$$

and coefficients

$$P_{0,n}^0, P_{0,n}^1, \ldots, P_{0,n}^n \quad \text{and} \quad P_{0,n}^n, P_{1,n}^{n-1}, \ldots, P_{n,n}^0.$$

Property A9. *Degree Elevation for Rational Bernstein Polynomials*

Degree elevation can be extended to rational Bernstein polynomials (see [54]). The nth order rational Bernstein polynomial given by Equation (6) can be rewritten as a rational Bernstein polynomial of order $n+1$ with weights

$$w_{i,n+1} = \frac{i}{n+1} w_{i-1,n} + \left(1 - \frac{i}{n+1}\right) w_{i,n}.$$

where $w_{0,n+1} = w_{0,n}$ and $w_{n+1,n+1} = w_{n,n}$, and coefficients

$$P_{i,n+1} = \frac{\frac{i}{n+1} w_{i,n} P_{i,n} + \left(1 - \frac{i}{n+1}\right) w_{i+1,n} P_{i+1,n}}{\frac{i}{n+1} w_{i,n} + \left(1 - \frac{i}{n+1}\right) w_{i+1,n}}.$$

References

1. Milford, M.; Anthony, S.; Scheirer, W. Self-driving vehicles: Key technical challenges and progress off the road. *IEEE Potentials* **2019**, *39*, 37–45. [CrossRef]
2. Mogili, U.R.; Deepak, B. Review on application of drone systems in precision agriculture. *Procedia Comput. Sci.* **2018**, *133*, 502–509. [CrossRef]
3. Koerhuis, R. Odd.Bot Robot Takes on Herbicide Free Weed Elimination. 2018 . Available online: https://www.futurefarming.com/Tools-data/Articles/2018/12/OddBot-robot-takes-on-herbicide-free-weed-elimination-375027E/ (accessed on 4 May 2020)
4. Lutz, M.; Meurer, T. Optimal Trajectory Planning and Model Predictive Control of Underactuated Marine Surface Vessels using a Flatness-Based Approach. *arXiv* **2021**, arXiv:2101.12730.
5. Ackerman, E. Startup Developing Autonomous Delivery Robots That Travel on Sidewalks. 2015. Available online: https://spectrum.ieee.org/automaton/robotics/industrial-robots/starship-technologies-autonomous-ground-delivery-robots (accessed on 4 May 2020)
6. Ackerman, E.; Koziol, M. In the Air with Zipline's Medical Delivery Drones. 2019. Available online: https://spectrum.ieee.org/robotics/drones/in-the-air-with-ziplines-medical-delivery-drones (accessed on 4 May 2020)
7. Lardinois, F. A First Look at Amazon's New Delivery Drone. 2019. Available online: https://techcrunch.com/2019/06/05/a-first-look-at-amazons-new-delivery-drone/ (accessed on 4 May 2020)
8. Johnson, A.; Hautaluoma, G. NASA's Ingenuity Mars Helicopter Succeeds in Historic First Flight. 2021. Available online: https://www.nasa.gov/press-release/nasa-s-ingenuity-mars-helicopter-succeeds-in-historic-first-flight (accessed on 18 June 2020).
9. Lumelsky, V.J.; Stepanov, A.A. Path-planning strategies for a point mobile automaton moving amidst unknown obstacles of arbitrary shape. *Algorithmica* **1987**, *2*, 403–430. [CrossRef]

10. Kamon, I.; Rivlin, E. Sensory-based motion planning with global proofs. *IEEE Trans. Robot. Autom.* **1997**, *13*, 814–822. [CrossRef]
11. McGuire, K.; de Croon, G.; Tuyls, K. A Comparative Study of Bug Algorithms for Robot Navigation. *Robot. Auton. Syst.* **2019**, *121*, 103261. [CrossRef]
12. Fiorini, P.; Shiller, Z. Motion planning in dynamic environments using the relative velocity paradigm. In Proceedings of the IEEE International Conference on Robotics and Automation, Atlanta, GA, USA, 2–6 May 1993; pp. 560–565.
13. Abe, Y.; Yoshiki, M. Collision avoidance method for multiple autonomous mobile agents by implicit cooperation. In Proceedings of the 2001 IEEE/RSJ International Conference on Intelligent Robots and Systems, Expanding the Societal Role of Robotics in the the Next Millennium (Cat. No. 01CH37180), Maui, HI, USA, 29 October–3 November 2001; Volume 3, pp. 1207–1212.
14. Large, F.; Sckhavat, S.; Shiller, Z.; Laugier, C. Using non-linear velocity obstacles to plan motions in a dynamic environment. In Proceedings of the 7th International Conference on Control, Automation, Robotics and Vision, ICARCV 2002, Singapore, 2–5 December 2002; Volume 2, pp. 734–739.
15. Wilkie, D.; Van Den Berg, J.; Manocha, D. Generalized velocity obstacles. In Proceedings of the 2009 IEEE/RSJ International Conference on Intelligent Robots and Systems, St. Louis, MO, USA, 10–15 October 2009; pp. 5573–5578.
16. Khatib, O. Real-time obstacle avoidance for manipulators and mobile robots. In *Autonomous Robot Vehicles*; Springer: New York, NY, USA, 1986; pp. 396–404.
17. Chen, Y.B.; Luo, G.C.; Mei, Y.S.; Yu, J.Q.; Su, X.L. UAV path planning using artificial potential field method updated by optimal control theory. *Int. J. Syst. Sci.* **2016**, *47*, 1407–1420. [CrossRef]
18. Orozco-Rosas, U.; Montiel, O.; Sepúlveda, R. Mobile robot path planning using membrane evolutionary artificial potential field. *Appl. Soft Comput.* **2019**, *77*, 236–251. [CrossRef]
19. Kavraki, L.E.; Svestka, P.; Latombe, J.C.; Overmars, M.H. Probabilistic roadmaps for path planning in high-dimensional configuration spaces. *IEEE Trans. Robot. Autom.* **1996**, *12*, 566–580. [CrossRef]
20. LaValle, S.M.; Kuffner, J.J., Jr. Randomized kinodynamic planning. *Int. J. Robot. Res.* **2001**, *20*, 378–400. [CrossRef]
21. Karaman, S.; Frazzoli, E. Incremental sampling-based algorithms for optimal motion planning. *arXiv* **2010**, arXiv:1005.0416.
22. Sleumer, N.; Tschichold-Gürmann, N. *Exact Cell Decomposition of Arrangements Used for Path Planning in Robotics*; Technical Report; ETH Zurich, Department of Computer Science: Zurich, Switzerland, 1999; Volume 329.
23. Cai, C.; Ferrari, S. Information-driven sensor path planning by approximate cell decomposition. *IEEE Trans. Syst. Man Cybern. Part B (Cybern.)* **2009**, *39*, 672–689.
24. Hart, P.E.; Nilsson, N.J.; Raphael, B. A formal basis for the heuristic determination of minimum cost paths. *IEEE Trans. Syst. Sci. Cybern.* **1968**, *4*, 100–107. [CrossRef]
25. Vasconcelos, J.V.R.; Brandão, A.S.; Sarcinelli-Filho, M. Real-Time Path Planning for Strategic Missions. *Appl. Sci.* **2020**, *10*, 7773. [CrossRef]
26. Likhachev, M.; Ferguson, D.I.; Gordon, G.J.; Stentz, A.; Thrun, S. Anytime Dynamic A*: An Anytime, Replanning Algorithm. In Proceedings of the International Conference on Automated Planning and Scheduling (ICAPS), Monterey, CA, USA, 5–10 June 2005; Volume 5, pp. 262–271.
27. Karaman, S.; Walter, M.R.; Perez, A.; Frazzoli, E.; Teller, S. Anytime motion planning using the RRT. In Proceedings of the 2011 IEEE International Conference on Robotics and Automation, Shanghai, China, 9–13 May 2011; pp. 1478–1483.
28. Mellinger, D.; Kumar, V. Minimum snap trajectory generation and control for quadrotors. In Proceedings of the 2011 IEEE International Conference on Robotics and Automation, Shanghai, China, 9–13 May 2011; pp. 2520–2525.
29. Schulman, J.; Ho, J.; Lee, A.X.; Awwal, I.; Bradlow, H.; Abbeel, P. Finding Locally Optimal, Collision-Free Trajectories with Sequential Convex Optimization. In Proceedings of the Robotics: Science and Systems, Berlin, Germany, 24–28 June 2013; Volume 9, pp. 1–10. [CrossRef]
30. Ratliff, N.; Zucker, M.; Bagnell, J.A.; Srinivasa, S. CHOMP: Gradient optimization techniques for efficient motion planning. In Proceedings of the 2009 IEEE International Conference on Robotics and Automation, Kobe, Japan, 12–17 May 2009; pp. 489–494.
31. Zucker, M.; Ratliff, N.; Dragan, A.D.; Pivtoraiko, M.; Klingensmith, M.; Dellin, C.M.; Bagnell, J.A.; Srinivasa, S.S. Chomp: Covariant hamiltonian optimization for motion planning. *Int. J. Robot. Res.* **2013**, *32*, 1164–1193. [CrossRef]
32. Kalakrishnan, M.; Chitta, S.; Theodorou, E.; Pastor, P.; Schaal, S. STOMP: Stochastic trajectory optimization for motion planning. In Proceedings of the 2011 IEEE International Conference on Robotics and Automation, Shanghai, China, 9–13 May 2011; pp. 4569–4574.
33. Tang, S.; Thomas, J.; Kumar, V. Hold or Take Optimal Plan (HOOP): A quadratic programming approach to multi-robot trajectory generation. *Int. J. Robot. Res.* **2018**, *37*, 1062–1084. [CrossRef]
34. Semenas, R.; Bausys, R.; Zavadskas, E.K. A Novel Environment Exploration Strategy by m-generalised q-neutrosophic WASPAS. *Stud. Inform. Control* **2021**, *30*, 19–28. [CrossRef]
35. Farouki, R.; Goodman, T. On the optimal stability of the Bernstein basis. *Math. Comput. Am. Math. Soc.* **1996**, *65*, 1553–1566. [CrossRef]
36. Cichella, V.; Kaminer, I.; Walton, C.; Hovakimyan, N. Optimal Motion Planning for Differentially Flat Systems Using Bernstein Approximation. *IEEE Control Syst. Lett.* **2018**, *2*, 181–186. [CrossRef]
37. Cichella, V.; Kaminer, I.; Walton, C.; Hovakimyan, N.; Pascoal, A. Bernstein approximation of optimal control problems. *arXiv* **2018**, arXiv:1812.06132.

38. Cichella, V.; Kaminer, I.; Walton, C.; Hovakimyan, N.; Pascoal, A.M. Optimal Multi-Vehicle Motion Planning using Bernstein Approximants. *IEEE Trans. Autom. Control* **2020**, *66*, 1453–1467. [CrossRef]
39. Kielas-Jensen, C.; Cichella, V. BeBOT: Bernstein polynomial toolkit for trajectory generation. In Proceedings of the 2019 IEEE/RSJ International Conference on Intelligent Robots and Systems (IROS), Macau, China, 3–8 November 2019; pp. 3288–3293.
40. Kielas-Jensen, C.; Cichella, V. BeBOT. Available online: https://github.com/caslabuiowa/BeBOT (accessed on 21 May 2020).
41. Bézier, P. Definition numerique des courbes et surface. *Automatisme* **1966**, *11*, 625–632.
42. Bézier, P. Définition numérique des courbes et surfaces (ii). *Automatisme* **1967**, *12*, 17–21.
43. Forrest, A.R. Interactive interpolation and approximation by Bézier polynomials. *Comput. J.* **1972**, *15*, 71–79. [CrossRef]
44. Chang, J.W.; Choi, Y.K.; Kim, M.S.; Wang, W. Computation of the minimum distance between two Bézier curves/surfaces. *Comput. Graph.* **2011**, *35*, 677–684. [CrossRef]
45. Gilbert, E.G.; Johnson, D.W.; Keerthi, S.S. A fast procedure for computing the distance between complex objects in three-dimensional space. *IEEE J. Robot. Autom.* **1988**, *4*, 193–203. [CrossRef]
46. Van Den Bergen, G. Proximity queries and penetration depth computation on 3d game objects. In Proceedings of the Game Developers Conference, San Jose, CA, USA, 20–24 March 2001; Volume 170.
47. Kielas-Jensen, C.; Cichella, V. Trajectory Generation Using Bernstein Polynomials (Bézier Curves). Available online: https://www.youtube.com/watch?v=e0SPt92W578 (accessed on 31 July 2019).
48. *Programming Language C Standard*; International Organization for Standardization: Geneva, Switzerland, 1999.
49. Patterson, A.; Lakshmanan, A.; Hovakimyan, N. Intent-aware probabilistic trajectory estimation for collision prediction with uncertainty quantification. In Proceedings of the 2019 IEEE 58th Conference on Decision and Control (CDC), Nice, France, 11–13 December 2019; pp. 3827–3832.
50. Hauser, J.; Saccon, A. A barrier function method for the optimization of trajectory functionals with constraints. In Proceedings of the 45th IEEE Conference on Decision and Control, San Diego, CA, USA, 13–15 December 2006; pp. 864–869.
51. Zhang, X.; Liniger, A.; Borrelli, F. Optimization-based collision avoidance. *IEEE Trans. Control. Syst. Technol.* **2020**, *29*, 972–983. [CrossRef]
52. Lee, B.G.; Park, Y. Distance for Bézier curves and degree reduction. *Bull. Aust. Math. Soc.* **1997**, *56*, 507–515. [CrossRef]
53. Prautzsch, H.; Kobbelt, L. Convergence of subdivision and degree elevation. *Adv. Comput. Math.* **1994**, *2*, 143–154. [CrossRef]
54. Farin, G. Algorithms for rational Bézier curves. *Comput.-Aided Des.* **1983**, *15*, 73–77. [CrossRef]

sensors

Article

A Distributed Algorithm for Real-Time Multi-Drone Collision-Free Trajectory Replanning

Bahareh Sabetghadam *, Rita Cunha and António Pascoal

Laboratory of Robotics and Engineering Systems (LARSyS), ISR/IST, University of Lisbon, 1649-004 Lisbon, Portugal; rita@isr.ist.utl.pt (R.C.); antonio@isr.ist.utl.pt (A.P.)
* Correspondence: bsabetghadam@isr.ist.utl.pt

Abstract: In this paper, we present a distributed algorithm to generate collision-free trajectories for a group of quadrotors flying through a common workspace. In the setup adopted, each vehicle replans its trajectory, in a receding horizon manner, by solving a small-scale optimization problem that only involves its own individual variables. We adopt the Voronoi partitioning of space to derive local constraints that guarantee collision avoidance with all neighbors for a certain time horizon. The obtained set of collision avoidance constraints explicitly takes into account the vehicle's orientation to avoid infeasiblity issues caused by ignoring the quadrotor's rotational motion. Moreover, the resulting constraints can be expressed as Bézier curves, and thus can be evaluated efficiently, without discretization, to ensure that collision avoidance requirements are satisfied at any time instant, even for an extended planning horizon. The proposed approach is validated through extensive simulations with up to 100 drones. The results show that the proposed method has a higher success rate at finding collision-free trajectories for large groups of drones compared to other Voronoi diagram-based methods.

Keywords: distributed trajectory generation; Voronoi diagram; multi-drone applications; real-time replanning

Citation: Sabetghadam, B.; Cunha, R.; Pascoal, A. A Distributed Algorithm for Real-Time Multi-Drone Collision-Free Trajectory Replanning. *Sensors* **2022**, *22*, 1855. https://doi.org/10.3390/s22051855

Academic Editor: Sašo Blažič

Received: 26 January 2022
Accepted: 24 February 2022
Published: 26 February 2022

1. Introduction

Trajectory generation is a key element for the execution of complex autonomous vehicle missions. It can be defined as the computational problem of finding a valid trajectory that guides a vehicle from an initial state to a given final state in an environment with static and/or moving obstacles. In most applications, the main concern, rather than just finding a feasible trajectory between the initial and final states, is to obtain the optimal trajectory with respect to a certain objective function. In such a setting, trajectory generation is formulated as an optimization problem with a cost function that quantifies the accomplishment of mission goals and objectives, and different types of constraints to ensure safety and feasibility of resulting trajectories.

With rapid advances in communication and computational technology, autonomous vehicles continue to take part in more complex missions, and even engage in teams of collaborating vehicles to take on increasingly demanding tasks. This necessitates incorporating intervehicle collision avoidance constraints in the optimization problem to guarantee that generated trajectories for a group of vehicles sharing a common workspace are collision-free. Therefore, for a large group of vehicles, the optimization problem would involve a large number of constraints and decision variables, and the computational cost of solving it centrally can be prohibitively high. To reduce the computational complexity, a multitude of distributed schemes, reviewed below, have been proposed for decomposing the optimization problem into smaller sub-problems that can be solved locally by each vehicle. The major challenge is to ensure that local decisions do also satisfy the coupling collision avoidance constraints. This is mainly addressed by exchanging information among the vehicles on their current states, future input sequences, etc. Depending on the communication strategy, the sub-problems might be solved sequentially or concurrently, with possibly several iterations of optimization and communication to achieve the required performance.

In [1], the collision avoidance constraint, usually expressed in terms of the two-norm of a relative position vector, is approximated by a set of linear constraints. The sub-problem for each vehicle is then formulated as a mixed-integer linear programming (MILP) that includes the vehicle's individual variables as well as variables of a subset of neighbors. This enables cooperation among vehicles by allowing a vehicle to make feasible perturbations to neighboring vehicles' decisions. The sub-problems are solved sequentially by each vehicle, and the algorithm iterates over the group of vehicles until convergence, during each cycle of a model predictive control (MPC) scheme.

Sequential convex programming (SCP)-based methods have also been used for solving distributed multiple vehicle trajectory generation problems [2,3]. The work in [4] addresses the infeasiblity of intermediate problems in decoupled-SCP methods, arising from convex approximation of collision avoidance constraints, i.e., linearizing them, and proposes incremental SCP (iSCP), which tightens collision constraints incrementally. Compared to sequential approaches in [5–7], that cast the trajectory of anterior vehicles as dynamic obstacles for a posterior vehicle, the methods proposed in [1,4] result in less constrained intermediate problem and faster convergence rate, yet, similar to most MPC-SCP-based methods, they would require the vehicles to exchange a full representation of their decisions to neighboring vehicles over a communication network.

The synchronous approach in [8] extends the distributed MPC (dMPC) scheme in [9] for formation control, based on alternating direction method of multipliers (ADMM), to problems with intervehicle collision avoidance constraints. These constraints are decoupled using separating hyperplanes, which enforces each vehicle to stay within one half-space of a time-varying plane over a certain time horizon. The resulting sub-problems are solved simultaneously by vehicles, while the normal vector and offset shared between a vehicle and a neighbor, for characterizing their separating hyperplane, are updated at each cycle of the dMPC, using the interchanged information about generated trajectories at the previous cycle.

In the decentralized trajectory planner proposed in [10], vehicles replan their trajectories asynchronously, independent of the planning status of other vehicles. At each iteration, a vehicle considers trajectories assigned to neighboring vehicles as constraints, and solves an optimization problem including as decision variables the normal and offset of planes that separate the outer polyhedral representation of its trajectory and those of its neighbors. A check–recheck scheme is then performed to ensure that the generated trajectory does not collide with trajectories other vehicles have committed to during the optimization time. Therefore, to guarantee deconfliction between vehicles, the planner requires a vehicle to broadcast its computed trajectory to its neighboring vehicles at the end of each replanning iteration.

The on-demand approach to local collision avoidance, proposed in [11], imposes constraints only at specific time instances when collisions between a vehicle and its neighbors are predicted. Predicting collisions along a time horizon, however, relies on an accurate knowledge of the neighbors' future actions which must be communicated at every sampling time. The dMPC scheme in [12] for distributed trajectory generation is based on this predict–avoid paradigm and an event-triggered replanning strategy, and has been shown to result in less conservative trajectories, but at the cost of voiding the collision avoidance guarantees for all time instances over the horizon. To capture the downwash effect of quadrotor's propellers, the collision avoidance constraint in [12] is modified with a diagonal scaling matrix, which approximates the quadrotor body with a translating ellipsoid elongated along the vertical axis, yet it ignores the quadrotor's rotational motion.

Reciprocal velocity obstacle (RVO) and its variants have been widely used in distributed collision avoidance [13–17]. At each time step, RVO [13] builds the set of all relative velocities, leading to a collision between a vehicle and its neighbors, and chooses a new constant velocity outside this set, and closest to the desired value, to avoid collisions. Therefore, RVO requires the position and velocity information to be communicated, or sensed, between nearby neighbors. Other variants, such as acceleration velocity obstacle (AVO), which addresses the instantaneous change of velocity in RVO by taking into account acceleration constraints, need further information such as acceleration to be interchanged. Reciprocally-rotating velocity obstacle (RRVO) [18] uses rotation information to mitigate deadlocks caused by symmetries of representing vehicles with translating discs in RVO. It relies on the assumption that neighbors may rotate equally (or equally opposite), bounded by a maximum value, to compute an approximation of swept areas for rotating polygon-shaped vehicles, and uses them for constructing velocity obstacles. A new velocity and rotation is then selected at each time step to avoid collisions.

Another approach to distributed collision avoidance is to construct the Voronoi diagram of the group of vehicles and generate the trajectory for each vehicle so that it is entirely within the vehicle's Voronoi cell [19–22]. Since Voronoi cells do not overlap, it can be guaranteed that the generated trajectories are collision-free. To consider the physical size of a vehicle, the modified Voronoi cell used in [23,24] retracts boundary hyperplanes of the general Voronoi cell by a safety radius for disc-shaped vehicles. At each sampling time, upon receiving the relative position information, trajectories are replanned to conform to the updated Voronoi diagram. The resulting sub-problems can be solved simultaneously, in a receding horizon manner, until the vehicles reach their final positions. The Voronoi-based approaches only require the vehicles to know relative positions to neighboring vehicles, and therefore are well suited to applications where vehicles only have relative position sensing and no communication network [25].

In this paper we develop a distributed trajectory generation framework, with low computation and communication demands, for multiple quadrotors flying in (relatively) close proximity to each other. We specifically address the shortcomings of approximating the drone body with a disc (or sphere) for generating feasible collision-free trajectories for large groups of drones. A sphere model, used in most existing distributed collision avoidance schemes, may be overly conservative in confined spaces since it invalidates trajectories whose feasibility depends on the consideration of the flight attitude. Instead, we model the drone body with an ellipsoid, and employ the Voronoi partitioning of space to derive local collision avoidance constraints that take into account the drone's real size and orientation. The same approach can be integrated into other distributed schemes that utilize separating hyperplanes for decoupling collision avoidance constraints. Yet the main reason for adopting Voronoi diagram is that using time-invariant boundary hyperplanes determined prior to solving a sub-problem, despite being more conservative, can significantly reduce communication and computational load, allowing for higher replanning rates. Incorporating the resulting set of constraints into sub-problems, solved by each vehicle, allows finding collision-free trajectories for guiding a group of drones through confined spaces by proper adjustment of attitude angles. In addition, the obtained set of constraints can be expressed as Bézier curves, and hence can be efficiently evaluated to guarantee that intervehicle collision avoidance requirements are met at any instant of time even over a long planning horizon.

In the proposed synchronous distributed scheme, each vehicle uses the position information of its neighbors, updated at each sampling time, and solves a sub-problem to generate its trajectory inside (a subset of) its Voronoi cell towards the closest point (in the cell) to its goal position. We present an efficient method to compute this point, which is needed to appropriately define the terminal constraint and cost in the sub-problem. A sequence of sub-problems are then solved in a receding-horizon manner until the vehicles reach their goal positions. The simulation results show that the proposed method has a higher success rate at finding collision-free trajectories for larger groups of quadrotors compared to other Voronoi diagram-based methods. In addition, it can effectively reduce the total flight time required to perform point-to-point maneuvers. Furthermore, the computation time of generating those trajectories satisfies timing constraints imposed by real-time applications.

The rest of this paper is organized as follows: In Section 2 we formulate the optimization sub-problem solved by each vehicle. In Section 2.1 we study the differentially flat system describing the drone equations of motion, and parameterize trajectories with Bézier curves. We derive the set of local collision avoidance constraints in Section 2.2, and present an efficient algorithm for finding the closest point in a Voronoi polytope to a goal position in Section 2.3. In Section 3 we obtain the continuity conditions between two adjacent Bézier curve segments, and present a method for evaluating inequalities in Bézier form without discretization. Finally, we provide simulation results in Section 4.

2. Problem Formulation

The multiple vehicle trajectory generation problem addressed in this paper can be defined as finding optimal trajectories that act as references to guide a group of vehicles from their initial positions to some desired final positions. The generated trajectories should jointly minimize a cost function, corresponding to the accomplishment of mission goals and objectives, and satisfy a set of local and coupling constraints, so that they are dynamically-feasible and collision-free. For N_v vehicles, this problem can be formulated as the following optimal control problem.

$$\begin{aligned} \min_{\substack{\mathbf{u}_i(.)\\ i\in[N_v]}} \; & \sum_{i\in[N_v]} J(\mathbf{x}_i(.), \mathbf{u}_i(.)) && (1)\\ s.t.\; & \dot{\mathbf{x}}_i(t) = f(\mathbf{x}_i(t), \mathbf{u}_i(t)) && \text{(Dynamics)}\\ & \mathbf{x}_i(0) = \mathbf{x}_{i,0} && \text{(Initial state)}\\ & \mathbf{x}_i(t_f) = \mathbf{x}_{i,f} && \text{(Final state)}\\ & c(\mathbf{x}_i(t), \mathbf{x}_j(t)) \leq 0 \quad j \in [N_v]\backslash\{i\} && \text{(collision avoidance)}\\ & \mathbf{x}_i(t) \in \mathcal{X}_i && \text{(State Constraints)}\\ & \mathbf{u}_i(t) \in \mathcal{U}_i && \text{(Input constraints)} \end{aligned}$$

where $[N_v] = \{1, \ldots, N_v\}$. The cost to be minimized is the sum of the vehicles' individual costs, J, given by the functional,

$$J[\mathbf{u}_i(.)] = \int_0^{t_f} L(\mathbf{x}_i, \mathbf{u}_i)dt \tag{2}$$

where $\mathbf{x}_i(t) \in \mathbb{R}^{n_x}$ and $\mathbf{u}_i(t) \in \mathbb{R}^{n_u}$ are the state and the input vectors of the vehicle's model described by an ODE, and $\mathbf{x}_{i,0}$ and $\mathbf{x}_{i,f}$ are the initial and final values of the state of the i-th vehicle, respectively. $\mathcal{X}_i$ and $\mathcal{U}_i$ denote the set of admissible states and inputs for the i-th vehicle derived from limits imposed by vehicle dynamics and the surrounding environment.

In order to reduce the computational complexity of solving (1) for larger N_v with increased numbers of constraints and variables, one can divide the problem into a set of small-scale sub-problems. Here, the sub-problems are formulated such that each involves only a vehicle's individual costs and constraints, and hence can be solved independently by the vehicle. The sub-problems must include constraints to ensure that the trajectory generated locally by a vehicle does satisfy the coupling collision avoidance constraints.

The key idea to ensure intervehicle collision avoidance is to decompose the space into non-overlapping regions, provided by a Voronoi diagram, and generate the trajectory for each vehicle such that it is entirely within its partition. The Voronoi diagram is updated at each sampling time according to the relative positions of vehicles, and a sequence of sub-problems is solved in a receding horizon manner until the vehicles reach their final positions. For the i-th vehicle, the problem that has to be solved at the time instant t_k can be formulated as

$$
\begin{aligned}
\min_{\mathbf{x}_{i,k}(.),\mathbf{u}_{i,k}(.)} \quad & J(\mathbf{x}_{i,k}(.),\mathbf{u}_{i,k}(.)) && (3)\\
s.t. \quad & \dot{\mathbf{x}}_{i,k}(t) = f(\mathbf{x}_{i,k}(t),\mathbf{u}_{i,k}(t)) && \text{(Dynamics)}\\
& \mathbf{x}_{i,k}(t_k) = \hat{\mathbf{x}}_{i,k} && \text{(Initial state)}\\
& \mathbf{x}_{i,k}(t) \in \mathcal{C}_{i,k}(\bar{\mathbf{x}}_k) && \text{(collision avoidance)}\\
& \mathbf{x}_{i,k}(t) \in \mathcal{X}_{i,k} && \text{(State Constraints)}\\
& \mathbf{u}_{i,k}(t) \in \mathcal{U}_{i,k} && \text{(Input constraints)}
\end{aligned}
$$

where $\mathbf{x}_{i,k}(t)$ and $\mathbf{u}_{i,k}(t)$ are the state and the input profiles of the vehicle over the time interval $[t_k, t_k + t_h]$, with t_h being the planning horizon, and $\hat{\mathbf{x}}_{i,k}$ denotes its state at the time instant t_k. The cost function in the above sub-problem is modified as

$$J[\mathbf{u}_{i,k}(.)] = \int_{t_k}^{t_k+t_h} L(\mathbf{x}_{i,k},\mathbf{u}_{i,k})dt + \phi(\mathbf{x}_{i,k}(t_k+t_h)) \tag{4}$$

where the second term is added to penalize the distance, at $t_k + t_h$, to the point in the Voronoi partition that is closest to the goal position.

In the optimization problem (3), $\mathcal{C}_{i,k}$ may denote the Voronoi partition assigned to the i-th vehicle. The Voronoi diagram is updated for each sub-problem according to the vehicles' configuration at each time instant t_k, i.e., $\bar{\mathbf{x}}_k = \{\hat{\mathbf{x}}_{j,k}\}_{j\in[N_v]}$. Since Voronoi partitions are disjoint and the assigned trajectory to each vehicle for the time horizon t_h is contained within its partition, it can be guaranteed that there is no collision between the trajectories over the time interval $[t_k, t_k + t_h]$.

The distributed trajectory generation framework is summarized in Algorithm 1. In Section 2.2, we study the Voronoi diagram for a group of vehicles and modify $\mathcal{C}_{i,k}$ to explicitly take into account the orientation while generating collision-free trajectories for multiple drones.

Algorithm 1 Distributed Trajectory Generation Framework

```
k = 0
x̂_{i,0} ← Initial position of the i-th vehicle
repeat
    Receive position information from neighbors
    Broadcast own position to neighbors
    Update Voronoi partition
    Compute the closest point in the Voronoi partition to
    the goal position      ▷ Section 2.3
    Set the cost function (4)
    Set the constraints    ▷ Sections 2.2 and 3.1
    Solve the optimization sub-problem
until x̂_{i,k} = x_{i,f}.
```

2.1. Quadrotor Model

In this paper, the simplified quadrotor equations of motion are described by

$$m\ddot{\mathbf{p}} = mg\mathbf{e}_3 + f, \tag{5}$$

where $\mathbf{p} \in \mathbb{R}^3$ is the position and m is the mass of the quadrotor. In addition, $g = 9.8\,\frac{\text{m}}{\text{s}^2}$ is the gravitational acceleration, and $\mathbf{e}_3 = [0\,0\,1]^T$. The first term on the right-hand side of (5) is gravity in the $\mathbf{z}_I$ direction, and the second term, $f \in \mathbb{R}^3$, is the thrust force aligned with the body's $\mathbf{z}$-axis.

$$f = -T\,{}^I\mathbf{z}_B \tag{6}$$

where $T \in \mathbb{R}$ is the net thrust, ${}^I\mathbf{z}_B = R\,{}^B\mathbf{z}_B = R\mathbf{e}_3$ is the body frame $\mathbf{z}$-axis expressed in $\{I\}$, and $R \equiv {}^I_B R \in SO(3)$ is the rotation matrix from the body frame $\{B\}$, centered at the quadrotor's center of gravity, to the fixed inertial frame $\{I\}$. For simplicity, we drop the superscript ^{I} and consider $\mathbf{z}_B = {}^I\mathbf{z}_B$. Figure 1 is a graphical representation of the quadrotor and the associated reference frames.

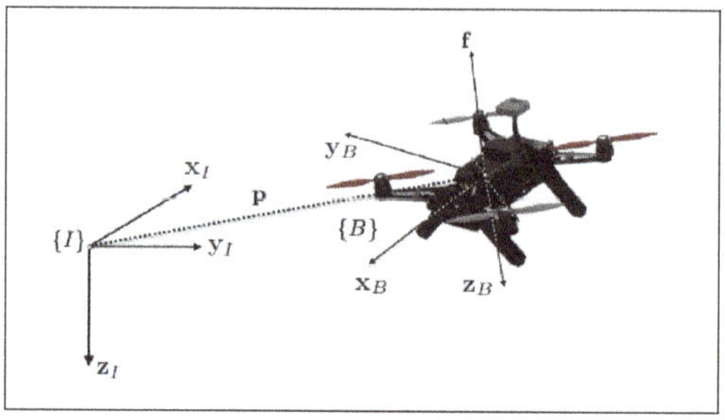

Figure 1. The quadrotor reference frames.

Trajectory Parametrization

The quadrotor dynamics (5) with the four inputs is differentially flat [26], and therefore the state and the input of the system can be expressed as functions of the flat outputs and a finite number of its derivatives. The position vector together with the yaw angle can be selected as flat outputs of the system. Here, $\mathbf{p} \in \mathbb{R}^3$ is parameterized as a Bézier curve, given by

$$\mathbf{p}(\tau) = \sum_{l=0}^{n} \bar{p}_l B_{l,n}(\tau), \tag{7}$$

where $\bar{p}_l \in \mathbb{R}^3$ are the control points, $B_{l,n}$ are Bernstein basis polynomials of degree n, and $\tau \in [0,1]$ is defined as

$$\tau = \frac{t}{t_f} \tag{8}$$

The linear velocity, $\mathbf{v} = \dot{\mathbf{p}}$, and linear acceleration, $\mathbf{a} = \ddot{\mathbf{p}}$, can be expressed as parametric Bézier curves through the first and second derivative of $\mathbf{p}$ with respect to time, yielding

$$\begin{aligned} \mathbf{v}(\tau) &= \sum_{l=0}^{n-1} \bar{v}_l B_{l,n-1}(\tau) \\ \mathbf{a}(\tau) &= \sum_{l=0}^{n-2} \bar{a}_l B_{l,n-2}(\tau) \end{aligned} \tag{9}$$

where the control points $\bar{v}_l$ and $\bar{a}_l$ are obtained as

$$\begin{aligned} \bar{v}_l &= \frac{n}{t_f}(\bar{p}_{l+1} - \bar{p}_l) & l &= 0,\dots,n-1 \\ \bar{a}_l &= \frac{n(n-1)}{t_f^2}(\bar{p}_{l+2} - 2\bar{p}_{l+1} + \bar{p}_l) & l &= 0,\dots,n-2 \end{aligned} \tag{10}$$

The thrust T and rotation matrix R can also be expressed as functions of the flat output and its derivatives. The net thrust T can be written as

$$T = m\|\ddot{\mathbf{p}} - g\mathbf{e}_3\|. \tag{11}$$

Assuming that the rotation matrix $R = [\mathbf{x}_B, \mathbf{y}_B, \mathbf{z}_B]$ is parameterized by the Z-Y-X Euler angles $\boldsymbol{\lambda} = [\phi, \theta, \psi]^T$ as

$$R = R_z(\psi)R_y(\theta)R_x(\phi), \tag{12}$$

then the columns of the rotation matrix are extracted from

$$\mathbf{z}_B = \frac{\mathbf{g}\mathbf{e}_3 - \ddot{\mathbf{p}}}{\|\mathbf{g}\mathbf{e}_3 - \ddot{\mathbf{p}}\|} \quad \mathbf{x}_B = \frac{\mathbf{r} \times \mathbf{z}_B}{\|\mathbf{r} \times \mathbf{z}_B\|}, \quad \mathbf{y}_B = \mathbf{z}_B \times \mathbf{x}_B \tag{13}$$

where the unit vector $\mathbf{r}$ is defined as

$$\mathbf{r} = [-\sin\psi, \cos\psi, 0]^T \tag{14}$$

The above equations declare that the vehicle's orientation can be fully determined from the second derivative of the trajectory and the yaw angle. As mentioned before, the yaw angle ψ is a component of the flat output, and therefore it can be controlled independently without affecting the trajectory generation. Using the differential flatness property of the system, trajectories consistent with dynamics can be planned in the space of flat outputs, where (5) is trivially satisfied and the original input and state constraints are transformed into constraints on the flat output and its derivatives.

2.2. Collision Avoidance

In this section, we present a Voronoi diagram-based approach to decoupling inter-vehicle collision avoidance constraints. Although the presence of obstacles, interpreted as non-decision-making agents, is not explicitly considered here, incorporating vehicle–obstacle collision avoidance constraints into the problem simply amounts to taking into account the obstacles' position when updating the Voronoi partition (step 6 of Algorithm 1).

The widely used approach in the literature to avoiding collisions with obstacles in the environment is to model the drone body as a sphere with radius $r_{\mathbf{D}}$, and then simply building the collision-free space, $\mathcal{C}_{free}$, by inflating the obstacles with a factor $r_{\mathbf{D}}$. As a result, collision-free trajectories can be obtained by enforcing the vehicle, which is now treated as a point in the space, to be inside $\mathcal{C}_{free}$ [27]. Considering now the collision avoidance between the i-th and j-th drones, the corresponding constraint can be derived similarly by

$$\|\mathbf{p}_i - \mathbf{p}_j\| \geq 2r_{\mathbf{D}} \tag{15}$$

where $\|.\|$ denotes the Euclidean distance. Ignoring the real shape and orientation of the drone, and approximating its body with a sphere, invalidates trajectories that are feasible upon considering the flight attitude. For this reason, the above approach might be too conservative for trajectory generation in confined spaces and can even fail to find feasible collision-free trajectories when multiple drones are involved.

Approximating the drone body with an ellipsoid, whose principal axes are aligned with the body frame axes, allows considering the drone orientation while inspecting for collisions against other vehicles. For the i-th drone, the ellipsoid, E_i, centered at the drone position $\mathbf{p}_i$, is given by

$$E_i \equiv \{\mathbf{p} \in \mathbb{R}^3 | \mathbf{p} = \mathbf{p}_i + R_i\Lambda\mathbf{w}, \|\mathbf{w}\| \leq 1\} \tag{16}$$

where Λ is a 3×3 diagonal matrix of the form

$$\Lambda = \begin{bmatrix} r_{\mathbf{D}} & 0 & 0 \\ 0 & r_{\mathbf{D}} & 0 \\ 0 & 0 & h_{\mathbf{D}} \end{bmatrix} \tag{17}$$

with $r_{\mathbf{D}}$ and $h_{\mathbf{D}}$ being the lengths of the principle semi-major and semi-minor axes, respectively. (See Figure 2).

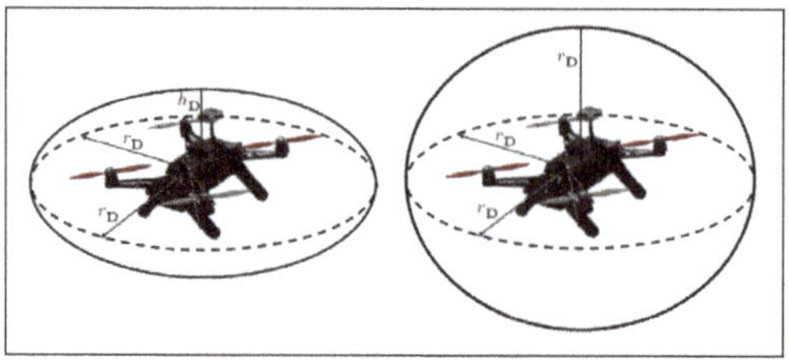

Figure 2. The quadrotor body can be represented as a sphere with radius $r_{\mathbf{D}}$ (**right**), or an ellipsoid aligned with the axes of the body frame (**left**). Approximating the drone body with an ellipsoid allows considering the quadrotor's rotational motion.

As proposed in [28], collision avoidance constraints for two ellipsoid-shaped drones can be derived using separating hyperplanes. Denoting by $\mathbf{a} \in \mathbb{R}^3$ and $b \in \mathbb{R}$ the normal vector and offset of a hyperplane, respectively, the separating hyperplane for E_i and E_j, defined as $H \equiv \{\mathbf{p}|\mathbf{a}^T\mathbf{p} - b = 0\}$, must satisfy

$$\begin{aligned} \mathbf{a}^T\mathbf{p} - b \leq 0 \quad \forall \mathbf{p} \in E_i \\ \mathbf{a}^T\mathbf{p} - b > 0 \quad \forall \mathbf{p} \in E_j \end{aligned} \tag{18}$$

Since

$$-\|\Lambda R^T\mathbf{a}\| \leq \mathbf{a}^T R \Lambda \mathbf{w} \leq \|\Lambda R^T\mathbf{a}\| \tag{19}$$

The set of inequalities (18) holds if, and only if,

$$\begin{aligned} \mathbf{a}^T\mathbf{p}_i - b \geq & \quad \|\Lambda R_i^T\mathbf{a}\| \\ \mathbf{a}^T\mathbf{p}_j - b \leq & \quad -\|\Lambda R_j^T\mathbf{a}\| \end{aligned} \tag{20}$$

Satisfying the set of constraints (20) will guarantee that there is no collision between the two ellipsoids E_i and E_j associated with the i-th and j-th drones, respectively.

For multiple vehicle trajectory generation, collision avoidance constraints, either in the form of the inequality constraint (15), for spheres, or the set of constraints (20), for ellipsoids, must be incorporated in the optimization problem for each pair of vehicles. As the number of vehicles involved in a mission grows, the resulting increase in the number of constraints would inevitably exacerbate the computational issues of finding collision-free trajectories in a centralized manner.

Distributed Collision Avoidance

Here, we propose a distributed approach to collision avoidance which takes into account the shape and orientation of a drone. The presented approach uses Voronoi partitioning of space and generates the trajectory of each vehicle such that it is entirely within (a subset of) the vehicle's Voronoi cell for a time interval t_h. Since Voronoi cells are pairwise disjoint, and each vehicle only moves inside its Voronoi cell, intervehicle collision avoidance is guaranteed for all future time before t_h.

Each Voronoi cell in an n-dimensional space is a convex polytope bounded by a number of $(n-1)$-dimensional convex polytopes. For a group of vehicles in three-dimensional space, the general Voronoi cell of the i-th vehicle is defined as

$$\mathcal{V}_i = \{\mathbf{p} \in \mathbb{R}^3 | \mathbf{p}_{ij}{}^T(\mathbf{p} - \frac{1}{2}(\mathbf{p}_i + \mathbf{p}_j)) \leq 0, \forall j \in [N_v] \backslash \{i\}\} \tag{21}$$

where $\mathbf{p}_{ij} = \mathbf{p}_j - \mathbf{p}_i$, and $\mathbf{p}_i$ and $\mathbf{p}_j$ are the position of the i-th and j-th vehicles at the current time instant. Note that $\mathcal{V}_i$ is the intersection of half-spaces corresponding to hyperplanes with $a = \mathbf{p}_{ij}$ and $b = \frac{1}{2}\mathbf{p}_{ij}^T(\mathbf{p}_i + \mathbf{p}_j)$. An arbitrary point in $\mathcal{V}_i$ is closer to the i-th vehicle than any other vehicle [22], i.e.,

$$\|\mathbf{p} - \mathbf{p}_i\| \leq \|\mathbf{p} - \mathbf{p}_j\|, \quad \forall \mathbf{p} \in \mathcal{V}_i \quad \& \quad j \neq i \tag{22}$$

The boundary of the Voronoi cell, $\partial\mathcal{V}_i$, is the union of multiple faces, each of which include points in the space that are equidistant to the i-th vehicle and a neighboring vehicle.

In order to account for the size of a vehicle, the buffered Voronoi cell (BVC) proposed in [24] retracts the boundary of the general Voronoi cell by a safety radius, so that if the vehicle's center is inside the BVC, its body, approximated by a sphere of radius $r_{\mathbf{D}}$, will be entirely within its Voronoi cell. The BVC of the i-th vehicle, denoted by $\bar{\mathcal{V}}_i$, is defined as

$$\bar{\mathcal{V}}_i = \{\mathbf{p} \in \mathbb{R}^3 | \quad \mathbf{p}_{ij}^T(\mathbf{p} - \frac{1}{2}(\mathbf{p}_j + \mathbf{p}_i)) + r_{\mathbf{D}}\|\mathbf{p}_{ij}\| \leq 0, \forall j \in [N_v] \backslash \{i\}\} \tag{23}$$

It can be easily shown that BVC is a subset of the general Voronoi cell, i.e., $\bar{\mathcal{V}}_i \subset \mathcal{V}_i$. In addition, for any two points $\mathbf{p}' \in \bar{\mathcal{V}}_i$ and $\mathbf{q}' \in \bar{\mathcal{V}}_j$, $\|\mathbf{p}' - \mathbf{q}'\| \geq 2r_{\mathbf{D}}$ holds. Therefore, the vehicles are guaranteed to avoid collisions due to the buffer region of $r_{\mathbf{D}}$ along $\partial\mathcal{V}_i$. Figure 3 shows the Voronoi diagram for 10 drones in a collision-free configuration and the BVC for two adjacent drones.

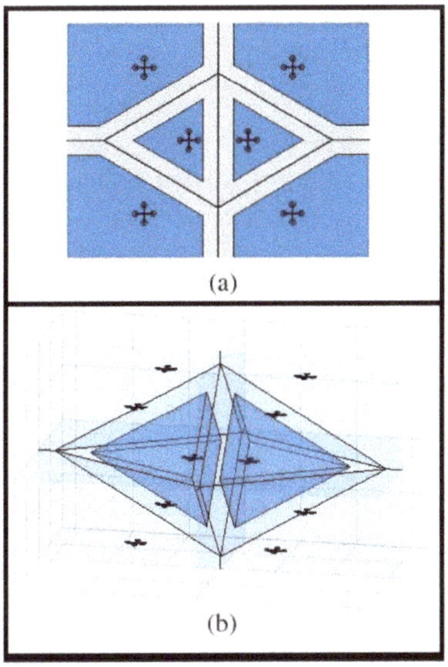

Figure 3. (**a**) The Voronoi diagram for six drones in 2D space. The Voronoi boundary edges are shown with solid black lines, and the buffered Voronoi cells are shaded in dark blue. (**b**) The Voronoi diagram for 10 drones in a collision-free configuration in 3D space. The Voronoi boundary $\partial\mathcal{V}$ is shaded in light blue, and the buffered Voronoi cells $\bar{\mathcal{V}}$ for two neighboring drones in the center are shown in dark blue.

The BVC is defined based on a symmetrical approximation of the vehicle's body with a translating disc. In order to reduce the conservatism and avoid infeasibility issues due to ignoring the real shape and orientation of the vehicle, we approximate the drone with an ellipsoid (16), and bearing in mind that

$$R\Lambda^2R^T = r_{\mathbf{D}}^2I + (h_{\mathbf{D}}^2 - r_{\mathbf{D}}^2)\mathbf{z}_B\mathbf{z}_B^T, \tag{24}$$

we propose $\mathcal{C}_i$ in problem (3) to be defined as

$$\mathcal{C}_i = \{(\mathbf{p}, \ddot{\mathbf{p}}) \in \mathbb{R}^6 | \mathbf{p}_{ij}^T(\mathbf{p} - \frac{1}{2}(\mathbf{p}_j + \mathbf{p}_i)) + \|\Lambda R^T \mathbf{p}_{ij}\| \leq 0, \forall j \in [N_v] \backslash \{i\}\} \quad (25)$$

If the trajectory of the i-th drone $\mathbf{p}_i(t)$ satisfies the above set of local collision avoidance constraints for all $t \in [t_k, t_k + t_h]$, then the ellipsoid representing the drone body is within the Voronoi cell for the entire time horizon; that is, the ellipsoid centered at $\mathbf{p}_i$ and aligned with the columns of R_i does not intersect the Voronoi boundary, stated mathematically

$$\|\partial E_i - \partial \mathcal{V}_i\| \geq 0 \quad (26)$$

Noting that

$$h_{\mathbf{D}} \|\mathbf{p}_{ij}\| \leq \|\Lambda R^T \mathbf{p}_{ij}\| \leq r_{\mathbf{D}} \|\mathbf{p}_{ij}\|, \quad (27)$$

it can be induced that

$$\bar{\mathcal{V}}_i(r_{\mathbf{D}}) \subset \text{proj}_{XYZ} \mathcal{C}_i \subset \bar{\mathcal{V}}_i(h_{\mathbf{D}}) \subset \mathcal{V}_i \quad (28)$$

where $\text{proj}_{XYZ}\mathcal{C}_i$ is the projection of $\mathcal{C}_i$ onto the three-dimensional subspace spanned by $\mathbf{e}_1$, $\mathbf{e}_2$, and $\mathbf{e}_3$.

Therefore, incorporating (25) into the optimization problem (3) will ensure that the generated trajectories are collision-free while alleviating infeasibility problems by taking orientations into account. Also, since $\mathbf{z}_B$ is fully obtained from $\ddot{\mathbf{p}}$ (13), the above set of local collision avoidance constraints can be expressed as constraints imposed on Bézier curves. Later, we present an efficient method for evaluating inequalities in Bézier form.

2.3. Finding the Closest Point to the Goal Position

As explained above, at each time instant the Voronoi cell $\mathcal{V}_i$ is updated according to the relative position of the i-th vehicle to other vehicles. The optimization problem (3) is then solved to generate a trajectory, for a time horizon t_h, that guides the vehicle towards the point in the cell closest to the goal position. This process is repeated until the vehicle reaches its final position. At each sampling time, the closest point must be found prior to solving the trajectory generation problem. Therefore, having an efficient scheme for finding the closest point is critically important to avoid long computational delays between updating the Voronoi cell and replanning the trajectory.

The point in a convex polytope that is closest to a query point $\mathbf{q}$ is either $\mathbf{q}$ itself or a point on the boundary of the polytope. A naive way to find the closest point in a convex polytope in a three-dimensional space, represented by $P = (\mathbb{F}, \mathbb{E}, \mathbb{V})$, where $\mathbb{F}$ is the set of faces, $\mathbb{E}$ is the set of edges, and $\mathbb{V}$ is the set of vertices, is to check the distance between $\mathbf{q} \in \mathbb{R}^3$ to all faces, edges, and vertices for finding the minimum. However, for complex polytopes, the computation time is not negligible.

The geometric approach proposed in [24] for a polygon in a two-dimensional space calculates the barycentric coordinates and an angle from the query point to the two vertices of each edge to find the closest point. Since this approach iterates over all edges, its computational complexity can significantly increase as the number of Voronoi neighbors of a vehicle increases. Here, we make use of the Gilbert–Johnson–Keerthi (GJK) distance algorithm and devise an approach that can efficiently determine whether the query point is inside the polytope, i.e., $\mathbf{q} \in P$, in which case the closest point is $\mathbf{q}$ itself. Otherwise, the presented algorithm returns the closest feature of P to $\mathbf{q}$, and the closest point can be obtained by projecting $\mathbf{q}$ onto it. The proposed approach is not limited to distance queries between a point and a polytope, and can also be used when the final constraint in (3) is relaxed to a small box or sphere around the goal position, and used in conjunction with a terminal cost term.

The GJK distance algorithm or simply GJK algorithm is an iterative algorithm that relies on a support mapping function to incrementally build simplices that are closer to the query point [29]. The algorithm has been extensively used for collision detection between generic convex shapes [30,31]. The original algorithm, however, can be used to compute the minimum distance, and also the respective pair of (closest) points, between two convex shapes [32].

In order to obtain the minimum distance between two general convex sets A and B, GJK approximates the closest point in the Minkowski difference of the two sets, $C = A - B$ to the origin, denoted by $\nu(C)$, with an iterative search. At each iteration, a simplex in C is constructed such that it is closer to the origin $\mathcal{O}$ than the simplex in the previous iteration. In $\mathbb{R}^3$, a simplex can be a point, a line, a triangle, or a tetrahedron with 1, 2, 3, and 4 affinely independent vertices.

GJK relies on the so-called support mappings to construct a new simplex. A support mapping function $s_C(\mathbf{d})$ of the convex set C maps the vector $\mathbf{d}$ to a point in the set, called the support point, according to

$$\mathbf{d}^T s_C(\mathbf{d}) = \max\{\mathbf{d}^T\mathbf{p}; \mathbf{p} \in C\} \tag{29}$$

At each iteration, a support point $\mathbf{w}_k = s_C(-\nu_k)$ is added as a vertex to the current simplex, indicated with V_k, where ν_k is the closest point of V_k to the origin, i.e., $\nu_k = \nu(\text{conv}(V_k))$. V_{k+1} is then updated such that it only contains the smallest set of vertices that supports $\nu_{k+1} = \nu(\text{conv}(V_k \cup \mathbf{w}_k))$, and earlier vertices that do not back ν_{k+1} are discarded [30].

It is proved in [30] that in each iteration the new ν is closer to the origin than the previous one, and thus the sequence of $\{\nu_k\}$ converges to the closest point of C to the origin. In addition, it is shown that

$$\|\nu_k - \nu(C)\|^2 \leq \|\nu_k\|^2 - \nu_k^T \mathbf{w}_k \tag{30}$$

which is used to construct the terminating condition of the GJK algorithm for general convex shapes.

The GJK algorithm, as explained above, depends heavily on the computation of ν_k to test the termination condition and to determine the search direction for finding the support point. In each iteration of the algorithm, ν_k must be computed with the *Johnson Distance Subalgorithm* [29] or more robust alternatives such as the signed volume method in [33]. Here, we exploit unique features of polytopes and propose a faster way to evolve simplices in the GJK algorithm without computing ν_k in each iteration.

For polytopes, GJK arrives at the actual $\nu(C)$ in a finite number of iterations [30]. The pseudocode in Algorithm 2 describes the GJK distance algorithm for a polygon P. In order to find the support point $\mathbf{w}_k$, we employ a search direction $\mathbf{d}_k \uparrow\downarrow \nu_k$, which is updated in each iteration of the algorithm with S1D, S2D, or S3D subroutines. To update $\mathbf{d}$ and V, these subroutines, summarized in Algorithms 3–5, inspect the Voronoi regions of the simplex for the one that contains the origin. Figure 4 shows the Voronoi regions of an m-simplex ($m = 1, 2, 3$) where the origin possibly lies. Once the Voronoi region containing the origin is found, $\mathbf{d}$ is determined as a vector from the associated vertex, edge, or face (of the simplex) pointing towards the origin.

The stop criterion for the conditional loop in Algorithm 1 is also constructed using the search direction, offered as

$$\mathbf{d}_k^T(\mathbf{w}_k - \mathbf{v}_1) \leq 0 \tag{31}$$

where $\mathbf{v}_1 \in V_k$. Considering that $\mathbf{d}_k^T(\nu_k - \mathbf{v}_1) = 0$, we can conclude that the above criterion, for deciding whether V_k represents the closest feature of P to the origin, is equivalent to the stop criterion in [30] for determining whether ν_k s the closest point, that is

$$\|\nu_k\|^2 - \nu_k^T \mathbf{w}_k \leq 0 \longleftrightarrow \mathbf{d}_k^T(\mathbf{w}_k - \mathbf{v}_1) \leq 0 \tag{32}$$

If the inequality (31) holds, then the closest feature of P to the origin can be determined from V_k. Figure 5 shows all possible outputs of Algorithm 2, which can be a vertex, an edge, or a face of P or a tetrahedron inside it, and $\nu(P)$ for each of the cases.

Algorithm 2 Compute the closest point of P to the origin

$\mathbf{v}$ = "Arbitrary point in vert(P)"
$\mathbf{d} = -\mathbf{v}$
$V = \{\mathbf{v}\}$
repeat
  $\mathbf{w} = s_P(\mathbf{d})$;
  if $\mathbf{d}^T(\mathbf{w} - \mathbf{v}_1) \leq 0$ **then**
    V represents the closest feature of P to $\mathcal{O}$
    return $\nu(V)$
  end if
  $V \leftarrow V \cup \mathbf{w}$;
  $[V, \mathbf{d}] \leftarrow \text{CallSmD}(V)$;[1]
until $|V| = 4$;
P contains $\mathcal{O}$
return $\nu = \mathcal{O}$

[1] One of the three subroutines S1D, S2D or S3D is called in accordance with $|V|$.

A support point of a convex polytope can also be computed efficiently. For a polytope P, the support point is a vertex of P, i.e., $s_P(\mathbf{d}) \in \text{vert}(P)$, and we can take $\mathbf{w} = s_P(\mathbf{d}) = s_{\text{vert}(P)}(\mathbf{d})$, that is,

$$\mathbf{d}^T s_P(\mathbf{d}) = \max\{\mathbf{d}^T\mathbf{v}; \mathbf{v} \in \text{vert}(P)\} \quad (33)$$

Therefore, for polytopes, the support point can be uniquely determined by simply scanning through the list of vertices for the vertex that is the most extreme in the search direction $\mathbf{d}$. Therefore, the computation time is linear in the number of vertices of P. For complex polytopes, the vertices adjacency information and the coherence between consecutive calls to support mapping functions can be exploited to find the support point with almost constant time complexity [30].

Algorithm 3 Sub-routine for $|V| = 2$

function S1D($\{\mathbf{v}_2, \mathbf{v}_1\}$) [1]
  if $\mathbf{v}_1^T\mathbf{v}_{12} \geq 0$ **then**
    $V \leftarrow \{\mathbf{v}_1\}$
    $\mathbf{d} \leftarrow -\mathbf{v}_1$
  else
    $V \leftarrow \{\mathbf{v}_2, \mathbf{v}_1\}$
    $\mathbf{d} \leftarrow -\mathbf{v}_{12} \times \mathbf{v}_1 \times \mathbf{v}_{12}$
  end if
end function

[1] The input is the ordered list of vertices, with $\mathbf{v}_1$ being the last added element to V_k.

1-simplex 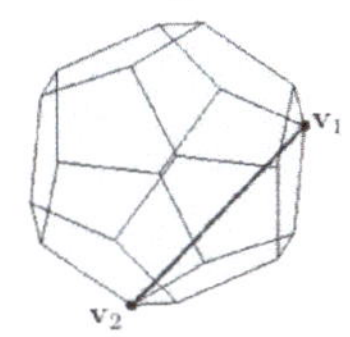	$\mathcal{V}_{(1)}$ $\mathcal{V}_{(1,2)}$
2-simplex 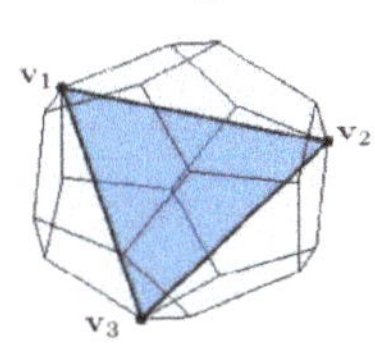	$\mathcal{V}_{(1)}$ $\mathcal{V}_{(1,2)}, \mathcal{V}_{(1,3)}$ $\mathcal{V}_{(1,2,3)}$
3-simplex 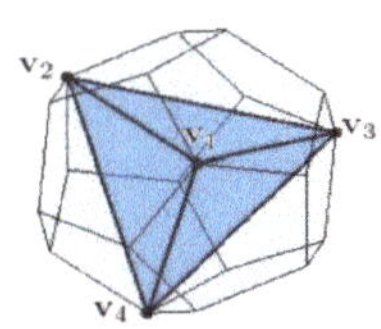	$\mathcal{V}_{(1)}$ $\mathcal{V}_{(1,2)}, \mathcal{V}_{(1,3)}, \mathcal{V}_{(1,4)}$ $\mathcal{V}_{(1,2,3)}, \mathcal{V}_{(1,2,4)}, \mathcal{V}_{(1,3,4)}$ $\mathcal{V}_{(1,2,3,4)}$

Figure 4. An m-simplex is linked to $2^{m+1} - 1$ Voronoi regions associated with its vertices, edges, faces, and volume. The list of 2^m Voronoi regions that can possibly contain the origin is given in this table. It should be noted that $\mathbf{v}_1$ is the latest vertex added to V.

Algorithm 4 Sub-routine for $|V| = 3$

function S2D ($\{\mathbf{v}_3, \mathbf{v}_2, \mathbf{v}_1\}$)
    $\mathbf{n}_{\mathbf{v}_3\mathbf{v}_2\mathbf{v}_1} = \mathbf{v}_{12} \times \mathbf{v}_{13}$
    if $\mathbf{v}_1^T(\mathbf{n}_{\mathbf{v}_3\mathbf{v}_2\mathbf{v}_1} \times \mathbf{v}_{12}) \geq 0$ **then**
        $[V, \mathbf{d}] \leftarrow \text{S1D}(\{\mathbf{v}_2, \mathbf{v}_1\})$
    else
        if ${\mathbf{v}_1}^T(\mathbf{v}_{13} \times \mathbf{n}_{\mathbf{v}_3\mathbf{v}_2\mathbf{v}_1}) \geq 0$ **then**
            $[V, \mathbf{d}] \leftarrow \text{S1D}(\{\mathbf{v}_3, \mathbf{v}_1\})$
        else
            if $\mathbf{v}_1^T \mathbf{n}_{\mathbf{v}_3\mathbf{v}_2\mathbf{v}_1} \geq 0$ **then**
                $V \leftarrow \{\mathbf{v}_3, \mathbf{v}_2, \mathbf{v}_1\}$
                $\mathbf{d} \leftarrow -\mathbf{n}_{\mathbf{v}_3\mathbf{v}_2\mathbf{v}_1}$
            else
                $V \leftarrow \{\mathbf{v}_3, \mathbf{v}_2, \mathbf{v}_1\}$
                $\mathbf{d} \leftarrow \mathbf{n}_{\mathbf{v}_3\mathbf{v}_2\mathbf{v}_1}$
            end if
        end if
    end if
end function

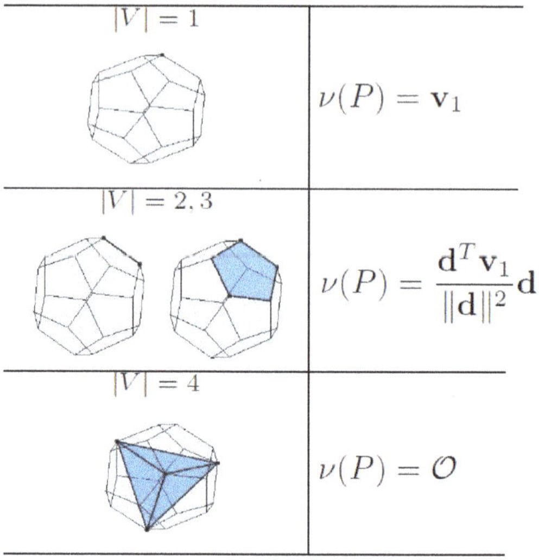

Figure 5. Examples of the closest feature of a polyhedron to a query point are shown above. Once the closest feature is obtained from Algorithm 1, the closest point, i.e., $\nu(P)$, can be determined, as shown above.

Algorithm 5 Sub-routine for $|V| = 4$

function S3D($\{\mathbf{v}_4, \mathbf{v}_3, \mathbf{v}_2, \mathbf{v}_1\}$)
    $\mathbf{n}_{\mathbf{v}_3\mathbf{v}_2\mathbf{v}_1} = \mathbf{v}_{12} \times \mathbf{v}_{13}$
    if $(\mathbf{v}_1^T \mathbf{n}_{\mathbf{v}_3\mathbf{v}_2\mathbf{v}_1})(\mathbf{v}_{14}^T \mathbf{n}_{\mathbf{v}_3\mathbf{v}_2\mathbf{v}_1}) \geq 0$ **then**
        $[V, \mathbf{d}] \leftarrow$ S2D($\{\mathbf{v}_3, \mathbf{v}_2, \mathbf{v}_1\}$)
    else
        $\mathbf{n}_{\mathbf{v}_4\mathbf{v}_3\mathbf{v}_1} = \mathbf{v}_{13} \times \mathbf{v}_{14}$
        if $(\mathbf{v}_1^T \mathbf{n}_{\mathbf{v}_4\mathbf{v}_3\mathbf{v}_1})(\mathbf{v}_{12}^T \mathbf{n}_{\mathbf{v}_4\mathbf{v}_3\mathbf{v}_1}) \geq 0$ **then**
            $[V, \mathbf{d}] \leftarrow$ S2D($\{\mathbf{v}_4, \mathbf{v}_3, \mathbf{v}_1\}$)
        else
            $\mathbf{n}_{\mathbf{v}_4\mathbf{v}_2\mathbf{v}_1} = \mathbf{v}_{12} \times \mathbf{v}_{14}$
            if $(\mathbf{v}_1^T \mathbf{n}_{\mathbf{v}_4\mathbf{v}_2\mathbf{v}_1})(\mathbf{v}_{13}^T \mathbf{n}_{\mathbf{v}_4\mathbf{v}_2\mathbf{v}_1}) \geq 0$ **then**
                $[V, \mathbf{d}] \leftarrow$ S2D($\{\mathbf{v}_4, \mathbf{v}_2, \mathbf{v}_1\}$)
            else
                $V \leftarrow \{\mathbf{v}_4, \mathbf{v}_3, \mathbf{v}_2, \mathbf{v}_1\}$
            end if
        end if
    end if
end function

3. Bézier Curves

3.1. Continuity Conditions

As explained before, at each sampling time, a trajectory, expressed as a parametric Bézier curve, is generated for the time horizon $[t_k, t_k + t_h]$, and the trajectory for the entire flight time $[0, T]$ is formed by joining segments of these Bézier curves end-to-end. The smoothness of the resulting composite trajectory must be guaranteed by enforcing continuity at the joining points of two consecutive segments up to a certain derivative. In the following, in order to derive conditions that address parameter continuity between consecutive curves, we assume that the time horizon is equal to $\Delta t_k = t_{k+1} - t_k$, which is not necessarily the same for all sub-problems. In practice, however, the time horizon is greater than Δt_k, in which case a Bézier curve describing the segment over the time interval $[t_k, t_{k+1}]$ can be obtained by subdividing $\mathbf{p}_k(.)$ at t_{k+1} with the de Casteljau's algorithm. For simplicity we drop the subscript $i \in [N_v]$.

The Bézier curve describing the trajectory over the time interval $[t_k, t_{k+1}]$ is defined as

$$\mathbf{p}_k(\tau_k) = \sum_{l=0}^{n_k} \bar{p}_{l,k} B_{l,n_k}(\tau_k), \tag{34}$$

Assuming that the global parameter t runs over the interval $[t_k, t_{k+1}]$, the local parameter τ_k is related to t by

$$0 \leq \tau_k = \frac{t - t_k}{t_{k+1} - t_k} \leq 1 \tag{35}$$

The parametric continuity condition, C^r, requires the r-th derivative and all lower derivatives of two consecutive segments to be equal at the joining point. In other words,

$$\frac{d^r \mathbf{p}_k(1)}{dt^r} = \frac{d^r \mathbf{p}_{k+1}(0)}{dt^r} \qquad r \in \{0, \ldots, \mathrm{r}\} \tag{36}$$

Zero-order parametric continuity, C^0, requires the endpoints of two consecutive curves, $\mathbf{p}_k(.)$ and $\mathbf{p}_{k+1}(.)$, to intersect at one endpoint, that is,

$$\mathbf{p}_k(1) = \mathbf{p}_{k+1}(0) \tag{37}$$

Since a Bézier curve is coincident with its control points at the two ends, i.e.,

$$\mathbf{p}_k(0) = \bar{p}_{0,k} \quad \mathbf{p}_k(1) = \bar{p}_{n_k,k}, \tag{38}$$

the position continuity condition (37) translates into

$$\bar{p}_{n_k,k} = \bar{p}_{0,k+1} \tag{39}$$

The first-order parametric continuity condition, C^1, for the k-th and $k+1$-th Bézier curves, can be obtained as

$$\begin{aligned} &\Delta t_{k+1} n_k (\bar{p}_{n_k,k} - \bar{p}_{n_k-1,k}) = \\ &\qquad \Delta t_k n_{k+1} (\bar{p}_{1,k+1} - \bar{p}_{0,k+1}) \end{aligned} \tag{40}$$

Finally, the k-th and $k+1$-th Bézier curves are C^2-continuous if

$$\begin{aligned} &\frac{\Delta t_{k+1}}{\Delta t_k} (\bar{p}_{n_k-1,k} - \bar{p}_{n_k-2,k}) \\ &\qquad + n_k(n_{k+1} - 1)\bar{p}_{n_k-1,k} + n_k \bar{p}_{n_k,k} = \\ &\frac{\Delta t_k}{\Delta t_{k+1}} (\bar{p}_{1,k+1} - \bar{p}_{2,k+1}) \\ &\qquad + n_{k+1}(n_k - 1)\bar{p}_{1,k+1} + n_{k+1}\bar{p}_{0,k+1} \end{aligned} \tag{41}$$

Higher-order parametric continuity conditions can be obtained likewise.

3.2. Evaluating Inequalities in Bézier Form

Parameterizing the trajectory with a Bézier curve converts the original infinite dimensional problem (3) into a semi-infinite optimization problem with a finite number of decision variables and an infinite number of constraints. The commonly used approach to obtaining a standard optimization problem is time gridding, which inspects satisfaction of constraints on a finite number of points only. Although this method is straightforward, it cannot guarantee that constraints are satisfied over the entire time interval. Using fine discretization can remedy this issue, but, it will increase the number of constraints as well as the computation time. Since all constraints involved in the trajectory generation problem addressed above can be expressed as Bézier curves, we can employ the method proposed in [33] to recast the semi-infinite optimization problem into one that is computationally tractable. As explained below this method exploits unique features of Bézier curves to efficiently evaluate constraints while avoiding problems associated with time gridding.

If $h(\tau)$ can be expressed as a Bézier curve, then any inequality constraint of the form $h(\tau) \leq 0, \tau \in [0,1]$ can be replaced by a finite set of constraints on its control points. More specifically, if $h(\tau)$ is defined as

$$h(\tau) = \sum_{l=0}^{n_h} \bar{h}_l B_{l,n_h}(\tau), \tag{42}$$

then from the *convex hull* property of Bézier curves we know that

$$h(\tau) \in CH(\bar{H}) \quad \tau \in [0,1] \tag{43}$$

where $CH(\bar{H}) = \{\alpha_0\bar{h}_0 + \cdots + \alpha_{n_h}\bar{h}_{n_h} | \alpha_0 + \ldots, \alpha_{n_h} = 1, \alpha_l \geq 0\}$ is the convex hull defined by the set of control points [34]. Thus, the inequality constraint $h(\tau) \leq 0$ holds if

$$\bar{h}_l \leq 0 \quad for \quad l = 0, \ldots, n_h. \tag{44}$$

This finite set of inequality constraints can ensure that the original inequality constraint is satisfied over the entire interval $[0,1]$. However, Ineqs. (44) might be conservative due to the existing gap between the control points $\bar{h}_l$ and the actual curve $h(\tau)$. This problem can be alleviated by refining the control polygon and finding closer control points to the curve using recursive subdivision of $h(\tau)$ with the de Casteljau's algorithm. The sequence of control polygons generated with repeated subdivision converges to the underlying Bézier curve [35]. Figure 6 shows a threefold subdivision of a cubic Bézier curve. Furthermore, the de Casteljau's algortithm allows refining the control polygon locally. Using recursive subdivsion of $h(\tau)$ to reduce the conservatism in the finite set of constraints results in an increase in the number of constraints; hence, a trade-off has to be made between the computational effort and the conservatism. Nevertheless, the optimization variables remain the same [36].

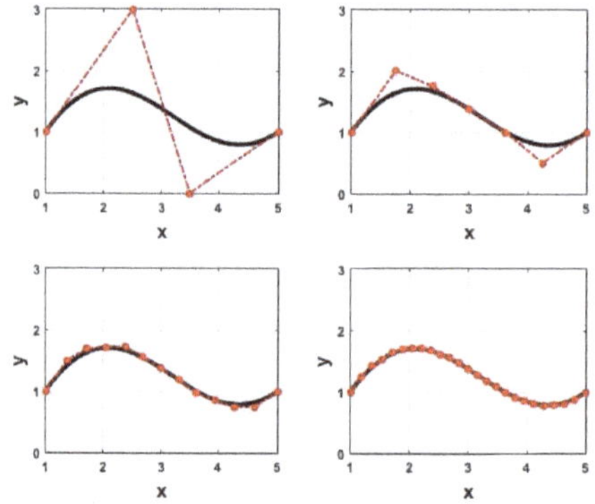

Figure 6. A cubic Bézier curve (**top left**) is subdivided into two Bézier curves of the same degree (**top right**) using the de Casteljau's algorithm. The control polygon generated by recursive subdivision converges to the original Bézier curve ref. [28]. Copyright 2021 IEEE. Successive refinement of the original control polygon after 2 (**bottom left**) and 3 (**bottom right**) subdivisions.

4. Simulation Results

In this section, the efficacy of the proposed method for generating feasible and collision-free trajectories in (vehicle-) dense environments are assessed through different simulation examples. We compare the resulting trajectories to those generated with the well-studied BVC approach. We specifically test the capability of the two methods to generate trajectories that ensure all drones involved in a simulation example reach their final positions, and compare the flight time, obtained with each of them, to complete point-to-point transition missions. We also present the recorded computation time for executing the proposed algorithm in this paper to emphasize its suitability for real-time applications.

In the simulations presented below, we assume all drones have the same size, and their BVC (23) is defined with the safety radius $r_D = 0.30$ m. To specify the set (25), we approximate the drone body with an oblate spheroid with $\Lambda = \text{diag}([0.30\text{ m}, 0.30\text{ m}, 0.11\text{ m}])$. In both methods, trajectories are parameterized with Bézier curves. Upper and lower bounds on the speed and acceleration are assumed to be $\pm 2.3\ \frac{\text{m}}{\text{s}}$ and $\pm 7.1\ \frac{\text{m}}{\text{s}^2}$ respectively. At each replanning step, the planner finds the closest point in the updated Voronoi cell to the goal position using the algorithm in Section 2.3. The computed point is then used to define the terminal cost term. The first term of the cost function in all subproblems is defined as $\int_0^1 \|\mathbf{p}_{i,k}^{(4)}(\tau)\|^2 d\tau$. The time horizon and the replanning step are also considered to be the same for both methods. The obtained solution at the previous replanning step is used to set the initial guess for the current sub-problem. We use FORCES Pro [37] to generate solvers for the resulting sub-problems. The sub-problems, involving the set of control points $\bar{p}_{i,k}$ as decision variables, can be reformulated to match the supported classes of problem in FORCES Pro. Here, all computations are executed on a single desktop computer, with 2.60 GHz i7-4510U CPU and 6.00 GB RAM; however, in practice, the resulting independent sub-problems can be solved in parallel.

As mentioned before in the paper, in Voronoi-based methods, a vehicle only requires the position information from its neighboring vehicles to generate its trajectory. Therefore, they are more suitable for implementation when vehicles have limited communication capability, and have to rely solely on onboard sensing. In reality, the position sensor noise can impact the planner performance, yet this is more pronounced when estimating other information, such as velocity, from noise-corrupted measurements is needed. Therefore, Voronoi-based planners are more robust when there is no communication network. Nevertheless, in the following simulations, we assume that accurate position information is available with no delay at the replanning time.

In the first example, we consider five drones flying from their initial positions to given final positions. This example is similar to one in [24] where a random offset is added to break the symmetry in the drones' initial and final configurations. Figure 7 (right) shows collision-free trajectories generated with the distributed scheme described above, with a replanning rate of 20 Hz. For this particular example, the resulting trajectories match those generated with BVC with a flight time of 11.6782 s. Figure 7 (left) shows collision-free trajectories obtained from the centralized solution, which delivers a total flight time of 9.4347 s, yet, while the central solution is obtained in 601 ms, the average computation time for solving the sub-problems in the decentralized scheme is only 49 ms.

In the next example, we consider 18 drones switching positions in a 3 m × 5 m × 2 m space, with a maximum speed and acceleration of $\pm 4.7\ \frac{\text{m}}{\text{s}}$ and $\pm 9.8\ \frac{\text{m}}{\text{s}^2}$, respectively. Figure 8a shows the initial and final configurations, and Figure 8b displays collision-free trajectories generated with the proposed distributed scheme in the paper implemented at 10 Hz. While both methods could find collision-free trajectories for guiding the team of drones from their initial positions to their goal positions, the flight time achieved with the proposed method is markedly shorter than the time obtained with BVC. We also performed a trial simulation with 34 drones in a similar configuration. Table 1 compares the success rate and the flight time to complete the transition using BVC and the proposed method.

In the third example, we consider 100 drones flying in an 8 m × 8 m × 3.5 m space. The initial and final positions for the drones are displayed with dot and square markers in Figure 9. We test both methods in 30 different trials. In each trial, final positions are randomly assigned to drones. A trial is considered successful if all drones could reach their final positions within the stipulated time. The proposed method with 23 successful trials and an average total flight of 1*s* outperforms the BVC with only 16 completed trials. It should be noted that using well-devised deadlock prevention strategies or loosening time constraints can improve the success rate of both methods. Figure 9 shows collision-free trajectories generated with the proposed method for one of the trials at different time steps. The average computation time for solving sub-problems in this example was around 115 milliseconds. In addition, compared to the geometric algorithm in [24], the closest point in a Voronoi cell to the goal position was obtained at least 10 times faster with the proposed algorithm in Section 2.3. The computation time for finding the closest point, and solving the optimization problem, depends on the number of neighboring drones (See Table 2 for recorded computation times in simulation examples with 18, 34, and 100 drones). In most applications, with typical Voronoi diagrams, the number of boundary planes, i.e., the number of Voronoi neighbors, is small. Thus, the proposed distributed algorithm is scalable to arbitrary numbers of vehicles.

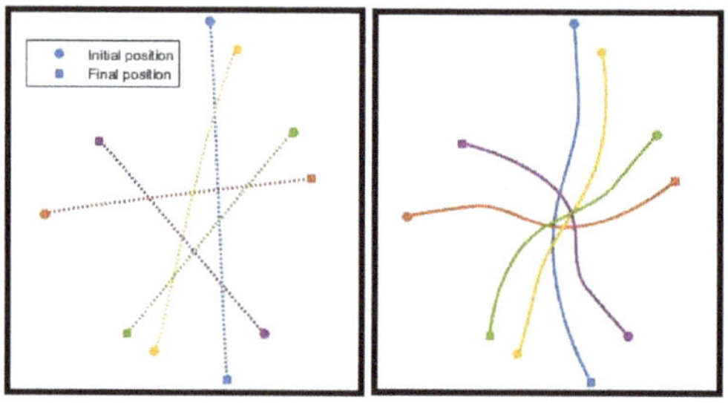

Figure 7. Comparing collision-free trajectories generated with the centralized solution (**left**) and the proposed decentralized approach (**right**) for five drones flying from their initial positions to given final positions. While the central solution yields a shorter flight time, its computation time is significantly longer than average time required to solve the sub-problems in the distributed method.

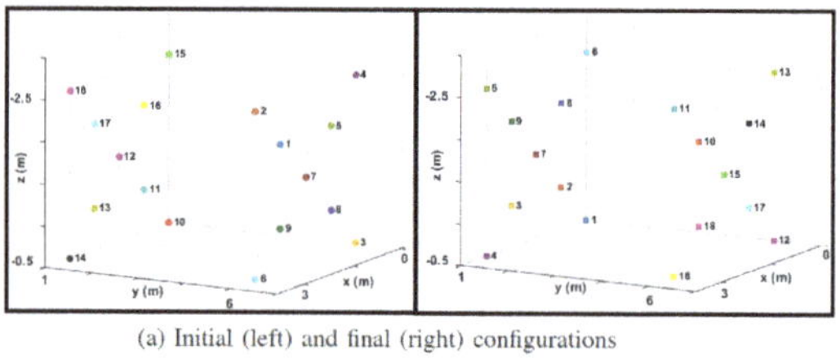

(a) Initial (left) and final (right) configurations

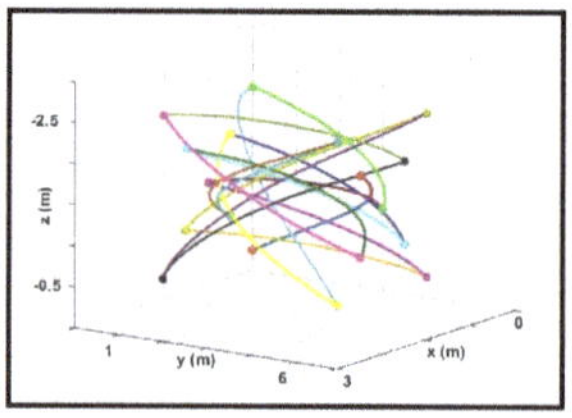

(b) Collision-free trajectories

Figure 8. (**a**) Initial (left) and final (right) position configurations for 18 drones. Each drone is assigned a unique color and a number next to it. (**b**) Collision−free trajectories for 18 drones switching their positions in a 3 m × 5 m × 2 m space. The total flight time for the drones to reach their final positions is 5.1 s using the proposed method, which is shorter than the 6.3 s flight time obtained with the BVC.

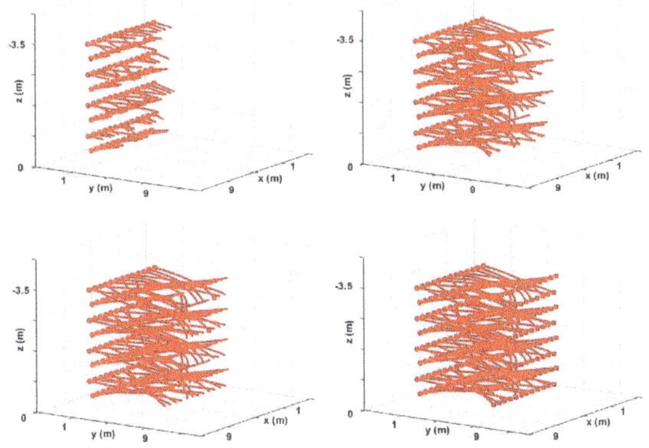

Figure 9. Collision−free trajectories for 100 drones flying from their initial positions (dots) to randomly specified final positions (squares) at different replanning steps.

Table 1. Comparing the number of successful trials and the average flight time achieved with the BVC and the proposed method in the paper.

Number of Drones	BVC		Proposed Method	
	Flight Time	Completed Trials	Flight Time	Completed Trials
18	6.812 s	5/5	5.327 s	5/5
34	8.105 s	7/10	6.625 s	8/10
100	14.573 s	16/30	11.462 s	23/30

Table 2. Recorded computation times for finding the closest point in a Voronoi cell to the goal position and solving the optimization problem in simulation examples with 18, 34, and 100 drones.

Number of Drones	Computation Time (ms)	
	Finding the Closest Point	Solving the Sub-Problem
18	<0.1	77.562
34	<0.1	98.330
100	0.171	121.633

5. Conclusions

In this paper, we introduce an efficient distributed algorithm for generating collision-free trajectories for multiple drones, taking into account their orientation. In order to avoid substantial communication between drones, we adopt Voronoi-based space partitioning and derive local constraints that guarantee collision avoidance with neighboring vehicles for an entire time horizon. We leverage Bézier curve properties to ensure that the set of collision avoidance constraints are satisfied at any time instant of the planning horizon. The same approach can be employed to obtain local collision avoidance constraints for the cases where the normal vector and offset of separating planes are time-varying parameters or decision variables of sub-problems. Yet, adopting Voronoi diagram with fixed planes for an entire planning horizon, though being conservative, results in simple, small sub-problems allowing for the trajectories to be replanned at a higher rate. We present different simulation results to highlight the scalability of the algorithm to large numbers of drones, and also its capability to generate less conservative trajectories with notably shorter flight times, compared to other Voronoi-based methods.

Our future work includes implementation and experimental validation of the algorithm for teams of drones. As we explain in the paper, at each time sample, upon receiving (or sensing) the new position information, a vehicle must find the closest point in its Voronoi cell to the goal position, and solve an optimization problem that uses the current state of the vehicle as the initial condition, to generate its trajectory for a certain time horizon. Although the time to compute the closest point is mainly negligible, the computation time to find the optimal solution can lead to a (significant) delay between updating the position information and executing the trajectory. Therefore, in practice, the computational delay must be explicitly considered to avoid performance degradation (or even failure) of the planner.

Author Contributions: Conceptualization and writing—original draft preparation, B.S.; Supervision and writing—review and editing, R.C. and A.P. All authors have read and agreed to the published version of the manuscript.

Funding: This work was partially funded by Fundação para a Ciência e a Tecnologia and FEDER funds through the projects UIDB/50009/2020 and LISBOA-01-0145-FEDER-031411. B. Sabetghadam acknowledges the support of Instituto Superior Técnico through scholarship BL216/2018/IST-ID.

Institutional Review Board Statement: Not applicable.

Informed Consent Statement: Not applicable.

Data Availability Statement: Not applicable.

Conflicts of Interest: The authors declare no conflict of interest.

References

1. Kuwata, Y.; How, J.P. Cooperative distributed robust trajectory optimization using receding horizon MILP. *IEEE Trans. Control Syst. Technol.* **2010**, *19*, 423–431. [CrossRef]
2. Augugliaro, F.; Schoellig, A.P.; D'Andrea, R. Generation of collision-free trajectories for a quadrocopter fleet: A sequential convex programming approach. In Proceedings of the IEEE/RSJ International Conference on Intelligent Robots and Systems, Vilamoura-Algarve, Portugal, 7–12 October 2012; pp. 1917–1922.
3. Morgan, D.; Chung, S.J.; Hadaegh, F.Y. Model predictive control of swarms of spacecraft using sequential convex programming. *J. Guid. Control Dyn.* **2014**, *37*, 1725–1740. [CrossRef]
4. Chen, Y.; Cutler, M.; How, J.P. Decoupled multiagent path planning via incremental sequential convex programming. In Proceedings of the IEEE International Conference on Robotics and Automation (ICRA), Seattle, WA, USA, 26–30 May 2015; pp. 5954–5961.
5. Kuwata, Y.; Richards, A.; Schouwenaars, T.; How, J.P. Distributed robust receding horizon control for multivehicle guidance. *IEEE Trans. Control Syst. Technol.* **2007**, *15*, 627–641. [CrossRef]
6. Chaloulos, G.; Hokayem, P.; Lygeros, J. Distributed hierarchical MPC for conflict resolution in air traffic control. In Proceedings of the American Control Conference, Baltimore, MD, USA, 30 June–2 July 2010; pp. 3945–3950.
7. Tedesco, F.; Raimondo, D.M.; Casavola, A.; Lygeros, J. Distributed collision avoidance for interacting vehicles: A command governor approach. *IFAC Proc. Vol.* **2010**, *43*, 293–298. [CrossRef]
8. Van Parys, R.; Pipeleers, G. Distributed model predictive formation control with inter-vehicle collision avoidance. In Proceedings of the 2017 11th Asian Control Conference (ASCC), Gold Coast, QLD, Australia, 17–20 December 2017; pp. 2399–2404.
9. Van Parys, R.; Pipeleers, G. Distributed MPC for multi-vehicle systems moving in formation. *Robot. Auton. Syst.* **2017**, *97*, 144–152. [CrossRef]
10. Tordesillas, J.; How, J.P. MADER: Trajectory planner in multiagent and dynamic environments. *IEEE Trans. Robot.* **2021**, *38*, 463–476. [CrossRef]
11. Luis, C.E.; Schoellig, A.P. Trajectory generation for multiagent point-to-point transitions via distributed model predictive control. *IEEE Robot. Autom. Lett.* **2019**, *4*, 375–382. [CrossRef]
12. Luis, C.E.; Vukosavljev, M.; Schoellig, A.P. Online trajectory generation with distributed model predictive control for multi-robot motion planning. *IEEE Robot. Autom. Lett.* **2020**, *5*, 604–611. [CrossRef]
13. Van den Berg, J.; Lin, M.; Manocha, D. Reciprocal velocity obstacles for real-time multi-agent navigation. In Proceedings of the IEEE International Conference on Robotics and Automation, Pasadena, CA, USA, 19–23 May 2008; pp. 1928–1935.
14. Van Den Berg, J.; Guy, S.J.; Lin, M.; Manocha, D. Reciprocal n-body collision avoidance. In *Robotics Research*; Springer: Berlin/Heidelberg, Germany, 2011; pp. 3–19.
15. van den Berg, J.; Guy, S.J.; Snape, J.; Lin, M.C.; Manocha, D. rvo2 Library: Reciprocal Collision Avoidance for Real-Time Multi-Agent Simulation. 2011. Available online: https://gamma.cs.unc.edu/RVO2/ (accessed on 24 January 2022).

16. Alonso-Mora, J.; Breitenmoser, A.; Beardsley, P.; Siegwart, R. Reciprocal collision avoidance for multiple car-like robots. In Proceedings of the IEEE International Conference on Robotics and Automation, Saint Paul, MN, USA, 14–18 May 2012; pp. 360–366.
17. Snape, J.; Van Den Berg, J.; Guy, S.J.; Manocha, D. The hybrid reciprocal velocity obstacle. *IEEE Trans. Robot.* **2011**, *27*, 696–706. [CrossRef]
18. Giese, A.; Latypov, D.; Amato, N.M. Reciprocally-rotating velocity obstacles. In Proceedings of the 2014 IEEE International Conference on Robotics and Automation (ICRA), Hong Kong, China, 31 May–7 June 2014; pp. 3234–3241.
19. Bortoff, S.A. Path planning for UAVs. In Proceedings of the 2000 American Control Conference (ACC), Chicago, IL, USA, 28–30 June 2000; pp. 364–368.
20. Garrido, S.; Moreno, L.; Abderrahim, M.; Martin, F. Path planning for mobile robot navigation using voronoi diagram and fast marching. In Proceedings of the IEEE/RSJ International Conference on Intelligent Robots and Systems, Beijing, China, 9–15 October 2006; pp. 2376–2381.
21. Bhattacharya, P.; Gavrilova, M.L. Roadmap-based path planning-using the voronoi diagram for a clearance-based shortest path. *IEEE Robot. Autom. Mag.* **2008**, *15*, 58–66. [CrossRef]
22. Cortes, J.; Martinez, S.; Karatas, T.; Bullo, F. Coverage control for mobile sensing networks. *IEEE Trans. Robot. Autom.* **2004**, *20*, 243–255. [CrossRef]
23. Bandyopadhyay, S.; Chung, S.J.; Hadaegh, F.Y. Probabilistic swarm guidance using optimal transport. In Proceedings of the 2014 IEEE Conference on Control Applications (CCA), Juan Les Antibes, France, 8–10 October 2014; pp. 498–505.
24. Zhou, D.; Wang, Z.; Bandyopadhyay, S.; Schwager, M. Fast, on-line collision avoidance for dynamic vehicles using buffered voronoi cells. *IEEE Robot. Autom. Lett.* **2017**, *2*, 1047–1054. [CrossRef]
25. Senbaslar, B.; Hönig, W.; Ayanian, N. Robust trajectory execution for multi-robot teams using distributed real-time replanning. In *Distributed Autonomous Robotic Systems*; Springer: Cham, Switzerland, 2019; pp. 167–181.
26. Mellinger, D.; Kumar, V. Minimum snap trajectory generation and control for quadrotors. In Proceedings of the IEEE International Conference on Robotics and Automation, Shanghai, China, 9–13 May 2011; pp. 2520–2525.
27. LaValle, S.M. *Planning Algorithms*; Cambridge University Press: Cambridge, UK, 2006.
28. Sabetghadam, B.; Cunha, R.; Pascoal, A. Trajectory Generation for Drones in Confined Spaces using an Ellipsoid Model of the Body. *IEEE Control Syst. Lett.* **2021**, *6*, 1022–1027. [CrossRef]
29. Gilbert, E.G.; Johnson, D.W.; Keerthi, S.S. A fast procedure for computing the distance between complex objects in three-dimensional space. *IEEE J. Robot. Autom.* **1988**, *4*, 193–203. [CrossRef]
30. Van Den Bergen, G. *Collision Detection in Interactive 3D Environments*; CRC Press: Boca Raton, FL, USA, 2003.
31. Ericson, C. *Real-Time Collision Detection*; CRC Press: Boca Raton, FL, USA, 2004.
32. Tereshchenko, V.; Chevokin, S.; Fisunenko, A. Algorithm for finding the domain intersection of a set of polytopes. *Procedia Comput. Sci.* **2013**, *18*, 459–464. [CrossRef]
33. Montanari, M.; Petrinic, N.; Barbieri, E. Improving the GJK algorithm for faster and more reliable distance queries between convex objects. *ACM Trans. Graph. (Tog)* **2017**, *36*, 1–7. [CrossRef]
34. Farouki, R.T. The Bernstein polynomial basis: A centennial retrospective. *Comput. Aided Geom. Des.* **2012**, *29*, 379–419. [CrossRef]
35. Cichella, V.; Kaminer, I.; Walton, C.; Hovakimyan, N.; Pascoal, A.M. Optimal Multivehicle Motion Planning Using Bernstein Approximants. *IEEE Trans. Autom. Control* **2020**, *66*, 1453–1467. [CrossRef]
36. Sabetghadam, B.; Cunha, R.; Pascoal, A. Real-time Trajectory Generation for Multiple Drones using Bézier Curves. *IFAC-PapersOnLine* **2020**, *53*, 9276–9281. [CrossRef]
37. Domahidi, A.; Jerez, J. *Forces Professional*; Embotech AG: Zürich, Switzerland, 2014–2019. Available online: https://embotech.com/FORCES-Pro (accessed on 22 June 2021).

sensors

MDPI

Article

A Local Planner for Accurate Positioning for a Multiple Steer-and-Drive Unit Vehicle Using Non-Linear Optimization

Henrik Andreasson [1,*], Jonas Larsson [2] and Stephanie Lowry [1]

[1] Centre for Applied Autonomous Sensor Systems (AASS), Örebro University, 701 82 Örebro, Sweden; stephanie.lowry@oru.se
[2] ABB Corporate Research, 722 26 Västerås, Sweden; jonas.larsson@se.abb.com
* Correspondence: henrik.andreasson@oru.se

Abstract: This paper presents a local planning approach that is targeted for pseudo-omnidirectional vehicles: that is, vehicles that can drive sideways and rotate on the spot. This local planner—MSDU–is based on optimal control and formulates a non-linear optimization problem formulation that exploits the omni-motion capabilities of the vehicle to drive the vehicle to the goal in a smooth and efficient manner while avoiding obstacles and singularities. MSDU is designed for a real platform for mobile manipulation where one key function is the capability to drive in narrow and confined areas. The real-world evaluations show that MSDU planned paths that were smoother and more accurate than a comparable local path planner Timed Elastic Band (TEB), with a mean (translational, angular) error for MSDU of (0.0028 m, 0.0010 rad) compared to (0.0033 m, 0.0038 rad) for TEB. MSDU also generated paths that were consistently shorter than TEB, with a mean (translational, angular) distance traveled of (0.6026 m, 1.6130 rad) for MSDU compared to (0.7346 m, 3.7598 rad) for TEB.

Keywords: local planning; optimal control; obstacle avoidance

Citation: Andreasson, H.; Larsson, J.; Lowry, S. A Local Planner for Accurate Positioning for a Multiple Steer-and-Drive Unit Vehicle Using Non-Linear Optimization. *Sensors* **2022**, *22*, 2588. https://doi.org/10.3390/s22072588

Academic Editor: Maysam Abbod

Received: 1 March 2022
Accepted: 24 March 2022
Published: 28 March 2022

1. Introduction

Omni-motion capability—the ability to drive in all directions, not just forwards—allows greater maneuverability for mobile robots and similar platforms [1,2]. A platform with restricted omni-motion capability must make more complex maneuvers to achieve goal positions [1], which can be difficult to plan [3], time-consuming, and even impossible in crowded or constrained physical environments [3]. Furthermore, greater mobile maneuverability allows a robot to achieve more accurate pose targets [4] which are essential for many industrial applications.

A fully holonomic platform has maximal omni-motion capability as such a platform can move freely in any direction [2], and fully holonomic motion can be achieved using omnidirectional wheels such as the Mecanum (or Swedish) wheel [5]. However, there are drawbacks to such a wheel configuration: these wheels are mechanically complex and the wheels are sensitive to non-smooth surfaces and thus require either clean, flat floors or complex suspension systems that ensure that the wheels are in contact with the ground [6]. For this reason, while fully holonomic systems are common in academic research [7], very few commercial platforms exist, although an exception is the Omnibot from KUKA [8].

Alternatively to a fully holonomic platform, a platform with a wheel configuration that uses multiple combined steer and drive wheels is an attractive option for an omni-motion robotic platform. It is not fully holonomic, but it has many advantages: it is robust to floor and environment conditions; it has good motion performance, with high acceleration, deacceleration, and turning capability; and it provides accurate wheel odometry, compared to the slip experienced by fully holonomic wheels.

However, a challenge with steerable wheels is the increased complexity in vehicle motion control. This is due to the resulting non-holonomic nature of the vehicle motion as well as the inverse kinematics equations possessing singularities. Therefore, specialized motion-planning algorithms are required to ensure that the robot motion planning and control optimizes the agility and other advantages of the combined steer and drive wheels, while ensuring that the planner produces feasible paths for the platform to follow, and avoids singularities.

Existing local motion-planning algorithms often provide generic paths that do not fully exploit the pseudo-omnidirectional capabilities of the platform [9] or the kinodynamic constraints [10]. Furthermore, many local motion-planning algorithms do not inherently handle obstacles [11], which can be critical when navigating in confined and narrow environments. Furthermore, local motion planners often try to solve the global planning problem, which is computationally inefficient or infeasible for long paths [10]. Alternatively, local motion planners which do not solve the global path planning problem can cause the platform to get stuck in local minima and fail to reach the goal position [12,13].

The contribution of this paper is a local motion planner that exploits the pseudo-omnidirectional capabilities of the platform to generate an efficient trajectory, allow precise positioning, and avoid singular configurations. It combines constraints from both vehicle dynamics and obstacles observed by the robot's perception module to ensure that the outputted trajectories are kinodynamically feasible as well as safe from obstacle collisions.

The local motion planner is designed to integrate with a global motion planner that provides an initial path estimate based on the robot's internal map of the environment. Coupling with a global planner avoids the need to address longer, time-consuming trajectory optimization steps or to use additional reasoning in the framework to select a shorter path [9], and ensures the platform avoids local minima. This global path estimate provides waypoints as sub-goals to the local planner. The local planner then computes the best local path and provides a control signal to the robot's low-level controller for navigation.

The local planner—denoted Multiple Steer Drive Unit Local Planner, or MSDU—is evaluated on a real-world robotic platform. It is demonstrated to provide more accurate positioning than existing local planners, achieving a smaller distance from the desired goal position in terms of both translation and angular displacement, while also taking a shorter path (in terms of both translational and angular distance) to the goal.

2. Related Work

This paper focuses on the problem of motion planning and trajectory optimization. The challenge of motion planning is to produce paths or trajectories for a robot to travel to a specified goal, where a path is a curvature defined as a continuous or discretized function and a trajectory also contains velocity information. Trajectory optimization is often difficult as obstacles and complex robot dynamics have to be considered. The motion planner needs to consider both kinematic and kinodynamic feasibility. A kinematically feasible output simply means that the output paths are possible to drive: for example, the paths have continuous curvature. A kinodynamically feasible output is one where the given velocity and acceleration bounds on the vehicle are met.

Local motion planners that handle short-term planning with a limited horizon are used in complex environments where obstacle avoidance is the key. One "classical" local planner is the Dynamic Window Approach (DWA) [14], which, given a goal and current sensory readings, searches for feasible velocities that will drive the robot towards the goal. Another approach that is commonly used in practice is based on the Elastic-Band (EB) planner [15]. However, these planners provides outputs that are not kinodynamically smooth and lack a velocity profile [10]. Other approaches based on potential fields have also been proposed [16].

The elastic band has been extended with an optimization framework TEB [9,17] and provides a local planner that utilize optimization, which provides a trajectory that is followed by a separate controller. TEB is designed for differential drive, car-like and full holonomic vehicles. In [9], a set of possible local trajectories were used blending the difference between global and local planning. TEB-based formulations have also been utilized for motion planning of manipulators [18].

There is a large variety of optimization-based approaches which, nowadays, are becoming more and more popular primarily thanks to the increased computational power available. Often the task of trajectory generation and tracking/control is separated into two different entities [19,20]. As trajectory optimization is essentially what is solved in the MPC scheme and for example, similar to the approach suggested here, Schoels et al. [10] is a mixture of these two as the proposed system both generates trajectories and performs the tracking. In general, a more complicated footprint in combination with limited dynamics of a platform makes both the global motion planning search part as well as the trajectory generation and tracking more difficult, hence the need to separate them.

As the output of a global planner is typically a path, we can also define approaches that generate trajectories to reach points along the route as "path following" approaches. A generic description of wheeled platforms and path following is described by Oftadeh et al. [11] which also contains evaluations of a combined steer and drive platform. One key difference here is how the controllers are coupled with the error directly between the current pose and corresponding static path and do not rely on horizon-based optimization, and the followed path is assumed to be free of obstacles. There are a large variety of path-following for cars, such as lateral control, which also is formulated for 4WS4WD (four-wheel-steering–four-wheel-driving) or 4WID (four-wheel-independent-drive) [21,22] and often focuses on more complex dynamics and stability at higher speeds and not on obstacle handling. Typically these car-like vehicles have a limitation on the amount of steering each wheel can undertake [23].

How to formulate obstacle handling efficiently into an optimization framework is of importance and is related to both how to represent the vehicle's footprint, but also how to represent the shape of the obstacles. Circular or ellipsoidal constraints is commonly used [24] but other techniques using, for example, convex polygons described through signed distance functions [10,24].

Depending on the kinematic configuration and footprint of the platform there are many different motion planners, such as RRT [25], probabilistic roadmaps [26] and lattice-based motion planners [19,27]. There are also many variants of the RRT planner, such as RRT* [28] which is provably asymptotically optimal, RRT-Blossom [29] which is designed for highly constrained environments or RRT*-UNF [30] which targets dynamic environments.

As the given platform has a rather simplistic footprint and good maneuverability, we exploit this to have a fast Dijkstra-based global planner that is fast to compute and therefore can be continuously re-invoked with an up-to-date representation containing dynamic and static obstacles.

3. System Design

This section introduces the system that the MSDU Local Planner is designed to integrate with, including the robotic platform, the available sensors, the existing planning and navigation capability and the low-level control functionality. The challenges associated with the pseudo-omnidirectional motion capabilities of the platform are presented in Section 4, which describes the geometry and kinematics of the combined steer and drive wheels, and how the velocity and control values can be suitably represented to provide constraints to the optimization formulation. Section 5 then introduces the MSDU Local Planner presented in this paper with Section 5.1 describing the formulation of the optimization problem.

3.1. The Platform

The robotic platform used in this work is presented in Figure 1. It has four steerable wheels located in each corner and is what is often termed as a "pseudo-omnidirectional" platform [16].

Figure 1. Robotic platform used in the evaluation. It has four combined steer and drive wheels located at each corner. The onboard computer is an Intel NUC i7.

The steerable wheels in the steerable wheel research platform are an in-house design and are based on the outrunner hub-motors design commonly used in commodified personal e-mobility vehicles. These motors are superior when it comes to the torque-to-cost ratio, which was utilized to realize a direct-drive (no gearbox) solution for both steering and driving motion. High-quality motion performance is obtained using high-accuracy position sensors and state-of-the-art motor control. The result is a compact, high-performance, reliable and very quiet actuation module.

The vehicle control is designed to receive independent linear velocities in the 2D plane and rotation (v_x, v_y, ω) . To ensure high-performance dynamics of the vehicle-motion reference tracking, a dynamic model is used to compute feedforward torque to joints in a prediction-closed loop correction scheme. A dedicated Kalman filter is employed to make the vehicle control tolerant to potentially relatively low frequency and jittery references coming from the motion planner in the transition to the high frequency real-time system. In the tests reported in the present paper, the vehicle control system is reading the communication bus for new references from the planner at 500 Hz.

The wheel odometry computation is designed to take advantage of the redundancy of this specific configuration of four steerable wheels such that it selects wheels to be included to compute the measured 2D pose based on the motion. For example, any wheel that is close to ICR is omitted from the odometry calculation. This results in an accurate and robust measured vehicle-speed estimation, which is communicated to the navigation system at 500 Hz.

3.2. Steerable Wheels

The wheel configuration of multiple combined steer and drive wheels units was selected as it can provide omni-motion capability, while also having other benefits relative to other comparable systems:

- *Payload capability.* A steerable wheel is straightforward to scale in payload.
- *Stability and stiffness.* A steerable wheel allows stiff or no suspension, enabling a minimum footprint for a given mobile manipulation application stiffness requirement.
- *Wheel odometry.* Steerable wheels enable accurate wheel odometry due to their limited slippage. For robots with more than two steerable wheels the system also has

inherent redundancy that enables diagnostics of vehicle motion estimation based on wheel motions.

- *Robustness.* Steerable wheels are relatively robust to floor and environment conditions, and vehicle motion control is maintained even if some wheels lose floor contact, as long as at least two wheels remain in contact with grip. The size and material of the wheel can be chosen to suit the application; for example, a larger- or smaller-sized wheel can be used, and the system can employ soft material for the tire to increase the contact area if pressure in the floor contact needs to be limited.
- *Motion performance.* Steerable wheels are capable of high acceleration, deceleration, and turning due to good ground grip and low steering inertia. Steerable wheels have high scalability of speed capacity, and low vehicle-direction isotropic friction losses or slippage.
- *Flexible wheel configuration.* Steerable wheels can be placed arbitrarily in the chassis for design flexibility, and the number of wheels can be arbitrarily chosen. This flexibility in the wheel configuration can be useful to allow for modular robot payload and motion performance capacity.

Note that having more than two combined steer and drive wheels will provide redundancy to the system that can be beneficial for better traction in general and better handling of non-flat surfaces.

3.3. Perception

The platform is equipped with a SICK TiM571 2D LiDAR, which is used for localization, navigation, and planning, as well as for obstacle detection. It is also used for building the map (see Figure 8) used by the robot to perform localization, navigation, and planning. The sensor characteristics taken from the data sheet are a systematic errors of ± 60 mm and statistical errors of ± 20 mm at ranges up to max range of 25 m with 90% remission and for 10% remission up to 6 m range the statistical error is ± 10 mm; however, all numbers are typical and dependent on ambient conditions.

3.4. Localization, Navigation, and Planning System

The platform is equipped with all systems required for navigation; localization, global planning and the proposed MSDU local planning (where MSDU stands for Multiple Steer-And-Drive Unit) that provides the control output. All system components exist as ROS [31] packages and the proposed local planner is compatible with the navigation stack [32]. When provided with a map of its environment, the robot can localize within its map, using its range sensor and odometry readings. As a localization module we utilize the AMCL package with only minor parametric changes; the motion model is set to "omni-corrected". The robot also has a global planner that can provide a path based on the robot's map of the environment and also incorporated LiDAR readings. The global planner we use is the planner provided in the navigation stack with default parameters. The output is a path containing a set of poses and does not contain any velocity information.

The platform footprint is encoded as two circles. One circle has an inscribed radius, which gives the smallest opening you could drive the robot through with a *specific* angle, and the second circle has a circumscribed radius which is the smallest opening you could drive the robot through given *any* angle. What this shows us is that the global planner does not fully factor in the shape of the platform and it is up to the local planner to handle narrow passages where the opening size is between the inscribed and circumscribed radii, as can be seen in Figure 16.

Furthermore, the global planner does not consider the full kinematics of the robot and the generated global path is non-smooth. However, the global planner does consider the surrounding using LiDAR data (through the local costmap) and the given map (global costmap) to rapidly compute paths that keeps a distance from obstacles and that find plans that are suitable from a global perspective.

The local planner presented in this paper is designed to integrate with the global planner and use the provided path to generate waypoints as input to the local planner, which then formulates a control output for the low-level wheel controller. Unlike the global planner, the local planner factors in a large number of constraints, including the kinematic constraints of the robot and a footprint that can handle different orientation (to ensure safe navigation through confined spaces). Furthermore, the local planner integrates real-time information from the LiDAR sensor to guarantee that the generated control actions provide motions that are collision free.

Figure 2 presents an overview of the different components of the system and how they interact. As can be seen, the local planner integrates information from many components, including the global planner and the sensor data from the LiDAR, to calculate and output a control signal to the robot.

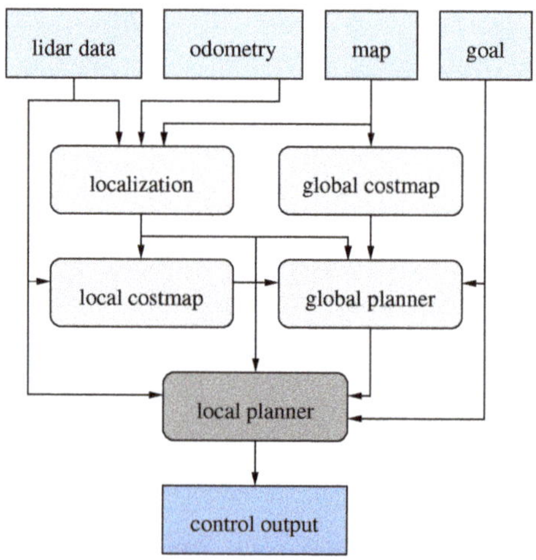

Figure 2. Overview of the different components of the localization, navigation and planning system and their interactions.

4. Geometry and Kinematics of the Combined Steer and Drive Wheels

This section provides an overview of the geometry and kinematics of the combined steer and drive wheels present on the robotic platform used in this paper. It outlines the physical limitations on the linear and angular velocities of the wheels, which will be used to provide the constraints for the optimization formulation in Section 5 below. Please note that we do not consider nor model any wheel slippage.

4.1. Definitions

Let the input control signal to the platform be denoted $\mathbf{u}_p = (\mathbf{v}, \omega)$ where $\mathbf{v} = (v_x, v_y)$ is the linear velocity (a 2D vector) and ω is the rotational velocity (a scalar). Using a 2D world reference frame $\{A\}$, let θ denote the orientation of the platform and x, y its 2D position (see Figure 3a). Then we have the following equations for the velocity of the platform in the world frame:

$$\begin{aligned} \dot{x}^{\{A\}} &= v_x^{\{A\}} \\ \dot{y}^{\{A\}} &= v_y^{\{A\}} \\ \dot{\theta}^{\{A\}} &= \omega \end{aligned} \tag{1}$$

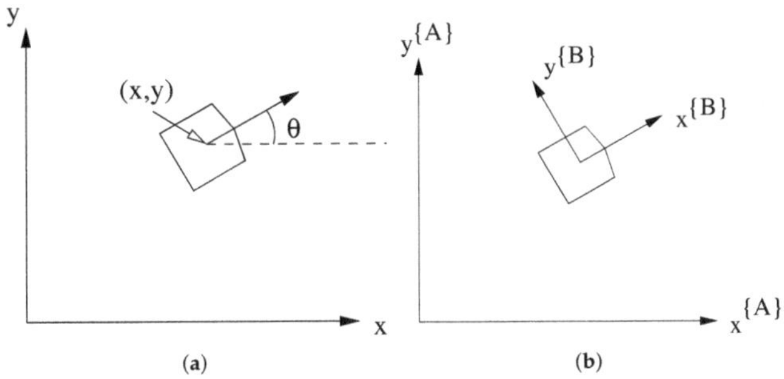

Figure 3. Platform pose and coordinate frames used. (**a**) The platform pose is denoted by position (x, y) and orientation θ in a world frame; (**b**) the relationship between the world coordinate frame $\{A\}$ and the platform coordinate frame $\{B\}$.

We also have the platform frame $\{B\}$, which has its origin at the center of the platform. The $\{B\}$ frame is defined so that the x direction will point "forward" (see Figure 3b). If the velocity is defined in the platform frame $\{B\}$, the resulting equations become:

$$
\begin{aligned}
\dot{x}^{\{A\}} &= v_x^{\{B\}} \cos(\theta) - v_y^{\{B\}} \sin(\theta) \\
\dot{y}^{\{A\}} &= v_x^{\{B\}} \sin(\theta) + v_y^{\{B\}} \cos(\theta) \\
\dot{\theta}^{\{A\}} &= \omega
\end{aligned} \tag{2}
$$

4.2. Wheel Configuration

On the platform there are four combined steer and drive wheels ($W_{1\ldots4}$). We denote for each wheel W_i the velocity v_i and the steering angle φ_i. The position of the wheel $\mathbf{p}_i^{\{B\}} = (p_x^i, p_y^i)$ is given in the platform frame. The velocity vector $\mathbf{v}_i$ is used to represent the combination of φ_i and v_i (see Figure 4).

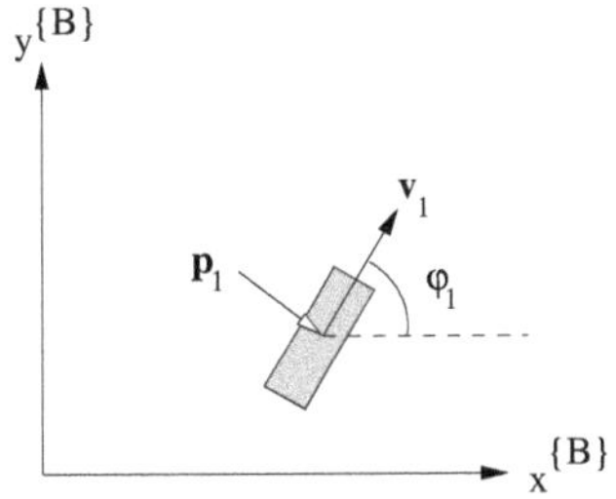

Figure 4. Wheel parameters: $\mathbf{p}_1$ indicates the position, $\mathbf{v}_1$ the velocity vector, $v_i = \|\mathbf{v}_i\|$ the velocity (scalar) and φ_1 is the steering (scalar).

4.3. The Effect of the Control Input on the Wheels

Given a control value $\mathbf{u} = (\mathbf{v}, \omega)$ to the platform, it is necessary is to compute the corresponding steering angle φ_i and drive wheel velocity v_i of each wheel.

We therefore seek a function f with the following property:

$$
(\varphi_i, v_i) = f(\mathbf{p}_i^{\{B\}}, \mathbf{v}^{\{B\}}, \omega). \tag{3}
$$

In this formulation, $\mathbf{v}$ (and $\mathbf{p}$) are expressed in the platform frame $\{B\}$.

Intuitively, function f consists of two different components. (1) $f_\mathbf{v}$ which handles linear motion and (2) f_ω which takes care of rotational motion. The function is a combination of the two: "$f = f_\mathbf{v} + f_\omega$".

4.3.1. Linear Velocity Component $f_{\mathbf{v}}$

Firstly, assume there is only a linear control component $\mathbf{v}$ (and $\omega = 0$). This would correspond to each wheel having the same steering angle ($\varphi_1 = \varphi_2 = \varphi_3 = \varphi_4$), given by the velocity vector $\mathbf{v}$ as:

$$\begin{aligned} (\varphi_i, v_i) &= f(\mathbf{p}_i, \mathbf{v}, 0) \\ \varphi_i &= \text{atan2}(v_y, v_x) \\ v_i &= \|\mathbf{v}\| \end{aligned} \tag{4}$$

Note that: (1) the steering angle is independent on the wheels' position $\mathbf{p}_{1\ldots4}$, and (2) when there is no velocity $\mathbf{v}$ given in this case, the steering angles $\varphi_{1\ldots4}$ can be arbitrarily set.

4.3.2. Rotational Velocity Component—f_ω

Now, assume that there is only a rotational velocity control component ω (and $\mathbf{v} = \mathbf{0}$). Given that $\omega \neq 0$ the steering angles can be computed as (see also Figure 5):

$$\begin{aligned} (\varphi_i, v_i) &= f(\mathbf{p}_i, \mathbf{0}, \omega) \\ \varphi_i &= \text{atan2}(p_y^i, p_x^i) + \frac{\pi}{2} \\ \mathbf{v}_i &= \boldsymbol{\omega} \times \mathbf{p}_i \\ v_i &= \|\mathbf{v}_i\| \end{aligned} \tag{5}$$

where for the $\times$ computation we have assumed that $\boldsymbol{\omega} = [0\ 0\ \omega]^T$ and $\mathbf{p}_i = [p_x^i\ p_y^i\ 0]^T$. Note that $\mathbf{v}_i$ is a vector (compared to v_i which only is a scalar).

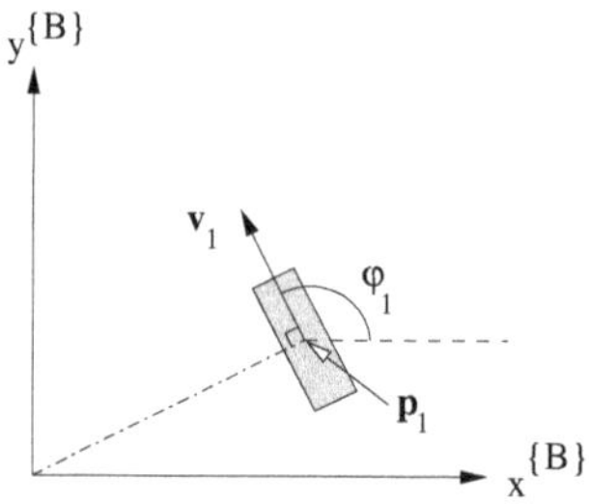

Figure 5. Wheel 1 parameters given that $\mathbf{v} = \mathbf{0}$, which enforces the steering wheel angle φ_1 to make the steering direction vector $\mathbf{v}_1$ to be orthogonal towards the origin of the frame.

Note that if there is a wheel located at the center of the platform frame ($p_x^i = p_y^i = 0$), this wheel angle is not defined and could in principle be set arbitrarily; note that this point is the Instantaneous Center of Rotation (ICR).

4.3.3. Combining Linear and Rotational Components

From the previous sections we have obtained a set of linear velocity vectors ($\mathbf{v}$ and $\mathbf{v}_{1\ldots4}$), where $\mathbf{v}$ originates from the linear velocity component $f_{\mathbf{v}}$ and $\mathbf{v}_{1\ldots4}$ from f_ω.

These velocity vectors are combined through vector summation as follows:

$$\begin{aligned} (\varphi_i, v_i) &= f(\mathbf{p}_i, \mathbf{v}, \omega) \\ \varphi_i &= \angle \boldsymbol{\omega} \times \mathbf{p}_i + \mathbf{v} \\ v_i &= \|\boldsymbol{\omega} \times \mathbf{p}_i + \mathbf{v}\| \end{aligned} \tag{6}$$

Note that if $\mathbf{v} = -\boldsymbol{\omega} \times \mathbf{p}_i$, we have the point of rotation (ICR) at $\mathbf{p}_i$. Note at this exact point the wheel can in principle be set arbitrarily, however, this point is singular as small changes to the ICR will require rapid changes to the angle φ. In short, this singular point can be avoided by ensuring that the ICR point is sufficiently distant from $\mathbf{p}_i$.

The ICR point is a function of the control $\mathbf{u}_p = (\mathbf{v}, \omega) = (v_x, v_y, \omega)$ as:

$$\begin{aligned} \text{ICR}_x &= -\frac{v_y}{\omega} \\ \text{ICR}_y &= \frac{v_x}{\omega} \end{aligned} \tag{7}$$

Note that if there is no rotational element ω, the ICR will be at infinity.

4.4. Representing the Control Action

Taken from the previous sections we have the following ODE:

$$\begin{aligned} \dot{x}^{\{A\}} &= v_x^{\{B\}} \cos(\theta) - v_y^{\{B\}} \sin(\theta) \\ \dot{y}^{\{A\}} &= v_y^{\{B\}} \sin(\theta) + v_y^{\{B\}} \cos(\theta) \\ \dot{\theta}^{\{A\}} &= \omega \end{aligned} \tag{8}$$

We need to incorporate the different wheel configurations; that is, to add the different wheels into the formulation. From the discussion above, there are limitations on the turning of each wheels as well as singular points. The key question here is how can we constrain the control to ensure that it is feasible. Equation (6) above is a function that satisfies the property stated in Equation (3):

$$(\varphi_i, v_i) = f(\mathbf{p}_i, \mathbf{v}, \omega) \tag{9}$$

That is, Equation (6) computes the orientation and velocity of each wheel. However, it is necessary to ensure that the velocity v_i of each wheel is within bounds and does not exceed a max value. Similarly there is also a limitation on the turning rate of each wheel $\dot{\varphi}_i$.

The first step is to find a more suitable representation of the control $\mathbf{u}_p$ that is better aligned with the wheel configuration such as the steering angle of the wheels. Therefore we utilize the following control variable as proposed in [33]:

$$\mathbf{u}_p = (v, \varphi, \omega) \tag{10}$$

where $v = \sqrt{v_x^2 + v_y^2}$ and $\varphi = \text{atan2}(v_y, v_x)$. In essence we change the base of the linear velocity to contain the linear velocity direction φ, as well as the linear velocity along this direction as v. As it is easy to convert between the different linear velocity representations ($v_x = v\sin(\varphi)$, $v_y = v\cos(\varphi)$) they are used interchangeably in this work.

If we now only have a linear velocity ($\omega = 0$) each steering wheel angle is given by φ, that is $\varphi_i = \varphi$ for each wheels i. Given this representation we can now add limitations on the change of the control $\dot{\varphi}$ to better reflect the limitation on the change in steering angle. One key situation is when approaching the goal: the control action must be limited, otherwise small changes in linear velocities would require extremely large changes in wheel steering angles $\dot{\varphi}_i$. Note that this is not the same problems as discussed above regarding ICR as when $\omega = 0$ the ICR is at the center of the platform and far from the wheels' locations $\mathbf{p}_i$.

Another benefit is that the change in linear velocity control representation is that the linear velocity speed is separated from the orientation components. This allows a more intuitive way of formulating an acceleration profile. Furthermore, a profile on the change of linear direction—which is highly influential of the steering wheel angles—can be formulated. Specifically, the following limitations are added on the control variables:

$$\begin{aligned} -\dot{v}^{\max} &\le \dot{v} \le \dot{v}^{\max} \\ -\dot{\varphi}^{\max} &\le \dot{\varphi} \le \dot{\varphi}^{\max} \\ -\dot{\omega}^{\max} &\le \dot{\omega} \le \dot{\omega}^{\max} \end{aligned} \tag{11}$$

Note that these are not directly connected to any physical limitation of the steer and drive wheels. For example, the change in steering wheel angle is often fast (in the platform used in the evaluation the changes are >20 rad/s). However, these limitations are useful for a smooth drive characteristic.

There is also a maximum velocity limitation on the drive wheel. This limitation is also explicitly considered and is detailed in the next section.

4.5. Limitations on the Control Action Due to Maximum Velocity of the Drive Wheels

The linear and angular velocity of the wheels are connected. Informally speaking, you cannot both drive quickly and turn quickly at the same time, because, as with differential drive platforms, the same wheel velocities v_i are utilized to both obtain the linear velocity $\mathbf{v}$ as well as the angular velocity ω of the platform. However, for a differential drive platform the ICR point is on the line that connects the left and right wheel, while for the platform used here the ICR can be set arbitrarily. The maximum rotational velocity a wheel can achieve without having any linear velocity (that is the ICR is given at (0, 0)) is given by:

$$\omega_i^{\max} = \frac{v_i^{\max}}{||\mathbf{p}_i||} \tag{12}$$

It is clear that the rotational speed is greatly dependent on the distance between the ICR and the wheels. For a differential driven platform with a linear velocity forward and turning right it will be the left wheel that will reach the $v_i^{\max}$ boundary first. Hence it is the wheel with the furthest distance from the ICR that will be the limitation. In principle for the platform at hand, it would be possible to have a larger linear velocity $(v_x, v_y) = (v, 0)$ (along the x-axis) than $(v_x, v_y) = (\sqrt{2}, \sqrt{2})$ as the distance between the ICR and the wheel furthest away is longer in the latter case.

In the proposed approach this difference is neglected and we consider a combined linear and rotational max boundary as follows:

$$\begin{aligned} -\omega^{\max} &\leq \omega - \frac{v}{d} \leq \omega^{\max} \\ -\omega^{\max} &\leq \omega + \frac{v}{d} \leq \omega^{\max} \end{aligned} \tag{13}$$

where $d = d_i = ||\mathbf{p}_i||$ which is assumed to be the same distance for all wheels i. In practice, this corresponds to always assuming the worst-case scenario where one wheel is always the furthest possible distance away from the ICR.

5. Defining the Optimization Problem

This section outlines the non-linear optimization problem that forms the basis of the MSDU Local Planner. The core objective of the MSDU Local Planner is to obtain a local feasible trajectory that drives the platform towards a goal. This goal is obtained from a global planner that is periodically updated (in the order of approx. 1 Hz). Hence, the local plan obtained does not have to consider getting stuck despite its local nature.

We here formulate a non-linear optimization problem that will generate a feasible trajectory. Due to the computational complexity that comes with non-linear solving, care has to be taken to formulate an optimization problem that is fast enough to solve. This will impact how the problem is formulated, as well as restricting the size of certain parameters; for example, the look-ahead distance and the sampling resolution. On the positive side, the non-linearity approach allows us to be much more free in how we select the objective function.

5.1. Problem Formulation

Two different approaches are used to steer the trajectory generation by the optimization. The first approach is to add factors into an objective function. The second is to post constraints on the different variables. Constraints are a very powerful and intuitive way to steer the optimization, but can be problematic if constraints are posted that simply cannot be satisfied. One example would be the goal pose $\mathbf{P}^{goal}$ that we would like to arrive at. Typically, it is not guaranteed that given other constraints on, for example, the acceleration, velocity, and turning speed limits, that we can reach the goal. Instead, we use the distance between the current pose and future poses to the goal pose in the objective function. If the goal cannot be reached, that is acceptable as no constraints are violated and at the same time the minimization of the cost function (that is, the distance to the goal) will drive the vehicle towards the goal.

The state $\mathbf{s}$ consists of the vehicle pose (x, y, θ) and linear velocity, direction and angular velocity (v, φ, ω) whereas the optimization control variables $\mathbf{u}$ are $(dv, d\varphi, d\omega)$. Note that the control action $\mathbf{u}_p$ used to drive the vehicle is actually part of the state, however, we will use an additional optimization variables of derivatives of the control values to fulfill additional requirements such as limiting the maximum acceleration permitted; see Equation (10).

The model of the vehicle dynamics ($\dot{\mathbf{s}} = f(\mathbf{s}, \mathbf{u})$) is described as:

$$\begin{aligned} \dot{x} &= v\cos(\varphi)\cos(\theta) - v\sin(\varphi)\sin(\theta) \\ \dot{y} &= v\cos(\varphi)\sin(\theta) + v\sin(\varphi)\cos(\theta) \\ \dot{\theta} &= \omega \\ \dot{v} &= dv \\ \dot{\varphi} &= d\varphi \\ \dot{\omega} &= d\omega \end{aligned} \tag{14}$$

Given the dynamics, we formulate a constrained optimal control problem (OCP):

$$\begin{aligned} \text{minimize}_{\mathbf{s},\mathbf{u}} \quad & \phi(T) + \int_0^T l(\mathbf{s}(t), \mathbf{u}(t))dt \\ \text{subject to} \quad & \mathbf{s}(0) = \hat{\mathbf{s}}_0 \\ & \dot{\mathbf{s}}(t) = f(\mathbf{s}(t), \mathbf{u}(t)), \quad t \in [0, T] \\ & h(\mathbf{s}(t), \mathbf{u}(t)) \leq 0, \quad t \in [0, T] \\ & d(\mathbf{s}(t), \mathbf{o}, \mathbf{c}) \leq 0, \quad t \in [0, T], \mathbf{o} \in \mathcal{O}, \mathbf{c} \in \mathcal{C} \end{aligned} \tag{15}$$

where T is the horizon length in seconds, $\phi(T)$ is the terminal cost, l is the cost for time t, $\hat{\mathbf{s}}_0$ is the initial state, f is the vehicle dynamics function (Equation (14)), h is the path constraints containing limits on inputs $\mathbf{u}$, such as max accelerations, as well as pure state constraints on $\mathbf{s}$, such as bounds on max velocities, and finally d provides a mean to ensure collision free state poses given a set of obstacle points $\mathcal{O}$ as well as a set of circles $\mathcal{C}$ representing the shape of the vehicle. Both the obstacle point $\mathbf{o} = [o_x, o_y]$ and the circle $\mathbf{c} = [c_x, c_y, c_R]$ are given in the vehicle frame where c_R is the radius of the circle.

To solve the OCP problem defined above we discretize it into a non-linear program (NLP) using multiple shooting [34]. The trajectory consists of vehicle states at discrete timestamps and holds N steps covering T seconds which gives us that each increment brings us $dt = \frac{T}{N}$ seconds into the future.

The discrete decision variable is $\zeta = \{\mathbf{s}_i, \mathbf{u}_i\}_{i=1}^{N}$, and the non-linear program is written as:

$$\begin{aligned} \text{minimize}_{\zeta} \quad & \phi(\zeta_N) + \sum_{k=0}^{N-1} l(\zeta_k) \\ \text{subject to} \quad & \mathbf{s}_0 = \hat{\mathbf{s}}_0 \\ & \mathbf{s}_{k+1} = F(\mathbf{s}_k, \mathbf{u}_k, dt), \quad k = 0 \ldots N-1 \\ & h(\mathbf{s}_k, \mathbf{u}_k) \leq 0, \quad k = 0 \ldots N \\ & d(\mathbf{s}_k, \mathbf{o}, \mathbf{c}) \leq 0, \quad k = 0 \ldots N, \mathbf{o} \in \mathcal{O}, \mathbf{c} \in \mathcal{C} \end{aligned} \tag{16}$$

where the objective function is described in Equation (19), F is the discrete model of the dynamics (see Equation (20)), the path constraints h are given in Equations (21) and (22) and finally the constraints to ensure collision-free motions d are described in Equation (24).

As we are primarily interested in the next control action to take, we follow the classical model-predictive control (MPC) scheme and use the obtained decision variables ζ to extract the next control action. Depending on inherent lag in the system, it is also possible to take not just the first control action available but to take a future one.

Because the problem is formulated as a standard non-linear program, it can be straightforwardly integrated into existing non-linear solvers. In this work, the formulation was implemented with CasADi [35] which here utilizes the Ipopt library [36] to solve the posted non-linear problem.

The objective and constraints are discussed further in the following sections.

5.2. Inputs and Outputs

As described above we continuously receive a global plan (at approximately 1 Hz) from which we extract the next local "goal" based on our current localization estimate, which is also provided continuously (at 50 Hz, where the localization system runs slower and is dependent on the translation and rotational distance, but is augmented with odometry readings which are obtained at 50 Hz). To simplify the formulation the local goal is converted into the robot frame $\{B\}$, which allows us to assume that we always start at pose (0, 0, 0). The continuous sensory input (at 10 Hz) is already provided in the robot frame $\{B\}$ as the sensory setup is located on the robot itself.

The controller or the local planner is queried at 10 Hz in which the latest received plan, localization estimate, and sensory data are used.

The output is the next control action to be executed $\mathbf{u}_p = (v_x, v_y, \omega)$.

5.3. Objectives

As discussed above, the force that drives the robot towards the goal $\mathbf{g} = (g_x, g_y, g_\theta)$ lies in the cost objective which contains the distance between the goal and each pose in the N-step long trajectory $(x, y, \theta)_{1\ldots N}$. The goal part to the objective is as:

$$\mathbf{J}^{goal} = \sum_{i=1}^{N} w_i^x (g_x - x_i)^2 + w_i^y (g_y - y_i)^2 + w_i^\theta (g_\theta - \theta_i)^2 \tag{17}$$

where we have different weighting factors $w_x^{1\ldots N}$, $w_y^{1\ldots N}$ and $w_\theta^{1\ldots N}$. These weights can be selected and tuned as needed, but in the evaluation presented in this paper all position weights were set to be the same. It is also possible to have a lower cost on the intermediate weights in the range $(1\ldots N-1)$ compared to the last terminal state weights w_N. The core idea is that we want to steer the optimization towards the goal as quickly as possible. Additional constraints to limit the velocities and accelerations were also added as discussed in Section 5.4 below.

Another cost relates to the magnitude of the control actions utilized. This was found particularly important to limit the amount of turning when driving the platform close to the goal. The cost on the decisions variables related to the derivatives of the generated control output is defined as:

$$\mathbf{J}^{control} = \sum_{i=1}^{N} w_i^{dv} (dv)^2 + w_i^{d\varphi} (d\varphi)^2 + w_i^{d\omega} (d\omega)^2 \tag{18}$$

Our objective is the sum of the above and can be rewritten as:

$$\phi(\zeta_N) + \sum_{k=0}^{N-1} l(\zeta_k) = \sum_{i=1}^{N} \mathbf{x}_i^T \mathbf{Q}_i \mathbf{x}_i + \sum_{i=1}^{N} \mathbf{u}_i^T \mathbf{R}_i \mathbf{u}_i \tag{19}$$

where $\mathbf{x}_i = [g_x - x_i, g_y - y_i, g_\theta - \theta_i]^T$ and $\mathbf{Q}_i$ together with $\mathbf{R}_i$ are diagonal weighting matrices.

5.4. Constraints

The constraints added relate to limitations of the vehicle, such as maximum drive velocities, but also define harder acceleration limits to obtain a softer and smoother driving characteristic. The cost objective defined in Section 5.3 pulls all states towards the goal, while the constraints limit the velocity profile. The constraints are also used to handle obstacle avoidance by the planner.

5.4.1. Velocity Profiles and Vehicle Constraints

The constraints utilized are to ensure strict boundaries as max velocities and accelerations are not exceeded. Constraints are also used to ensure that the generated trajectory is consistent; that is, that the states' progression by applying the corresponding control action will bring the current state to the next. In principle it would be sufficient to provide the initial state and the generated control action to derive the trajectory, but this work utilizes a multiple-shooting [34] approach where we keep track of the states in each iteration which is useful as some constraints added contains state information (note that the control output sent to the vehicle is defined in the state).

Given a control action $\mathbf{u} = [dv, d\varphi, d\omega]$ the next state is computed by integrating the vehicle dynamics f specified in Equation (14), using Runge–Kutta RK4 for the integration. The equality constraint used to make sure the state integrated with control is consistent with the next state $\mathbf{s}_{k+1} = F(\mathbf{s}_k, \mathbf{u}_k, dt)$ is defined by :

$$\begin{aligned} k_1 &= f(\mathbf{s}_k, \mathbf{u}_k) \\ k_2 &= f(\mathbf{s}_k + k_1\frac{dt}{2}, \mathbf{u}_k) \\ k_3 &= f(\mathbf{s}_k + k_2\frac{dt}{2}, \mathbf{u}_k) \\ k_4 &= f(\mathbf{s}_k + k_3 dt, \mathbf{u}_k) \\ \mathbf{s}_{k+1} &= \mathbf{s}_k + (k_1 + 2k_2 + 2k_3 + k_4) \cdot dt \end{aligned} \quad (20)$$

To limit the control actions as well as the maximum velocities, the following non-equality constraints are added; also see Equation (11):

$$\begin{aligned} -dv^{\max}dt &\le dv \le dv^{\max}dt \\ -d\varphi^{\max}dt &\le d\varphi \le d\varphi^{\max}dt \\ -d\omega^{\max}dt &\le d\omega \le d\omega^{\max}dt \end{aligned} \quad (21)$$

To ensure that the maximum drive velocity on each wheel is not exceeded, the following constraints are utilized:

$$\begin{aligned} -\omega^{\max} &\le \omega - v/d \le \omega^{\max} \\ -\omega^{\max} &\le \omega + v/d \le \omega^{\max} \end{aligned} \quad (22)$$

Here, d is the distance between the centre of the platform and the wheels, which is assumed to be the same for all wheels.

The above two sets of constraints are used on all control variables in the optimization.

5.4.2. Obstacle Constraints

As one of the objectives is to drive in confined areas it is necessary to have good spatial resolution of obstacles. Rather than using a grid-based representation that would discretize the environment and hence lower the available resolution, instead we use range readings directly to have as high resolution as possible and to receive quick feedback. This requires that we have a good overview of the surrounding of the platform at all times. In essence, an obstacle $\mathbf{o} = [o_x, o_y]$ is a point in 2D.

Note that the global path planner will, to a great extent, handle obstacle avoidance while providing the global path. However, the obstacle constraints are necessary to safely drive at close proximity to obstacles.

From the raw range readings we process the data to only extract the key LiDAR readings. This is essential as each constraint will add additional complexity to the optimization problem at hand. How the obstacle constraints are formulated is also greatly affected by the representation used to model the platform. Due to the square shape of the platform the orientation is essential. For example, to drive through a narrow passage straight ahead along the x-axis, the orientation θ must be kept close to the following angles $(0, \pi/2, \pi, 3\pi/4)$ as the required width would be approximately $\sqrt{2}$ larger if we instead had an orientation of $\pi/4$.

As the optimization problem can be altered on the fly, there are two different shapes that are used. One is used for driving, and thus we want the platform to drive with a specific forward direction, and the other is used for close-proximity driving when reaching goals. In the first scenario we define the footprint as two circles, one slightly forward and one slightly backwards. The front circle and back circle will stick out and make the footprint form in a clear direction, which is the best way to pass a narrow passage. The front circle will act as a "plow" that divides obstacles on either side. For close proximity near a goal, we instead approximate the shape of the robot better using five circles, one center and the remainder used to cover the corners of the platform.

To approximate the platform using only a circle would simplify the computation of the constraint to be the following:

$$\begin{aligned} \sqrt{(x_k - o_x)^2 + (y_k - o_y)^2} &\geq R \\ \forall\, \mathbf{o} &\in \mathcal{O} \end{aligned} \tag{23}$$

where R is the radius of the circle. This constraint is added both for the different states $k = 1 \ldots N$ as well as for all obstacles $\mathbf{o} \in \mathcal{O}$, which gives us in total of $N \times |\mathcal{O}|$ where $|\mathcal{O}|$ is number of obstacles.

Circles that have have their centres at an offset ($\mathbf{c} = [c_x, c_y, c_R]$) also require that the platform orientation θ is considered. These obstacle constraints are formulated as:

$$\begin{aligned} \sqrt{\Big(x_k + c_x \cos(\theta_k) - c_y \sin(\theta_k) - o_x\Big)^2 + \Big(y_k + c_x \sin(\theta_k) + c_y \cos(\theta_k) - o_y\Big)^2} &\geq c_R \\ \forall\, \mathbf{o} \in \mathcal{O}, \forall\, \mathbf{c} &\in \mathcal{C} \end{aligned} \tag{24}$$

which gives us the total number of constraints added to be $N \times |\mathcal{O}| \times |\mathcal{C}|$, where $|\mathcal{O}|$ and $|\mathcal{C}|$ is the number of obstacles and circles respectively.

One difficulty not addressed up to this point is that each obstacle $\mathbf{o}$ is directly derived from sensory data and hence is subject to various sets of noise. If we are operating the robot close to an object it might happen that in one iteration the obstacles are on the "correct" side of the boundary, whereas in the next iteration they are not and the optimization problem is infeasible. To handle this situation we avoid constraints using the first state variable ($k = 0$) as this state is an initial condition and it is not possible to alter it in the optimization. In principle, one could also remove collision checks on further states ahead that then could handle larger noise values or cause the robot to back off when other dynamic entities such as people within the collision boundaries. As the trajectory is discretized based on dt, ignoring any step apart from $k = 0$ is not recommended especially if dt is large as not all actions effect the state and its collision boundaries will be incorporated and checked. If there is no feasible solution the vehicle is stopped; in practice this can happen if you quickly place an obstacle into the collision constraint zone, then the vehicle will simply stay still until this obstacle is removed.

Another approach to handling the noisy LiDAR data would be to incorporate the obstacle avoidance as part of the objective function. One option here would then be to utilize sigmoid functions that give a high cost if the the obstacle is within the collision boundary and a low one if it is not. One problem with this is that from an optimization point of view you also want an objective that is smooth (i.e., has a nice derivative) which from a numerical and practical point of view could be challenging. Due to this, we instead opted for using constraints on the obstacle avoidance and we found this approach to be more intuitive.

5.4.3. Constraints to Avoid Singularities

As described in Section 4.3.3 there is a singular configuration when the ICR is very close to the wheel position which makes small changes in linear and angular velocities $(\mathbf{v}, \omega)$ to create large changes in the steering angle of that wheel. To avoid control actions that are given in these singular regions, one option is to adopt the following set of constraints (one for each wheel position $\mathbf{p}_{1\ldots4}$):

$$||\text{ICR} - \mathbf{p}_i|| \leq R \tag{25}$$

where R is the boundary radius that the ICR should not enter. However, while this constraints does seem to address the problem, the time complexity involved in adding these constraints was simply not worth it compared to using a more simplistic approach. Instead of adding this limitation into the optimization problem we can simply adjust the corresponding command sent to the platform (or have the platform do this internally) by adjusting the command $\mathbf{u} = (\mathbf{v}, \omega)$ to make sure that the above criteria $||\text{ICR} - \mathbf{p}_{1\ldots4}||$ holds by adjusting the control values; see Figure 6.

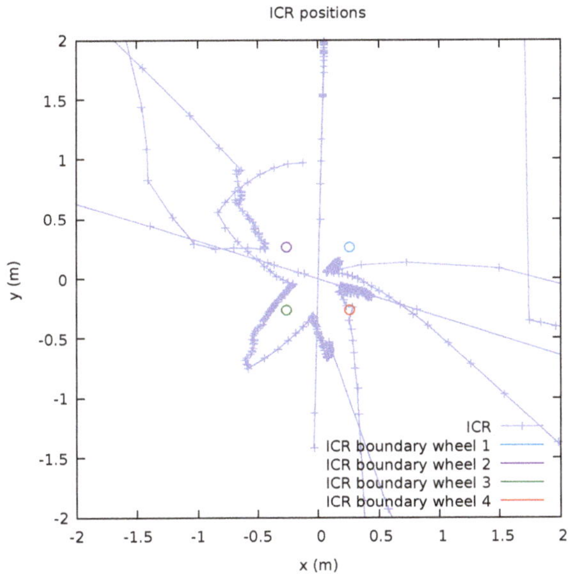

Figure 6. An example of the locations of ICR when moving around the platform, where the line represents the the consecutive changes of ICR locations. One can see the effect of ω (Equation (7)) that a relatively large change in ICR happens when there is low rotational velocity ω. In this example, we can see that the ICR was close to the boundary of wheel 4 (red) and probably was adjusted to stay outside the boundary. Note that to smoothly move from pure rotational velocity to linear velocity the ICR point has to incrementally move from the center passing in between the ICR constraint regions, therefore it is important to make the ICR constraint region not larger than needed.

This can be achieved quickly by first checking if the ICR is within any of the bounds. If not the command is valid and can be directly sent to the low level controller. If it is within any of the bounds we have the option of changing either the linear or angular velocity component or both before sending the command. Here we adopted the approach to adjust the linear **v** component and to keep the rotational velocity ω. One benefit of doing this at a higher level compared to just sending any control commands and letting the low-level controller handle it is that the the control values that will be used in the forward model will be more correct. It should also be noted that the optimization problem formulated here is to be used as a tracking-based controller as the key input is the next local goal pose, rather than a trajectory to follow, which makes the system robust to large disturbances.

Performing the navigation task though the doorway, as depicted in Figure 16 using ICR as constraints into the optimization had an average optimization time of 46 ± 29 ms compared to 29 ± 3 ms if ICR were instead corrected after optimization as described in the previous paragraph. It is worth noting the large variance which indicates that optimization problem is rather ill-posed but yet possible to solve.

5.4.4. Initial Conditions

As mentioned above, the initial state pose is assumed to be $(0,0,0)$ and the local goal **g** is given in the robot frame $\{B\}$ along with the extracted obstacles $\mathbf{o}_{1\ldots O}$ from LiDAR data. One part that has not been addressed is how the remaining state variables, that from a control perspective is actually used to command the platform at hand, are determined. Looking at the state Equation (14), we also have the linear velocity v, the linear velocity direction φ and the angular velocity ω. One key question is how to obtain and provide these values.

As the platform is non-holonomic by nature, since there it takes some time in order to turn the steering wheels, we cannot directly set any arbitrary φ value. Nor can we directly set an arbitrary rotational velocity ω and expect an instantaneous response. One approach could be to define that the robot always starts moving in one direction and then use this as an initial condition, but this would limit the advantages of the platform that you can start moving in an arbitrary direction or to start by a rotation. Instead of modeling the time taken for the robot to rotate its wheels which corresponds to the commanded linear and angular velocity, we at startup directly look at the feedback from the system that contains these velocities, see Figure 7. Without doing so our internal velocity prediction step will assume that the velocity is greater than the robot platform can deliver. When the vehicle has started to drive we then can utilize our acceleration models used within the optimization to predict more velocities.

As the MPC scheme to a great extent solves very similar problems from one iteration to the next, we also utilize a "warm-start", which is to provide the previous solution to the optimization problem as an initial start configuration. Performing the navigation task though the doorway as depicted in Figure 16 with warm start had an average optimization time of 29 ± 3 ms compared to 34 ± 5 ms without warm start, a reduction of approximately 10–20%.

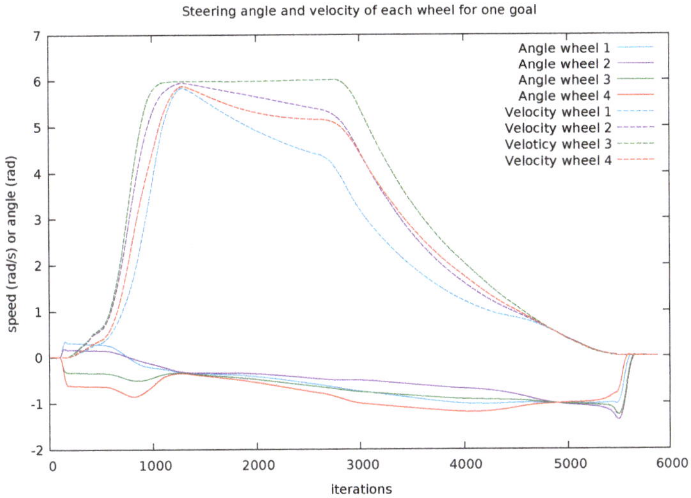

Figure 7. An example output driving from a start pose to a goal pose showing the heading and velocity of all four steer and drive units. In the beginning, the steering wheels all have an angular heading of zero. To allow the vehicle to start driving at an arbitrary direction, we utilize the feedback given by the platform to ensure that all wheels are aligned before the velocity is applied. When the vehicle has reached its goal the heading angles is set back to zero by the internal controller.

5.5. Perception

As mentioned in previous sections, the amount of constraints per added obstacles can be rather high and depends on the footprint representation and the horizon length N. To keep the amount of obstacle points $\mathbf{o}$ low we have to do some "clever" filtering of the range data. An example on filtered data is depicted in Figure 15. At a higher level, we have a perception system that is based on a previously built occupancy map which is overlaid with LiDAR data. This occupancy map is used to generate the global path. As we use this global path to extract the local path we do have some obstacle-avoidance systems in place. The global path planner has two main parameters that are of importance. First, to simplify the global motion planning the platform's footprint is approximated with a circle, to allow the robot to traverse narrow areas. The circle does not enclose the full actual footprint but ensures that the vehicle can pass given that its orientation is suitable. The second parameter is the distance from any object or occupied cell that will cause the planned motion to have additional cost. Given an empty environment and a wall, this distance will ensure that the generated plan will always keep this distance away from the wall, unless the goal is given closer towards the wall. These two parameters ensure that the global plan finds traversable areas in constrained areas while it makes sure that the the robot keep some distance in areas where there is space.

From the LiDAR we get a set of range readings that can be transformed into a corresponding 3D coordinate depending on sensor placement and internal parameters. Around the yaw rotation in the robot frame, a sector is defined where the closest reading that is considered to be an obstacle is stored (for example, readings that are sampled from the floor or objects that are higher than the actual height of the robot are ignored). To simplify processing the order, the sector is sorted based on the corresponding yaw angle. From this set of sectors we would like to extract the O closest obstacles. From all sectors we choose the closest reading as the first obstacle. It would be likely that the next closest obstacle is from a sector close by, but instead of looking solely at the second closest we make sure that the distance between the selected readings is at least a threshold distance apart. By doing so we can drastically reduce the amount of obstacles while still providing a set that can reasonably well represent the spatial layout of the environment. These obstacle readings will be continuously updated at the next iteration so obstacles that were filtered away might now be visible. Another aspect is that the global plan provided will keep a path with a distance to obstacles. The obstacles here will be used as constraint in the optimization framework, meaning that they will only impact the control actions taken if the obstacles are within the specified collision bounds. This requires that the obstacle is sufficiently close to the robot.

6. Experiments and Analysis

The MSDU Local Planner was evaluated in a number of real-world experiments on the robotic platform. The evaluation was performed in a research lab, and initially the robot was manually driven through the environment to build a global map using the LiDAR sensors (see Figure 8 for the created map). This map was used for planning and localization. The map was not updated during the experiments, which took place over several weeks, even though minor modifications occurred within the environment, with obstacles appearing and disappearing. Nonetheless, the system was able to successfully localize and plan.

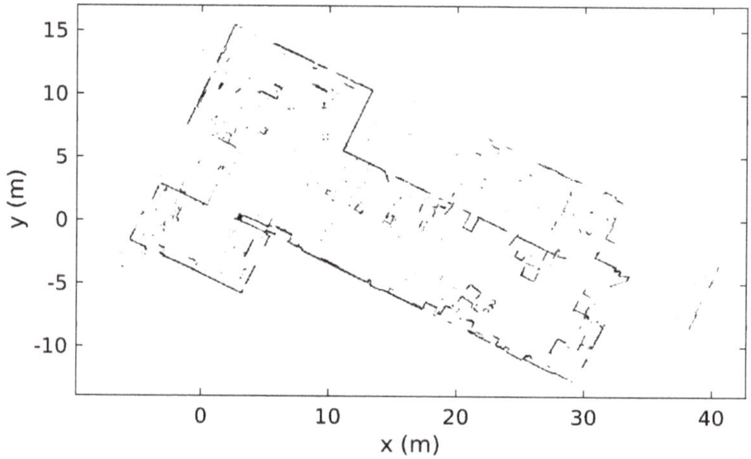

Figure 8. The map generated by the robot and used for localization, navigation, and planning. The robot was manually driven around the environment to create the map before the experiments took place.

The experiments performed evaluated the efficiency and accuracy of the paths planned by the MSDU Local Planner, and compared its performance to the TEB Local Planner [9], a widely used local planner. Further experiments evaluated the performance of MSDU in specific scenarios relating to particularly challenging environment aspects, such as narrow gaps and doorways.

The first evaluation compared the paths generated by MSDU against TEB. The goal positions were randomly generated in the environment, and the robot drove between the goals positions using MSDUṪhe experiment was repeated using TEB.

The termination criteria used are the distance between the current pose and goal as well as the maximum allowed velocity. If both these criteria are met the robot is considered to have reached the goal. In order to give as fair comparison as possible, these parameters were tweaked so that for both planners all goals were reachable within reasonable time and without causing excessive turning in the end. It should also be noted that the TEB Local Planner is not targeting pseudo-omnidirectional platform specifically and the parameters used were empirically found to produce outputs that the platform at hand worked well with. This comparison should therefore primarily be seen as a base-line of how an off-the-shelf planner can perform.

The generated paths were evaluated according to the following criteria:

- Accuracy: how closely did the final robot pose match the desired goal pose? To evaluate the accuracy, the displacement (both translational and angular) between the final robot pose and the desired goal pose was measured for each goal.
- Efficiency: how far did the robot need to travel to achieve the desired goal pose? To evaluate this measure, the distance traveled between the robot's start pose and final pose was measured for each goal. Both translational and angular distance were measured. It was decided that distance traveled was a fairer measure of efficiency for the path planners rather than time taken, as the time taken was too dependent on external parameters to the planners such as the maximum allowed velocity.

The localization performance was evaluated using the robot's pre-generated map and pose.

The experiment was performed across both "short" and "long" paths. Short paths are particularly challenging for the local motion planners, as they have to take tighter, more confined routes to the goal position. Here, "short" paths had a direct distance between starting and ending positions of less than 1 m. A total of 69 short paths were tested, with a distance normally distributed around a mean of 0.55 m, with a maximum distance of 0.98 m and a minimum distance of 0.12 m. The longer, less challenging paths had a mean distance between starting and ending positions of 2.9 m, with a maximum distance of 3.8 m and a minimum distance of 2.3 m. The experiments are summarized in Table 1.

Table 1. Summary of accuracy and precision experiments.

Accuracy and Precision Experiments	Number of Repetitions	Mean Distance (m)	Max Distance (m)	Min Distance (m)
Long paths	20	2.9 m	2.3 m	3.8 m
Short paths	69	0.55 m	0.12 m	0.98 m

Repeatability experiments were also performed, where the robot traveled repeatedly between two goals. The first repeatability experiment was over a long path between two goals 2.4 m apart, while the second repeatability experiment was over a short path between two goals 0.42 m apart. The robot traveled to each goal position 20 times. The repeatability experiments are summarized in Table 2.

Table 2. Summary of repeatability experiments.

Repeatability Experiments	Number of Repetitions	Distance between Goals (m)
Long paths	20 per goal (40 total)	2.4 m
Short paths	20 per goal (40 total)	0.42 m

7. Results

This section presents the experimental evaluation of the MSDU Local Planner. Section 7.1 presents the quantitative experiments evaluating the paths generated by MSDU and TEB. Sections 7.2 and 7.3 present qualitative descriptions of the performance of MSDU in narrow and confined spaces. Finally, Section 7.4 provides a brief summary of the main conclusions from the path evaluation experiments.

7.1. Accuracy and Efficiency of Planned Paths

Figure 9 presents the error (both translational and angular) between the final robot pose and the desired goal pose for each goal, for both MSDU and TEB for "long" (2.3–3.8 m) paths. Some key metrics are also summarized in Table 3. It can be seen (Figure 9a) that both planners achieved a high accuracy in terms of linear position. However, MSDU typically has a smaller translational error than TEB: MSDU has a mean error of 0.0019 m from the goal while TEB has a mean error of 0.0031 m from the goal position.

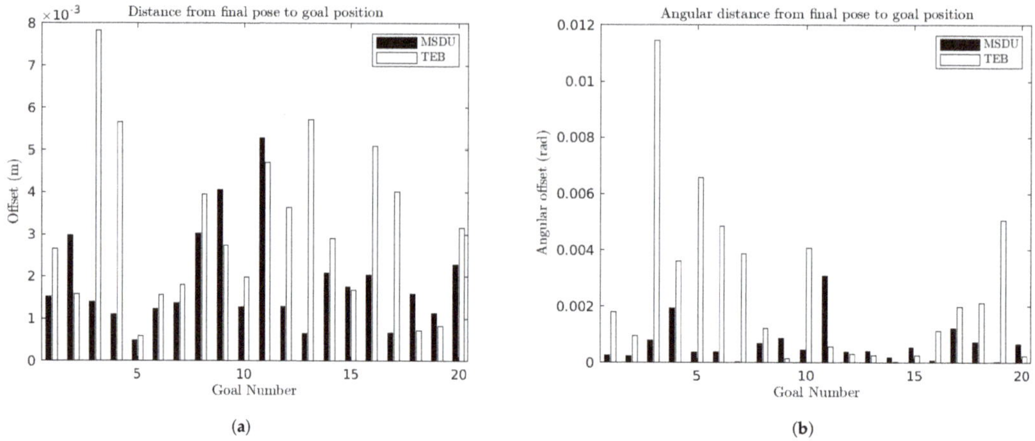

Figure 9. Distance between final robot pose and desired goal position for the MSDU Local Planner (black) and the TEB Local Planner (white) over long paths. (**a**) Translational error between the final robot pose and the desired goal position; (**b**) angular error between the final robot pose and the desired goal position.

Similar results can be seen when it comes to angular error (see Figure 9b and Table 3). MSDU has a mean angular error of 0.0007 rad while TEB has a mean angular error of 0.0025 rad.

Table 3. Summary of performance of MSDU and TEB for long paths.

Translational Error (m)	**MSDU**	**TEB**
Mean (m)	0.0019	0.0031
Standard deviation (m)	0.0012	0.0019
Max (m)	0.0053	0.0078
Min (m)	0.0005	0.0006
Angular error (rad)		
Mean (rad)	0.0007	0.0025
Standard deviation (rad)	0.0007	0.0029
Max (rad)	0.0031	0.0114
Min (rad)	4×10^{-5}	4×10^{-5}

Figure 10 and Table 4 present the distance traveled by the robot to the final goal position. It can be seen that TEB consistently takes longer paths to the goal than MSDU. This is particularly the case for angular distance, where in the worst case TEB travels 20 times the angular distance of MSDU (Goal 18).

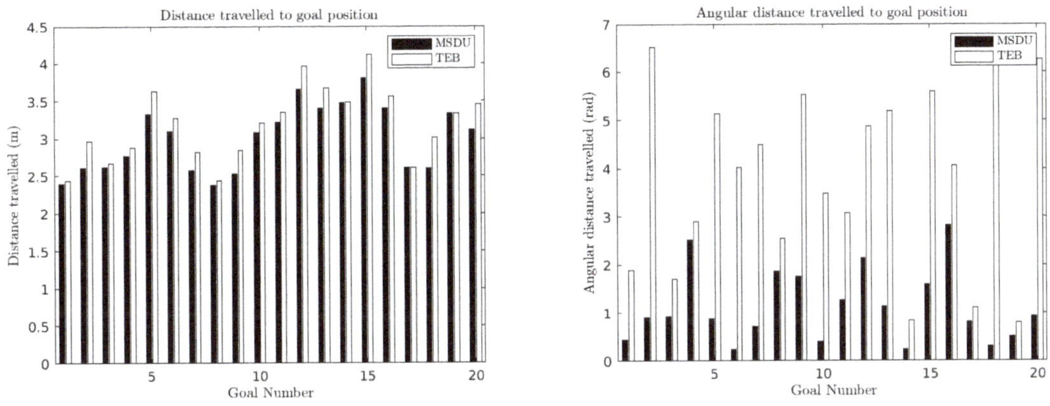

(a) (b)

Figure 10. Distance traveled between the starting and final robot pose for the MSDU Local Planner (black) and the TEB Local Planner (white) over long paths. (**a**) Translational distance traveled between the starting and final robot pose (**b**) Angular distance traveled between the starting and final robot pose.

Table 4. Summary of distances traveled by MSDU and TEB for long paths.

Translational Distance Traveled (m)	**MSDU**	**TEB**
Mean (m)	3.0011	3.1877
Standard deviation (m)	0.4437	0.4823
Max (m)	3.8077	4.1208
Min (m)	2.3772	2.4355
Angular distance traveled (rad)		
Mean (rad)	1.1690	3.8198
Standard deviation (rad)	0.7646	1.8938
Max (rad)	2.1898	6.5075
Min (rad)	0.2373	0.7898

The results for the short paths can be seen in Figure 11 and Table 5. The first thing to note is that the short paths are more challenging for both planners than the long paths: the mean (translational, angular) error increases from (0.0019 m, 0.0007 rad) on the long paths to (0.0028 m , 0.0010 rad) on the short paths for MSDU and from (0.0031 m, 0.0025 rad) on the long paths to (0.0033 m, 0.0038 rad) on the short paths for TEB. However, MSDU continues to outperform TEB in terms of accuracy.

For the short paths, the distance traveled to each goal pose is displayed in Figure 12 and Table 6. The translational distance traveled is longer for the TEB in 97% of the experiments (67 out of 69). In the cases where the translational distance traveled by TEB is shorter, the MSDU path is no more than 1.07 times as long as the TEB path (0.51 m to 0.48 m), while when the TEB path is longer; in the worst case it is 1.78 times as long as MSDU (0.98 m to 0.55 m).

As with the long paths, the angular distance traveled by TEB is consistently longer than that traveled by MSDU: in the worst case TEB takes an angular distance 51 times as long as that of MSDU (Goal 47).

These results show that the MSDU Local Planner can achieve final robot poses that are both closer to the desired goal pose while traveling shorter distances both in terms of translational and angular displacement.

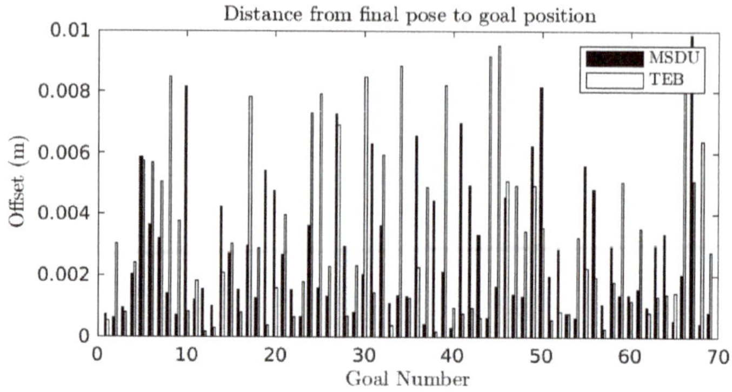

(a)

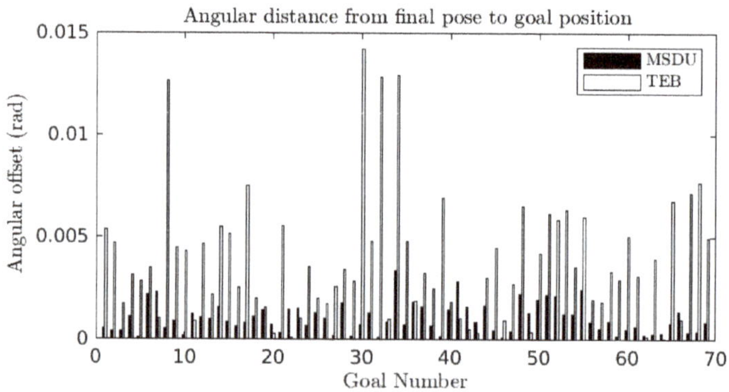

(b)

Figure 11. Distance between final robot pose and desired goal position for the MSDU Local Planner (black) and the TEB Local Planner (white) over short (<1 m) paths. (**a**) Translational error between the final robot pose and the desired goal position; (**b**) angular error between the final robot pose and the desired goal position.

Table 5. Summary of performance of MSDU and TEB for short paths (<1 m).

Translational Error (m)	**MSDU**	**TEB**
Mean (m)	0.0028	0.0033
Standard deviation (m)	0.0022	0.0028
Max (m)	0.0099	0.0095
Min (m)	0.0003	0.0002
Angular error (rad)		
Mean (rad)	0.0010	0.0038
Standard deviation (rad)	0.0007	0.0031
Max (rad)	0.0033	0.0142
Min (rad)	5×10^{-5}	1×10^{-5}

To understand why the distances traveled differ so much between planners, we display a number of the trajectories taken by the robot. We display the trajectories for the first four short goals in Figure 13. In each figure, the goal is marked by the black circle, the trajectory planned by MSDU is displayed in black and the trajectory planned by TEB is displayed in red. It is important to note that the trajectories, even when generally smooth, appear to have a slightly "jagged" appearance. This does not represent the actual path of the robot, but is due to the localization process, where the dead-reckoning estimate based on odometry is updated periodically by the AMCL localization process, thus creating small apparent "jumps" in the path.

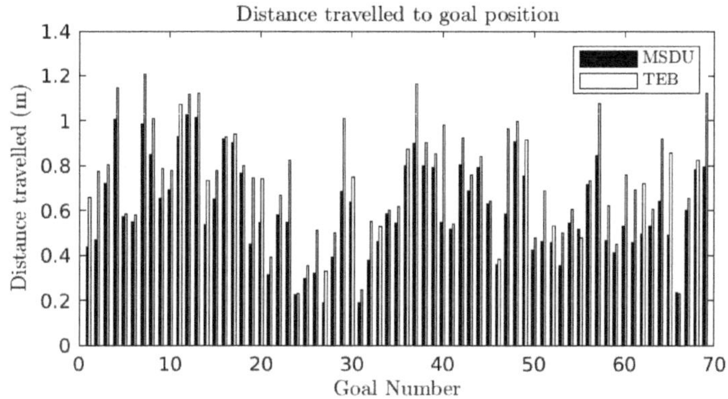

(**a**)

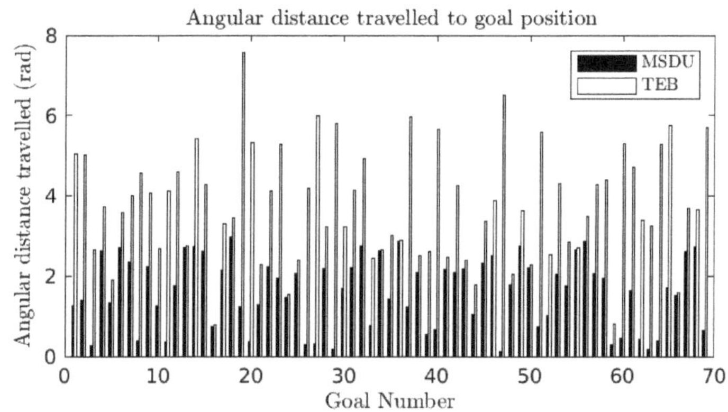

(**b**)

Figure 12. Distance traveled between the starting and final robot pose for the MSDU Local Planner (black) and the TEB Local Planner (white) over short (<1 m) paths. (**a**) Translational distance traveled between the starting and final robot pose; (**b**) angular distance traveled between the starting and final robot pose.

Table 6. Summary of distances traveled by MSDU and TEB for short paths.

Translational Distance Traveled (m)	MSDU	TEB
Mean (m)	0.6026	0.7346
Standard deviation (m)	0.2114	0.2390
Max (m)	1.0260	1.2085
Min (m)	0.1869	0.2822
Angular distance traveled (rad)		
Mean (rad)	1.6130	3.7598
Standard deviation (rad)	0.8756	1.4064
Max (rad)	2.9687	7.5681
Min (rad)	0.1260	0.7815

However, it is still possible to see the deviation between the paths planned by MSDU and those planned by TEB. TEB often takes a wider, more sweeping path—as one might see from a more car-like vehicle—and sometimes (as in the case of Goal 2) must do some complicated corrections close to the goal. In comparison, MSDU takes a more direct path, and very little correction occurs close to the goal.

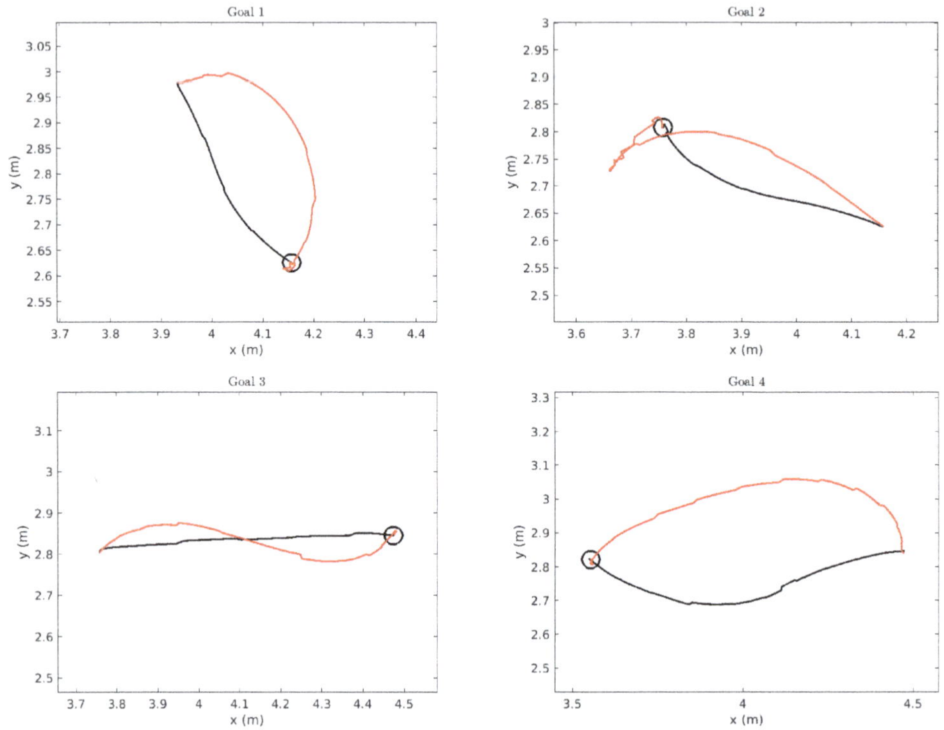

Figure 13. Trajectories taken by the robot for each goal. The goal is denoted by the black circle. The position of the robot using the MSDU Local Planner is displayed in black, and the robot position by the TEB Local Planner is displayed in red. Note that discrepancy in the curves are due to the localization update.

The repeatability of the planners is displayed in Figure 14 below, for both the long trajectory (Figure 14a) and the short trajectory (Figure 14b). The trajectory of the robot using MSDU is displayed in black, and the trajectory of the robot using TEB is displayed in red. The qualitative difference between the two planners' behavior can be clearly seen. The quantitative comparison is presented in Tables 7 and 8. A key distinction between the two planners is that in each case the distance MSDU travels to Goal 1 is very similar to the distance it travels to Goal 2, while TEB takes quite different-length paths between the two goals for the short paths. This can also be observed in Figure 14b.

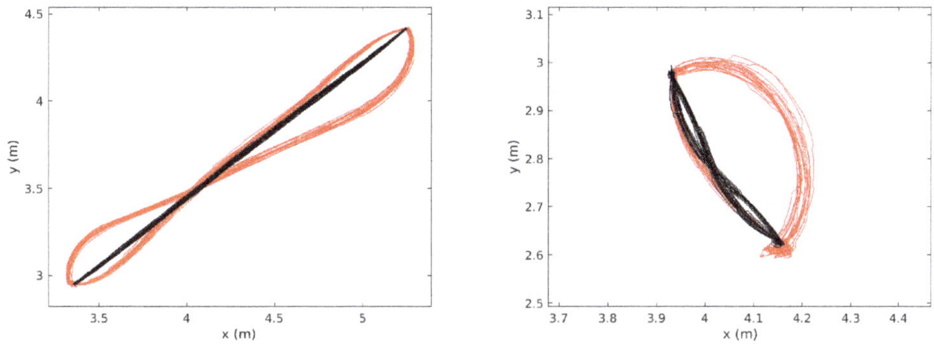

(a) (b)

Figure 14. Trajectories taken by the robot during the repeatability experiment. The trajectory of the robot using MSDU is displayed in black, and the robot trajectory using TEB is displayed in red. (**a**) Long path (2.4 m between the two goals); (**b**) short path (0.42 m between the two goals).

Table 7. Repeatability of performance of MSDU and TEB on long paths.

Goal Pose Error	**MSDU (Mean, Std)**	**TEB (Mean, Std)**
Goal 1 translational error (m)	(0.0021, 0.0014)	(0.0057, 0.0027)
Goal 2 translational error (m)	(0.0027, 0.0015)	(0.0030, 0.0014)
Goal 1 angular error (rad)	$(5 \times 10^{-4}, 6 \times 10^{-4})$	(0.0040, 0.0025)
Goal 2 angular error (rad)	$(6 \times 10^{-4}, 4 \times 10^{-4})$	(0.0034, 0.0030)
Distance Traveled		
Goal 1 translational (m)	(2.3971, 0.0039)	(2.4384, 0.0072)
Goal 2 translational (m)	(2.3960, 0.0047)	(2.7068, 0.0213)
Goal 1 angular (rad)	(0.4414, 0.0038)	(1.9345, 0.0452)
Goal 2 angular (rad)	(0.4398, 0.0050)	(5.4977, 0.0399)

Table 8. Repeatability of performance of MSDU and TEB on short paths.

Goal Pose Error	**MSDU (Mean, Std)**	**TEB (Mean, Std)**
Goal 1 translational error (m)	(0.0025, 0.0018)	(0.0029, 0.0014)
Goal 2 translational error (m)	(0.0033, 0.0018)	(0.0028, 0.0017)
Goal 1 angular error (rad)	$(6 \times 10^{-4}, 4 \times 10^{-4})$	(0.0035, 0.0022)
Goal 2 angular error (rad)	$(7 \times 10^{-4}, 5 \times 10^{-4})$	(0.0051, 0.0042)
Distance Traveled		
Goal 1 translational (m)	(0.4418, 0.0087)	(0.6915, 0.0241)
Goal 2 translational (m)	(0.4467, 0.0087)	(0.4581, 0.0079)
Goal 1 angular (rad)	(1.2668, 0.0016)	(5.0739, 0.0161)
Goal 2 angular (rad)	(1.2663, 0.0013)	(1.3190, 0.0212)

7.2. Local Planning Capabilities

To give an intuitive understanding of what types of problem the local planner can address, and to also give some illustration regarding how obstacles are extracted and how collision constraints are formed, we performed the following task. When the robot was located in a doorway, we performed a 180° turn on the spot. As the global planner does not have a notion of the shape of the robot, it just gave the goal location directly. However, due to the shape of the vehicle it could not turn on the spot without causing a collision between the robot and the door frame; instead it had to move out of the doorway and there perform the rotation before entering the doorway again. This is a problem that the optimization-based local planner addressed out of the box and snapshots of this turn are depicted in Figure 15.

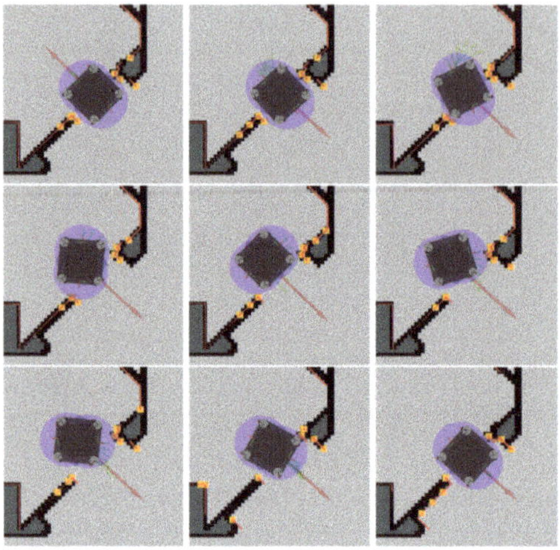

Figure 15. Set of snapshots showing the capability of the local planner to do a 180° turn inside a narrow doorway where there is not enough space to turn on the spot. Top left is the start configuration and bottom right is the goal configuration. In each picture, the red arrow depicts the current goal. The green arrows depict the future horizon states $(x, y, \theta)_{1...N}$. The blue circles depicts the collision circles $\mathbf{c}_{1...C}$. The yellow squares is the extracted obstacle points $\mathbf{o}_{1...O}$. The red smaller dots are the LiDAR points.

7.3. Passing a Doorway

To illustrate the connection between the global planner and the proposed local planner when passing through a doorway, a sequence of images are depicted in Figure 16. The local goal orientation is computed by the directions given by the consecutive points before and after the current local goal point.

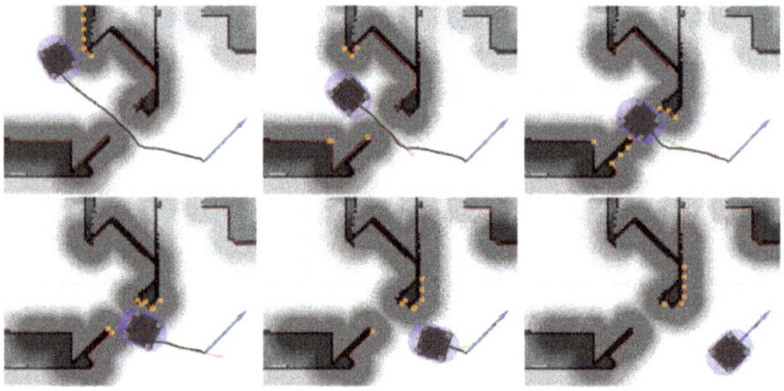

Figure 16. Set of snapshots illustrating the interaction between the global planner (the planned path is depicted as a black line and the goal is light-blue). The red arrow depicts the current local goal that the proposed local planner is driving towards, the green arrows depict the future horizon states (also see the caption in Figure 15. Due to the look ahead, the optimization allows for a smoother driven path compared to the relatively non-smooth and jerky path provided by the global path planner. The additional spacing between the obstacle points and the robot when there is enough space can be seen in the top left figure as the cost utilized by the global planner (illustrated with different gray levels) contains an additional offset. By setting this offset to at least corresponding to a single collision radius of the platform, that the platform can be in any orientation makes the obstacle constraints only to be active in areas where there is simply not enough space. Additionally, this additionally safety margin is very useful as the robot would otherwise drive very close to obstacles as the obstacle constraints do not have any other cost associated with them based on distance.

7.4. Summary of Results

The experiments demonstrate that MSDU consistently generated paths for the vehicle that were more accurate than those planned by TEB. Over the long paths in the experimental evaluation, MSDU had a mean translational error of 0.0019 m while TEB had a mean translational error of 0.0031 m. The mean angular error for MSDU was 0.0007 rad compared to 0.0025 rad for TEB. Over the (more challenging) short paths, MSDU had a mean translational error of 0.0028 m while TEB had a mean translational error of 0.0033 m. The mean angular error for MSDU was 0.0010 rad compared to 0.0038 rad for TEB.

The MSDU paths were also shorter than the TEB paths. Over the long paths, MSDU had a mean translational distance of 3.0011 m while TEB had a mean translational distance of 3.1877 m. Over the short paths, MSDU had a mean translational distance of 0.6026 m while TEB had a mean translational distance of 0.7346 m. In terms of angular distance traveled, MSDU had a mean angular distance traveled of 1.1690 rad over the long paths and 1.6130 rad over the short paths, while TEB had a mean angular distance traveled of 3.8198 rad over the long paths and 3.7598 rad over the short paths. In the worst-case scenario the TEB path traversed an angular distance 51 times as long as that of MSDU.

8. Conclusions

This work presents a local planner for mobile autonomous platforms with combined steer and drive wheels. It uses a non-linear optimization formulation to handle the kinematic characteristics of the wheel configuration, the physical limitations of the platform, and the obstacles in the environment, to generate a trajectory that generates smooth and efficient paths to a goal, and achieves highly accurate goal pose positions.

The local planner is tightly integrated into a full localization, navigation and motion-planning pipeline, and utilizes a global planner that is continuously re-executed to give an up-to-date global feasible path with respect to the latest readings and map updates. This enables the local planner to look exclusively at local problems and provide an online control signal based on immediate sensor and motion data, while the global planner resolves problems that require a broader global knowledge of the environment.

An interesting future research direction is how dynamic obstacles in the environment—such as people, or other autonomous robots—could be managed by the system. An interesting and non-trivial question to ask is in which layer of the system the management of dynamic obstacles should be included. For example, if we obtain a prior knowledge of typical movement in the environment then this would be suitable to use when providing the global plan. Dynamic obstacles can also be relatively straightforwardly formulated as constraints connected to the predicted horizon if we have a model of how these obstacles move in the future.

In the current formulation, the actual steering wheel direction is not explicitly modeled. It would be of interest to study how to incorporate these in an efficient way. If we can look at the steering angle and the rate of change directly as constraint many detours taken in the current approach, such as taking care of the ICR, would indirectly be handled. One problem is that with a set of four steer-and-drive wheels there is a significant amount of redundancy that is far from straightforward to put together as an optimization problem that both gives reasonable outputs and is also sufficiently fast to solve.

Author Contributions: The individual contributions of the authors are as follows. Conceptualization, H.A. and J.L.; methodology, validation, investigation, draft preparation and writing, H.A., S.L. and J.L.; software and formal analysis, H.A. and S.L.; supervision, project administration and funding acquisition, H.A. All authors have read and agreed to the published version of the manuscript.

Funding: This research was funded by FORMAS (Swedish Research Council for Sustainable Development), grant number 2019-02264.

Institutional Review Board Statement: Non applicable.

Informed Consent Statement: Non applicable.

Data Availability Statement: Not applicable.

Conflicts of Interest: The authors declare no conflicts of interest.

References

1. Siciliano, B.; Sciavicco, L.; Villani, L.; Oriolo, G. *Robotics: Modelling, Planning and Control*; Advanced Textbooks in Control and Signal Processing; Springer: London, UK, 2010.
2. Siegwart, R.; Nourbakhsh, I.; Scaramuzza, D. *Introduction to Autonomous Mobile Robots*, 2nd ed.; Intelligent Robotics and Autonomous Agents series; MIT Press: Cambridge, MA, USA, 2011.
3. Li, Z.; Canny, J. *Nonholonomic Motion Planning*; The Springer International Series in Engineering and Computer Science; Springer: New York, NY, USA, 2012.
4. Kavraki, L.E.; LaValle, S.M. Motion Planning. In *Springer Handbook of Robotics*; Siciliano, B., Khatib, O., Eds.; Springer International Publishing: Cham, Switzerland, 2016; pp. 139–162. [CrossRef]
5. Ilon, B.E. Wheels for a Course Stable Selfpropelling Vehicle Movable in Any Desired Direction on the Ground or Some Other Base. U.S. Patent 3,876,255, 8 April 1975.
6. Campion, G.; Chung, W. Wheeled Robots. In *Springer Handbook of Robotics*; Siciliano, B., Khatib, O., Eds.; Springer: Berlin/Heidelberg, Germany, 2008; pp. 391–410. [CrossRef]
7. Mišković, D.; Milić, L.; Čilag, A.; Berisavljević, T.; Gottscheber, A.; Raković, M. Implementation of Robots Integration in Scaled Laboratory Environment for Factory Automation. *Appl. Sci.* **2022**, *12*, 1228. [CrossRef]
8. Sprunk, C.; Lau, B.; Pfaffz, P.; Burgard, W. Online generation of kinodynamic trajectories for non-circular omnidirectional robots. In Proceedings of the 2011 IEEE International Conference on Robotics and Automation, Shanghai, China, 9–13 May 2011; pp. 72–77. [CrossRef]
9. Rösmann, C.; Hoffmann, F.; Bertram, T. Integrated online trajectory planning and optimization in distinctive topologies. *Robot. Auton. Syst.* **2017**, *88*, 142–153. [CrossRef]
10. Schoels, T.; Palmieri, L.; Arras, K.O.; Diehl, M. An NMPC Approach using Convex Inner Approximations for Online Motion Planning with Guaranteed Collision Avoidance. In Proceedings of the 2020 IEEE International Conference on Robotics and Automation (ICRA), Paris, France, 31 May–31 August 2020; pp. 3574–3580. [CrossRef]
11. Oftadeh, R.; Ghabcheloo, R.; Mattila, J. Universal Path-Following of Wheeled Mobile Robots: A Closed-Form Bounded Velocity Solution. *Sensors* **2021**, *21*, 7642. [CrossRef] [PubMed]
12. Hu, J.; Cheng, C.; Wang, C.; Zhao, C.; Pan, Q.; Liu, Z. An Improved Artificial Potential Field Method Based on DWA and Path Optimization. In Proceedings of the 2019 IEEE International Conference on Unmanned Systems (ICUS), Beijing, China, 17–19 October 2019; pp. 809–814. [CrossRef]
13. Orthey, A.; Frész, B.; Toussaint, M. Motion Planning Explorer: Visualizing Local Minima Using a Local-Minima Tree. *IEEE Robot. Autom. Lett.* **2020**, *5*, 346–353. [CrossRef]
14. Fox, D.; Burgard, W.; Thrun, S. The dynamic window approach to collision avoidance. *IEEE Robot. Autom. Mag.* **1997**, *4*, 23–33. [CrossRef]
15. Quinlan, S.; Khatib, O. Elastic bands: Connecting path planning and control. In Proceedings of the IEEE International Conference on Robotics and Automation, Atlanta, GA, USA, 2–6 May 1993; Volume 2, pp. 802–807. [CrossRef]
16. Connette, C.P.; Parlitz, C.; Hagele, M.; Verl, A. Singularity avoidance for over-actuated, pseudo-omnidirectional, wheeled mobile robots. In Proceedings of the 2009 IEEE International Conference on Robotics and Automation, Kobe, Japan, 12–17 May 2009; pp. 4124–4130. [CrossRef]
17. Rösmann, C.; Feiten, W.; Woesch, T.; Hoffmann, F.; Bertram, T. Trajectory modification considering dynamic constraints of autonomous robots. In Proceedings of the 7th German Conference on Robotics ROBOTIK 2012, Munich, Germany, 21–22 May 2012; pp. 1–6.
18. Magyar, B.; Tsiogkas, N.; Deray, J.; Pfeiffer, S.; Lane, D. Timed-Elastic Bands for Manipulation Motion Planning. *IEEE Robot. Autom. Lett.* **2019**, *4*, 3513–3520. [CrossRef]
19. Andersson, O.; Ljungqvist, O.; Tiger, M.; Axehill, D.; Heintz, F. Receding-Horizon Lattice-Based Motion Planning with Dynamic Obstacle Avoidance. In Proceedings of the 2018 IEEE Conference on Decision and Control (CDC), Miami, FL, USA, 17–19 December 2018; pp. 4467–4474. [CrossRef]
20. Andreasson, H.; Saarinen, J.; Cirillo, M.; Stoyanov, T.; Lilienthal, A.J. Fast, continuous state path smoothing to improve navigation accuracy. In Proceedings of the 2015 IEEE International Conference on Robotics and Automation (ICRA), Seattle, WA, USA, 26–30 May 2015; pp. 662–669.
21. Zhang, Z.; Yang, C.; Zhang, W.; Xu, Y.; Peng, Y.; Chi, M. Motion Control of a 4WS4WD Path-Following Vehicle: Further Studies on Steering and Driving Models. *Shock Vib.* **2021**, *2021*, 8830841. [CrossRef]
22. Chen, J.; Shuai, Z.; Zhang, H.; Zhao, W. Path Following Control of Autonomous Four-Wheel-Independent-Drive Electric Vehicles via Second-Order Sliding Mode and Nonlinear Disturbance Observer Techniques. *IEEE Trans. Ind. Electron.* **2021**, *68*, 2460–2469. [CrossRef]
23. Guo, J.; Luo, Y.; Li, K. An Adaptive Hierarchical Trajectory Following Control Approach of Autonomous Four-Wheel Independent Drive Electric Vehicles. *IEEE Trans. Intell. Transp. Syst.* **2018**, *19*, 2482–2492. [CrossRef]

24. Helling, S.; Roduner, C.; Meurer, T. On the Dual Implementation of Collision-Avoidance Constraints in Path-Following MPC for Underactuated Surface Vessels. In Proceedings of the 2021 American Control Conference (ACC), New Orleans, LA, USA, 26–28 May 2021; pp. 3366–3371. [CrossRef]
25. LaValle, S.M. Rapidly-exploring random trees: A new tool for path planning. The Annual Research Report. 1998. Avaliable online: http://msl.cs.illinois.edu/~lavalle/papers/Lav98c.pdf (accessed on 1 March 2022).
26. Kavraki, L.; Svestka, P.; Latombe, J.C.; Overmars, M. Probabilistic roadmaps for path planning in high-dimensional configuration spaces. *IEEE Trans. Robot. Autom.* **1996**, *12*, 566–580. [CrossRef]
27. Pivtoraiko, M.; Kelly, A. Fast and feasible deliberative motion planner for dynamic environments. In Proceedings of the ICRA Workshop on Safe Navigation in Open and Dynamic Environments: Application to Autonomous Vehicles, Kobe, Japan, 12–17 May 2009.
28. Karaman, S.; Frazzoli, E. Sampling-based algorithms for optimal motion planning. *Int. J. Robot. Res.* **2011**, *30*, 846–894. [CrossRef]
29. Kalisiak, M.; van de Panne, M. RRT-blossom: RRT with a local flood-fill behavior. In Proceedings of the 2006 IEEE International Conference on Robotics and Automation, 2006 (ICRA 2006), Orlando, FL, USA, 15–19 May 2006; pp. 1237–1242. [CrossRef]
30. Adiyatov, O.; Varol, H.A. A novel RRT*-based algorithm for motion planning in Dynamic environments. In Proceedings of the 2017 IEEE International Conference on Mechatronics and Automation (ICMA), Takamatsu, Japan, 6–9 August 2017; pp. 1416–1421. [CrossRef]
31. Quigley, M.; Gerkey, B.; Conley, K.; Faust, J.; Foote, T.; Leibs, J.; Berger, E.; Wheeler, R.; Ng, A. ROS: An open-source Robot Operating System. In Proceedings of the ICRA Workshop on Open Source Software, Kobe, Japan, 12–17 May 2009.
32. Marder-Eppstein, E.; Berger, E.; Foote, T.; Gerkey, B.; Konolige, K. The Office Marathon: Robust Navigation in an Indoor Office Environment. In *Proceedings of the* International Conference on Robotics and Automation, Anchorage, AK, USA, 3–7 May 2010.
33. Connette, C.P.; Pott, A.; Hägele, M.; Verl, A.W. Control of an pseudo-omnidirectional, non-holonomic, mobile robot based on an ICM representation in spherical coordinates. In Proceedings of the 2008 47th IEEE Conference on Decision and Control, Cancun, Mexico, 9–11 December 2008; pp. 4976–4983.
34. Bock, H.; Plitt, K. A Multiple Shooting Algorithm for Direct Solution of Optimal Control Problems. *IFAC Proc. Vol.* **1984**, *17*, 1603–1608. [CrossRef]
35. Andersson, J.A.E.; Gillis, J.; Horn, G.; Rawlings, J.B.; Diehl, M. CasADi—A software framework for nonlinear optimization and optimal control. *Math. Program. Comput.* **2019**, *11*, 1–36. [CrossRef]
36. Wächter, A.; Biegler, L.T. On the implementation of an interior-point filter line-search algorithm for large-scale nonlinear programming. *Math. Program.* **2006**, *106*, 25–57. [CrossRef]

sensors

MDPI

Article

Fast Adaptation of Manipulator Trajectories to Task Perturbation by Differentiating through the Optimal Solution

Shashank Srikanth [1], Mithun Babu [1], Houman Masnavi [2], Arun Kumar Singh [2,*], Karl Kruusamäe [2] and Krishnan Madhava Krishna [1]

[1] Robotics Research Center, KCIS, IIIT Hyderabad, Hyderabad 500032, India; s.shashank2401@gmail.com (S.S.); mithunbabu1141995@gmail.com (M.B.); mkrishna@iiit.ac.in (K.M.K.)

[2] Institute of Technology, University of Tartu, 50090 Tartu, Estonia; houman.masnavi@ut.ee (H.M.); karl.kruusamae@ut.ee (K.K.)

* Correspondence: arun.singh@ut.ee

Abstract: Joint space trajectory optimization under end-effector task constraints leads to a challenging non-convex problem. Thus, a real-time adaptation of prior computed trajectories to perturbation in task constraints often becomes intractable. Existing works use the so-called warm-starting of trajectory optimization to improve computational performance. We present a fundamentally different approach that relies on deriving analytical gradients of the optimal solution with respect to the task constraint parameters. This gradient map characterizes the direction in which the prior computed joint trajectories need to be deformed to comply with the new task constraints. Subsequently, we develop an iterative line-search algorithm for computing the scale of deformation. Our algorithm provides near real-time adaptation of joint trajectories for a diverse class of task perturbations, such as (i) changes in initial and final joint configurations of end-effector orientation-constrained trajectories and (ii) changes in end-effector goal or way-points under end-effector orientation constraints. We relate each of these examples to real-world applications ranging from learning from demonstration to obstacle avoidance. We also show that our algorithm produces trajectories with quality similar to what one would obtain by solving the trajectory optimization from scratch with warm-start initialization. Most importantly, however, our algorithm achieves a worst-case speed-up of 160x over the latter approach.

Keywords: manipulation; task perturbation; optimization; control

Citation: Srikanth, S.; Babu, M.; Masnavi, H.; Kumar Singh, A.; Kruusamäe, K.; Krishna, K.M. Fast Adaptation of Manipulator Trajectories to Task Perturbation by Differentiating through the Optimal Solution. *Sensors* **2022**, *22*, 2995. https://doi.org/10.3390/s22082995

Academic Editor: Gregor Klancar

Received: 1 March 2022
Accepted: 6 April 2022
Published: 13 April 2022

1. Introduction

A change in task-specification is often unavoidable in real-world manipulation problems. For example, consider a scenario where a manipulator is handing over an object to a human. The robot's estimate of the goal position can change as it executes its prior computed trajectories. Consequently, it needs to quickly adapt its joint motions to reach the new goal position. In this paper, we model motion planning as a parametric optimization problem wherein the task specifications are encoded in the parameters. In this context, adaptation to a new task requires re-computing the optimal joint trajectories for the new set of parameters. This is a computationally challenging process as the underlying cost functions in typical manipulation tasks are highly non-linear and non-convex [1]. Existing works leverage the so-called warm-starting technique where prior computed trajectories are used as initialization for the optimization solvers [2]. However, our extensive experimentation with off-the-shelf optimization solvers such as Scipy-SLSQP [3] show it is not sufficient for real-time adaptation of joint trajectories to task perturbations.

1.1. Main Idea

The proposed work explores an alternate approach based on differentiating the optimal solution with respect to the problem parameters, hereafter referred to as the Argmin differentiation [4]. To understand this further, consider the following constrained optimization problem over variable $\boldsymbol{\xi}$ (e.g., joint angles) and parameter vector $\mathbf{p}$ (e.g., end-effector position).

$$\boldsymbol{\xi}^*(\mathbf{p}) = \arg\min f(\boldsymbol{\xi}, \mathbf{p}) \tag{1}$$

$$g_i(\boldsymbol{\xi}, \mathbf{p}) \leq 0, \forall i = 1, 2, \ldots n \tag{2}$$

$$h_j(\boldsymbol{\xi}, \mathbf{p}) = 0, \forall j = 1, 2, \ldots m \tag{3}$$

The optimal solution $\boldsymbol{\xi}^*$ satisfies the following Karush–Kuhn Tucker (KKT) conditions.

$$\nabla f(\boldsymbol{\xi}^*, \mathbf{p}) + \sum_i \lambda_i \nabla g_i(\boldsymbol{\xi}^*, \mathbf{p}) + \sum_j \mu_j \nabla h_j(\boldsymbol{\xi}^*, \mathbf{p}) - 0 \tag{4a}$$

$$g_i(\boldsymbol{\xi}^*, \mathbf{p}) \leq 0, \forall i \tag{4b}$$

$$h_j(\boldsymbol{\xi}^*, \mathbf{p}) = 0 \tag{4c}$$

$$\lambda_i \geq 0, \lambda_i g_i(\boldsymbol{\xi}^*, \mathbf{p}) = 0, \forall i. \tag{4d}$$

The gradients in (4a) are taken with respect to $\boldsymbol{\xi}$. The variables λ_i, μ_j are called the Lagrange multipliers. Now, consider a scenario where the optimal solution $\boldsymbol{\xi}^*$ for the parameter $\mathbf{p}$ needs to be adapted for the perturbed set $\overline{\mathbf{p}} = \mathbf{p} + \Delta\mathbf{p}$. As mentioned earlier, one possible approach is to resolve the optimization with $\boldsymbol{\xi}^*$ as the warm-start initialization. Alternately, for $\Delta\mathbf{p}$ with a small magnitude, an analytical perturbation model can be constructed. To be precise, we can compute the first-order differential of the r.h.s. of (4a)–(4d) to obtain analytical gradients in the following form [5–7].

$$(\nabla_{\mathbf{p}}\boldsymbol{\xi}^*, \nabla_{\mathbf{p}}\lambda_i, \nabla_{\mathbf{p}}\mu_i^*) = \mathbf{F}(\boldsymbol{\xi}^*, \mathbf{p}, \lambda_i, \mu_j) \tag{5}$$

Multiplying the gradients with $\Delta\mathbf{p}$ gives us an analytical expression for the new solution and Lagrange multipliers corresponding to the perturbed parameter set [7].

1.2. Contribution

Algorithmic Contribution: A critical bottleneck in using the gradient map of the form (5) to compute perturbed solutions is that the mapping between $\Delta\mathbf{p}$ and λ_i is highly discontinuous. In other words, even a small $\Delta\mathbf{p}$ can lead to large changes in the so-called active-set of the inequality constraints. Thus it becomes necessary to develop additional active-set prediction mechanisms [7]. In this paper, we bypass this complication by instead focusing on the parametric optimization with only bound constraints on the variable set. Argmin differentiation of such problems has a simpler structure, which we leverage to develop a line-search based algorithm to incrementally adopt joint trajectories to larger changes in the parameter/tasks. To give some example of "large perturbation", our algorithm can adapt the joint trajectories of Franka–Panda arm to a perturbation of up to 30 cm in the goal position. This is almost 30% of the workspace of the Franka arm.

Application Contribution: For the first time, we apply the Argmin differentiation concept to the problem of joint trajectory optimization for the manipulators under end-effector task constraints. We consider a diverse class of cost functions to handle (i) perturbations in joint configurations or (ii) end-effector way-points in orientation-constrained end-effector trajectories. We present an extensive benchmarking of our algorithm's performance as a function of the perturbation magnitude. We also show that our algorithm outperforms the warm-start trajectory optimization approach in computation time by several orders of magnitude while achieving similar quality as that measured by task residuals and smoothness of the resulting trajectory.

1.3. Related Works

The concept of Argmin differentiation has been around for a few decades, although often under the name of sensitivity analysis [8,9]. However, of late it has seen a resurgence, especially in the context of end-to-end learning of control policies [10,11]. Our proposed work is more closely related to those that use Argmin differentiation for motion planning or feedback control. In this context, a natural application of Argmin differentiation is in bi-level trajectory optimization where the gradients of the optimal solution from the lower level are propagated to optimize the cost function at the higher level. This technique has been applied to both manipulation and navigation problems in existing works [6,12]. Alternately, Argmin differentiation can also be used for the correction of prior-computed trajectories [7,13].

To the best of our knowledge, we are not aware of any work that uses Argmin differentiation for the adaptation of task-constrained manipulator joint trajectories. The closest to our approach is [5] that uses it to accelerate the inverse kinematics problem. Along similar lines, [7] considers a very specific example of perturbation in the end-effector goal position. In contrast to these two cited works, we consider a much more diverse class of task constraints. Furthermore, our formulation also has important distinctions with [7] at the algorithmic level. Authors in [7] use the log-barrier function for including inequality constraints as penalties in the cost function. In contrast, we note that in the context of the task-constrained trajectory optimization considered in this paper, the joint angle limits are the most critical. The velocity and acceleration constraints can always be satisfied through time-scaling based pre-processing [14]. Thus, by choosing a way-point parametrization for the joint trajectories, we formulate the underlying optimization with just box constraints on the joint angles. This, in turn, allows us to treat this constraint through simple projection (Line 4 in Algorithm 1) without disturbing the structure of the cost function and the resulting Jacobian and Hessian matrices obtained by Argmin differentiation.

Algorithm 1 Line-Search Based Joint Trajectory Adaptation to Task Perturbation

1: Initialize ${}^{k}\boldsymbol{\xi}^{*}$ as the solution for the prior parameter ${}^{k}\mathbf{p}$, the Hessian ${}^{k}\nabla^{2}_{\xi}f({}^{k}\boldsymbol{\xi},\mathbf{p})$, the gradient $\nabla_{\xi,p_i}f({}^{k}\boldsymbol{\xi},\mathbf{p})$, and ${}^{k}\Delta\mathbf{p}=\overline{\mathbf{p}}-{}^{k}\mathbf{p}$

2: **while** $\eta > 0$ **do**

$$\max \eta \tag{6a}$$

$$f({}^{k}\boldsymbol{\xi}^{*}(\mathbf{p}+\eta\Delta\mathbf{p}),\mathbf{p}+\Delta\mathbf{p}) \leq f({}^{k}\boldsymbol{\xi}^{*},\mathbf{p}+\Delta\mathbf{p}) \tag{6b}$$

3:

$${}^{k+1}\boldsymbol{\xi}^{*} = {}^{k}\boldsymbol{\xi}^{*} + \eta\nabla_{\mathbf{p}}\boldsymbol{\xi}^{*}\Delta^{k}\mathbf{p} \tag{7}$$

4:

$${}^{k+1}\boldsymbol{\xi}^{*} = Project(\xi_{lb},\xi_{ub}) \tag{8}$$

5: Update ${}^{k+1}\overline{\mathbf{p}} = ForwardRoll({}^{k+1}\boldsymbol{\xi}^{*})$

6: Update ${}^{k+1}\Delta\mathbf{p} = \overline{\mathbf{p}} - {}^{k+1}\mathbf{p}$.

7: Update Hessian $\nabla^{2}_{\xi}f({}^{k+1}\boldsymbol{\xi},{}^{k+1}\mathbf{p})$.

8: Update Jacobian $\nabla_{\xi,p}f({}^{k+1}\boldsymbol{\xi},{}^{k+1}\mathbf{p})$

9: **end while**

2. Proposed Approach

2.1. Symbols and Notations

We will use lower case normal font letters to represent scalars, while bold font variants will represent vectors. Matrices are represented by upper case bold fonts. The subscript t will be used to denote the time stamp of variables and vectors. The superscript T will represent the transposing of a matrix.

2.2. Argmin Differentiation for Unconstrained Parametric Optimization

We consider the optimal joint trajectories to be the solution of the following bound-constrained optimization with parameter $\mathbf{p}$.

$$\boldsymbol{\xi}^{*}(\mathbf{p}) = \arg\min_{\xi} f(\boldsymbol{\xi},\mathbf{p}) \tag{9a}$$

$$\xi_{lb} \leq \boldsymbol{\xi} \leq \xi_{ub} \tag{9b}$$

We are interested in computing the Jacobian of $\boldsymbol{\xi}^*(\mathbf{p})$ with respect to $\mathbf{p}$. If we ignore the bound-constraints for now, we can follow the approach presented in [4] to obtain them in the following form.

$$\nabla_{\mathbf{p}}\boldsymbol{\xi} = -(\nabla^2_{\boldsymbol{\xi}} f(\boldsymbol{\xi},\mathbf{p}))^{-1}[\nabla_{\xi,p_1} f(\boldsymbol{\xi},\mathbf{p}), \quad \dots \quad \nabla_{\xi,p_n} f(\boldsymbol{\xi},\mathbf{p})] \tag{10}$$

Using (10), we can derive a local model for the optimal solution corresponding to a perturbation $\Delta\mathbf{p}$ as

$$\boldsymbol{\xi}^*(\overline{\mathbf{p}}) = \boldsymbol{\xi}^*(\mathbf{p}) + \nabla_{\mathbf{p}}\boldsymbol{\xi}^* \overbrace{(\overline{\mathbf{p}} - \mathbf{p})}^{\Delta\mathbf{p}}, \tag{11}$$

Intuitively, (11) signifies a step of length $\Delta\mathbf{p}$ along the gradient direction. However, for (11) to be valid, the step-length needs to be small. In other words, the perturbed parameter $\overline{\mathbf{p}}$ needs to be in the vicinity of $\mathbf{p}$. Although it is difficult to mathematically characterize the notion of "small", in the following, we attempt a practical definition based on the notion of optimal cost.

Definition 1. *A valid $|\Delta\mathbf{p}|$ is one that satisfies the following relationship*

$$f(\boldsymbol{\xi}^*(\overline{\mathbf{p}} = \mathbf{p} + \Delta\mathbf{p}), \mathbf{p} + \Delta\mathbf{p}) \leq f(\boldsymbol{\xi}^*, \mathbf{p} + \Delta\mathbf{p}) \tag{12}$$

The underlying intuition in (12) is that the perturbed solution should lead to a lower cost for the parameter $\mathbf{p} + \Delta\mathbf{p}$ as compared to $\boldsymbol{\xi}^*$ for the same perturbed parameter.

2.3. Line Search and Incremental Adaption

Algorithm 1 couples the concept from the definition (11) with a basic line-search to incrementally adapt (11) to a large $\Delta\mathbf{p}$. The algorithm begins by initializing the optimal solution ${}^k\boldsymbol{\xi}$ and the parameter ${}^k\mathbf{p}$ with prior values for iteration $k = 0$. These variables are then used to initialize the Hessian and Jacobian matrices. The core computations takes place in line 2, wherein we compute the least amount of scaling that needs to be done to step length ${}^k\Delta\mathbf{p} = {}^k\overline{\mathbf{p}} - \mathbf{p}$ to guarantee a reduction in the cost. At line 3, we update the optimal solution based on step-length $\eta^k\Delta\mathbf{p}$ obtained in line 2, followed by a simple projection at line 4 to satisfy the minimum and maximum bounds. At line 5, we perform the called forward roll-out of the solution to update the parameter set. For example, if the parameter $\mathbf{p}$ models position of the end-effector at the final time instant of a trajectory, then line 5 computes how close the ${}^{k+1}\boldsymbol{\xi}^*$ takes the end-effector to the perturbed goal position $\overline{\mathbf{p}}$. On lines 7 and 8, we update the Hessian and the Jacobian matrices based on the updated parameter set and optimal solution.

3. Task Constrained Joint Trajectory Optimization

This section formulates various examples of the task-constrained trajectory optimization problem and uses the previous section's results for optimal adaptation of joint trajectories under task perturbation. To formulate the underlying costs, we adopt the way-point parametrization and represent the joint angles at time t as $\mathbf{q}_t$. Furthermore, we will use $(\mathbf{x}_e(\mathbf{q}_t), \mathbf{o}_e(\mathbf{q}_t))$ to describe the end-effector position and orientation in terms of Euler angles, respectively.

3.1. Orientation Constrained Interpolation between Joint Configurations

The task here is to compute an interpolation trajectory between a given initial $\mathbf{q}_0$ and a final joint configuration $\mathbf{q}_m$ while maintaining a specified orientation $\mathbf{o}_d$ for the end-effector at all times. We model it through the following cost function.

$$\sum_t f_s(\mathbf{q}_{t-k:t}) + \left\| \begin{matrix} \mathbf{q}_{t_1} - \mathbf{q}_0 \\ \mathbf{q}_{t_m} - \mathbf{q}_m \end{matrix} \right\|_2^2 + \sum_t \|\mathbf{o}_e(\mathbf{q}_t) - \mathbf{o}_d\|_2^2 \tag{13}$$

The first term the cost function models smoothness in terms of joint angles from $t-k$ to t [15]. For example, for $k=1$, the smoothness is defined as the first-order finite difference of the joint positions at subsequent time instants. Similarly, $k=2,3$, will model higher order smoothness through second and third-order finite differences respectively. We consider all three finite-differences in our smoothness cost term. The second term ensures that the interpolation trajectory is close to the given initial and final points. The final term in the cost function maintains the required orientation of the end-effector.

We can shape (13) in the form of (9a) by defining $\boldsymbol{\xi} = (\mathbf{q}_{t_1}, \mathbf{q}_{t_2}, \dots \mathbf{q}_{t_m})$. The bounds will correspond to the maximum and minimum limits on the joint angles at each time instant. We define the parameter set as $\mathbf{p} = (\mathbf{q}_0, \mathbf{q}_m)$. That is, we are interested in computing the adaptation when either or both of $\mathbf{q}_0$ and $\mathbf{q}_m$ gets perturbed.

Applications

Adaptation of $\boldsymbol{\xi}^*$ of (13) for different $\mathbf{q}_0, \mathbf{q}_m$ has applications in learning from demonstration setting where the human just provides the information about the initial and/or final joint configuration, and the manipulator then computes a smooth interpolation trajectory between the boundary configurations by adapting a prior computed trajectory.

Figure 1 presents an example of adaptation discussed above. The prior computed trajectory is shown in blue. This is then adapted to two different final joint configurations. The trajectory computed through Algorithm 1 is shown in green, while that obtained by resolving the optimization problem (with warm-starting) is shown in red.

3.2. Orientation-Constrained Trajectories through Way-Points

The task in this example is to make the end-effector move though given way-points while maintaining the orientation at $\mathbf{o}_d$. Let $\mathbf{x}_{d_t}$ represent the desired way-point of the end-effector at time t. Thus, we can formulate the following cost function for the current task.

$$\sum_t f_s(\mathbf{q}_{t-k:t}) + \sum_t \|\mathbf{o}_e(\mathbf{q}_t) - \mathbf{o}_d\|_2^2 + \sum_t \|\mathbf{x}_e(\mathbf{q}_t) - \mathbf{x}_{d_t}\|_2^2 \tag{14}$$

The first two terms in the cost function are the same as the previous example. The changes appear in the final term which minimizes the l_2 norm of the distance of the end-effector with the desired way-point. The defintion of $\boldsymbol{\xi}$ remains the same as before. However, the parameter set is now defined as $\mathbf{p} = (\mathbf{x}_{d_1}, \mathbf{x}_{d_2}, \dots \mathbf{x}_{d_m})$.

Application

Collision Avoidance As shown in Figure 2, a key application of the adaptation problem discussed above is in collision avoidance. A reactive planner such as [16] can provide new via-points for the manipulator to avoid collision. Our Algorithm 1 can then use the cost function (14) to adapt the prior trajectory shown in blue to that shown in green. For comparison, the trajectory obtained with resolve of the trajectory optimization is shown in red.

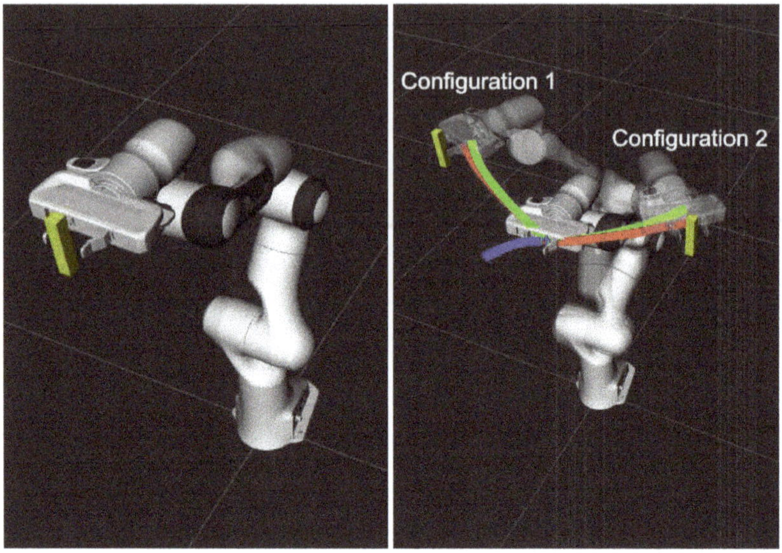

Figure 1. Prior trajectory shown in blue is used to adapt the joint motions to move towards two different final joint configurations while maintaining the horizontal orientation of the end-effector at all times.

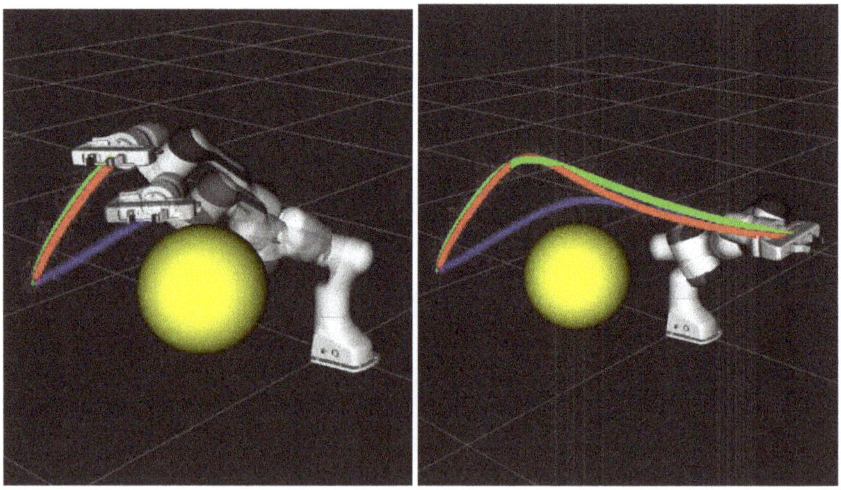

Figure 2. Collision avoidance by perturbing the mid-point of the prior computed end-effector trajectory.

Human–Robot Handover: Algorithm 1 with cost function (14) also finds application in human–robot handover tasks. An example is shown in Figure 3, where the manipulator adapts the prior trajectory (blue) to a new estimate of the handover position. As before, the trajectory obtained with Algorithm 1 is shown in green, while the one shown in red corresponds to a re-solve of the trajectory optimization with warm-start initialization.

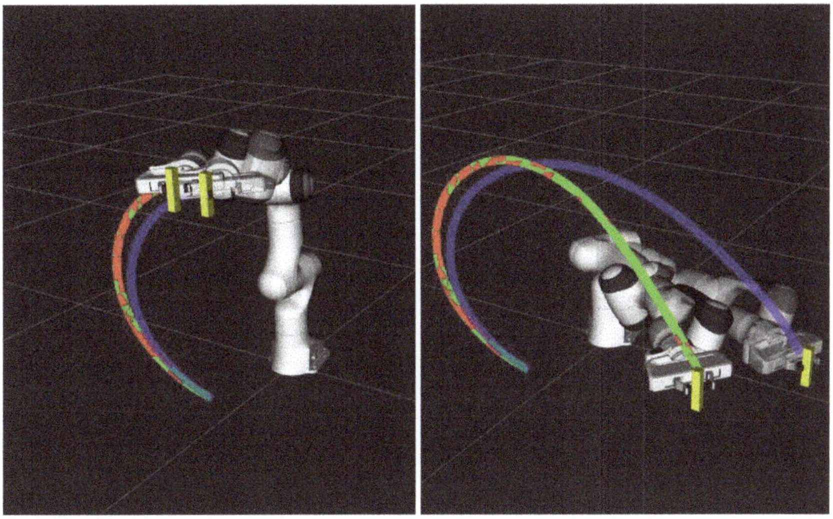

Figure 3. Perturbation in the final position of the end-effector.

4. Benchmarking

4.1. Implementation Details

The objective of this section is to compare the trajectories computed by Algorithm 1 with that obtained by re-solving the trajectory optimization for the perturbed parameters with warm-start initialization. We consider the same three benchmarks presented in Figures 1–3 implemented on a 7dof Franka Panda Arm, but for a diverse range of perturbations magnitude. For each benchmark, we created a data set of 180 trajectories by generating random perturbations in the task parameters. For the benchmark of Figure 1, the parameters are the joint angles, but in the following we use the forward kinematics to derive equivalent representation for the parameters in terms of end-effector position values.

Each joint trajectory is parameterized by a 50-dimensional vector of way-points. Thus, the underlying task constrained trajectory optimization involves a total of 350 variables. We use Scipy-SLSQP [3] to obtain the prior trajectory and also to re-solve the trajectory optimization for the perturbed parameters. We did our implementation in Python using Jax-Numpy [17] to compute the necessary Jacobian and Hessian matrices. We also used the just-in-time compilation ability of JAX to create an on-the-fly compiled version of our codes. The line-search in Algorithm 1 (line 2) was done through a parallelized search over a set of discretized η values. The entire implementation was done on a 32 GB RAM i7-8750 desktop with RTX 2080 GPU (8GB). To foster further research in this field and ensure reproducibility, we open-source our implementation for review at https://rebrand.ly/argmin-planner (First released on 15 September 2020).

4.2. Quantitative Results

Orientation Metric: For this analysis, we compared the pitch and roll angles at each time instant along trajectories obtained with Algorithm 1 and the resolving approach. Specifically, we computed the maximum of the absolute difference (or L_∞ norm) of the two orientation trajectories. The yaw orientation in all these benchmarks was a free variable and is thus not included in the analysis. The results are summarized in Figures 4–6. The histogram plot in these figures are generated for the medium perturbation ranges (note the figure legends). For the Figure 1 benchmark related to cost function (13), Figure 4 shows that all the trajectories obtained by Algorithm 1 have L_∞ norm of the orientation difference less than 0.1 rad. For the benchmark of Figure 2, which we recall involves perturbing the via-point of the end-effector trajectory, the histograms of Figure 5 show similar trends. All the trajectories computed by Algorithm 1 managed a similar orientation difference. For the benchmark of Figure 3 pertaining to the perturbation of the final position, 69.41% of the trajectories obtained by Algorithm 1 managed to maintain a orientation difference of 0.1 rad with the resolving approach.

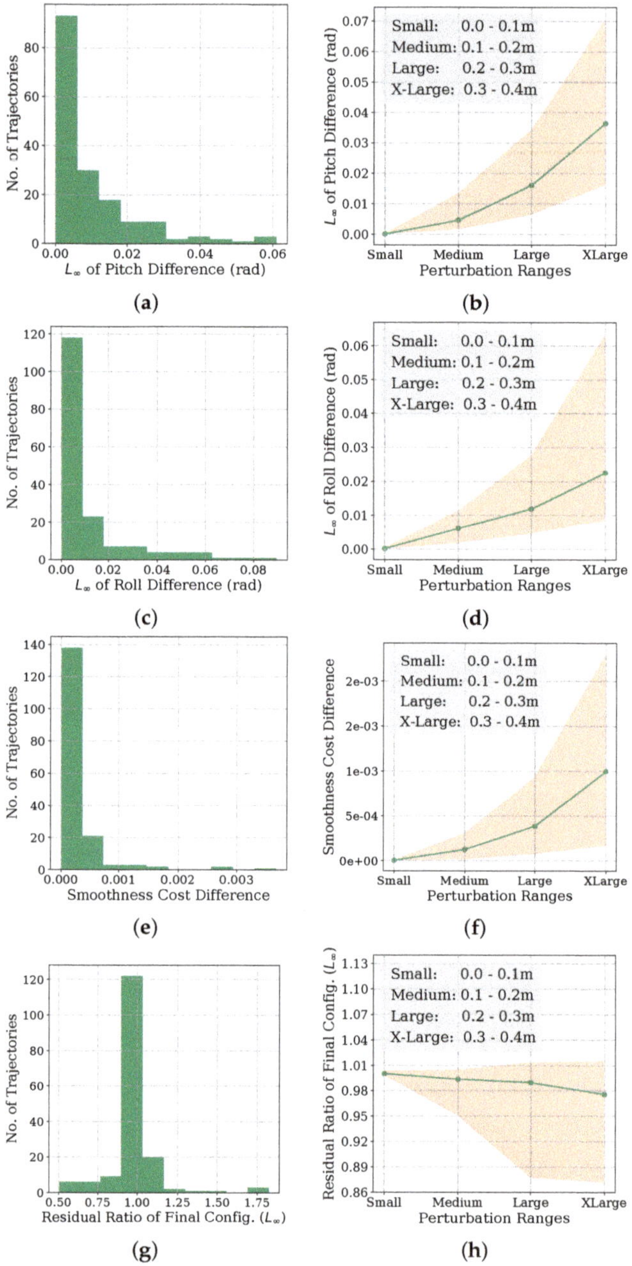

Figure 4. Performance of Algorithm 1 for different perturbation ranges on the benchmark of Figure 1 that involves perturbing the final joint configuration (recall cost function (13)). Note that the perturbation in the final joint is converted to position values by forward kinematics. The (**a**,**c**,**e**,**g**) column shows the histogram of orientation, smoothness and task residual ratio metrics for the medium range perturbation. The (**b**,**d**,**f**,**h**) column quantifies the metrics for different perturbation ranges.

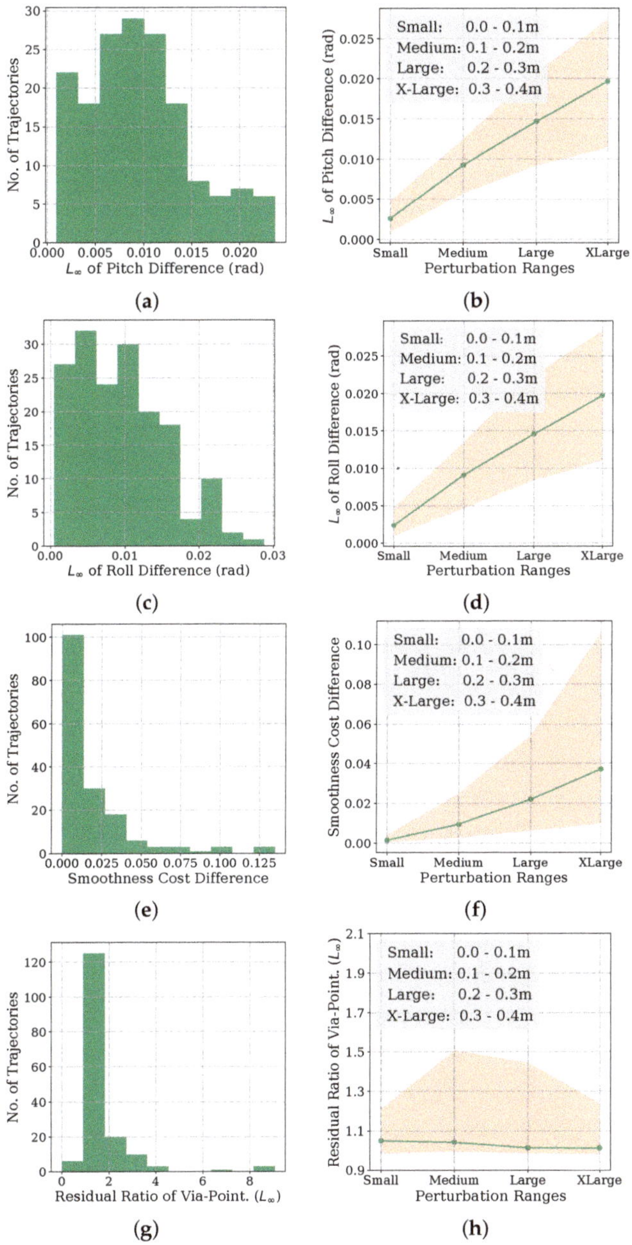

Figure 5. Performance of Algorithm 1 for different perturbation ranges on the benchmark of Figure 2 that involves perturbing the via-point of the end-effector trajectory (recall cost function (14)). The (**a**,**c**,**e**,**g**) and (**b**,**d**,**f**,**h**) columns show similar benchmarking as those of Figure 4.

Task residuals ratio metric: For this analysis, we compare the task residual between trajectories obtained from Algorithm 1 and the resolving approach. For example, for the benchmark of Figure 1, we want the manipulator final configuration to be close to the specified value (recall cost (13)) while maintaining the desired orientation at each time instant. Thus, we compute the L_∞ residual of $\mathbf{q}_t - \mathbf{q}_m$ for Algorithm 1 and compare it with that obtained from the resolving approach. Now, as previously, and to be consistent with the other benchmarks, we convert the residual of the joint angles to position values through forward kinematics. Similar analysis follow for the other benchmarks as well. For the ease of exposition, we divide the task residual of Algorithm 1 by that obtained with the resolving approach. A ratio greater than 1 implies that the former led to a higher task residual than the latter and vice-versa. Similarly, a ratio closer to 1 implies that both the approaches performed equally well.

The results are again summarized in Figures 4–6. From Figure 4, we notice that 97.05% of trajectories have a residual ratio less than 1.2. For the experiment involving via-point perturbation in Figure 5, the performance drops to 62.50% for the same value of residual ratio. Meanwhile, as shown in Figure 6, around 82.94% of the trajectories have a residual ratio less than 1.2 in the case of the final position perturbation benchmark of Figure 3.

Velocity Smoothness Metric: For this analysis, we computed the difference in the velocity smoothness cost (L_2 norm of first-order finite difference) between the trajectories obtained with Algorithm 1 and the resolving approach. The results are again summarized in Figures 4–6. For all the benchmarks, in around 65% of the examples, the difference was less than 0.05. This is 35% of the average smoothness cost observed across all the trajectories from both the approaches.

Scaling with Perturbation Magnitude: The line plots in Figures 4–6 represent the first quartile, median and the third quartile of the three metrics discussed above for different perturbation ranges.

For the benchmark of Figure 1, trajectories from Algorithm 1 maintains an orientation difference of less than 0.1 rad, with the trajectories of the resolving approach for perturbations as large as 40 cm. The difference in smoothness cost for the same range is also small, with the median value being in the order of 10^{-3}. The median task residuals achieved by Algorithm 1 is only 2% higher than that obtained by the resolving approach. For the benchmark of Figure 2, the performance remains same on the orientation metric, but the median difference in smoothness cost and task residual ration increases to 0.04 and 9% for the largest perturbation range. The benchmark of Figure 6 follows a similar trend in orientation and smoothness metric, but performs significantly worse in task residuals. For the largest perturbation range, Algorithm 1 leads to 50% higher median task residuals. However, importantly, for perturbation up to 30 cm, the task residual ratio is close to 1, suggesting that Algorithm 1 performed as well as the resolving approach for these perturbations.

Computation Time: Table 1 contrasts the average timing of our Algorithm 1 with the approach of resolving the trajectory optimization with warm-start initialization. As can be seen, our Argmin differentiation based approach provides a worst-case speed up of 160x on the benchmark of Figure 3. For the rest of the benchmarks, this number varies between 500 to 1000. We believe that this massive gain in computation time offsets whatever little performance degradation in terms of orientation, smoothness, and task residual metric that Algorithm 1 incurs compared to re-solving the problem using warm-start. Note that the high computation time of the re-solving approach is expected, given that we are solving a difficult non-convex function over a long horizon of 50 steps resulting in 350 decision variables. Even highly optimized planners like [1] show similar timings on closely related benchmarks [18].

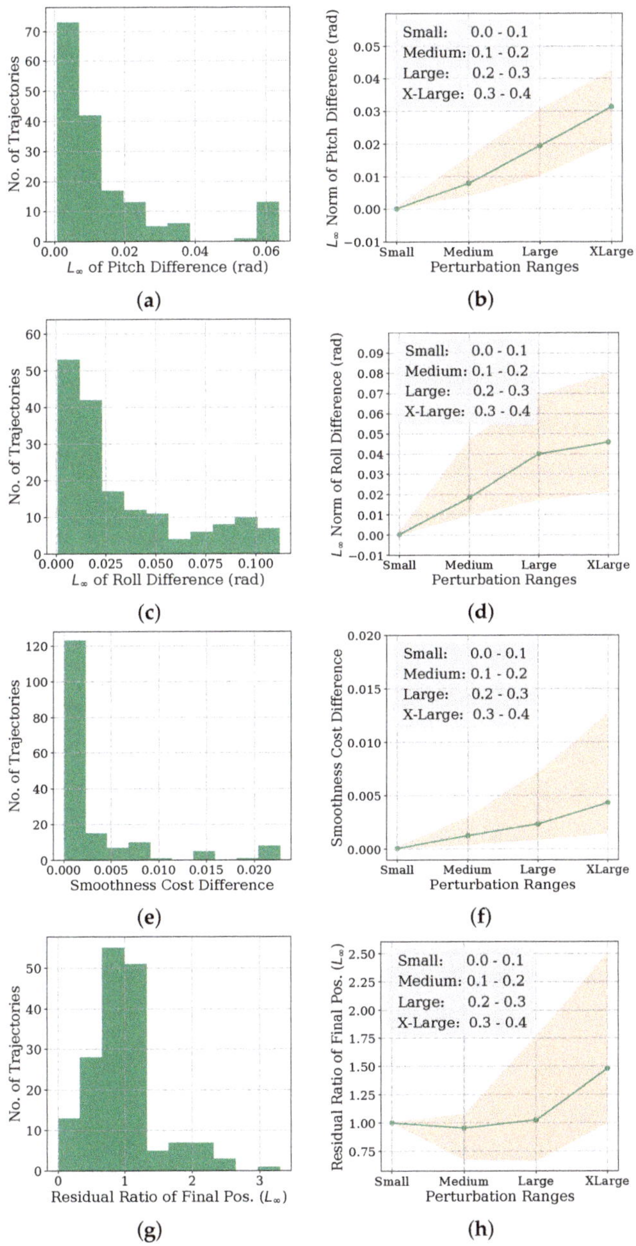

Figure 6. Performance of Algorithm 1 for different perturbation ranges on the benchmark of Figure 6 that involves perturbing the final end-effector position. (recall cost function (14)). The (**a**,**c**,**e**,**g**) and (**b**,**d**,**f**,**h**) columns show similar benchmarking as those of Figure 4.

Table 1. Computation times comparison between Algorithm 1 and resolving trajectory optimization approach on three benchmarks.

Benchmarks	SciPy-SLSQP		Our Algorithm 1
	Wall Time (s)	Wall Time w/o Jacobian and Function Evaluation Overhead (s)	Wall Time (s)
Final Configuration Perturbation (Figure 1)	43.91	41.09	**0.039**
Via Point Perturbation (Figure 2)	53.05	34.74	**0.09**
Final Position Perturbation (Figure 3)	35.91	29.09	**0.18**

5. Conclusions and Future Work

We presented a fast, near real-time algorithm for adapting joint trajectories to task perturbation as high as 40 cm in the end-effector position, almost half the radius of the Franka Panda arm's horizontal workspace used in our experiments. By consistently producing trajectories similar to those obtained by resolving the trajectory optimization problem but in a small fraction of a time, our Algorithm 1 opens up exciting possibilities for reactive motion control of manipulators in applications like human–robot handover.

Our algorithm is easily extendable to other kind of manipulators. The only requirement is that we should know the forward kinematics of the manipulator. This would allow us to get the algebraic expressions for functions $\mathbf{o}_e(\mathbf{q})$ and $\mathbf{x}_e(\mathbf{q})$ in cost function (13) and (14), respectively. In our implementation, we derived the forward kinematics and $\mathbf{o}_e(\mathbf{q})$ and $\mathbf{x}_e(\mathbf{q})$ through the DH representation of the manipulator. The DH table is available for many commercial manipulators, e.g., UR5e besides the Franka Panda Arm used in our simulation.

Our algorithm does not depend on any specific sensing modality. For example, in collision avoidance applications, we assume that obstacle information is used by some higher level planners that provides intermediate collision-free points to the manipulator, which then uses the ArgMin differentiation to replan its prior trajectories.

There are several ways to improve our algorithm. First, the joint bounds can also be included as penalties in the cost function itself, in addition to being handled by projection (Line 11 in Algorithm 1). This would ensure that the gradient and Hessian of the optimal cost is aware of the joint limit bounds. Second, we can consider a low dimensional polynomial representation of the trajectories. For example, the joint trajectories can be represented by a 10th order Bernstein polynomial with the coefficients acting as the variables of the optimization problem. This would drastically reduce the computation cost of obtaining the Hessian of the optimal cost as compared to current way-point paramaterization of the joint trajectory that requires around 50 variables to represent one joint trajectory.

In future works, we will extend our formulation to problems with dynamic constraints, such as torque bounds. We conjecture that by coupling the way-point parametrization with a multiple-shooting like approach, we can retain the constraints as simple box-bounds on the decision variables and consequently retain the computational structure of the Algorithm 1. We are also currently evaluating our algorithm's performance on applications such as autonomous driving.

Author Contributions: Conceptualization, S.S., M.B., A.K.S. and K.M.K.; Methodology, S.S., M.B., A.K.S. and K.M.K.; Software, S.S., M.B. and H.M.; Validation, S.S., M.B. and H.M.; formal analysis, S.S. and M.B.; investigation, S.S., M.B. and A.K.S.; resources, S.S., M.B. and A.K.S.; data curation, S.S., M.B. and H.M.; writing—original draft preparation, A.K.S., S.S., M.B., K.K. and K.M.K.; writing—review and editing, S.S., M.B., K.K. and K.M.K.; visualization, S.S., M.B. and H.M.; supervision, A.K.S. and K.M.K.; project administration, A.K.S. and K.M.K.; funding acquisition, A.K.S. and K.K. All authors have read and agreed to the published version of the manuscript.

Funding: The work was supported in part by thw European Social Fund via the "ICT programme" measure, by grant PSG753 from the Estonian Research Council, by AI & Robotics Estonia (AIRE), the Estonian candidate for European Digital Innovation Hub, funded by the Ministry of Economic Affairs and Communications in Estonia.

Data Availability Statement: Codes to reproduce the results are available at https://rebrand.ly/argmin-planner (accessed on 1 August 2020).

Conflicts of Interest: The authors declare no conflict of interest. The funders had no role in the design of the study; in the collection, analyses, or interpretation of data; in the writing of the manuscript, or in the decision to publish the results.

References

1. Berenson, D.; Srinivasa, S.S.; Ferguson, D.; Kuffner, J.J. Manipulation planning on constraint manifolds. In Proceedings of the 2009 IEEE International Conference on Robotics and Automation, Kobe, Japan, 12–17 May 2009; pp. 625–632.
2. Lembono, T.S.; Paolillo, A.; Pignat, E.; Calinon, S. Memory of motion for warm-starting trajectory optimization. *IEEE Robot. Autom. Lett.* **2020**, *5*, 2594–2601. [CrossRef]
3. Virtanen, P.; Gommers, R.; Oliphant, T.E.; Haberland, M.; Reddy, T.; Cournapeau, D.; Burovski, E.; Peterson, P.; Weckesser, W.; Bright, J.; et al. SciPy 1.0: Fundamental algorithms for scientific computing in Python. *Nat. Methods* **2020**, *17*, 261–272. [CrossRef] [PubMed]
4. Gould, S.; Fernando, B.; Cherian, A.; Anderson, P.; Cruz, R.S.; Guo, E. On differentiating parameterized argmin and argmax problems with application to bi-level optimization. *arXiv* **2016**, arXiv:1607.05447.
5. Hauser, K. Learning the problem-optimum map: Analysis and application to global optimization in robotics. *IEEE Trans. Robot.* **2016**, *33*, 141–152. [CrossRef]
6. Tang, G.; Sun, W.; Hauser, K. Time-Optimal Trajectory Generation for Dynamic Vehicles: A Bilevel Optimization Approach. In Proceedings of the 2019 IEEE/RSJ International Conference on Intelligent Robots and Systems (IROS), Macau, China, 3–8 November 2019; pp. 7644–7650.
7. Reiter, A.; Gattringer, H.; Müller, A. Real-time computation of inexact minimum-energy trajectories using parametric sensitivities. In Proceedings of the International Conference on Robotics in Alpe-Adria Danube Region, Torino, Italy, 21–23 June 2017; Springer: Berlin/Heidelberg, Germany, 2017; pp. 174–182.
8. Geffken, S.; Büskens, C. Feasibility refinement in sequential quadratic programming using parametric sensitivity analysis. *Optim. Methods Softw.* **2017**, *32*, 754–769. [CrossRef]
9. Pirnay, H.; López-Negrete, R.; Biegler, L.T. Optimal sensitivity based on IPOPT. *Math. Program. Comput.* **2012**, *4*, 307–331. [CrossRef]
10. Amos, B.; Jimenez, I.; Sacks, J.; Boots, B.; Kolter, J.Z. Differentiable MPC for end-to-end planning and control. In *Advances in Neural Information Processing Systems*; MIT Press: Cambridge, MA, USA, 2018; pp. 8289–8300.
11. Agrawal, A.; Barratt, S.; Boyd, S.; Stellato, B. Learning convex optimization control policies. In Proceedings of the 2nd Conference on Learning for Dynamics and Control, PMLR, Berkeley, CA, USA, 11–12 June 2020; pp. 361–373.
12. Landry, B.; Lorenzetti, J.; Manchester, Z.; Pavone, M. Bilevel Optimization for Planning through Contact: A Semidirect Method. *arXiv* **2019**, arXiv:1906.04292.
13. Kalantari, H.; Mojiri, M.; Dubljevic, S.; Zamani, N. Fast l_1 model predictive control based on sensitivity analysis strategy. *IET Control. Theory Appl.* **2020**, *14*, 708–716. [CrossRef]
14. Pham, Q.C. A general, fast, and robust implementation of the time-optimal path parameterization algorithm. *IEEE Trans. Robot.* **2014**, *30*, 1533–1540. [CrossRef]
15. Toussaint, M. A tutorial on Newton methods for constrained trajectory optimization and relations to SLAM, Gaussian Process smoothing, optimal control, and probabilistic inference. In *Geometric and Numerical Foundations of Movements*; Springer: Berlin/Heidelberg, Germany, 2017; pp. 361–392.
16. Flacco, F.; Kröger, T.; De Luca, A.; Khatib, O. A depth space approach to human-robot collision avoidance. In Proceedings of the 2012 IEEE International Conference on Robotics and Automation, Guangzhou, China, 11–14 December 2012; pp. 338–345.
17. Bradbury, J.; Frostig, R.; Hawkins, P.; Johnson, M.J.; Leary, C.; Maclaurin, D.; Wanderman-Milne, S. JAX: Composable Transformations of Python+NumPy Programs. 2018. Available online: http://github.com/google/jax (accessed on 1 August 2020).
18. Qureshi, A.H.; Dong, J.; Baig, A.; Yip, M.C. Constrained Motion Planning Networks X. *arXiv* **2020**, arXiv:2010.08707.

sensors

MDPI

Article

An Improved Rapidly-Exploring Random Trees Algorithm Combining Parent Point Priority Determination Strategy and Real-Time Optimization Strategy for Path Planning

Lijing Tian, Zhizhuo Zhang, Change Zheng *, Ye Tian, Yuchen Zhao, Zhongyu Wang and Yihan Qin

School of Technology, Beijing Forestry University, Beijing 100083, China; T7190221@bjfu.edu.cn (L.T.); zhangzhizhuo@bjfu.edu.cn (Z.Z.); tytoemail@bjfu.edu.cn (Y.T.); zhaoyuchen@bjfu.edu.cn (Y.Z.); wangzhongyu@bjfu.edu.cn (Z.W.); linxinyi031@bjfu.edu.cn (Y.Q.)

* Correspondence: zhengchange@bjfu.edu.cn

Abstract: In order to solve the problems of long path planning time and large number of redundant points in the rapidly-exploring random trees algorithm, this paper proposed an improved algorithm based on the parent point priority determination strategy and the real-time optimization strategy to optimize the rapidly-exploring random trees algorithm. First, in order to shorten the path-planning time, the parent point is determined before generating a new point, which eliminates the complicated process of traversing the random tree to search the parent point when generating a new point. Second, a real-time optimization strategy is combined, whose core idea is to compare the distance of a new point, its parent point, and two ancestor points to the target point when a new point is generated, choosing the new point that is helpful for the growth of the random tree to reduce the number of redundant points. Simulation results of 3-dimensional path planning showed that the success rate of the proposed algorithm, which combines the strategy of parent point priority determination and the strategy of real-time optimization, was close to 100%. Compared with the rapidly-exploring random trees algorithm, the number of points was reduced by more than 93.25%, the path planning time was reduced by more than 91.49%, and the path length was reduced by more than 7.88%. The IRB1410 manipulator was used to build a test platform in a laboratory environment. The path obtained by the proposed algorithm enables the manipulator to safely avoid obstacles to reach the target point. The conclusion can be made that the proposed strategy has a better performance on optimizing the success rate, the number of points, the planning time, and the path length.

Keywords: rapidly-exploring random trees; manipulator; priority determination; real-time optimization; path planning

Citation: Tian, L.; Zhang, Z.; Zheng, C.; Tian, Y.; Zhao, Y.; Wang, Z.; Qin, Y. An Improved Rapidly-Exploring Random Trees Algorithm Combining Parent Point Priority Determination Strategy and Real-Time Optimization Strategy for Path Planning. *Sensors* **2021**, *21*, 6907. https://doi.org/10.3390/s21206907

Academic Editor: Baochang Zhang

Received: 22 September 2021
Accepted: 14 October 2021
Published: 18 October 2021

Publisher's Note: MDPI stays neutral with regard to jurisdictional claims in published maps and institutional affiliations.

1. Introduction

Whether for mobile robots such as AGV (automated guided vehicle) carts working in automated workshops, or robotic arms such as agricultural manipulators in the field, the core of automation for autonomous robots is path planning. Path planning is the process of finding an obstacle-free path from an initial position to a target position in a known or partially known environment [1].

Path planning is one of the most important research focuses of robots. Chinthaka Premachandra et al. completed the robot's path planning in an indoor environment by a self-localization method through baseboard recognition and image processing [2]. Wenzhou Chen et al. used distributed sonar sensors to calculate the distance between the receiver and the generator in real time to control the moving path of the robot [3]. Chinthaka Premachandra et al. proposed a hybrid aerial-terrestrial robot system to help UAVs avoid obstacles during the movement [4]. Path planning algorithms can usually be divided into three types. The first type is the bionic-based path planning algorithm [5], of which the ant colony algorithm is a common one and has the advantages of robustness and environmental adaptability, but its convergence speed is slow and very easy to fall into the local optimum. The second type is the map-based path planning algorithm, of which the A* algorithm (Optimal A-algorithm), with the optimal surrogate and prognostic functions, is a commonly used one and has the advantages of heuristic search and the obtained path is optimal, but its planning time is long and not applicable to high-dimensional space [6]. The third type is the sampling-based path planning algorithm [7], of which the most commonly used one is the RRT (rapidly-exploring random trees) algorithm [8]. As an efficient path-planning method in a multi-dimensional space, the RRT algorithm uses an initial point as the root point and generates a random extended tree by randomly sampling and adding leaf points. When a leaf point in the random tree contains a target point or enters a target region, a path from the initial point to the target point, consisting of tree points, can be found in the random tree. The RRT algorithm is the most popular path planning algorithm due to the rapidness, probabilistic completeness, and good scalability [9–13].

However, the RRT algorithm also has many disadvantages. Among the main disadvantages of the RRT algorithm, one is that the whole random tree needs to be traversed to search the parent point in the process of new point generation, which consumes a lot of computation time. The other is the large number of redundant points generated during the path generation process. To address the above problems, this paper proposes an improved RRT algorithm based on the PPD strategy (the strategy of parent point priority determination) to speed up the path planning, and further optimizes the efficiency of the algorithm by incorporating the RO strategy (real-time optimization) on this basis. The PPD strategy would shorten the path planning time and the RO strategy would reduce the number of redundant points. Finally, MATLAB-based three-dimensional comparison simulation experiments were conducted, and the experimental results showed that the proposed algorithm has a faster planning speed and can generate fewer redundant points, which has a better performance compared with other improved algorithms.

The rest of this paper is organized as follows. Section 2 introduces the related work and Section 3 presents the proposed RRT algorithm in three parts including the RRT algorithm, the PPD strategy, and the RO strategy. Section 4 shows the simulation results of the proposed algorithm in three-dimensional space and the experiments using IRB1410 in the laboratory. In the following, a discussion is presented in Section 5. Finally, Section 6 presents the conclusions of this paper.

2. Related Work

The RRT algorithm has been widely used in the field of robot motion and path planning. However, the paths obtained are not optimal, mainly because of several aspects such as path planning time, the number of points, and the path length. To solve these shortcomings, many improved algorithms based on the RRT algorithm have been proposed to promote the path-planning efficiency. LaValle and Kuffner proposed a bidirectional extended random tree algorithm [14] that generated two random trees from the starting point and the target point simultaneously, expanding them in space separately. This algorithm used a greedy strategy to reduce the number of iterations in the path generation process. Sertac and Emilio proposed an asymptotically optimal RRT* algorithm (an improved algorithm for progressively optimizing path length by reselecting parent point) [15], which changed the selection of the parent point and used a cost function to select the point with the smallest cost in the neighborhood of the extended point as the parent point, thus reducing the cost of path generation and improving the search efficiency. Jordan et al. borrowed the RRT-Connect (bidirectional extended random tree) algorithm idea and proposed a bidirectional extended RRT* algorithm, namely B-RRT* (bidirectional version of RRT*) algorithm [16]. Wang Kun et al. proposed a two-way extended RRT* algorithm for heuristic search, which reduced the number of iterations to a certain extent [17]. Jordan proposed the B-RRT* algorithm [18], which used the strategy of reselecting the parent point and rewiring two trees to speed up the algorithm convergence. Qureshi et al. added the heuristic strategy to the B-RRT* algorithm and proposed the IB-RRT* (Intelligent bidirectional-RRT*) algorithm [16]. Qureshi et al. combined the artificial potential field method with the RRT* algorithm to improve the convergence speed of the algorithm [19]. Barfoot proposed the Inform-RRT* algorithm [20] to narrow the search range and speed up the convergence of the algorithm on the basis of obtaining feasible paths. Mashayekhi et al. proposed the Informed-RRT*-Connect algorithm [21], which used a bidirectional tree to quickly find the initial path before using a subset of heuristics to directly sample to accelerate convergence, and the heuristic algorithm performed better in the improved RRT*-Connect algorithm. While the RRT* algorithm and its improved algorithm helped to reduce the path length, their planning times were several times longer than that of the RRT algorithm. Although the RRT-Connect algorithm and its improved algorithm had a slight reduction in the number of redundant points and planning time, the optimization effect was not significant and the path length was much longer than that of the RRT algorithm.

The above improvement algorithms have different advantages. In this paper, we focused on both the path planning time and the number of redundant points. Two different improved strategies for each of the two aspects combined are proposed.

3. Methods

3.1. The Rapidly-Exploring Random Tree Algorithm (RRT)

The rapidly-exploring random tree algorithm is a probability-complete global path planning algorithm that obtains path points by random sampling in the search space and then achieving a feasible path from the start point to the goal point. The specific process is shown in Algorithm 1.

Algorithm 1. RRT algorithm.

a. Initialize the random tree P_{init}.
b. Select a random point P_{rand} in the search space.
c. Traverse the random tree and find the closest point to P_{rand} in the random tree, named P_{parent}.
d. Intercept the step length ρ along the direction from P_{near} to P_{rand} to get a new point P_{new}.
e. Repeat the above steps b–d until the target point P_{goal} is added to the random tree.

The random tree expansion diagram for the RRT algorithm is shown in Figure 1. The RRT algorithm generates new points by random sampling in the workspace. In the random tree expansion process, searching P_{parent} requires traversing the entire random tree, a process that takes a lot of time when the random tree grows relatively large, which in turn leads to a slow path-planning speed of the algorithm. The sampling method of the RRT algorithm is highly random, which results in a large number of redundant points. For these problems, two improved strategies are proposed in this paper.

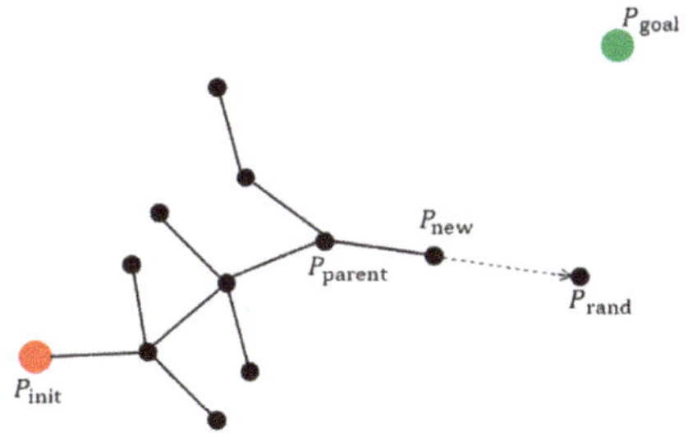

Figure 1. Random tree expansion diagram for the RRT algorithm. The red circle indicates the starting point and the green circle indicates the target point. Black circle indicates the path point, black solid line indicates the path, and black dashed line with arrow indicates the current expansion direction of the random tree. P_{rand} denotes the randomly sampled point, P_{parent} denotes the parent point, and P_{new} denotes the new point.

3.2. The Strategy of Parent Point Priority Determination (PPD)

In order to save time in traversing the whole random tree in the process of determining the parent point, this paper proposes a strategy of parent point priority determination to simplify this process and thus shorten the path planning time. The strategy of parent point priority determination is an improved strategy based on the RRT algorithm, whose core idea is to prioritize the parent point of the next new point before random sampling. Compared with the RRT algorithm, the improved algorithm based on the PPD strategy saves time in finding the P_{parent} and therefore speeds up the execution of the algorithm. The specific process is shown in Algorithm 2, where D_{new} and D_{parent} denote the distance from the new point to the target point and the distance from the parent point to the target point, respectively, which can be calculated by Equation (1).

$$\begin{cases} D_{new} = \sqrt{\left(P_{new(x)} - P_{goal(x)}\right)^2 + \left(P_{new(y)} - P_{goal(y)}\right)^2} \\ D_{parent} = \sqrt{\left(P_{parent(x)} - P_{goal(x)}\right)^2 + \left(P_{parent(y)} - P_{goal(y)}\right)^2} \end{cases} \quad (1)$$

Algorithm 2. PPD-RRT algorithm.

a. Initialize the random tree P_{init}.
b. Set the point P_{init} as the parent point P_{parent} of the next expansion.
c. Get four random points $P_{rand1} \sim P_{rand4}$ on the circumference of the circle with the parent point P_{parent} as the center and the step length ρ as the radius.
d. Select the closest point to the target point in $P_{rand1} \sim P_{rand4}$ as the random point P_{rand}.
e. Connect parent point P_{parent} to the random point P_{rand}, the random point P_{rand} is the new point P_{new}.
f. Use Equation (1) to calculate D_{new} and D_{parent} respectively, and choose the one which is closer to the target point as the parent point P_{parent} for the next expansion.
g. Repeat the above steps c–f until the target point P_{goal} is added to the random tree.

The random tree expansion diagram of the improved algorithm based on the PPD strategy is shown in Figure 2. In the process of generating new points, as the random tree becomes larger and there are more and more points in the random tree, traversing the random tree to search the parent point consumes a lot of computational time. In this article, the parent point is determined before the new point is generated, which can greatly save the path-planning time, and the larger the random tree gets, the more obvious this effect becomes.

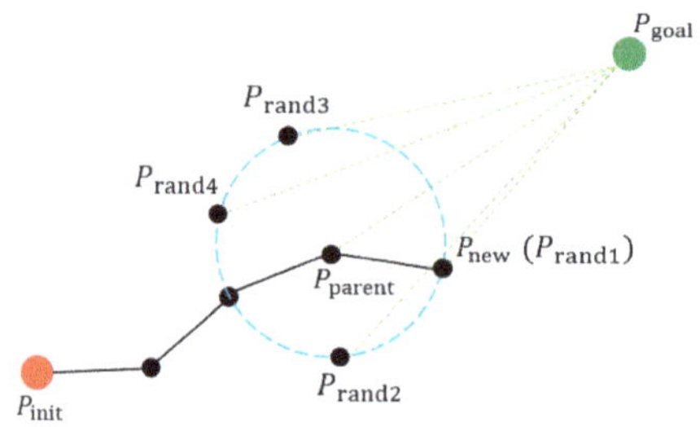

Figure 2. Random tree expansion diagram of the improved algorithm based on the PPD strategy. The red circle indicates the starting point, and the green circle indicates the target point. The black circle indicates the path point, the solid black line indicates the path. and the green dashed line indicates the distance from the point to the target point. The blue dashed line indicates the circumference of the circle with the parent point as the center and the step length ρ as the radius, which is the random sampling space. $P_{rand1} \sim P_{rand4}$ denote random sampling points, P_{parent} denotes parent point, and P_{new} denotes the new point.

3.3. The Strategy of Real-Time Optimization (RO)

The improved algorithm based on the PPD strategy can greatly reduce the path planning time, but the number of path points still needs to be optimized. Therefore, a strategy of real-time optimization was proposed in this paper. We calculated the distance D_{new} between the new point and the target point, and if D_{new} can satisfy both Equations (2) and (3), the generation of the new point is considered to beneficial to the growth of the random tree. The core idea of the real-time optimization strategy is to determine whether the new point has a positive impact on the subsequent growth of the random tree immediately after the new point is generated, as shown in the process of Algorithm 3. The real-time optimization of the RO policy can effectively reduce the number of redundant points of the random tree.

Algorithm 3. PPRO-RRT algorithm.

a. Initialize the random tree P_{init}.
b. Set the point P_{new} by the exploration process of PPD-RRT algorithm.
c. Calculate the distance from the new point, the parent point and its two nearest ancestor points to the target point according to Equation (2), respectively.
d. Determine whether the new point P_{new} is reserved according to Equation (3), and if it is satisfied, then it is reserved.
e. Repeat the above steps b–d until the target point P_{goal} is added to the random tree.

$$\begin{cases} D_{ancestor1,2} = \sqrt{(P_{ancestor1,2(x)} - P_{goal(x)})^2 + (P_{ancestor1,2(y)} - P_{goal(y)})^2} \\ D_{parent} = \sqrt{(P_{parent(x)} - P_{goal(x)})^2 + (P_{parent(y)} - P_{goal(y)})^2} \end{cases} \tag{2}$$

$$(D_{ancestor1} > D_{new}) \; or \; (D_{ancestor2} > D_{new}) \; or \; (D_{parent} > D_{new}) \tag{3}$$

The random tree expansion diagram of the improved algorithm based on the RO strategy is shown in Figure 3. Once a new point is generated, Equations (2) and (3) are used to decide the point to be left. If this new point does not contribute to the growth of the random tree, it is rejected, which avoids growing more redundant points from that point. Therefore, the number of redundant points is greatly reduced by a real-time judgment that prevents the generation of more redundant points.

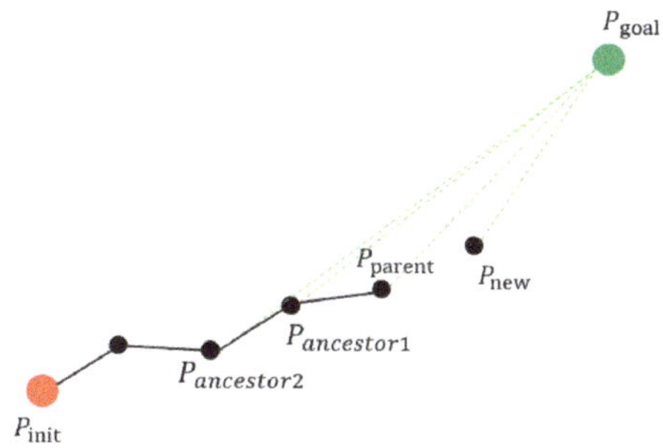

Figure 3. Random tree expansion diagram of the improved algorithm based on the RO strategy. The red circle indicates the starting point, and the green circle indicates the target point. The black circle indicates the path point, the solid black line indicates the path, and the green dashed line indicates the distance from the point to the target point. $P_{ancestor1}$~$P_{ancestor2}$ denote the ancestor points, P_{parent} denotes the parent point, and P_{new} denotes the new point.

4. Experiment and Analysis

The PPD strategy can effectively reduce the path planning time, and the RO strategy can reduce the number of redundant points. In order to further evaluate the performance of the PPD-RRT algorithm and the PPRO-RRT (parent point priority determination-real-time optimization-RRT) algorithm, we will conduct simulation experiments in 3-dimensional space for the above improved algorithm and the existing improved algorithm to verify the high-dimensional reliability and efficiency of the improved algorithm in this paper.

4.1. Simulation Experiments

The simulation experimental platform was configured with MATLAB 2019b, 64-bit Windows 10, processor Inter(R) Core (TM) i7-10700F CPU @ 2.90 GHz, and 16 GB of memory. The experimental simulation area was a 100 × 100 × 100 cube area with the starting position [5, 5, 5] and the target position [95, 95, 95]. The experimental environment was designated as a simple and complex one according to the number of obstacles [22]. Figures 4 and 5 show the simulation results of various algorithms in simple and complex environments.

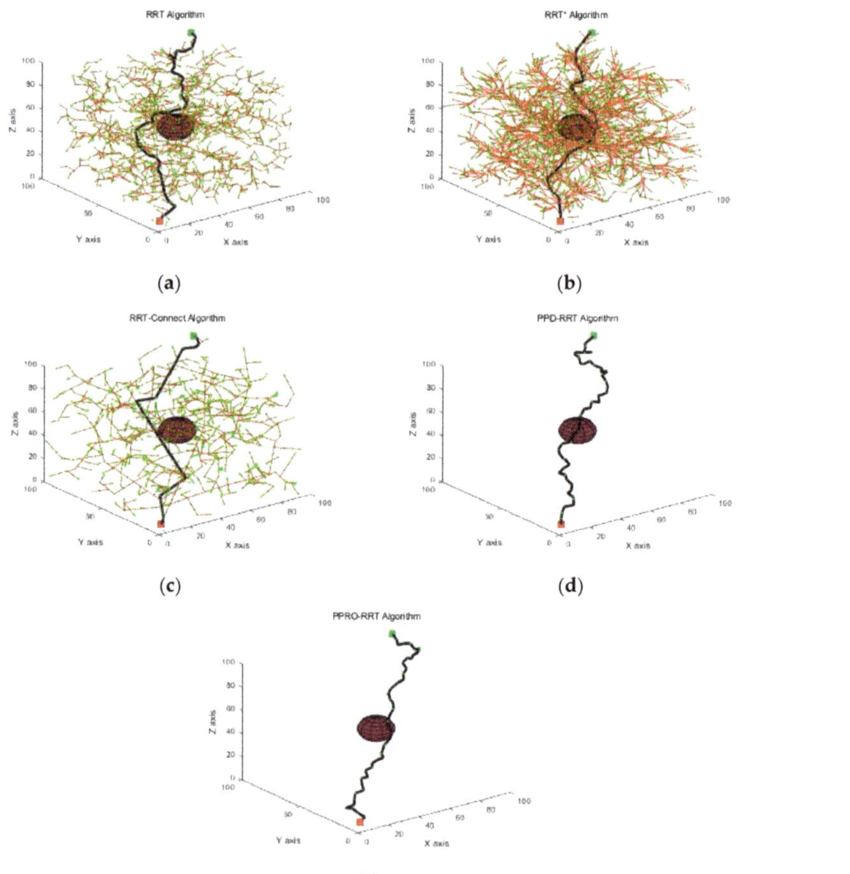

Figure 4. Paths planned by various algorithms in a simple environment. The red cube indicates the starting position and the green cube indicates the target position. The red sphere indicates the obstacles in the environment, the green dot indicates the points obtained by sampling during the path planning process, the red solid line indicates the branches of the random tree, and the black solid line indicates the feasible paths obtained. (**a**) RRT algorithm; (**b**) RRT* algorithm; (**c**) RRT-Connect algorithm; (**d**) PPD-RRT algorithm; (**e**) PPRO-RRT algorithm.

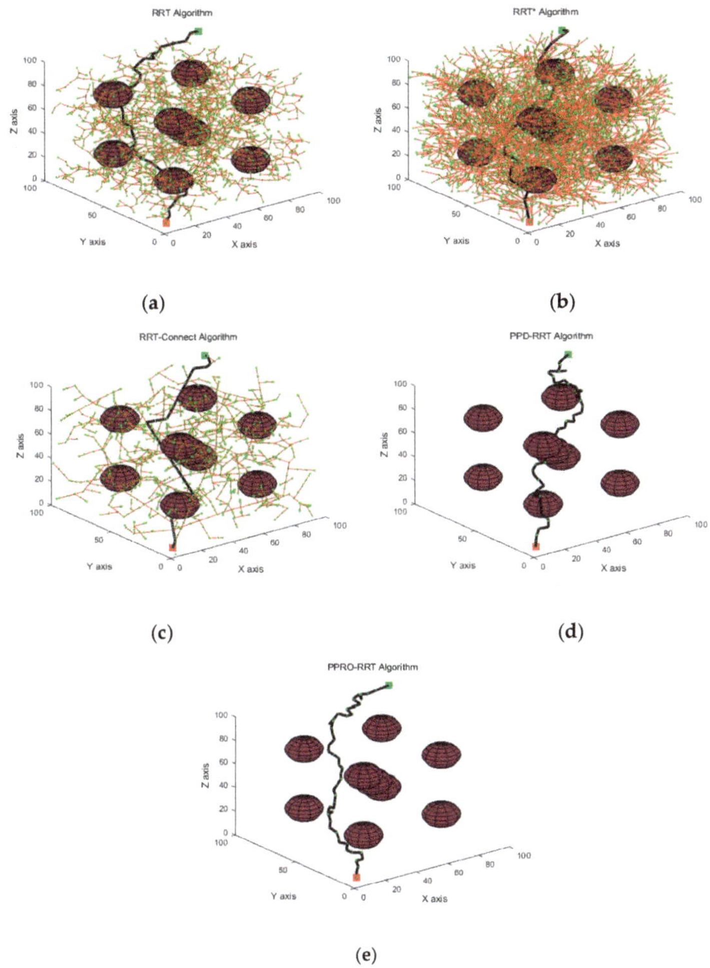

Figure 5. Paths planned by various algorithms in a complex environment. The red cube indicates the starting position and the green cube indicates the target position. The red sphere indicates the obstacles in the environment, the green dot indicates the points obtained by sampling during the path planning process, the red solid line indicates the branches of the random tree, and the black solid line indicates the feasible paths obtained. (**a**) RRT algorithm; (**b**) RRT* algorithm; (**c**) RRT-Connect algorithm; (**d**) PPD-RRT algorithm; (**e**) PPRO-RRT algorithm.

The simulation results in Figures 4 and 5 showed that there was a significant reduction in the number of sampling points because the PPRO-RRT algorithm incorporated the idea of PPD and RO.

4.2. Results and Analysis

In order to avoid the influence of the randomness of the path planning algorithm on the experimental results, each method was tested for 5000 iterations in the same environment, and the upper limit of the number of single iterations was set to 1000. The proposed algorithm was compared with other algorithms in terms of the success rate, the number of path points, the path planning time, and the path length. The results of each experiment are shown in Tables 1–4.

Table 1. The success rate of various algorithms in two kinds of experimental environments.

	RRT	RRT*	RRT-Connect	PPD-RRT	PPRO-RRT
Simple environments	5.12%	5.68%	49.70%	99.80%	100.00%
Complex environments	3.94%	4.62%	51.88%	99.48%	99.99%

Table 1 indicates the success rate of various path planning algorithms in two kinds of experimental environments. The experimental results showed that the success rates of the PPD-RRT and PPRO-RRT path planning algorithms were significantly higher than those of the other algorithms. The success rates of the PPD-RRT and PPRO-RRT algorithms were 99.80% and 100.00% in simple environments and 99.48% and 99.99% in complex environments, respectively. The success rate improvement was significant compared with other algorithms. The PPRO-RRT algorithm achieved success rate increases of 94.88% and 96.05% compared with the RRT algorithm; 94.32% and 95.37% compared with the RRT* algorithm; and 50.30% and 48.11% compared with the RRT-Connect algorithm. With a set number of iterations of 1000, the PPRO-RRT algorithm greatly enhances the success rate of the path planning algorithm.

Table 2. The number of random tree points of various algorithms in two kinds of experimental environments.

	RRT	RRT*	RRT-Connect	PPD-RRT	PPRO-RRT
Simple environments	831	829	446	153	56
Complex environments	834	823	447	152	55

Table 2 indicates the number of random tree points obtained by various path planning algorithms. For each of the algorithms, the number of random tree points hardly varied with the complexity of the environment. Compared with the RRT algorithm, the number of random tree points of the PPD-RRT algorithm was reduced by approximately 81.59% and 81.76% in two kinds of experimental environments, and after adding the RO idea, the number of points was reduced by about 93.25% and 93.32%. Compared to the RRT* and RRT-Connect algorithms, the number of random tree points of the PPRO-RRT algorithm was reduced by 93.23% and 87.42% in the simple environments and 93.23% and 87.55% in the complex environments, respectively.

Table 3. The path planning time(s) of various algorithms in two kinds of experimental environments.

	RRT	RRT*	RRT-Connect	PPD-RRT	PPRO-RRT
Simple environments	0.0240	0.0704	0.0318	0.0024	0.0017
Complex environments	0.0388	0.1318	0.0461	0.0044	0.0033

Table 3 indicates the path planning time of different algorithms in two kinds of experimental environments. For each of the proposed algorithms, the path planning time was approximately reduced by one order of magnitude. In the simple experimental environment, the path planning time of the PPD-RRT and PPRO-RRT algorithms was 90% and 92.92% shorter than that of the RRT algorithm, respectively. In the complex experimental environment, the path planning time of the PPD-RRT and PPRO-RRT algorithms was 88.66% and 91.49% shorter than that consumed by the RRT algorithm, respectively. As the RRT* algorithm and the RRT-Connect algorithm both require a longer path planning time than the RRT algorithm, the PPD-RRT algorithm and the PPRO-RRT algorithm were far superior to the RRT* algorithm and the RRT-Connect algorithm in terms of path planning time. Compared with the RRT* and RRT-Connect algorithms, the path planning time of the PPRO-RRT algorithm was reduced by 97.59% and 94.65% in the simple environments and 97.05% and 92.84% in the complex environments, respectively.

Table 4. The path length [23] of various algorithms in two kinds of experimental environments.

	RRT	RRT*	RRT-Connect	PPD-RRT	PPRO-RRT
Simple environments	221.42	190.15	231.34	241.95	203.96
Complex environments	225.00	195.22	240.49	237.67	202.48

Table 4 indicates the path length of the various algorithms in the two kinds of experimental environments. Compared with the RRT algorithm, the path length of the PPD-RRT algorithm became longer in two kinds of experimental environments due to the restricted sampling area. After adding the RO idea, the path length of the PPRO-RRT algorithm was reduced by 7.89% and 10.01% in simple environments and complex environments. While the path length of the PPRO-RRT algorithm was longer than that of the RRT* algorithm, the difference was not significant and was far superior to the RRT* algorithm in other aspects. Compared with the RRT-Connect algorithm, the path length of the PPRO-RRT algorithm was reduced by 11.84% and 15.81% in simple environments and complex environments.

From the above analysis, the PPRO-RRT algorithm can converge at a small number of iterations, and the success rate reaches close to 100% with a set upper limit of 1000 iterations. The PPRO-RRT algorithm saves the process of searching parent points, which greatly saves the path planning time; the real-time judgment of whether new points should be retained reduces the generation of more redundant points. We can conclude that the PPRO-RRT algorithm has significant advantages over the RRT algorithm in terms of success rate, number of points, planning time, and path length.

4.3. Experiments on the Manipulator

To demonstrate the feasibility and effectiveness of the algorithm proposed in this paper, an experimental platform for path planning was built in a laboratory environment using the IRB1410 robot arm. A random position (1.289 m, 0.013 m, 0.873 m) was selected as the position of an obstacle in the manipulator workspace. In this scene, the path point of the manipulator end-effector is obtained by the PPRO-RRT algorithm, and then the manipulator is controlled by the program procedure from the starting point (1.241 m, 0.156 m, 0.703 m) around the obstacle to the target point (1.166 m, −0.087 m, 0.921 m), and the sequence of the path is shown in Figure 6.

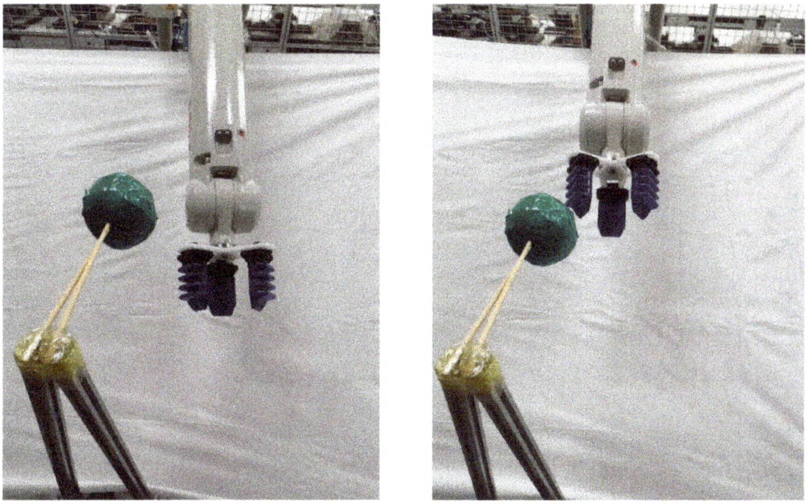

(a) (b)

Figure 6. *Cont.*

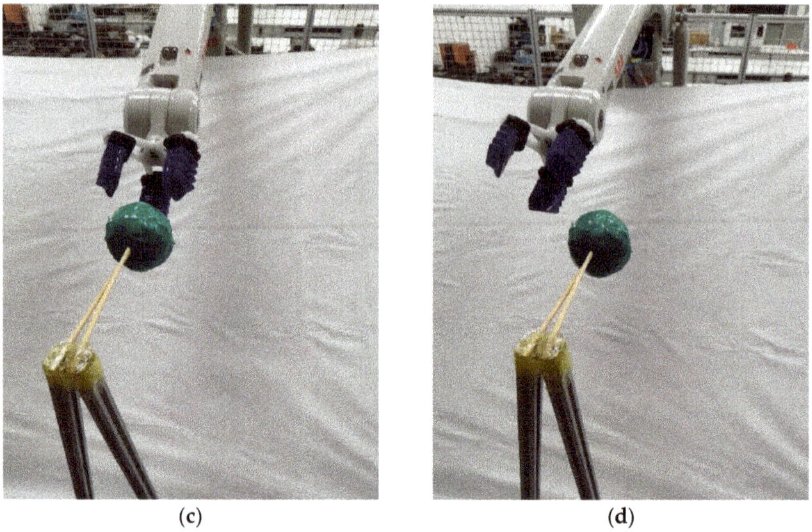

Figure 6. Path sequence with the PPRO-RRT algorithm in obstacle avoidance. (**a**–**d**) represent different states of the manipulator and obstacle at each moment.

In this project, the individual joint angles of the robot arm change, as shown in Figure 7, from which we can see that the individual joint angles change smoothly.

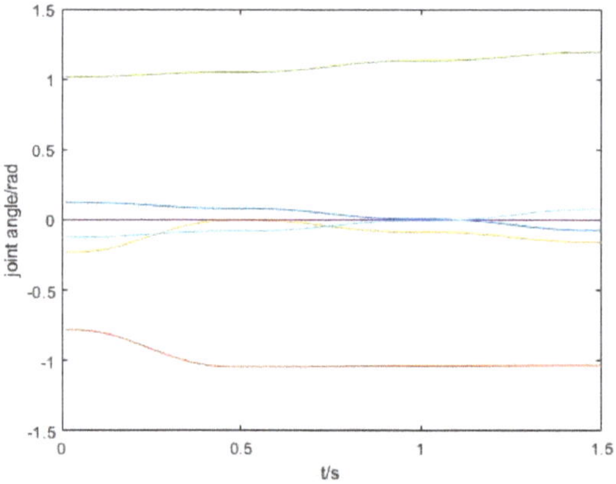

Figure 7. Variation of each joint angle during the process of obstacle avoidance.

5. Discussion

In this paper, simple and complex environments were chosen to verify the feasibility of the proposed algorithm. The experimental results showed that the PPRO-RRT algorithm had better performance in both environments than the RRT algorithm, the RRT* algorithm, and the RRT-Connect algorithm. However, in the field conditions, the position of the obstacles may change or there may be narrow gaps between the obstacles. The algorithm in this paper does not fit in the above scene, which creates some limitations on the path planning of the manipulator. Therefore, the following research will focus on enhancing the applicability of the algorithm to a wider range of scene.

In addition, in the real environment, the manipulator may collide with the obstacle, which is because the actual size of the manipulator needs to be considered during the experiment in the real scene, which would make the collision detection algorithm more complicated and increase the running time of the algorithm. In order to simplify the collision detection algorithm, the links of the manipulator are abstracted as lines and the obstacle is inflated. The actual size of the manipulator is added to the obstacle so that the complex process of collision detection is transformed into a problem of the position relationship between a line and a sphere. Experimental results showed that with the proposed algorithm in this paper, the manipulator could safely reach the target point from the starting point.

6. Conclusions

In order to simplify the complex process of determining the parent point that needs to traverse the random tree and reduce the number of redundant points in the random tree, a PPRO-RRT algorithm, combined the PPD strategy with the RO strategy, was proposed in this paper. The following conclusions were drawn through experiments and comparative analysis: (1) the PPD strategy significantly speeds up path planning, and the RO strategy reduces the number of redundant points; (2) the PPRO-RRT algorithm outperforms the other improved RRT algorithms for both simple and complex environments; (3) the experimental results by IRB1410 shows that the manipulator can safely avoid obstacles using the PPRO-RRT algorithm. However, compared with the RRT* algorithm, the algorithm proposed in this paper may produce longer paths and would be modified in the future in terms of shorter path lengths. The improved algorithm in this paper is relatively unsuitable for the scene with narrow channels. In terms of practical applications, the improved algorithm in this paper can be applied to the path planning problems of intelligent vehicles, manipulators, UAVs, etc. to speed up their path planning.

Author Contributions: Conceptualization, Z.Z. and Y.T.; Methodology, L.T., C.Z., Z.Z. and Y.T.; Software, L.T.; Validation, L.T., Y.T. and Z.Z.; Formal analysis, L.T., Y.Z. and Z.W.; Investigation, L.T., Y.T. and Z.Z.; Resources, L.T. and Y.Z.; Data curation, L.T.; Writing—original draft preparation, L.T.; Writing—review and editing, L.T., Y.Z. and Z.W.; Visualization, L.T., Y.Q. and Z.Z.; Supervision, C.Z., Y.T., Y.Z. and Y.Q.; Project administration, C.Z., Y.Q. and Z.W.; Funding acquisition, C.Z. All authors have read and agreed to the published version of the manuscript.

Funding: This research was funded by the National Natural Science Foundation of China (grant number 31971668) and the Beijing College Students' Innovation and Entrepreneurship Training Program (grant s20201002114).

Conflicts of Interest: The authors declare no conflict of interest.

References

1. Kavraki, L.E.; Švestka, P.; Latombe, J.C.; Overmars, M.H. Probabilistic roadmaps for path planning in high-dimensional configuration spaces. *IEEE Trans. Robot. Autom.* **1996**, *12*, 566–580. [CrossRef]
2. Premachandra, C.; Murakami, M.; Gohara, R.; Ninomiya, T.; Kato, K. Improving landmark detection accuracy for self-localization through baseboard recognition. *Int. J. Mach. Learn. Cybern.* **2017**, *8*, 1815–1826. [CrossRef]
3. Chen, W.; Xu, J.; Zhao, X.; Liu, Y.; Yang, J. Separated Sonar Localization System for Indoor Robot Navigation. *IEEE Trans. Ind. Electron.* **2020**, *68*, 6042–6052. [CrossRef]
4. Premachandra, C.; Otsuka, M.; Gohara, R.; Ninomiya, T.; Kato, K. A study on development of a hybrid aerial terrestrial robot system for avoiding ground obstacles by flight. *IEEE/CAA J. Autom. Sin.* **2018**, *6*, 327–336. [CrossRef]
5. Luo, Q.; Wang, H.; Zheng, Y.; He, J. Research on path planning of mobile robot based on improved ant colony algorithm. *Neural Comput. Appl.* **2020**, *32*, 1555–1566. [CrossRef]
6. Dolgov, D.; Thrun, S.; Montemerlo, M.; Diebel, J. Path planning for autonomous vehicles in unknown semi-structured environments. *Int. J. Rob. Res.* **2010**, *29*, 485–501. [CrossRef]
7. Karaman, S.; Frazzoli, E. Sampling-based algorithms for optimal motion planning. *Int. J. Rob. Res.* **2011**, *30*, 846–894. [CrossRef]
8. LaValle, S.M.; Kuffner, J.J. Randomized kinodynamic planning. *Int. J. Rob. Res.* **2001**, *20*, 378–400. [CrossRef]
9. Lau, B.; Sprunk, C.; Burgard, W. Kinodynamic motion planning for mobile robots using splines. In Proceedings of the 2009 IEEE/RSJ International Conference on Intelligent Robots and Systems, St. Louis, MO, USA, 10–15 October 2009; pp. 2427–2433. [CrossRef]

10. Islam, F.; Nasir, J.; Malik, U.; Ayaz, Y.; Hasan, O. RRT*-Smart: Rapid convergence implementation of RRT* towards optimal solution. In Proceedings of the 2012 IEEE International Conference on Mechatronics and Automation, Chengdu, China, 5–8 August 2012; pp. 1651–1656. [CrossRef]
11. Frazzoli, E. Incremental Random Sampling Algorithms for Optimal Motion Planning. *Proc. Robot. Sci.* **2010**, 1–8. Available online: http://sertac.scripts.mit.edu/web/wp-content/papercite-data/pdf/C21.pdf (accessed on 3 May 2010).
12. Du, M.; Chen, J.; Zhao, P.; Liang, H.; Xin, Y.; Mei, T. An improved RRT-based motion planner for autonomous vehicle in cluttered environments. In Proceedings of the 2014 IEEE International Conference on Robotics and Automation (ICRA), Hong Kong, China, 31 May–7 June 2014. [CrossRef]
13. Barfoot, T. Informed RRT*: Optimal Sampling-based Path Planning Focused via Direct Sampling of an Admissible Ellipsoidal Heuristic. In Proceedings of the 2014 IEEE/RSJ International Conference on Intelligent Robots and Systems, Chicago, IL, USA, 14–18 September 2014. Available online: https://www.google.com/%0Apapers3://publication/uuid/FD516565-4AD0-4704-809E-5EBDBC4886C9 (accessed on 13 October 2021).
14. Kuffner, J.J.; La Valle, S.M. RRT-connect: An efficient approach to single-query path planning. In Proceedings of the Proceedings 2000 ICRA. Millennium Conference. IEEE International Conference on Robotics and Automation. Symposia Proceedings (Cat. No. 00CH37065), San Francisco, CA, USA, 24–28 April 2000; pp. 995–1001. [CrossRef]
15. Adiyatov, O.; Varol, H.A. A novel RRT∗-based algorithm for motion planning in Dynamic environments. In Proceedings of the 2017 IEEE International Conference on Mechatronics and Automation (ICMA), Takamatsu, Japan, 6–9 August 2017; pp. 1416–1421. [CrossRef]
16. Qureshi, A.H.; Ayaz, Y. Intelligent bidirectional rapidly-exploring random trees for optimal motion planning in complex cluttered environments. *Rob. Auton. Syst.* **2015**, *68*, 1–11. [CrossRef]
17. Ying, Y.; Li, Z.; Ruihong, G.; Yisa, H.; Haiyan, T.; Junxi, M. Path planning of mobile robot based on Improved RRT Algorithm. In Proceedings of the 2019 Chinese Automation Congress (CAC), Hangzhou, China, 22–24 November 2019; Volume 39, pp. 4741–4746. [CrossRef]
18. Jordan, M.; Perez, A. Optimal Bidirectional Rapidly-Exploring Random Trees; Computer Science and Artificial Intelligence Laboratory Technical Report. 2013. Available online: https://dspace.mit.edu/bitstream/handle/1721.1/79884/MIT-CSAIL-TR-2013-021.pdf (accessed on 15 August 2013).
19. Qureshi, A.H.; Ayaz, Y. Potential functions based sampling heuristic for optimal path planning. *Auton. Robot.* **2016**, *40*, 1079–1093. [CrossRef]
20. Mashayekhi, R.; Idris, M.Y.I.; Anisi, M.H.; Ahmedy, I.; Ali, I. Informed RRT∗-Connect: An Asymptotically Optimal Single-Query Path Planning Method. *IEEE Access* **2020**, *8*, 19842–19852. [CrossRef]
21. Chen, Q.; Jiang, H.; Zheng, Y. A review of fast extended random tree algorithms for robot path planning. *Comput. Eng. Appl.* **2019**, *55*, 10–17.
22. Dai, J.; Li, Z.; Li, Y.; Zhao, J. Robot path planning based on improved Informed-RRT* algorithm. *J. Henan Univ. Technol. (Nat. Sci. Ed.)* **2021**, 1–11. Available online: http://kns.cnki.net/kcms/detail/41.1384.n.20210608.1446.004.html (accessed on 28 June 2021).
23. Mohammed, H.; Romdhane, L.; Jaradat, M.A. RRT*N: An efficient approach to path planning in 3D for Static and Dynamic Environments. *Adv. Robot.* **2020**, *35*, 168–180. [CrossRef]

sensors

MDPI

Article

Multi-Ship Control and Collision Avoidance Using MPC and RBF-Based Trajectory Predictions

Myron Papadimitrakis [1], Marios Stogiannos [1], Haralambos Sarimveis [2] and Alex Alexandridis [1,*]

[1] Department of Electrical and Electronic Engineering, University of West Attica, 12241 Aigaleo, Greece; m.papadimitrakis@uniwa.gr (M.P.); mstogia@uniwa.gr (M.S.)
[2] School of Chemical Engineering, National Technical University of Athens, 15780 Zografou, Greece; hsarimv@central.ntua.gr
* Correspondence: alexx@uniwa.gr; Tel.: +30-210-5381571

Abstract: The field of automatic collision avoidance for surface vessels has been an active field of research in recent years, aiming for the decision support of officers in conventional vessels, or for the creation of autonomous vessel controllers. In this paper, the multi-ship control problem is addressed using a model predictive controller (MPC) that makes use of obstacle ship trajectory prediction models built on the RBF framework and is trained on real AIS data sourced from an open-source database. The usage of such sophisticated trajectory prediction models enables the controller to correctly infer the existence of a collision risk and apply evasive control actions in a timely manner, thus accounting for the slow dynamics of a large vessel, such as container ships, and enhancing the cooperation between controlled vessels. The proposed method is evaluated on a real-life case from the Miami port area, and its generated trajectories are assessed in terms of safety, economy, and COLREG compliance by comparison with an identical MPC controller utilizing straight-line predictions for the obstacle vessel.

Keywords: autonomous vessels; collision avoidance; model predictive control; radial basis function networks; trajectory optimization

Citation: Papadimitrakis, M.; Stogiannos, M.; Sarimveis, H.; Alexandridis, A. Multi-Ship Control and Collision Avoidance Using MPC and RBF-Based Trajectory Predictions. *Sensors* **2021**, *21*, 6959. https://doi.org/10.3390/s21216959

Academic Editors: Baochang Zhang, Reza Ghabcheloo and Antonio M. Pascoal

Received: 3 August 2021
Accepted: 16 October 2021
Published: 20 October 2021

1. Introduction

In the last two decades, research on automatic collision avoidance and optimal path planning for surface vessels has intensified, driven by the ever-growing density of maritime traffic in narrow waterways, such as gulfs, ports, and canals [1]. Motivated by the design of autonomous surface vehicles (ASV) controllers, but also aiming for the decision support of officers on watch in conventional vessels [2], control and optimization tools that ensure the safety and the cost effectiveness of navigational actions are being intensively developed. These tools are perceptive of the surrounding environment through arrays of sensors, radars, and other positioning and communication aids. In this context, the automatic identification system (AIS) encompasses most of the aforementioned technologies in order to gather positioning and other vessel data. The already vast AIS comprises an ever-expanding worldwide maritime trajectory dataset, which is made available by vessels, port authorities, and other platforms in charge of efficient and safe maritime path planning. Given the fact that the majority of vessel accidents are related to erroneous handling rather than equipment failure or environmental conditions [3], these tools aim to phase out the human officers on watch as vessel controllers, or at least augment their navigational decision-making using optimization- and prediction-based methods.

The formulation of the trajectory optimization problem used in collision avoidance controllers must take multiple aspects of vessel navigation into account, while being perceptive of their surrounding environment in real time. The generation of control actions that will result in a trajectory remaining sufficiently clear from any stationary or moving objects is not the sole objective; an efficient controller should also ensure the economy of the control actions, as well as the adherence to the collision avoidance rules, commonly known as the COLREGs [4]. Multiple collision avoidance controllers have been proposed that fulfil the aforementioned specifications; in [5], a hierarchical multiobjective optimization problem is formulated, which generates an intermediate waypoint for the controlled vessel while accounting for the good seamanship rules. In [6], a fuzzy-Bayesian collision avoidance controller is formulated capable of addressing multiple obstacle vessels at once. In [7], optimal trajectories for the collision avoidance problem are generated using a B-Spline-based search algorithm. Lastly, in [8], a collision avoidance controller utilizes a probabilistic method in order to infer the one-step-ahead position of obstacle vessels, while also accounting for non-COLREG-compliant obstacle vessels.

In general, it has been observed that controllers that are not model-based can have trouble incorporating crucial aspects of the trajectory optimization problem, thus compromising practicality. Without a working model of the controlled vessel, its maneuvering capabilities cannot be easily included in the formulation, and neither can the effect of environmental conditions be quantified [9,10]. For these reasons, model predictive control (MPC) emerges as an effective control method for the problem at hand because it utilizes a model of the plant in order to compute an optimal control trajectory based on the predicted trajectory of other ships in the vicinity. As a framework, MPC can account for the uncertainties of both the utilized model of the plant and the trajectory prediction models of other ships, while also incorporating all possible control objectives (such as navigational risk, course smoothness, or deviation from the original path) in a single cost function. Some collision avoidance controllers based on MPC have been proposed in the literature; a robust MPC controller utilizing straight-line obstacle vessel trajectory predictions is proposed in [9], capable of COLREG compliance and handling of multiple obstacles. In [11], motion planning for an autonomous vessel using a sampling-based MPC method takes place. In [12], an MPC controller for the collision avoidance task is built by approximating the behavior of an LQR controller, thus ensuring asymptotic stability of the system. In [13], a neural network used to approximate the MPC response for the generation of COLREG compliant trajectories for multi ship encounters is presented. In addition, MPC has been integrated in distributed control frameworks of multi-ship schemes; for example, a distributed MPC scheme has been employed for a multi-vessel formation controller with collision avoidance capabilities [14], or for the robust distributed control of multiple vessels operating for the inter-terminal transport of containers [15].

It becomes apparent that for the scope of the collision avoidance task, information about the future trajectories of other ships plays a central role. Prevalent in non-data driven methodologies already used for the vessel trajectory prediction (VTP) problem is the first principles-based modeling technique [16], carrying a number of significant shortcomings, such as their inherent complexity, which has a greater negative impact due to the fact that the model is usually employed multiple times within the duration of each MPC sample. In order to simplify the solution of the employed kinematic differential equations and facilitate the real-time prediction of future states, these types of models are usually created using several assumptions that try to approximate real-world conditions, but also make the final model far less accurate. Therefore, one should employ a more sophisticated, data-driven approach for the creation of effective trajectory prediction models that are included in MPC controllers. Machine learning has answered the call of producing highly accurate models, which may be easily integrated in predictive frameworks through the use of black-box modeling, and more specifically, artificial neural network (NN) approaches [17]. NNs employ different architectures in order to remap the original non-linear problem to a higher-dimensional input space and approximate its dynamics utilizing standard functions. In this context, various NN techniques have been successfully utilized in control frameworks solving the vessel trajectory prediction problem.

Feedforward NN architectures, most commonly represented by the multilayer perceptrons (MLPs), have been employed to solve the vessel trajectory prediction problem as in [18,19], where MLP NNs are trained using the well-established backpropagation algorithm outperforming rival methodologies, i.e., linear models and Kalman filters. In [19], a real AIS dataset gathered from the confined space of a river waterway is used to approximate the vessel dynamics in such environments. Backpropagation has been the baseline of more efficient training methods as in [20], where different computational intelligence approaches like differential evolution, genetic algorithms, and swarm-based techniques are used to modify the original backpropagation algorithm in order to create more accurate feedforward NN models. Other NN architectures, like support vector machines (SVMs), have been employed in conjunction with computational intelligence optimization techniques, i.e., the particle swarm optimization (PSO) algorithm, on AIS datasets to solve the vessel trajectory prediction problem [21]. In most cases, the inherent abilities of NN architectures that can meet the standard of high accuracy are limited to a one-step ahead prediction horizon, in the sense that multi-step ahead predictions would require an approximation of unknown future states to be made and present an error enlarged through propagation to the end of the prediction horizon. Such an error would become critically high after a small number of steps, rendering the control framework useless.

To overcome this problem, long-term trajectory prediction approaches have been devised with the inclusion of memory features, such as the recurrent neural networks (RNNs), with their most notable representative, i.e., the long short-term memory (LSTM) NNs already used in the context of the vessel trajectory prediction problem [22–25]. Besides trajectory modeling and prediction in open waters, advances have also been made in crowded port waters as in [26], where another modification of the RNNs, namely the bidirectional gated recurrent unit, is used to address the vessel trajectory prediction problem, outperforming standard NN methods in such scenarios. Gated recurrent units are promising candidates for predicting the collective behavior of vessel fleets [27].

Radial basis function (RBF) networks form a unique NN architecture belonging to the general feedforward NN category. RBFs differ from other NN architectures, having simpler structures, employing faster training algorithms, and usually producing more accurate models than MLPs. Within the context of vessel trajectory prediction, RBFs have been integrated in control frameworks by approximating unknown vessel parameters [28–30]. Recently, RBFs have been applied on real AIS data in order to produce highly accurate models for one-step and multi-step ahead predictions [31], showing their potential in being integrated to receding horizon control methodologies.

Remarkably, in the collision avoidance research literature, there are no instances where the multi-step-ahead trajectory prediction of moving obstacles is addressed in such a systematic manner; usually these trajectories are either known a priori, or there are no obstacle ships present whatsoever. An exception occurs in [9], where straight line trajectory predictions are employed, based on estimates of current course and speed for the moving obstacle. In this work, a multi-ship MPC controller utilizing RBF obstacle trajectory prediction models trained using real AIS data is presented for the collision avoidance task. The main contributions of this work are as follows: first, we introduce a novel MPC scheme for collision avoidance control, where nonlinear data-driven models are used to predict the trajectories of obstacle ships; to the authors' best knowledge, this is the first such instance in the literature. The previous state-of-the-art approach of using straight-line obstacle trajectory predictions may have yielded satisfactory approximation results in open sea case studies, where ships are expected to travel in a straight line, but is of limited practical use for the cases of narrow gulfs, ports, or canals, where ships are expected to maneuver while navigating through, thus resulting to nonlinear trajectories. Secondly, the aforementioned nonlinear models are trained using an AIS data-driven methodology, which, once again, constitutes an original approach in the context of vessel collision avoidance. Using real trajectories as training data increases the practicality of the proposed scheme, since the resulting trajectory prediction models capture the dynamics of real vessels. The proposed method is tested in a collision avoidance case study at the Miami port area, and its performance is illustrated by the comparison with an MPC controller employing straight-line obstacle prediction models, which corresponds to the current state-of-the-art approach [9].

The paper is structured as follows. In Section 2, the AIS-data-driven methodology for the creation of the RBF trajectory prediction models is presented. In Section 3, some preliminaries on maritime collision avoidance and optimal trajectory generation are described, and later, the proposed method is presented. In Section 4, the case study based on the port of Miami is outlined, and the simulation results are discussed in depth. Lastly, in Section 5, concluding remarks are made.

2. AIS-Data-Based Trajectory Prediction Models

2.1. Radial Basis Function Neural Networks

RBF NNs have been successfully employed to approximate nonlinear system dynamics in order to predict future system states in numerous diverse applications [32,33]. Their success can be mainly attributed to their structure, which is simpler when compared to other NN architectures, as they comprise a single hidden layer, attached linearly to the network output. This property not only allows for using very fast training algorithms, but also makes RBF NNs suitable for integration in MPC schemes, as (a) it facilitates the controller design [34], and (b) helps to solve the MPC online optimization problem in shorter computational times [35], thus rendering such schemes applicable to systems with fast dynamics [36]. Another property that makes RBF NNs a popular modeling method in predictive control schemes is related to their increased approximation capabilities [37,38]. Indeed, notwithstanding their simple structure, RBF NNs have proven to be more accurate in modeling various nonlinear systems when compared to other machine learning methods, including MLPs, support vector machines (SVMs), random forests, etc. [39]. Especially within the context of the vessel modeling and control problems, RBFs have already shown great potential in modeling unknown vessel parameters [28,29], in order to create models capable to be integrated in trajectory tracking control algorithms. Furthermore, in a recent publication [31], it has been shown that RBF NNs trained with the fuzzy means (FM) algorithm outperform other data-driven techniques such as MLPs when modeling vessel trajectories based on AIS data.

Training an RBF NN is a process consisting mainly of two phases. The first phase is performed by applying a clustering technique on the training dataset in order to identify the centerpoint and optimized parameters of a number of radially symmetric basis functions called nodes. The incorporation of radial basis functions (e.g., Gaussian, quadratic, etc.) is the first main difference between RBFs and other feedforward NNs. The linear combination of these nodes produces the output prediction of the network. Finding the node weights is the goal of the second phase, a problem trivially solved by least squares.

A typical RBF network can be seen in Figure 1. The structure comprises three distinct layers, the first of which is the input layer and has the sole purpose of distributing the N inputs to the L nodes of the hidden layer. The second point differentiating RBFs from other architectures is the existence of only one hidden layer of N-dimensional nodes. In order to produce a prediction $\hat{y}(k)$ given an input datapoint $\mathbf{x}(k)$, at first the Euclidean norm is used to calculate the activity $\mu_l(\mathbf{x}(k))$ for every node l by using the difference between the k-th input vector $\mathbf{x}(k)$ and the l-th node center $\hat{\mathbf{u}}_l$, such that

$$\mu_l(\mathbf{x}(k)) = \|\mathbf{x}(k) - \hat{\mathbf{u}}_l\| = \sqrt{\sum_{i=1}^{N}\left(\mathbf{x}_i(k) - \hat{\mathbf{u}}_{i,l}\right)^2}, k = 1, \ldots, K \quad (1)$$

The activity acts as input to the free parameters of each node according to the chosen RBF. The hidden node response vector $\mathbf{z}(k)$ for the k-th datapoint is given by

$$\mathbf{z}(k) = [g(\mu_1(\mathbf{x}(k))), g(\mu_2(\mathbf{x}(k))), \ldots, g(\mu_L(\mathbf{x}(k)))] \quad (2)$$

where g corresponds to the chosen activation function. Note that in this work, a thin plate spline function is employed

$$g(\mu_l(\mathbf{x}(k))) = \mu_l^2(\mathbf{x}(k)) \cdot \log \mu_l(\mathbf{x}(k)) \quad (3)$$

due to the fact that it is an established choice as an RBF kernel producing models of high accuracy [40], but also because there are no tunable parameters other than the actual input to the function. Such parameters would require optimization techniques to be included in the training process, e.g., employment of the Gaussian function would need kernel width optimization, which is usually performed by cpu-intensive iterative algorithms or suboptimal trial-and-error techniques.

Finally, the network's prediction is calculated by linearly combining the hidden note responses such that

$$\hat{y}(k) = \mathbf{z}(k) \cdot \mathbf{w} \quad (4)$$

where $\mathbf{w}$ is a vector containing the node weights.

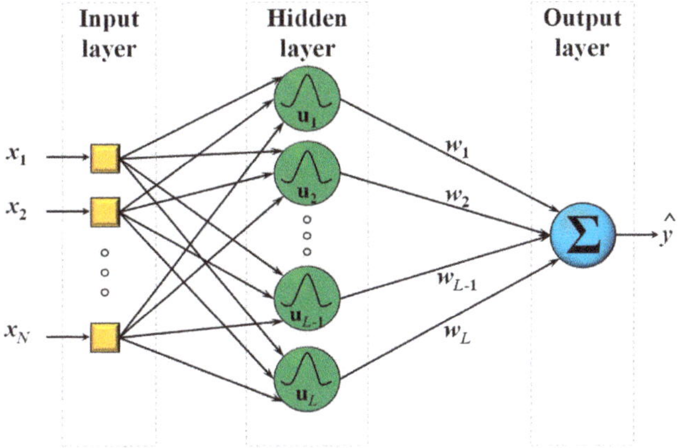

Figure 1. A typical radial basis function network structure using Gaussian RBFs.

2.2. The Fuzzy Means (Fm) Training Algorithm

For a given training dataset where $\mathbf{Y}$ denotes the real outputs, and after the hidden node responses $\mathbf{Z}$ are formulated, the weight vector $\mathbf{w}$ can be trivially calculated by least squares in matrix form

$$\mathbf{w}^T = \mathbf{Y}^T \cdot \mathbf{Z} \cdot \left(\mathbf{Z}^T \cdot \mathbf{Z}\right)^{-1} \quad (5)$$

thus completing the second training phase in one easy step.

The first phase of training requires a clustering algorithm to be applied to the training dataset, in which case the fuzzy means (FM) algorithm is a great candidate for this task [39]. Following the notation of the previous example, let us suppose a system with N normalized input variables x_i. At first, each input variable domain must be partitioned into s 1-D fuzzy sets (FS). Each fuzzy subspace $\mathbf{A}^l$, where $l = 1, 2, \ldots, s^N$, is formed by the selected s^N fuzzy sets according to the respective input variable. This process creates a N-dimensional grid, where all intersection points, also called nodes, are candidates to become RBF centerpoints. The FM algorithm undertakes the task of selecting the appropriate candidate nodes that will be assigned as RBF centers. To perform the selection procedure, a membership function

$$\mu_{\mathbf{A}^l}(\mathbf{x}(k)) = \begin{cases} 1 - d_r^l(\mathbf{x}(k)), & if\ d_r^l(\mathbf{x}(k)) \leq 1 \\ 0, & if\ otherwise \end{cases} \quad (6)$$

determines whether an input vector lies within the domain of an RBF centered around a candidate node. In the simple case where all input variable spaces are equally partitioned, the following distance metric can be used to assign N-dimensional spherical domains to each candidate node

$$d_r^l(\mathbf{x}(k)) = \sqrt{\sum_{i=1}^{N} \left(a_{i,ji}^l - x_i(k)\right)^2 / \sqrt{N}\delta a} \tag{7}$$

where $\mathbf{x}(k)$ is the k-th input vector, $a_{i,ji}^l$ is the centerpoint of fuzzy subspace $\mathbf{A}^l$, and δa is the sphere radius, which is the same for each input. The FM algorithm uses a fast non-iterative procedure to find a subset of the subspaces, so that the final RBF NN's hidden layer comprises only the fuzzy sets, which sufficiently cover all training datapoints, in the sense that each datapoint is included in at least one fuzzy set. For an in-depth description of the FM algorithm, the reader may refer to [39].

2.3. Data Preprocessing

Best modeling practices mandate that a training dataset should be error- and noise-free, a case that is far from truth when using data from AIS transceivers [41–45]. AIS data are irregularly sampled and contain heavy noise, missing data, and erroneous values. Thus, before employing any modeling technique, rigorous preprocessing is in order.

The Marine Cadastre service (www.marinecadastre.gov, accessed on 25 July 2021) has been the source of all data used in this work. MarineCadastre.gov is a is a service that gathers and publicly provides AIS data to marine planning initiatives. In this work, data from all days between 1 July 2019 and 30 June 2020 have been included and filtered to keep vessels sailing an area around the port of Miami covered by the geolocation rectangle defined by the latitudes of 25.720° through 25.840° and the longitudes of −80.145° through −80.042°. To conform to the initial assumption of similar size and similar dynamics, we allowed only cargo ships sailing on engine power into the dataset, further filtering the dataset to yield a total of 180 vessels.

To address the problems of sample irregularity, noise, and erroneous values, the dataset was resampled to 120 s, which was deemed enough to capture the high inertia dynamics of large cargo ships. The interpolation technique applied on the data to perform the resampling was the Akima piecewise cubic interpolation [46], which is quite effective on geolocation data, performing a mild denoising as well. A heuristic that rejects very far-off outlier values due to GPS errors was also applied. The trajectories were split in data samples, with each one containing ten consecutive vessel positions. Note that each trajectory's starting point should be the last point of the previous one resulting in an overlap of one point, but this final position will be used as the model's output, so no actual overlapping exists within the input data. The resampling and splitting process yielded a total of about 14 k samples from 3.1 k resampled trajectories of the initial 180 vessels. Algorithm 1 depicts the step-by-step procedure of preprocessing.

Algorithm 1 Preprocessing Stage

1: Process each entry in the common dataset so that it contains only the following: Vessel ID, timestamp, latitude, and longitude. Reject all other information.
2: Sort dataset by vessel ID and sort each vessel data by date.
3: Apply resampling and outlier filtering on the data of each vessel to achieve a resampling of 120 s.
4: Split vessel data into trajectories containing ten consecutive vessel positions each.
5: Create final preprocessed dataset, which should contain the vessel ID and final 10-position trajectories.

2.4. Modeling Procedures

AIS transceiver equipped vessels are able to record and exchange timestamped information, including geolocation, speed, direction, vessel identification, and specifications data. Vessel trajectory prediction algorithms integrated with collision avoidance techniques can incorporate trajectory prediction models in order to identify imminent threats and navigate safely and efficiently within heavily crowded port areas or open seas.

Let us suppose an available AIS dataset, comprising an arbitrary number of T^v trajectories for a total of V vessels, where $v = 1, 2, \ldots, V$. Let us also suppose that the included trajectories contain an arbitrary number of $K^{v,t}$ AIS messages $AISm_k^{v,t}$ (timestamped geolocation and other data). In this work, for simplicity reasons, we employ the following format in AIS messages

$$AISm_k^{v,t} = \left\{ \begin{array}{ccc} Ts_k^{v,t} & y_k^{v,t} & x_k^{v,t} \end{array} \right\} \tag{8}$$

where $k = 1, 2, \ldots, K^{v,t}$, and $Ts_k^{v,t}$ denotes the message timestamp, while $y_k^{v,t}$ and $x_k^{v,t}$ are the respective latitude and longitude contained in the k-th AIS message for the t-th trajectory of the v-th vessel. The fact that there are unknown parameters, e.g., the state and controls of the vessels, prohibits the use of kinematics in calculating future vessel states. Nevertheless, the vessel dynamics exist in the information hidden within the dataset and can be extracted and, in most cases, approximated by using a black-box modeling technique such as RBF NNs. We can assume that a common underlying pattern exists in the dynamics of same-size vessels executing similar maneuvers, for example, when approaching or leaving a port, when berthing, when crossing waterway paths, etc. Thus, if a suitable dataset of sufficient size is made available, an RBF NN can be trained to perform one-step-ahead predictions about a vessel's future geolocation by using past AIS messages as seen in the following equations

$$\left\{ \begin{array}{c} \Delta \hat{y}_{k+1}^{v,t} \\ \Delta \hat{x}_{k+1}^{v,t} \end{array} \right\} = RBF\ NN\left(\begin{array}{ccc} AISm_k^{v,t} & \ldots & AISm_{k-N}^{v,t} \end{array} \right) \tag{9}$$

$$\begin{array}{c} \hat{y}_{k+1}^{v,t} = y_k^{v,t} + \Delta \hat{y}_{k+1}^{v,t} \\ \hat{x}_{k+1}^{v,t} = x_k^{v,t} + \Delta \hat{x}_{k+1}^{v,t} \end{array} \tag{10}$$

where N is the number of past AIS messages given as inputs to the RBF NN.

The accuracy and simple structures of FM-trained RBF NNs make them ideal to be integrated in multi-step-ahead predictive control formulations. Receding horizon techniques require models of very high accuracy due to the inevitable error enlargement through propagation. This effect appears when the incorporated model is expected to recurrently make future predictions based on its own previous output throughout the prediction horizon. The problem is further worsened with the increase of the prediction horizon and can ultimately drive the control algorithm to failure. Another important point to be noted is that the linear combination used to produce an RBF NN's prediction is a simple and very fast calculation, a fact that benefits model predictive control (MPC) frameworks [36], which require an optimization problem to be solved iteratively and expect a significant number of model predictions to be made in order to converge to the optimal solution at each timestep.

Delta values of the last position of each sample were used as the model's output, while the first nine positions were the model's input

$$\left\{ \begin{array}{c} \Delta \hat{y}_{k+1}^{v,t} \\ \Delta \hat{x}_{k+1}^{v,t} \end{array} \right\} = RBF\ NN\left(\begin{array}{ccccc} y_k^{v,t} & x_k^{v,t} & \ldots & y_{k-8}^{v,t} & x_{k-8}^{v,t} \end{array} \right) \tag{11}$$

The $\Delta \hat{y}_{k+1}^{v,t}$ and $\Delta \hat{x}_{k+1}^{v,t}$ values may be added to the last input position to calculate the final predicted vessel position. Based on the above procedure, the results of the modeling process produced an RBF model of very high accuracy [31]. The step-by-step procedure of the modeling stage can be seen in Algorithm 2.

Note that the number of past inputs was determined after a trial-and-error procedure, where several RBF models were trained using a different number of inputs. After testing inputs in the range of 3 to 15 past vessel positions, data obtained on model performance showed that using less than nine inputs produced models with reduced prediction accuracy, while using more than nine inputs increased the model's complexity without any significant accuracy gain compared to the model using nine inputs.

Algorithm 2 Modeling Stage

1: Load final preprocessed dataset.
2 : Replace the final value of all included 10 −
position trajectories with the respective delta value according to $\left\{ \begin{array}{c} \Delta y_{10}^{v,t} \\ \Delta x_{10}^{v,t} \end{array} \right\} = \left\{ \begin{array}{c} y_{10}^{v,t} - y_9^{v,t} \\ x_{10}^{v,t} - x_9^{v,t} \end{array} \right\}$,
so that each trajectory sample is in the form
$\left[\begin{array}{ccccccccc} y_1^{v,t} & x_1^{v,t} & y_2^{v,t} & x_2^{v,t} & \ldots & y_9^{v,t} & x_9^{v,t} & \Delta y^{v,t} & \Delta x^{v,t} \end{array} \right]$.
3: Randomly permute the trajectory samples of each vessel.
4: Split the trajectory samples of each vessel into training, validation, and testing subsets (in this work a 50%–5%–25% percentage split is used). Do this so that all vessels contribute to all three subsets according to the chosen splitting.
5: Merge all subset samples, e.g., all training samples of all vessels together in one single dataset that will be used for training. Do the same for the validation and testing subsets.
6: Normalize the inputs and outputs of the training subset. Apply the normalization coefficients to the validation and testing subsets.
7 : Apply the fuzzy means algorithm on the training and validation dataset using the nine first sets of $y^{v,t}$ − $x^{v,t}$ values as inputs and the last set of $\Delta y^{v,t} - \Delta x^{v,t}$ values as output.
8: Final model is in the form of Equation (11).

Moreover, a series of tests has been performed by the recurrent application of this model based on a horizon of 5 timesteps for all trajectories of the testing subset, where, at each successive timestep, the model had to use an increasing number of its own previous predictions. As the model uses more of its past predictions, accuracy decreases due to the enlargement of the propagated prediction error. Such a test can provide intuition on the models' ability to be incorporated in receding horizon predictive frameworks. The quality metrics used for these tests were the root mean squared error (RMSE) and the root mean squared haversine formula distance (RMSHFD). The haversine formula is commonly used to measure great circle distances on spherical surfaces. Table 1 presents the performance metrics obtained after the recurrent application of the chosen model in order to make predictions for the full length of the trajectories included in the testing subset of the training procedure. Mean RMSE values for the two outputs of the model, namely the latitude and longitude, are provided in degrees, wherein can be seen that the error lies in the order of 1.5 thousandth of a degree. The mean RMSHFD metric shows the respective error margin in meters when combining the two model outputs to get the actual predicted future vessel position for all tested trajectories. More details on the modeling procedure for the one-step ahead models, including detailed results and comparison with other machine learning approaches, can be found in [31].

Table 1. Performance metrics of the produced RBF NN model.

	RBF NN	
	Latitude (y)	Longitude (x)
Mean RMSE (deg)	$1439{\cdot}10^{-6}$	$1567{\cdot}10^{-6}$
	Best combined RBF models	
Mean RMSHFD (m)	1200	

3. MPC for Multi-Ship Collision Avoidance

3.1. Preliminaries on Maritime Collision Avoidance and Trajectory Generation

The objective of maritime collision avoidance is the generation of a risk-free trajectory that the controlled vessel should follow. A well-defined and effective method of assessing collision risk in the near future is the closest point of approach (CPA). Stemming from the concept of the CPA, two metrics are defined: time to CPA (TCPA) and distance to CPA (DCPA) (see Figure 2). A discussion regarding the quick calculation of TCPA and DCPA using the line-of-sight (LOS) distance between the controlled vessel and the obstacle ship is presented in [5]. These metrics depict the urgency of the collision danger of vessel i with another vessel j as well as its magnitude, and by specifying lowest acceptable thresholds d_{min} and t_{min} concerning the minimum DCPA and minimum TCPA, respectively, one can construct a risk cost function, as presented in [5].

$$f_{r,ij} = \begin{cases} \exp\left(a_0\left(d_{min} - DCPA(T_i, T_j) + t_{min} - TCPA(T_i, T_j)\right)\right) - 1, \; if \\ \quad DCPA(T_i, T_j) \le d_{min} \;\; and \;\; TCPA(T_i, T_j) \le t_{min} \\ \quad 0, \;\; if \; otherwise \end{cases} \tag{12}$$

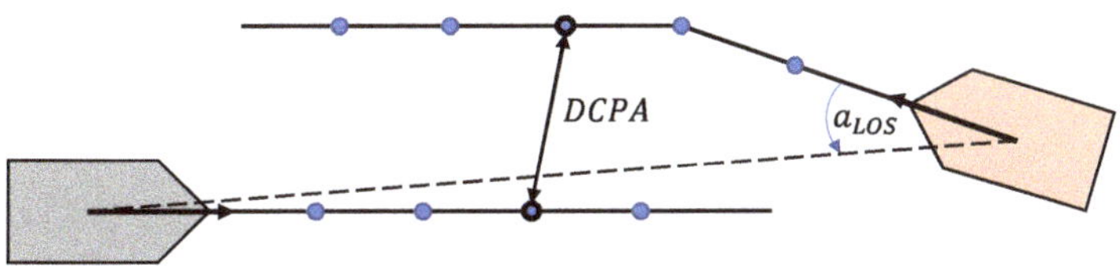

Figure 2. Illustration of the CPA metrics, as well as the LOS angle concept.

Here, a_0 is a scaling parameter, and $\boldsymbol{T}_i$ denotes the trajectory matrix containing the x-y position of the i vessel for every timestep

$$\boldsymbol{T}_i = \begin{bmatrix} x_1 & y_1 \\ \vdots & \vdots \\ x_n & y_n \end{bmatrix}. \tag{13}$$

By combining TCPA and DCPA, the spatial-temporal nature of a maritime collision risk with vessel i is successfully reflected. The physical interpretation of Equation (12) is that a candidate trajectory with larger minimum distance from an obstacle ship occurring at an earlier time will always be safer than a path with a smaller minimum distance and/or earlier time of occurrence. Common values for t_{min} and d_{min} are 10 min and 0.6 nm; because the present paper is concerned with collision avoidance in busy waterways such as ports, a lower d_{min} value of 0.4 nm is used. In any case, Equation (12) can be readily incorporated in the cost function of an MPC optimization problem formulation.

A second item in the domain of trajectory generation is efficiency. Vessels should strive to not deviate too much from their original course when addressing a collision risk with another vessel. The efficiency of the generated trajectory $\boldsymbol{T}_i$ for vessel i can be reflected by calculating the sum of absolute deviations from the original trajectory $\boldsymbol{T}_{OG,i}$

$$f_{d,i} = \|\boldsymbol{T}_{OG,i} - \boldsymbol{T}_i\|. \tag{14}$$

Next, an important requirement to be fulfilled when addressing the problem of collision avoidance are the COLREGs [4]. The implementation of the COLREGs restricts the domain of possible candidate paths according to the type of encounter, for example 'head-on,' 'crossing,' and 'overtaking.' Head-on vessels should pass each other on the port side, while a vessel crossing from the starboard side should be given the right of way. A visual depiction of the encounter rules takes place in Figure 3. Multiple approaches for the modeling of the COLREG rules have been made in the literature [2,5,8], although these are usually concerned with a one-step-ahead calculation. However, for the case of an MPC controller, in order to ensure COLREG compliance for a candidate trajectory, all of its waypoints must be taken into account. By assuming that the LOS angle is increasing in the anti-clockwise direction, one needs to evaluate whether the LOS angles of each sequential trajectory timestep position are increasing monotonically, in order to confirm the compliance of the trajectory for the 'head-on' and 'give-way' situations.

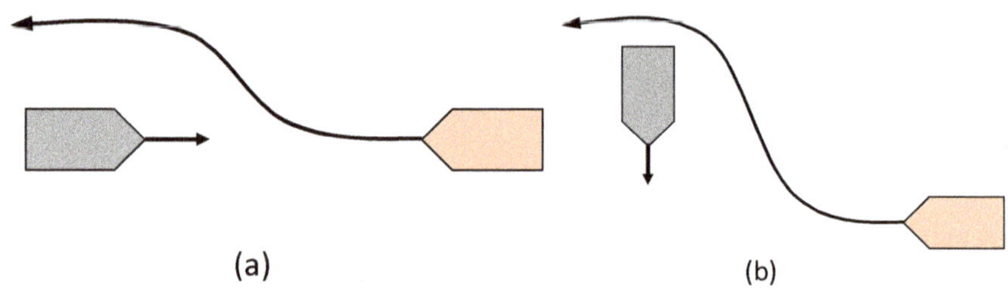

Figure 3. (**a**) A head-on situation between two ships; (**b**) a crossing situation between two ships (give-way); the orange ship must give way to the crossing ship on its starboard side.

The idea is depicted in Figure 4, where a head-on encounter between vessels i and j occurs; here, the LOS angles for trajectory $\boldsymbol{T}_i$ monotonically increase, and therefore, it is deemed as compliant. In contrast, the monotonically decreasing LOS angles of the $\boldsymbol{T}_i'$ trajectory confirm its non-compliance as per the COLREGs intentions. A penalty for non-compliance of a vessel i encountering a vessel j in a 'head-on' or 'give-way' situation can therefore be formulated,

$$P_{ij} = \begin{cases} 1, \ if \ \boldsymbol{a}_{LOS_{ij}} \searrow \\ 0, \ if \ otherwise \end{cases}, \tag{15}$$

where $\boldsymbol{a}_{LOS_{ij}}$ is the LOS angle vector, calculated for each trajectory point of encountering vessel i and the current position of encountered vessel j.

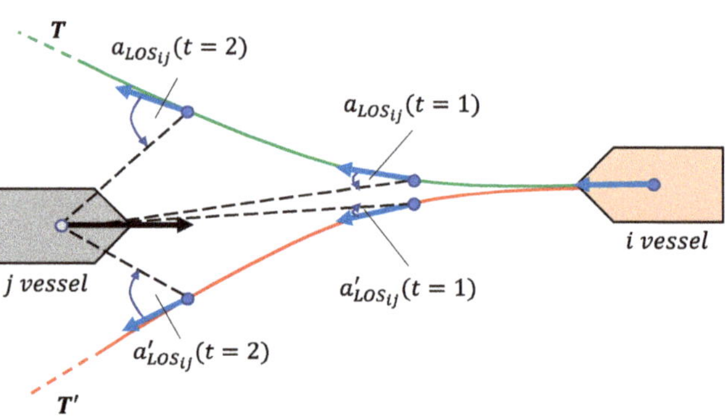

Figure 4. Head-on situation between vessels i and j; the LOS angle can be used to assess the COLREG compliance of a candidate trajectory.

Next, the generated vessel trajectory, apart from being safe and COLREG compliant, should also take into account the maneuvering capabilities of the controlled vessel, i.e., it should be guaranteed that the trajectory is technically possible to be tracked by the vessel. The feasible search domain of the trajectory optimization problem can be constructed by a purely geometric approach in the case of a one-step-ahead calculation, such as in [5], where the design variables are the vessel's next position and course. However, the extension of this geometric approach to multiple-steps-ahead requires the application of nonlinear constraints that would bound every sequential vessel position with its previous one, in order to enforce technical feasibility. For this reason, a model-based approach is preferred. The Nomoto models constitute a class of vessel course models that are tailored for this task, and have been widely adopted, not only for the design of collision avoidance schemes [47], but also for path tracking controllers [48]. The 1st order linear Nomoto model is shown as follows:

$$T_s \frac{d\omega}{dt} + \omega = K_s a. \tag{16}$$

Here, ω is the angular velocity of the vessel, while a is the control input to the vessel's rudder. The maneuvering capabilities of the vessel are reflected by the T_s and K_s constants, called time constants and rudder gain constants, respectively, while typical values are in the [0.5, 2] range for both. Solving the differential Equation (16) by assuming constant rudder angle input for a t time interval, the 1st order linear Nomoto model can be discretized as follows [8]:

$$\Delta\theta(t) = K_s\, a\left(t - T_s + T_s \exp\left(\frac{t}{T_s}\right)\right) \tag{17}$$

Here, $\Delta\theta$ is the course change that would occur if a control input of a was applied and held for a time period of t. By setting this time period t as the discretization interval Δt, a course model can be used to create a discrete vessel position model as follows

$$\begin{aligned} \theta_{k+1} &= \theta_{k-1} + \Delta\theta_k(a_k) \\ x_{k+1} &= x_k + \cos(\theta_{k+1}) V_k\, \Delta t \\ y_{k+1} &= y_k + \sin(\theta_{k+1}) V_k\, \Delta t \end{aligned} \tag{18}$$

Here, θ_k, x_k, y_k is the current course, horizontal displacement, and vertical displacement according to a global reference frame, respectively, while V_k is the vessel velocity. The discretization interval Δt can be set according to the simulation resolution required. Equation (18) constitute a discrete position model L_i for the i-th vessel,

$$\boldsymbol{x}_i(k+1) = L_i(\boldsymbol{u}_i,\, \boldsymbol{x}_i(k)), \tag{19}$$

with input vector $\boldsymbol{u}_i = \begin{bmatrix} a & V \end{bmatrix}$ and state vector $\boldsymbol{x}_i = \begin{bmatrix} \theta & x & y \end{bmatrix}$. By evaluating the discrete vessel position model L_i for $\{1, 2, \ldots, n\}$ consecutive timesteps, where n the total timesteps, a trajectory $\boldsymbol{T}_i$ can be created for the i-th vessel, as shown in Equation (13).

3.2. Collision Avoidance with Mpc and Obstacle Trajectory Prediction Models

The MPC framework has demonstrated its aptitude in handling the uncertainties and nonlinearities of the collision avoidance problem multiple times in the literature [9,49]; however, no other works have incorporated a nonlinear data-driven obstacle trajectory prediction model in their formulation. In MPC, the optimal moves of the controlled vessels are calculated for multiple steps ahead by solving a constrained optimization problem, with constraints in real time, for each controller sample time t_{cst}. The cost function of the optimization problem is constituted by two horizons, namely the prediction horizon h_p and the control horizon h_c; the first accounts for the total discrete timesteps ahead that the model can be evaluated, while the second for the number of timesteps that the control variables can be modified. Given a set of controlled vessels $V_c = \{1, 2, \ldots, N_c\}$ and a set of non-controlled or obstacle vessels $V_o = \{1, 2, \ldots, N_o\}$ where N_c and N_o are the total number of controlled and non-controlled vessels, respectively, the optimization problem's cost function can be formulated as the summation of all the cost functions of the respective controlled vessels for the kth timestep:

$$\begin{gathered} \min_{U(k)} \textstyle\sum_{i \in V} C_i(\boldsymbol{X}_c(k), \boldsymbol{X}_o(k)) \\ s.t.\ \boldsymbol{U}_u \leq \boldsymbol{U}(k) \leq \boldsymbol{U}_l \\ \boldsymbol{N}(\boldsymbol{X}_c(k), \boldsymbol{X}_o(k)) \geq \mathrm{d}_e \\ \boldsymbol{P}(\boldsymbol{X}_c(k), \boldsymbol{X}_o(k)) = 0 \end{gathered} \tag{20}$$

Here, $\boldsymbol{U}(k)$ is the input matrix and is created by the horizontal concatenation of the input vectors of all controlled vessels V_c, up until the control horizon h_c:

$$\boldsymbol{U}(k) = \begin{bmatrix} u_1(k) & u_1(k+1) & \cdots & u_1(k+h_c-1) \\ \vdots & & \ddots & \vdots \\ u_{N_c}(k) & u_{N_c}(k+1) & \cdots & u_{N_c}(k+h_c-1) \end{bmatrix}. \tag{21}$$

Next $\boldsymbol{X}_c(k)$ and $\boldsymbol{X}_o(k)$ are the controlled and non-controlled vessel state matrices, respectively, and are created by the horizontal concatenation of the state vectors of all controlled and non-controlled vessels V_c and V_o, respectively, up to the prediction horizon h_p.

$$\begin{aligned} \boldsymbol{X}_c(k) &= \begin{bmatrix} x_{c,1}(k) & x_{c,1}(k+1) & \cdots & x_{c,1}(k+h_p-1) \\ \vdots & & \ddots & \vdots \\ x_{c,N_c}(k) & x_{c,N_c}(k+1) & \cdots & x_{c,N_c}(k+h_p-1) \end{bmatrix} \\ \boldsymbol{X}_o(k) &= \begin{bmatrix} x_{o,1}(k) & x_{o,1}(k+1) & \cdots & x_{o,1}(k+h_p-1) \\ \vdots & & \ddots & \vdots \\ x_{o,N_o}(k) & x_{o,N_o}(k+1) & \cdots & x_{o,N_o}(k+h_p-1) \end{bmatrix} \end{aligned} \tag{22}$$

For simplicity, because consecutive states $x_i(k)$ up to $x_i(k+h_p-1)$ constitute a single trajectory $\boldsymbol{T}_i(k)$, one can write $\boldsymbol{X}_c(k)$ and $\boldsymbol{X}_o(k)$ as the concatenation of the trajectories of the respective vessel sets V_c, V_o as per Equation (13):

$$\boldsymbol{X}_c(k) = \begin{bmatrix} \boldsymbol{T}_{c,1}(k) \\ \vdots \\ \boldsymbol{T}_{c,\,N_c}(k) \end{bmatrix} \quad \boldsymbol{X}_o(k) = \begin{bmatrix} \boldsymbol{T}_{o,1}(k) \\ \vdots \\ \boldsymbol{T}_{o,N_o}(k) \end{bmatrix}. \tag{23}$$

Next, $C_i(k)$ is the cost function of the i-th controlled vessel, formulated as follows:

$$C_i(k) = F_i(\boldsymbol{X}(k)) + a_G G^2(\boldsymbol{U}_i(k)) \tag{24}$$

Here, $\boldsymbol{X}(k)$ is the vertical concatenation of the two state matrices $\boldsymbol{X}_c(k)$, $\boldsymbol{X}_o(k)$, containing the trajectories of all vessels $V = V_c \bigcup V_o$

$$\boldsymbol{X}(k) = \begin{bmatrix} \boldsymbol{T}_{c,1}(k) & \ldots & \boldsymbol{T}_{c,\,N_c}(k) & \boldsymbol{T}_{o,1}(k) & \ldots & \boldsymbol{T}_{o,N_o}(k) \end{bmatrix}'. \tag{25}$$

The cost function is comprised by two terms F_i and G, each concerned with the prediction and control horizon, respectively. The presence of G term, weighted by the a_G parameter, encourages the smoothness of the control actions and, consequently, the generated trajectories of the controlled vessels

$$G(\boldsymbol{U}_i(k)) = \sum_{j=1}^{h_c-1} \|\boldsymbol{U}_{i,j+1}(k) - \boldsymbol{U}_{i,j}(k)\|. \tag{26}$$

Term F_i consolidates the collision avoidance and course keeping objectives, and is specific to the i-th vessel

$$F_i(\boldsymbol{X}(k)) = a_r \sum_{j \in V \setminus i} \left(f_{r,ij}{}^2(\boldsymbol{X}_i(k), \boldsymbol{X}_j(k)) \right) \frac{1}{|V \setminus i|} + a_d\, f_{d,i}{}^2(\boldsymbol{X}_i(k)). \tag{27}$$

In Equation (26), $f_{r,ij}$ is the collision risk between the i-th and the j-th vessel, as calculated using their respective trajectories $\boldsymbol{X}_i(k)$, $\boldsymbol{X}_j(k)$ by applying Equation (12), and $f_{d,i}$ is the deviation from the original trajectory $\boldsymbol{T}_{OG,\,i}$, as expressed in Equation (14). Both terms are weighted by the a_r and a_d weighting parameters, respectively. Since we are concerned with the safety of the generated trajectory throughout the whole prediction horizon, the mean collision risk from all vessels in $V\backslash i$ is evaluated, in contrast to other approaches [5], where only the maximum collision risk at time k is minimized. This way, all possible collision risks are addressed and reduced simultaneously, thus avoiding the adverse possibility of evading one collision risk and increasing another. Moreover, the reason that risk avoidance is used as a control objective in Equation (27) and not as a hard optimization constraint is to ensure that the MPC optimization problem of Equation (20) will not fail in the case of the existence of an inescapable collision risk; as shown in Equation (12), risk is a function of distance to CPA, meaning that the controller will continue to attempt to maximize that distance, thus fulfilling the control intention in such an encounter. However, in order to guarantee that collisions will be avoided, one more constraint to the MPC optimization problem is added by setting an emergency distance d_e (where $d_e < d_{min}$); to be more specific, the vector $\boldsymbol{N}$ contains the DCPAs of all controlled vessels V_c, which are required to be above the emergency distance.

At this point, it must be noted that since the state matrix $\boldsymbol{X}(k)$ consolidates all vessel trajectories, controlled and non-controlled alike, a degree of cooperation is induced between the respective controlled vessels V_c. Lastly, returning to the optimization problem denoted in Equation (20), the $\boldsymbol{U}(k)$ input matrix is bounded by the upper and lower matrices $\boldsymbol{U}_u$, $\boldsymbol{U}_l$, respectively. The vector $\boldsymbol{P}$ contains the COLREG non-compliance penalties for the controlled vessels V_c as calculated in Equation (15), which are required to be zero via an equality constraint.

The next item to be addressed regarding the MPC formulation is the used model that maps the input variables $\boldsymbol{U}$ to the state variables of the controlled vessels $\boldsymbol{X}_c$. Here, the 1st order linear Nomoto model is used, as described in Equation (18), with the addition of input noise that accounts for modeling error e and environmental parameters:

$$\begin{aligned} \boldsymbol{x}_{c,i}(k+1) &= \hat{L}_i(\boldsymbol{u}_i,\ \boldsymbol{x}_{c,i}(k)),\ \ where\ i \in V_c \\ \hat{L}_i(\boldsymbol{u}_i, \boldsymbol{x}_{c,i}(k)) &= L_i(\boldsymbol{u}_i + e(\boldsymbol{u}_i),\ \boldsymbol{x}_{c,i}(k)), where\ e(\boldsymbol{u}_i) = \boldsymbol{u}_i\ G(0, \sigma^2). \end{aligned} \tag{28}$$

Here, G is a random variable sampled from a Gaussian distribution with a standard deviation of σ. Finally, the state matrix of the non-controlled vessels $\boldsymbol{X}_o(k)$ is assumed to be unknown for the scope of this research, and thus, an estimation is required, based on past positions. For this task, the RBF prediction model presented in Section 2.3 is employed for each non-controlled vessel j, and its trajectory for the k-th timestep is estimated using its past nine positions:

$$\hat{\boldsymbol{T}}_{o,j}(k) = RBF(\boldsymbol{x}_{o,j}(k), \boldsymbol{x}_{o,j}(k-1), \ldots, \boldsymbol{x}_{o,j}(k-9)),\ \ where\ j \in V_o. \tag{29}$$

Next, in order to alleviate a possible computational burden for the MPC optimization problem, an important assumption should be made. The formulation of the control scheme as-presented would give rise to a high-dimensional search space for the MPC optimization problem, thus greatly hindering its effective solution. It is assumed then that all vessels retain their initial speed, with the only controllable variable being the vessel's rudder angle; this way, the total number of control variables is reduced by half. This approach to the collision avoidance problem has occurred in the literature [5], and is not simplistic for two reasons: first, good seafaring practice dictates that course change maneuvers are preferred over speed ones, not only because they conserve energy, but also because they better emphasize the intentions of the vessel to outside observers, such as other vessels in the vicinity. Second, since large container ships will be examined in the scope this case study, their large longitudinal inertia [48] confirms the assumption that the speed remains almost constant during the timeframe of a typical collision avoidance maneuvering scenario.

Therefore, for the scope of this paper, the input matrix $\boldsymbol{U}$ at timestep k is formulated as follows:

$$\boldsymbol{U}(k) = \begin{bmatrix} a_1(k) & a_1(k+1) & \cdots & a_1(k+h_c-1) \\ \vdots & & \ddots & \vdots \\ a_{N_c}(k) & a_{N_c}(k+1) & \cdots & a_{N_c}(k+h_c-1) \end{bmatrix}, \tag{30}$$

where $a_i(k)$ is the rudder angle of vessel i at timestep k.

Having defined all aspects of the MPC optimization problem, a reiteration of the challenges of the collision avoidance control problem and how they are addressed by the controller is in order: firstly, the goal of the control design is to generate trajectories for the controlled vessels that are risk-free (Equation (12)), smooth (Equation (26)), COLREG-compliant (Equation (15)), and do not deviate from the original course (Equation (14)). Possible collision risks are assessed by utilizing trajectory predictions for non-controlled (obstacle) vessels in the vicinity. The controllable variables are the rudder angles of the vessels (vessel speed is considered constant), while a discrete 1st order Nomoto model (Equation (28)) is used for the modeling of the vessel dynamics, which was also infused with a noise signal for the purpose of accounting for uncertainties and environmental factors. The aforementioned vessel dynamics model has been compared to its higher-order nonlinear counterparts in [50], and it was shown that vessel course inaccuracies occur only for high yaw rates. Given the fact that the proposed collision avoidance method is concerned with large vessels with slow dynamics, the used vessel dynamics model is deemed adequate for the case. In addition, MPC has shown to be robust against model uncertainties or input noise [36]. Finally, the constraints that must be adhered to when searching for the optimal solution (Equation (20)) are the technical bounds on the controlled variables (i.e., maximum and minimum rudder angles) and the COLREG compliance of the result trajectory.

3.3. Control Framework

Having presented the proposed MPC controller, this section describes its integration within a general control framework. As shown in Figure 5, the framework is comprised by an offline and an online process. The offline process corresponds to the RBF trajectory prediction model training, using data from a specific area of interest (for example, a port)—naturally then, it could be undertaken by the port authority. The online process corresponds to the real-time control of autonomous vessels in the presence of obstacle vessels in the area of interest. The MPC collision avoidance controller, as described in Section 3.2, is integrated here and is supplied with real time trajectory predictions of all obstacle vessels in order to calculate the optimal control actions for the controlled vessels. Since the RBF trajectory prediction model has been trained offline in the port authority premises, it is sensible to place the MPC controller there too, and to communicate the computed control actions per control timestep via a communications link with the controlled vessels. Figure 6 demonstrates this concept.

The MPC optimization problem described in Equation (20) is solved using the sequential quadratic programming (SQP) algorithm, which involves iterative calls to the objective function [35]. As shown in Figure 5, the integration of the MPC controller in the control framework requires the calculation of the obstacle vessel trajectory predictions for every controller timestep. Therefore, two main sources of computational complexity arise: the first is the evaluation of the RBF trajectory prediction model, which is shown to be in the order of magnitude of milliseconds [31], meaning that multiple obstacle vessels can be accounted for by the control scheme. The second is the solution of the optimization problem (Equation (20)) by the SQP algorithm, which is known to converge quickly and with few objective function calls [51]. It is concluded that a typical controller timestep duration, comprised by the two aforementioned sources, will not exceed the order of magnitude of seconds, which is considered reasonable given the slow dynamics of large vessels.

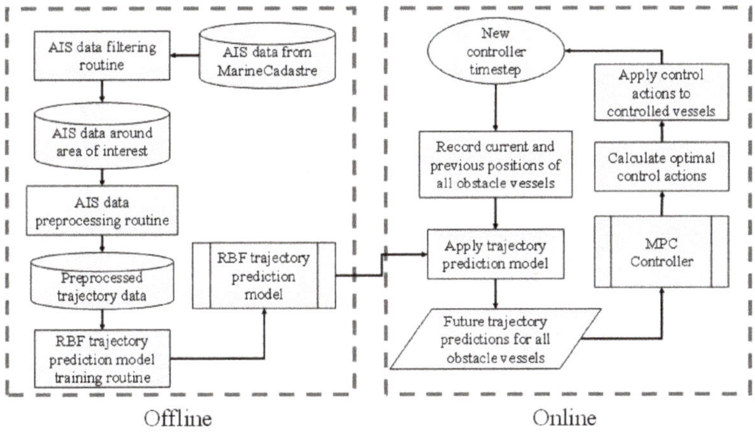

Figure 5. The proposed control framework.

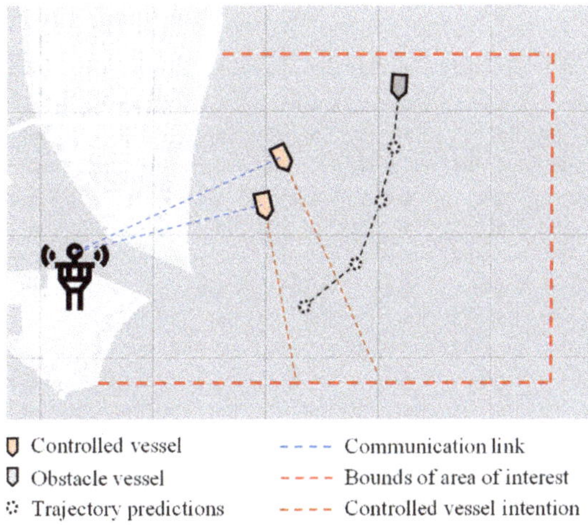

Figure 6. Communications within the proposed control framework.

4. Case Study

In this section, the performance of the proposed multi-ship MPC controller is assessed using real-life obstacle ship trajectories, which were sourced and preprocessed as described in Section 2.4. In order to underline the importance of using sophisticated trajectory prediction models in the context of collision avoidance controller design, the proposed method is compared to an MPC controller that uses straight-line predictions for the trajectories of obstacle ships based on their current course and speed [9]. To this end, two crossing scenarios are examined, while performance indicators of the generated trajectories are extracted and discussed in detail. The simulations were coded and executed on Matlab 2020b, on a computer with an Intel i7 processor and 16 GB RAM. The simulation sample time is 30″. Lastly, the tuning and parameters of the methods are shown in Table 2, while the vessel parameters are shown in Table 3.

Table 2. MPC tuning parameters.

Parameter	Description	Value
t_{cst}	Controller sample time	$1'$
h_p	Control horizon	5
h_c	Prediction horizon	15
a_0	Risk function scaling parameter	3
a_r	Risk term weighting parameter	1
a_d	Course deviation term weighting parameter	0.05
a_G	Control action smoothness term weighting parameter	5

Table 3. Vessel parameters.

Parameter	Description	Value
d_{min}	Minimum allowable DCPA for risk calculation	750 m
d_e	Emergency distance	200 m
t_{min}	Minimum allowable TCPA for risk calculation	$10'$
K_s	Rudder gain constant	0.5
T_s	Rudder time constant	2

4.1. Multi-Ship Collision Avoidance Control for the Miami Port

For this case study, two controllable vessels are chosen, moving in parallel to each other and encountering an obstacle vessel moving into the port of Miami. For the performance evaluation of the two controllers, two scenarios are created; the first contains a head-on encounter type, while the second an overtaking maneuver that changes into a crossing encounter as time progresses. In the first scenario, the two controlled vessels are leaving the port of Miami at a course of 110°, when they encounter a single obstacle on their starboard side, which, in turn, is looking to enter the port. In the second scenario, the two controlled vessels are overtaking an obstacle vessel on her port side when, suddenly, she turns port-side in order to enter the port of Miami, crossing into their intended path. The challenge posed by the two scenarios is that the two controllable vessels should maintain a safe distance between each other and the obstacle vessel, while also navigating smoothly and without unnecessary deviation from their original course. It should also be noted that the obstacle vessel is non-controllable and, therefore, follows a predetermined path, without considering other vessels.

The response of the MPC controller utilizing straight-line prediction models (hereby referred to as 'MPC-SLP') for the first scenario is shown in the left column of the subfigures within Figure 7 for the 3-, 9-, 10-, and 16.5-min timesteps. The response of the proposed MPC controller utilizing RBF prediction models (hereby referred to as 'MPC-RBFP') for the same scenario and same time instances are shown in the right column of the subfigures within Figure 8. Next, the responses of MPC-SLP and MPC-RBFP for the second scenario are shown in the left and right subfigure columns of Figure 8, respectively, for the 6-, 12-, 13.5-, and 17-min timesteps. In the aforementioned response figures, the red and blue dotted lines denote the original, undisturbed trajectory for controlled vessels 1 & 2, respectively, while the black dotted line shows the predetermined path that the obstacle ship will follow as the simulation progresses. Next, the red and blue dashed lines denote the trajectory that the controlled vessels intend to follow, as calculated by the current MPC iteration, while the black dashed line shows the current trajectory prediction of the obstacle ship, as utilized by the MPC controller. The grey dashed circles have a radius of d_{min} and denote the safe ship domain for the two controlled vessels; should any vessel enter another's domain at any time, a collision risk arises. Lastly, the red-colored and blue-colored rectangles mark the controlled vessels 1 & 2 positions, respectively, while the grey rectangle marks the obstacle ship's position; it should be noted that the markers are not to-scale with the real dimensions of the vessels, since they have been enlarged for graphical convenience.

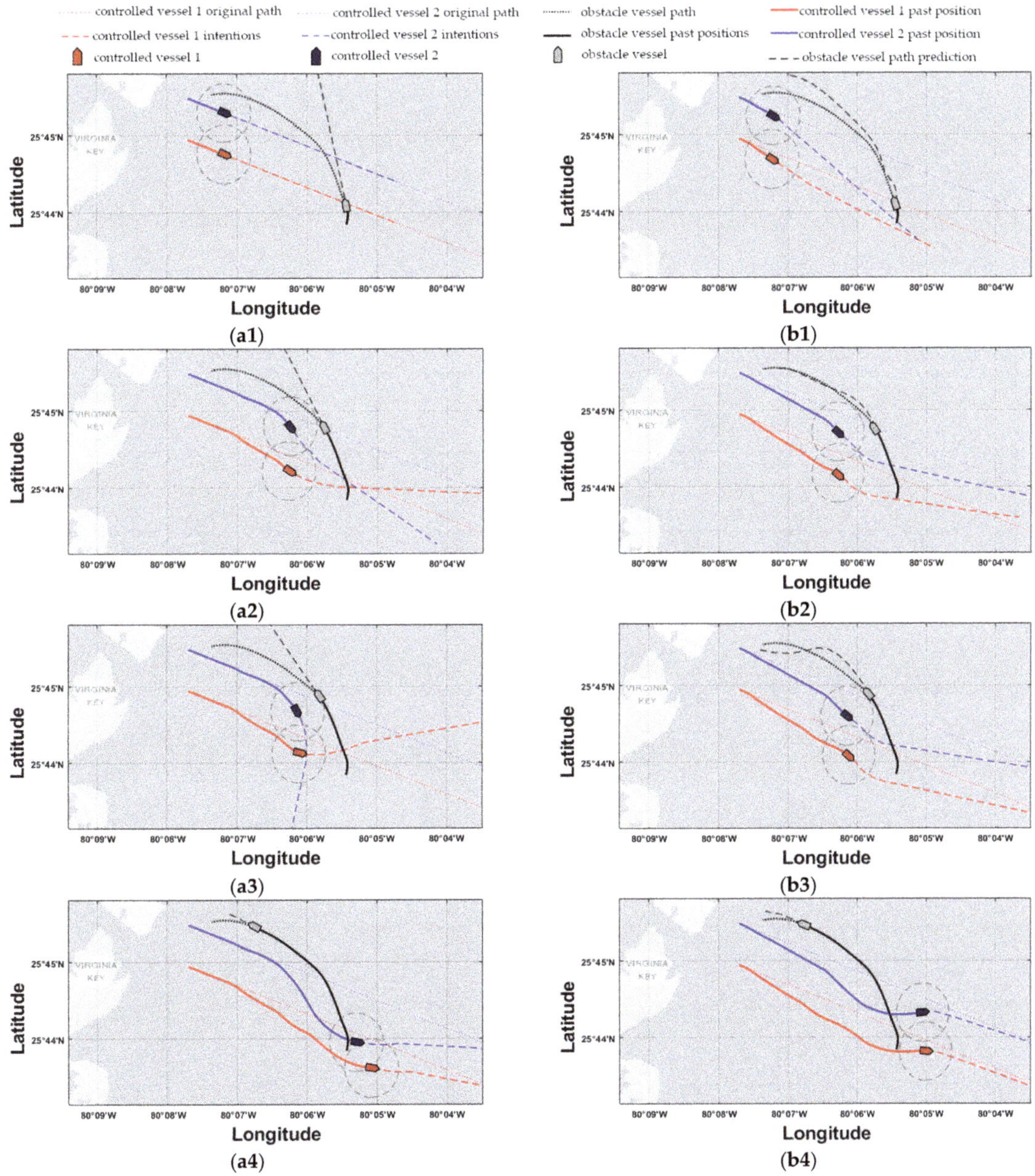

Figure 7. Scenario 1. The left subfigure column refers to the MPC-SLP scheme, while the right to the MPC-RBFP scheme. Subfigures (**a1**,**b1**) refer to time instance 3′, (**a2**,**b2**) to 9′, (**a3**,**b3**) to 10′, and finally, (**a4**,**b4**) to 16.5′.

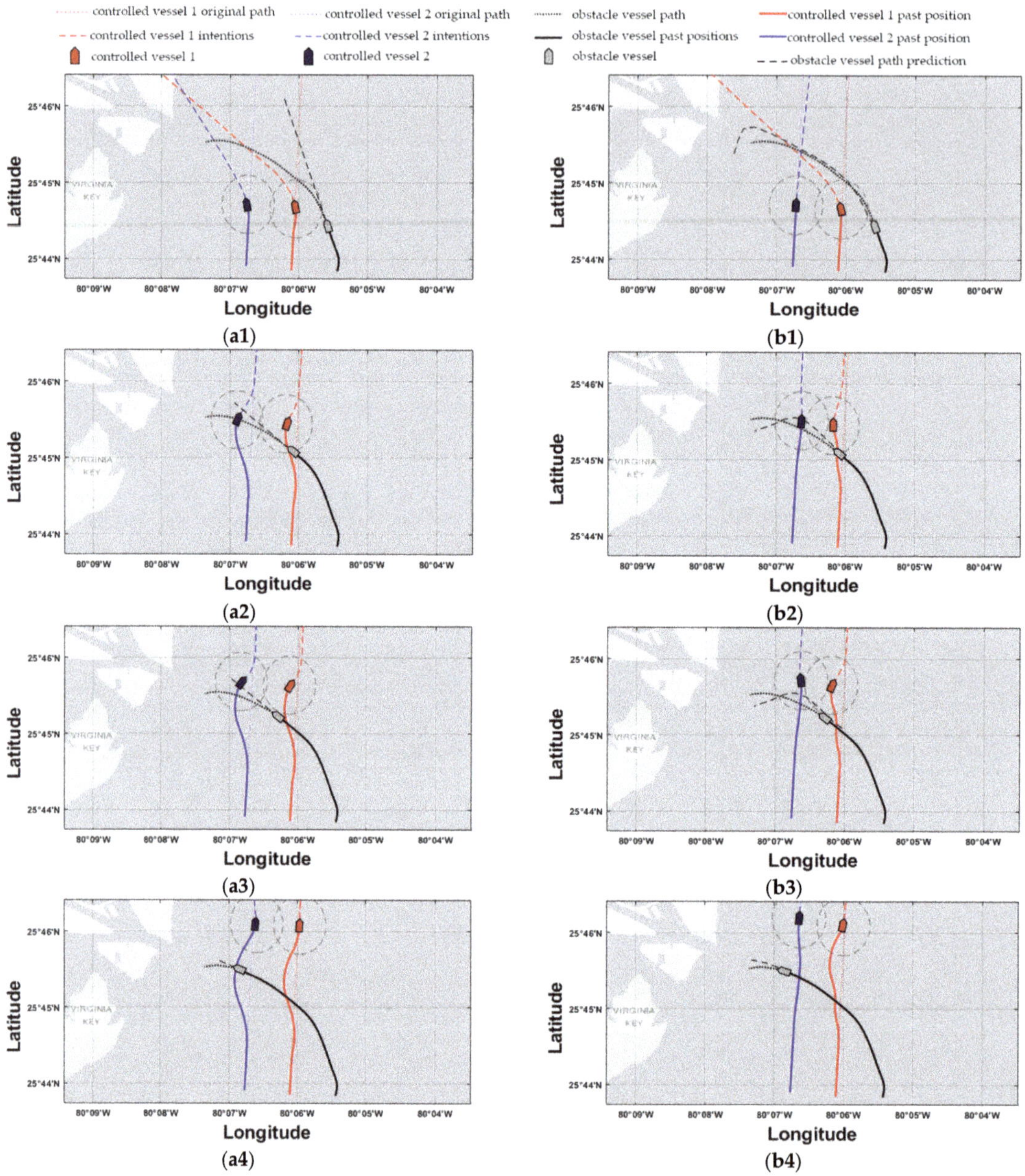

Figure 8. Scenario 2. The left subfigure column refers to the MPC-SLP scheme, while the right to the MPC-RBFP scheme. Subfigures (**a1**,**b1**) refer to time instance $6'$, (**a2**,**b2**) to $12'$, (**a3**,**b3**) to $13.5'$, and finally, (**a4**,**b4**) to $17'$.

4.2. Discussion

Firstly, in order to assist the discussion in this subsection, distance plots are generated for the controlled vessels that are in closest proximity with the obstacle ship for each scenario (see Figure 9). In addition, the performance metrics for each controller in each scenario are shown in Table 4.

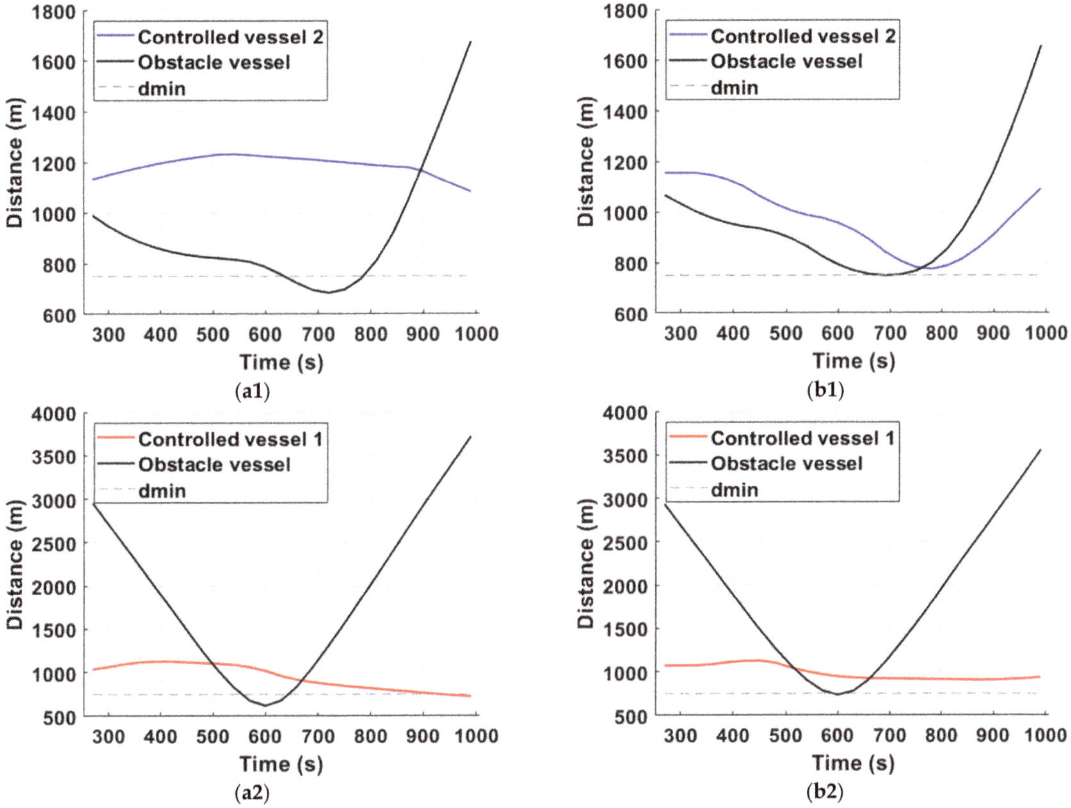

Figure 9. Distance plots for scenarios 1 & 2. The left subfigure column refers to the MPC-SLP scheme, while the second to the MPC-RBFP scheme. Note that for scenario 1 (**a1**,**b1**) and for scenario 2 (**a2**,**b2**), the MPC-SLP violates the lower limit on distance from CPA; therefore, its trajectories are deemed unsafe.

Table 4. Performance metrics for the generated trajectories of the MPC-RBFP and MPC-SLP schemes for the two simulation scenarios.

		Scenario 1		Scenario 2	
s	**Controlled Vessel**	**MPC-RBFP**	**MPC-SLP**	**MPC-RBFP**	**MPC-SLP**
Course deviations [(1)]	1	$1.31 \cdot 10^4$	$2.21 \cdot 10^4$	$0.658 \cdot 10^4$	$0.521 \cdot 10^4$
	2	$1.49 \cdot 10^4$	$2.85 \cdot 10^4$	$0.404 \cdot 10^4$	$0.529 \cdot 10^4$
Control action smoothness [(2)]	1	307.35	476.59	242.12	167.85
	2	290.94	424.43	92.47	128.41
Risk of trajectory [(3)]	1	0	$4.032 \cdot 10^6$	0	0
	2	0	0	0	$3.949 \cdot 10^6$
Cost of trajectory [(4)]	1	$9.05 \cdot 10^6$	$1.62 \cdot 10^{13}$	$2.63 \cdot 10^6$	$1.49 \cdot 10^6$
	2	$2.63 \cdot 10^6$	$4.15 \cdot 10^7$	$8.58 \cdot 10^5$	$1.55 \cdot 10^{13}$

[(1)] As calculated by Equation (14). [(2)] As calculated by Equation (26). [(3)] As calculated by Equation (12). [(4)] As calculated by Equation (24).

For the head-on encounter of scenario 1, the correct trajectory prediction of the obstacle ship proves vital for the success of the proposed scheme. Considering timestep 3 (see Figure 7(a1,b1)), the MPC-RBFP scheme is already applying evasive control actions, since the correct inference of the general direction of the obstacle ship has given rise to a possible collision risk in the near future. In contrast, the MPC-SLP controller does not apply any control actions yet, because, based on the straight-line prediction model that it utilizes, the obstacle vessel will continue north and, thus, remain well clear of the controlled vessels. For the same reason, it takes MPC-SLP another $5'$ minutes in order to correctly assess the collision risk and apply decisive control actions, but by then it is too late; by timestep $9'$ (see Figure 7(a3,b3)), controlled vessel 2 reaches its CPA with the obstacle ship, with a DCPA of 680 m for controlled vessel 2, well below the acceptable minimum distance d_{min}, as shown in Figure 9(a1). In contrast, the MPC-RBFP controller generates a smooth, safe, and consistent trajectory, owed to the correct trajectory prediction of the obstacle vessel. Not only does it reach an acceptable DCPA of 751 m for controlled vessel 2, but it also manages to apply consistent control actions and not significantly deviate from the original course, as shown in Table 4.

Next, the performance of the two controllers is assessed in an overtaking/crossing encounter in scenario 2. Here, the effect of the used trajectory prediction models is once again eminent: At timestep 6 (see Figure 8(a1,b1)), MPC-RBFP calculates a sharp control move to port-side for controlled vessel 1 in anticipation of the obstacle ship's crossing towards the port of Miami; in contrast, MPC-SLP applies a lower rate of steering for controlled vessel 1, because the straight-line trajectory prediction places its CPA with the obstacle ship at a later time instance. This failure to correctly place the CPAs has adverse effects on vessel 2 trajectory too, since it is displaced unnecessarily to the left in false anticipation of a collision risk. In addition, the obstacle ship crosses into the domain of controlled vessel 1 (see Figure 8(a2)) once it changes course towards the Miami port at timestep $8'$. On the other hand, the MPC-RBFP scheme places controlled vessel 1 in a better position to narrowly evade the breach of its safe domain (see Figure 8(b2)) throughout the simulation. This performance is owed to the trajectory that the RBF model generated for the obstacle vessel, placing its predicted CPA much closer to the real CPA for both controlled vessels. Moreover, it should be noted that for scenario 2, unnecessary deviations from the original course are avoided for controlled vessel 2, as indicated by the total deviation values in Table 4. In general, the proposed method achieves a lower overall cost for the generated trajectories, as shown in Table 4, while obtaining a certain degree of cooperation between the two vessels, where one makes way for the other in anticipation of their upcoming evasive maneuvers. Moreover, the results show that the proposed method exhibits robust characteristics against environmental effects, which are modeled as input noise in the vessel dynamic model for the scope of the simulations, while accounting for COLREGs. Lastly, the average CPU time evaluation of the MPC calculation was recorded as 7s for both scenarios, which is well within the allocated simulation controller timestep t_{cst} of 60 s, proving, in fact, that the proposed method is scalable to a greater number of controlled and obstacle vessels.

5. Conclusions

In this paper, a multi-ship MPC controller utilizing RBF obstacle ship trajectory prediction models trained on real AIS data is proposed for the collision avoidance task in busy ports or waterways. The proposed method is compared to an MPC controller using straight-line obstacle ship trajectory prediction models for a real simulation case for the port of Miami. The simulations have shown that the incorporation of a trajectory prediction model with a moderate degree of accuracy greatly benefits the performance of a collision avoidance controller; this is due to the fact that a collision risk can be detected earlier and in time, so that it can be accommodated by the slow maneuvering dynamics of larger vessels such as container ships. Moreover, this early detection enables the planning of a more economic trajectory for the controlled vessels and enables their better cooperation.

Author Contributions: The authors' individual contributions, are summarized below: M.P.: Conceptualization; Data curation; Formal analysis; Investigation; Methodology; Software; Validation; Writing—original draft, Writing—review and editing. M.S.: Conceptualization; Data curation; Investigation; Methodology; Software; Visualization; Writing—original draft, Writing—review and editing. H.S.: Conceptualization; Formal analysis; Supervision; Writing—original draft, Writing—review and editing. A.A.: Conceptualization; Formal analysis; Funding acquisition; Methodology; Project administration; Resources; Supervision; Writing—original draft, Writing—review and editing. All authors have read and agreed to the published version of the manuscript.

Funding: This research is co-financed by Greece and the European Union (European Social Fund-ESF) through the Operational Programme «Human Resources Development, Education and Lifelong Learning 2014–2020» in the context of the project "Cooperative distributed adaptive model predictive control methods using computational intelligence" (MIS 5050291).

Institutional Review Board Statement: Not applicable.

Informed Consent Statement: Not applicable.

Data Availability Statement: In this work, the publicly available AIS dataset provided by the Marine Cadastre service (www.marinecadastre.gov, accessed on 25 July 2021) has been used for trajectory modeling purposes.

Conflicts of Interest: The authors declare no conflict of interest. The funders had no role in the design of the study; in the collection, analyses, or interpretation of data; in the writing of the manuscript, or in the decision to publish the results.

References

1. Zaman, M.B.; Kobayashi, E.; Wakabayashi, N.; Khanfir, S.; Pitana, T.; Maimun, A. Fuzzy FMEA model for risk evaluation of ship collisions in the Malacca Strait: Based on AIS data. *J. Simul.* **2014**, *8*, 91–104. [CrossRef]
2. Zhang, J.; Zhang, D.; Yan, X.; Haugen, S.; Guedes Soares, C. A distributed anti-collision decision support formulation in multi-ship encounter situations under COLREGs. *Ocean Eng.* **2015**, *105*, 336–348. [CrossRef]
3. Puisa, R.; Lin, L.; Bolbot, V.; Vassalos, D. Unravelling causal factors of maritime incidents and accidents. *Saf. Sci.* **2018**, *110*, 124–141. [CrossRef]
4. IMO COLREG. *Convention on the International Regulations for Preventing Collisions at Sea, 1972*; IMO: London, UK, 2003.
5. Hu, L.; Naeem, W.; Rajabally, E.; Watson, G.; Mills, T.; Bhuiyan, Z.; Raeburn, C.; Salter, I.; Pekcan, C. A multiobjective optimization approach for COLREGs-Compliant path planning of autonomous surface vehicles verified on networked bridge simulators. *IEEE Trans. Intell. Transp. Syst.* **2020**, *21*, 1167–1179. [CrossRef]
6. Perera, L.P.; Carvalho, J.P.; Guedes Soares, C. Intelligent ocean navigation and fuzzy-Bayesian decision/action formulation. *IEEE J. Ocean. Eng.* **2012**, *37*, 204–219. [CrossRef]
7. Zhang, X.; Wang, C.; Chui, K.T.; Liu, R.W. A Real-Time Collision Avoidance Framework of MASS Based on B-Spline and Optimal Decoupling Control. *Sensors* **2021**, *21*, 4911. [CrossRef] [PubMed]
8. Wang, T.; Wu, Q.; Zhang, J.; Wu, B.; Wang, Y. Autonomous decision-making scheme for multi-ship collision avoidance with iterative observation and inference. *Ocean Eng.* **2020**, *197*, 106873. [CrossRef]
9. Johansen, T.A.; Perez, T.; Cristofaro, A. Ship collision avoidance and COLREGS compliance using simulation-based control behavior selection with predictive hazard assessment. *IEEE Trans. Intell. Transp. Syst.* **2016**, *17*, 3407–3422. [CrossRef]
10. Aguiar, A.P.; Pascoal, A.M. Dynamic positioning and way-point tracking of underactuated AUVs in the presence of ocean currents. *Int. J. Control* **2007**, *80*, 1092–1108. [CrossRef]
11. Caldwell, C.; Dunlap, D.; Collins, E. Motion planning for an autonomous Underwater Vehicle via Sampling Based Model Predictive Control. In Proceedings of the Oceans 2010 MTS/IEEE Seattle, Seattle, WA, USA, 20–23 September 2010; Volume 670–671, pp. 1370–1377.
12. Taherian, S.; Halder, K.; Dixit, S.; Fallah, S. Autonomous Collision Avoidance Using MPC with LQR-Based Weight Transformation. *Sensors* **2021**, *21*, 4296. [CrossRef]
13. Xie, S.; Garofano, V.; Chu, X.; Negenborn, R.R. Model predictive ship collision avoidance based on Q-learning beetle swarm antenna search and neural networks. *Ocean Eng.* **2019**, *193*, 106609. [CrossRef]
14. Chen, L.; Hopman, H.; Negenborn, R.R. Distributed model predictive control for vessel train formations of cooperative multi-vessel systems. *Transp. Res. Part C Emerg. Technol.* **2018**, *92*, 101–118. [CrossRef]
15. Zheng, H.; Negenborn, R.R.; Lodewijks, G. Robust Distributed Predictive Control of Waterborne AGVs-A Cooperative and Cost-Effective Approach. *IEEE Trans. Cybern.* **2018**, *48*, 2449–2461. [CrossRef]
16. Von Ellenrieder, K.D. Stable Backstepping Control of Marine Vehicles with Actuator Rate Limits and Saturation *. *IFAC-PapersOnLine* **2018**, *51*, 262–267. [CrossRef]
17. Haykin, S. *Neural Networks: A Comprehensive Foundation*, 3rd ed.; Prentice-Hall, Inc.: Upper Saddle River, NJ, USA, 1999; Volume 13, ISBN 0131471392.
18. Zhou, H.; Chen, Y.; Zhang, S. Ship Trajectory Prediction Based on BP Neural Network. *J. Artif. Intell.* **2019**, *1*, 29–36. [CrossRef]

19. Xu, T.; Liu, X.; Yang, X. Ship trajectory online prediction based on BP neural network algorithm. In Proceedings of the Proceedings-2011 International Conference of Information Technology, Computer Engineering and Management Sciences, ICM 2011, Penang, MY, USA, 13–14 June 2011; Volume 1, pp. 103–106.
20. Ma, S.; Liu, S.; Meng, X. Optimized BP neural network algorithm for predicting ship trajectory. In Proceedings of the 2020 IEEE 4th Information Technology, Networking, Electronic and Automation Control Conference, ITNEC 2020, Chongqing, China, 12–14 June 2020; Institute of Electrical and Electronics Engineers Inc.: Piscataway, NJ, USA, 2020; pp. 525–532.
21. Liu, X.; He, W.; Xie, J.; Chu, X. Predicting the Trajectories of Vessels Using Machine Learning. In Proceedings of the 2020 5th International Conference on Control, Robotics and Cybernetics, CRC 2020, Wuhan, China, 16–18 October 2020; Institute of Electrical and Electronics Engineers Inc.: Piscataway, NJ, USA, 2020; pp. 66–70.
22. Li, W.; Zhang, C.; Ma, J.; Jia, C. Long-term vessel motion predication by modeling trajectory patterns with AIS data. In Proceedings of the ICTIS 2019-5th International Conference on Transportation Information and Safety, Liverpool, UK, 14–17 July 2019; Institute of Electrical and Electronics Engineers Inc.: Piscataway, NJ, USA, 2019; pp. 1389–1394.
23. Ding, M.; Su, W.; Liu, Y.; Zhang, J.; Li, J.; Wu, J. A Novel Approach on Vessel Trajectory Prediction Based on Variational LSTM. In Proceedings of the 2020 IEEE International Conference on Artificial Intelligence and Computer Applications, ICAICA 2020, Dalian, China, 27–29 June 2020; Institute of Electrical and Electronics Engineers Inc.: Piscataway, NJ, USA, 2020; pp. 206–211.
24. Tampakis, P.; Chondrodima, E.; Pikrakis, A.; Theodoridis, Y.; Pristouris, K.; Nakos, H.; Petra, E.; Dalamagas, T.; Kandiros, A.; Markakis, G.; et al. Sea Area Monitoring and Analysis of Fishing Vessels Activity: The i4sea Big Data Platform. In Proceedings of the Proceedings-IEEE International Conference on Mobile Data Management, Versailles, France, 30 June–3 July 2020; Institute of Electrical and Electronics Engineers Inc.: Piscataway, NJ, USA, 2020; pp. 275–280.
25. Forti, N.; Millefiori, L.M.; Braca, P.; Willett, P. Prediction oof Vessel Trajectories from AIS Data Via Sequence-To-Sequence Recurrent Neural Networks. In Proceedings of the ICASSP, IEEE International Conference on Acoustics, Speech and Signal Processing-Proceedings, Barcelona, Spain, 4–8 May 2020; Institute of Electrical and Electronics Engineers Inc.: Piscataway, NJ, USA, 2020; pp. 8936–8940.
26. Wang, C.; Ren, H.; Li, H. Vessel trajectory prediction based on AIS data and bidirectional GRU. In Proceedings of the Proceedings-2020 International Conference on Computer Vision, Image and Deep Learning, CVIDL 2020, Chongqing, China, 1–12 July 2020; Institute of Electrical and Electronics Engineers Inc.: Piscataway, NJ, USA, 2020; pp. 260–264.
27. Tritsarolis, A.; Chondrodima, E.; Tampakis, P.; Pikrakis, A. Online Co-movement Pattern Prediction in Mobility Data. *arXiv* **2021**, arXiv:2102.08870.
28. Zhu, G.; Du, J.; Kao, Y. Robust adaptive neural trajectory tracking control of surface vessels under input and output constraints. *J. Frankl. Inst.* **2020**, *357*, 8591–8610. [CrossRef]
29. Zhang, C.; Wang, C.; Wei, Y.; Wang, J. Robust trajectory tracking control for underactuated autonomous surface vessels with uncertainty dynamics and unavailable velocities. *Ocean Eng.* **2020**, *218*, 108099. [CrossRef]
30. Li, C.; Zhao, Y.; Wang, G.; Fan, Y.; Bai, Y. Adaptive RBF neural network control for unmanned surface vessel course tracking. In Proceedings of the 6th International Conference on Information Science and Technology, ICIST 2016, Dalian, China, 6–8 May 2016; Institute of Electrical and Electronics Engineers Inc.: Piscataway, NJ, USA, 2016; pp. 285–290.
31. Stogiannos, M.; Papadimitrakis, M.; Sarimveis, H.; Alexandridis, A. Vessel Trajectory Prediction Using Radial Basis Function Neural Networks. In Proceedings of the 2021 IEEE 19th International Conference on Smart Technologies (EUROCON), Lviv, Ukraine, 6–8 July 2021.
32. Alexandridis, A.; Stogiannos, M.; Papaioannou, N.; Zois, E.; Sarimveis, H. An inverse neural controller based on the applicability domain of RBF network models. *Sensors* **2018**, *18*, 315. [CrossRef]
33. Yang, Q.; Ye, Z.; Li, X.; Wei, D.; Chen, S.; Li, Z. Prediction of flight status of logistics uavs based on an information entropy radial basis function neural network. *Sensors* **2021**, *21*, 3651. [CrossRef]
34. Bhartiya, S.; Whiteley, J.R. Factorized approach to nonlinear MPC using a radial basis function model. *AIChE J.* **2001**, *47*, 358–368. [CrossRef]
35. Alexandridis, A.; Sarimveis, H.; Ninos, K. A Radial Basis Function network training algorithm using a non-symmetric partition of the input space-Application to a Model Predictive Control configuration. *Adv. Eng. Softw.* **2011**, *42*, 830–837. [CrossRef]
36. Stogiannos, M.; Alexandridis, A.; Sarimveis, H. Model predictive control for systems with fast dynamics using inverse neural models. *ISA Trans.* **2018**, *72*, 161–177. [CrossRef] [PubMed]
37. Han, H.G.; Wu, X.L.; Qiao, J.F. Real-time model predictive control using a self-organizing neural network. *IEEE Trans. Neural Netw. Learn. Syst.* **2013**, *24*, 1425–1436. [CrossRef] [PubMed]
38. Wang, T.; Gao, H.; Qiu, J. A Combined Adaptive Neural Network and Nonlinear Model Predictive Control for Multirate Networked Industrial Process Control. *IEEE Trans. Neural Netw. Learn. Syst.* **2016**, *27*, 416–425. [CrossRef] [PubMed]
39. Alexandridis, A.; Chondrodima, E.; Giannopoulos, N.; Sarimveis, H. A fast and efficient method for training categorical radial basis function networks. *IEEE Trans. Neural Netw. Learn. Syst.* **2017**, *28*, 2831–2836. [CrossRef] [PubMed]
40. Karamichailidou, D.; Kaloutsa, V.; Alexandridis, A. Wind turbine power curve modeling using radial basis function neural networks and tabu search. *Renew. Energy* **2021**, *163*, 2137–2152. [CrossRef]
41. Tu, E.; Zhang, G.; Rachmawati, L.; Rajabally, E.; Huang, G. Bin Exploiting AIS Data for Intelligent Maritime Navigation: A Comprehensive Survey from Data to Methodology. *IEEE Trans. Intell. Transp. Syst.* **2018**, *19*, 1559–1582. [CrossRef]

42. Last, P.; Bahlke, C.; Hering-Bertram, M.; Linsen, L. Comprehensive Analysis of Automatic Identification System (AIS) Data in Regard to Vessel Movement Prediction. *J. Navig.* **2014**, *67*, 791–809. [CrossRef]
43. Zhang, L.; Meng, Q.; Xiao, Z.; Fu, X. A novel ship trajectory reconstruction approach using AIS data. *Ocean Eng.* **2018**, *159*, 165–174. [CrossRef]
44. Fu, P.; Wang, H.; Liu, K.; Hu, X.; Zhang, H. Finding Abnormal Vessel Trajectories Using Feature Learning. *IEEE Access* **2017**, *5*, 7898–7909. [CrossRef]
45. Emmens, T.; Amrit, C.; Abdi, A.; Ghosh, M. The promises and perils of Automatic Identification System data. *Expert Syst. Appl.* **2021**, *178*, 114975. [CrossRef]
46. Akima, H. A Method of Bivariate Interpolation and Smooth Surface Fitting Based on Local Procedures. *Commun. ACM* **1974**, *17*, 18–20. [CrossRef]
47. Zhang, J.; Yan, X.; Chen, X.; Sang, L.; Zhang, D. A novel approach for assistance with anti-collision decision making based on the International Regulations for Preventing Collisions at Sea. *Proc. Inst. Mech. Eng. Part M J. Eng. Marit. Environ.* **2012**, *226*, 250–259. [CrossRef]
48. Fossen, T.I. *Handbook of Marine Craft Hydrodynamics and Motion Control*; Wiley: Chichester, UK, 2011; ISBN 9781119991496.
49. Zheng, H.; Negenborn, R.R.; Lodewijks, G. *Trajectory Tracking of Autonomous Vessels Using Model Predictive Control*; IFAC: Cape Town, South Africa, 2014; Volume 19, ISBN 9783902823625.
50. Zhu, M.; Hahn, A.; Wen, Y.Q.; Bolles, A. Identification-based simplified model of large container ships using support vector machines and artificial bee colony algorithm. *Appl. Ocean Res.* **2017**, *68*, 249–261. [CrossRef]
51. Byrd, R.H.; Gilbert, J.C.; Nocedal, J. A trust region method based on interior point techniques for nonlinear programming. *Math. Program. Ser. B* **2000**, *89*, 149–185. [CrossRef]

sensors

MDPI

Article

Universal Path-Following of Wheeled Mobile Robots: A Closed-Form Bounded Velocity Solution †

Reza Oftadeh [1,*], Reza Ghabcheloo [2] and Jouni Mattila [2]

1 Department of Computer Science and Engineering, Texas A&M University, College Station, TX 77840, USA
2 Department of Automation Technology and Mechanical Engineering, Tampere University, 33720 Tampere, Finland; reza.ghabcheloo@tuni.fi (R.G.); jouni.mattila@tuni.fi (J.M.)
* Correspondence: reza.oftadeh@tamu.edu

† This paper is an extended version of our paper published in Oftadeh, R.; Ghabcheloo, R.; Mattila, J. A time-optimal bounded velocity path-following controller for generic Wheeled Mobile Robots. In Proceedings of the 2015 IEEE International Conference on Robotics and Automation (ICRA), Seattle, WA, USA, 26–30 May 2015, pp. 676–683.

Abstract: This paper presents a nonlinear, universal, path-following controller for Wheeled Mobile Robots (WMRs). This approach, unlike previous algorithms, solves the path-following problem for all common categories of holonomic and nonholonomic WMRs, such as omnidirectional, unicycle, car-like, and all steerable wheels. This generality is the consequence of a two-stage solution that tackles separately the platform path-following and wheels' kinematic constraints. In the first stage, for a mobile platform divested of the wheels' constraints, we develop a general paradigm of a path-following controller that plans asymptotic paths from the WMR to the desired path and, accordingly, we derive a realization of the presented paradigm. The second stage accounts for the kinematic constraints imposed by the wheels. In this stage, we demonstrate that the designed controller simplifies the otherwise impenetrable wheels' kinematic and nonholonomic constraints into explicit proportional functions between the velocity of the platform and that of the wheels. This result enables us to derive a closed-form trajectory generation scheme for the asymptotic path that constantly keeps the wheels' steering and driving velocities within their corresponding, pre-specified bounds. Extensive experimental results on several types of WMRs, along with simulation results for the other types, are provided to demonstrate the performance and the efficacy of the method.

Keywords: wheeled mobile robots; path-following; nonholonomic constraints

Citation: Oftadeh, R.; Ghabcheloo, R.; Mattila, J. Universal Path-Following of Wheeled Mobile Robots: A Closed-Form Bounded Velocity Solution. *Sensors* **2021**, *21*, 7642. https://doi.org/10.3390/s21227642

Academic Editor: Subhas Mukhopadhyay

Received: 9 October 2021
Accepted: 9 November 2021
Published: 17 November 2021

1. Introduction

Wheeled Mobile Robots (WMRs) form a significant subset of Unmanned Ground Vehicles (UGVs). The continuing demand for more advanced and autonomous UGVs entails more reliable and higher-performance motion controllers for WMRs. The multitude of motion controllers proposed for mobile robots, especially those with nonholonomic constraints, may be roughly categorized into three groups [1]: point stabilization [2], trajectory tracking [3], and path-following [4]. Typically, the path-following approach is used under a decoupled control architectures [5], where a path-planner provides the desired geometric path. Then, a path-following module, while considering the temporal and other intrinsic constraints of the system, maneuvers the robot toward the planned path and steers it so that it indefinitely follows the path.

The path-following algorithms are classified into several branches, including, but not limited to, Optimal Control approaches [6,7], Feedback Linearizion methods [8], Line of Sight guidance laws [9], Pure Pursuit techniques [10], and Vector Fields methods [11] (see [12] for a thorough review). The majority of these algorithms incorporate a concept that goes by many names, including "Virtual Target Point", "Carrot", and "Rabbit". In this concept, a virtual point is selected and moved along the desired path, while a tracking controller, also called a guidance controller, makes the robot follow and converge to that point. The differences among and between those branches mainly occur in the design of the guidance controller, the method for selecting the virtual point, and the its motion strategy along the path.

1.1. Related Work

This work belongs to a category of path-followers that parametrize (using the path's natural parameter) a virtual position for the mobile robot on the desired path. A nonlinear guidance controller is employed for the robot to track the virtual point based on an error space projected on the path through the path's Serret–Frenet frame.

Early notable works in this field were conducted by [13] for the car-like WMRs, by [14] for the unicycle types, and by [15] for both unicycles and WMRs with two steering wheels. However, the projection scheme of this approach, which selects the path's closest point to the robot as the virtual point, would result in singularities that, in turn, would impose stringent initial conditions on the system and on the drivable path curvatures. This drawback was overcome by having the path's natural parameter as an auxiliary state and, therefore, deriving a control law for its progression rate along the path [16]. There have been various extensions of the original problem, such as covering uncertainties [17], actuator saturations [18], obstacle avoidance [19] and an extension to aerial [20], marine [21] and articulated frame [22] vehicles. Comprehensive experimental results of some of these algorithms have been provided by [23]. However, the majority of recent studies on the path-following of WMRs consider only a special type of WMR: usually the unicycle type, with some exceptions, such as [24]. Nevertheless, there is no unified solution for the path-following of WMRs.

Aside from the general difficulties in designing a universal path-follower for WMRs, this paper tackles several challenges in the design of motion controllers that are *specific* to certain types of WMRs. Most notable is the presence of singularities, both inherently [25] and in the representation of the configuration space [26] of WMRs with active steering wheels. While several types of WMR possess steering wheels, those singularities are a major issue for WMRs that utilize two or more actively steered standard wheels. Due to their kinematic configuration, such WMRs are called nonholonomic omnidirectional robots [27,28] or pseudo-omnidirectional robots [29,30]. They have recently gained a significant level of popularity and are now being used in a wide range of practical fields, including service robotics (PR2 [31], Care-O-bot [32], Rollin' Justin [33]), space robotics (Mars-Exo-Rover [30]), agricultural applications [34,35], among others [36].

The common way of formulating the configuration space of such WMRs is with the notion of Instantaneous Center of Rotation (ICR) [37]. As the ICR moves closer to a wheel axis, the driving velocity of that wheel decreases, while the curvature of the wheel's footprint, and, hence, its steering velocity, unboundedly increases. When the ICR coincides with the wheel, its steering angle becomes undefined and singular. To circumvent such singular configurations, many proposed solutions rely on numerical methods to plan singularity-free ICR trajectories in velocity space [25]. Others treat the neighborhood of the singularities as obstacles and solve a navigation problem [29,38]. However, in all of those methods, considerable portions of the configuration space are avoided, thus reducing the maneuverability of the platform. Furthermore, when the robot is required to follow a desired path and heading profile, ICR position has already been determined and, therefore, none of those approaches are suitable for path-following problems.

1.2. Contributions and Organization

The contributions of this paper are as follows:

- This study solves the path-following problem for all WMRs categories in which their wheels roll without skidding. To the best knowledge of the authors, this is the first study that coherently solves the path-following problem with this level of generality.

- Unlike other path-followers, in this design, the control signals and the resultant vector field of closed-loop equations of motion are *linearly proportional* to the base speed (In this paper, speed exclusively refers to the magnitude of velocity vectors.). In fact, the controller acts as a feedback path-planner that minimizes the Lyapunov function of errors as its corresponding cost function.
- The kinematic constraints of all types of wheels are rigorously derived in their most general form. We derive and prove sufficient conditions for a path-following controller that simplify the kinematic and nonholonomic constraints into explicit relations between the speed of the base and that of the wheels.
- Based on this framework, we present a *closed-form* solution for the speed of the WMR's base so that all the wheels' steering and driving speeds remain within their respective bounds. We show that the solution is time-optimal, because it provides a bang-bang velocity profile in which, at each time step, at least one of the wheels runs at its maximum speed.
- This solution allows WMRs with active steering wheels to get close to, and even pass, their singular configurations by regulating the speed of the WMR and the steering velocities of the wheels. Hence, this method expands the allowable configuration space of such robots and allows them to exploit their whole maneuverability.

This paper is the culmination of several earlier studies presented by the authors in [39–41]. Compared to [41], this paper has several novelties. Section 3 is new, in which the controller in [41] is generalized into a generic parametrized form and we demonstrate that the controller serves as an example of such generic form. Moreover, the results of [40] are coherently included to address the singularities of WMRs with steering wheels. The majority of the equations, especially the kinematic constraints, are derived in a more general form and are presented in compact matrix format that is further consistent with the formulation of WMR constraints in the literature. A new set of comprehensive experimental and simulation results in a more complex scenario is presented, with further explanations and remarks.

This paper is organized as follows. In Section 2, we formally define the WMR that is the focus of this paper and define the corresponding path-following problem. Next, in Section 3, we present the path-following solution in a parametrized generic form. In Section 4, the variables in the parametrized model are meticulously derived and categorized for different types of wheels and WMRs. We explain how the solution solves the problem of singularities for WMRs with steering wheel in Section 4.3 and, finally, Section 5 covers the implementation results and provides extensive experimental data for three types of WMRs, and simulation results for the other two.

2. Problem Description

WMRs are classified based on the seminal work of [42] into five kinematically feasible categories. An ordered pair (δ_m, δ_s) is assigned to each category the degree of mobility and degree of steerability of the WMR, respectively. The number of wheels, their types, and their arrangements determine δ_m, and δ_s and, therefore, the WMRs' category. δ_m represents the dimension of the tangent space of the configuration space, while δ_s corresponds to the ability to change (steer) the basis of the tangent space by means of steering wheels. The degree of maneuverability defined as $\delta_M = \delta_m + \delta_s$ then, analogous to degrees of freedom for mechanism, determines the total mobility of the WMR.

2.1. WMR Architecture and Definitions

Definition 1 (WMR). *The WMR considered in this paper is equipped with n wheels, which are attached to a main body called the* base. *It belongs to and possesses the minimum actuated wheels of one of those five kinematically feasible categories of WMRs, and the actuators provide velocity and position control. Each wheel is of one of the following types:* fixed wheels, standard steerable wheels, off-centered steerable wheels (Caster wheels), Swedish wheels. *Furthermore,* *the following assumptions hold. The WMR traverses on a flat and horizontal plane. The base and the wheels are rigid, the tires are non-deformable, their contact surface with the ground can be approximated with a point, and there is no mechanical constraint for the steering of the steerable wheels; hence, they are free-turn. However, we will discuss the case with limited steering angle separately in Section 5.3.*

Figure 1, depicts a schematic view of a WMR and the desired path. Additionally, Table 1 lists the definitions of the corresponding parameters and variables. We define the base current posture $\boldsymbol{\mathcal{X}}$, as

$$\boldsymbol{\mathcal{X}} = \begin{bmatrix} \boldsymbol{q}^{\mathrm{T}} & \theta_b \end{bmatrix}^{\mathrm{T}}, \tag{1}$$

where $\boldsymbol{q}$ is the position vector of point Q (the origin of body frame $\boldsymbol{\mathcal{B}}$) and θ_b is the heading angle. The base linear velocity at Q is $v\hat{\boldsymbol{v}}$ with v being the base speed and $\hat{\boldsymbol{v}}(\psi_v)$ being the direction of the velocity as a function of the linear velocity angle ψ_v. Moreover, based on Definition 1, for a general WMR, several types of wheels may be connected to the base. As shown in Figure 1, the connection point of the ith wheel is $\boldsymbol{L}_i$ and the velocity of the wheel is $v_i\hat{\boldsymbol{v}}_i$, in which v_i is the driving speed and $\hat{\boldsymbol{v}}_i(\phi_i)$ is the direction of wheels heading, which, in turn, is a function of wheel's steering angle ϕ_i.

The desired path $\boldsymbol{P}_d$ is a 2D and bounded-curvature regular curve on the horizontal plane. This is defined by the vector-valued function $\boldsymbol{P}_d(s) : [0, L_d] \to \mathbb{R}^2$, where s and L_d are the natural parametrization (arc length) and the length of $\boldsymbol{P}_d$, respectively. The desired heading function $\theta_d(s) : [0, L_d] \to \mathbb{R}$ of class C^3 determines the base's desired heading θ_d. Similar to Equation (1), the WMR desired posture, $\boldsymbol{\mathcal{X}}_d$, is defined as

$$\boldsymbol{\mathcal{X}}_d = \begin{bmatrix} \boldsymbol{P}_d^{\mathrm{T}} & \theta_d \end{bmatrix}^{\mathrm{T}}. \tag{2}$$

Furthermore, the path $\boldsymbol{P}_a(\lambda)$ is an asymptotic path to $\boldsymbol{P}_d$. It is defined by $\boldsymbol{P}_a(\lambda) : [0, L_a] \to \mathbb{R}^2$, where λ and L_a are the natural parametrization (arc length) and length of the path, respectively, with its tangent angle denoted by ψ_a. Correspondingly, the function $\theta_a(\lambda) : [0, L_d] \to \mathbb{R}$ is an asymptotic angle from the base current heading θ_b, to the desired heading θ_d. We will later show that $\boldsymbol{P}_a(\lambda)$ and $\theta_a(\lambda)$ are solutions to autonomous differential equations with the current pose of the base as the initial conditions.

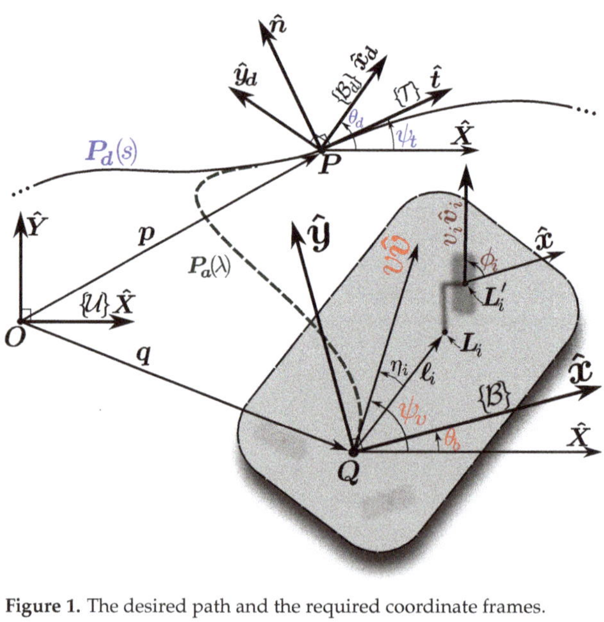

Figure 1. The desired path and the required coordinate frames.

Table 1. Parameters and Variables.

	Coordinate Frames and Their Basis
$\mathcal{U}\{\hat{X}, \hat{Y}\}$	The inertial frame with its origin $\boldsymbol{O}$
$\mathcal{B}\{\hat{x}, \hat{y}\}$	The body frame attached to the WMR's base $\boldsymbol{Q}$
$\mathcal{B}_v\{\hat{v}, \hat{u}\}$	The base velocity coordinate frame at $\boldsymbol{Q}$
$\mathcal{T}\{\hat{t}, \hat{n}\}$	The Frenet–Serret frame of the desired path, $\boldsymbol{P}_d$, at $\boldsymbol{P}$
$\mathcal{B}_d\{\hat{x}_d, \hat{y}_d\}$	The desired heading coordinate frame at $\boldsymbol{P}$
	Paths' Parameters and Variables
s	Natural parameter(arc length) of the desired path, $\boldsymbol{P}_d$
$\kappa_d(s)$	The curvature of the desired path, $\boldsymbol{P}_d$ ($\lvert\kappa_d\rvert < \kappa \neq \infty$)
λ	Natural parameter(arc length) of the asymptotic path, $\boldsymbol{P}_a$
	Position Vectors
$\boldsymbol{p} \triangleq$	$\begin{bmatrix} p_x & p_y & 0 \end{bmatrix}^{\mathrm{T}}$; the position vector of the virtual target point, $\boldsymbol{P}$, with respect to $\boldsymbol{O}$
$\boldsymbol{q} \triangleq$	$\begin{bmatrix} q_x & q_y & 0 \end{bmatrix}^{\mathrm{T}}$; the position vector of the base, $\boldsymbol{Q}$, with respect to $\boldsymbol{O}$
$\boldsymbol{\ell}_i$	The position of ith wheel attachment point, $\boldsymbol{L}_i$, to the base with respect to $\boldsymbol{Q}$
	Angles
θ_b	The WMR's heading angle that defines the body frame $\mathcal{B}$
θ_d	The desired heading angle; defines the frame $\mathcal{B}_d$ at $\boldsymbol{P}$
ψ_v	The angle of the base linear velocity direction, $\hat{v}$
ψ_d	The desired angle for the base linear velocity direction, $\hat{v}$
$\psi_{vb} \triangleq$	$\psi_v - \theta_b$
ψ_t	The tangent angle; defines the desired path tangent vector $\hat{t}$
ϕ_i	The ith wheel steering angle
η_i	The angle between the base linear velocity $v\hat{v}$ and the ith wheel position vector, $\boldsymbol{\ell}_i$
	Others
$\hat{v}$	The direction vector of the base linear velocity at $\boldsymbol{Q}$
v	The speed of the base at $\boldsymbol{Q}$
$\hat{v}_i$	The direction vector of the ith wheel linear velocity
v_i	The driving speed of the ith wheel
$\omega_b \triangleq$	$\dot{\theta}_b$; the base angular velocity
$\omega_v \triangleq$	$\dot{\psi}_v$; the angular rate of $\hat{v}$
$\omega_{vb} \triangleq$	$\omega_v - \omega_b$

2.2. Problem Formulation

As shown in Figure 1, the virtual target point on $\boldsymbol{P}_d$ is denoted as $\boldsymbol{P}$. It is determined by s, which is set as an *auxiliary* state with $\dot{s}$ being its corresponding control signal. Along with s, we define error state variables as

$$\mathcal{S} = (x_e, y_e, \theta_e, \psi_e) \tag{3a}$$

$$\mathcal{S}^* = (x_e, y_e, \theta_e), \tag{3b}$$

where,

$$\begin{bmatrix} x_e \\ y_e \end{bmatrix} = {}^{\mathcal{U}}\boldsymbol{R}_{\mathcal{T}}^{-1}(\boldsymbol{q} - \boldsymbol{p}) \tag{4a}$$

$$\theta_e = \theta_d - \theta_b, \tag{4b}$$

$$\psi_e = \psi_d(s, \boldsymbol{\mathcal{S}}^*) - \psi_v, \tag{4c}$$

and ${}^{\mathcal{U}}\boldsymbol{R}_{\mathcal{T}}$, equivalent to $\boldsymbol{R}(\psi_t)$, is the rotation matrix from frame $\boldsymbol{\mathcal{T}}$ to frame $\boldsymbol{\mathcal{U}}$. In the above, the position error signals, x_e and y_e, are measured along $\hat{\boldsymbol{t}}$ and $\hat{\boldsymbol{n}}$, respectively, while θ_e independently represents the heading error. The signal ψ_d, as a function of $\boldsymbol{\mathcal{S}}^*$, is the desired heading that is determined by the controller and basically generates a suitable approach angle to $\boldsymbol{P}_d(s)$.

The time derivation of Equations (4a) to (4c) yields to the open-loop equations of motion

$$\dot{x}_e = \dot{s}(\kappa_d(s)y_e - 1) + v\cos(\psi_t - \psi_v) \tag{5a}$$

$$\dot{y}_e = -\dot{s}\kappa_d(s)x_e - v\sin(\psi_t - \psi_v) \tag{5b}$$

$$\dot{\theta}_e = \frac{d\theta_d}{ds}\dot{s} - \omega_b \tag{5c}$$

$$\dot{\psi}_e = \dot{\psi}_d - \omega_v, \tag{5d}$$

in which the angular velocity of the frame $\boldsymbol{\mathcal{B}}$, is $\omega_b \triangleq \dot{\theta}_b$ and, similarly, $\omega_v \triangleq \dot{\psi}_v$. We assemble all of the above formulations into the following definition for the WMR's base.

Definition 2 (Base Path-Following). *Base path-following error dynamics is a system with the full states $(s, \boldsymbol{\mathcal{S}})$ or the reduced states $(s, \boldsymbol{\mathcal{S}}^*)$, and with the control inputs $\boldsymbol{\mathcal{C}}$, which are defined as*

$$\boldsymbol{\mathcal{C}} = (\dot{s}, \omega_b, \omega_v). \tag{6}$$

The dynamics of this system is given by Equations (5a) to (5d), in which the base speed v is seen as an exogenous input.

Some further notes are given here for the reduced states, $\boldsymbol{\mathcal{S}}^*$. For the majority of WMRs, except a special case, it is possible to incorporate a rather simpler scheme by dismissing ψ_e in (4c) as an error state ($\psi_e = 0 \quad \forall t$), therefore controlling the base velocity direction $\hat{\boldsymbol{v}}(\psi_v)$ directly. In the reduced states $\boldsymbol{\mathcal{S}}^*$, ψ_v acts as a control signal by directly setting $\psi_v = \psi_d(s, \boldsymbol{\mathcal{S}}^*)$. In Section 4.2, we will detail the types of WMRs and conditions under which the choice of $\boldsymbol{\mathcal{S}}^*$ is not possible.

In what follows, we formally define the problem that is the focal point of this paper.

Problem 1 (WMR Bounded Velocity Path-Following). *Given the desired path $\boldsymbol{P}_d(s)$ and heading profile $\theta_d(s)$ derive feedback control laws for the wheels' driving v_i and steering inputs ϕ_i (or their rates $\dot{\phi}_i$) such that:*

- *(I) Path-Following: The velocity frame $\boldsymbol{\mathcal{B}}_v$ converges and follows the tangent frame $\boldsymbol{\mathcal{T}}$; that is, error signals x_e and y_e, remain bounded and converge to zero (See Equation (4a)).*
- *(II) Heading Control: The body frame $\boldsymbol{\mathcal{B}}$ converges and follows $\boldsymbol{\mathcal{B}}_d$; that is, the heading error signal θ_e, remains bounded and converges to zero (See (4b)).*
- *(III) Bounded Velocity: v_i and $\dot{\phi}_i$ should not exceed their corresponding predefined limits.*

We solve the above problem in two stages. In the first stage, we place our focus solely on the base path-following defined by Definition 2, its stability and features. In the second stage, we focus on the kinematic constraints between the wheels and the base, the details of which are given in Section 3.2. We utilize these constraints to solve sub-problems (I) and (II) of Problem 1 by mapping the intermediary control inputs $\boldsymbol{\mathcal{C}}$, to the wheels' variables.

Furthermore, those mappings are used to solve the bounded velocity problem (sub-problem (III)) by selecting an appropriate v that bounds the driving and steering velocities.

At first glance, this approach is solely applicable to holonomic omnidirectional WMRs ($\delta_m = 3$). However, this treatment is a rather general and we will show that, for each category of WMRs, a proper subset of the above intermediary control sets along with a pertinent choice of body origin Q results in a feasible solution that abides by the kinematic limitations of that category.

3. WMR Path-Following: The Generic Form

In the following, Section 3.1 focuses on a generic form of $\boldsymbol{C}$ for the base path-following that has some unique features. Then, in Section 3.3, those features are utilized to present a parametrized version of the WMRs' kinematic constraints, which, in turn, are incorporated into a closed-form solution for Problem 1.

3.1. Base Path-Following

Proposition 1. *For the base path-following system defined by Definition 2, assume that some exists feedback control laws exist for $\boldsymbol{C}$ that render the origin of the error space asymptotically stable and they are in the generic form of*

$$\dot{s} = s'(s, \boldsymbol{\mathcal{S}})v \tag{7a}$$
$$\omega_b = \kappa_b(s, \boldsymbol{\mathcal{S}})v \tag{7b}$$
$$\omega_v = \kappa_v(s, \boldsymbol{\mathcal{S}})v \tag{7c}$$
$$\psi_d = \psi_d(s, \boldsymbol{\mathcal{S}}^*), \tag{7d}$$

in which s', κ_b, and κ_v are functions of only *s and error states, $\boldsymbol{\mathcal{S}}$. Then, at any given time $t \geq 0$, the closed-loop equations of motion result in a set of differential equations for an asymptotic path $\boldsymbol{P}_a(\lambda)$ and a heading $\theta_a(\lambda)$, with the initial conditions being $\boldsymbol{q}(t)$ and $\theta_b(t)$, respectively. In other words, $\boldsymbol{\mathcal{X}}(t)$, the base posture at the time t, is*

$$\boldsymbol{\mathcal{X}}(t) = \begin{bmatrix} \boldsymbol{P}_a^T(\lambda = 0) & \theta_a(\lambda = 0) \end{bmatrix}^T, \tag{8}$$

and as λ increases, $\begin{bmatrix} \boldsymbol{P}_a^T(\lambda > 0) & \theta_a(\lambda > 0) \end{bmatrix}^T$ asymptotically converges toward the desired posture $\boldsymbol{\mathcal{X}}_d$.

Proof. To prove the above proposition, notice that $||\dot{\boldsymbol{q}}|| = v$ and a geometric variable λ exist, where $\dot{\lambda} = v$. Consequently, Equation (7a) becomes $\dot{s} = s'\dot{\lambda}$. Based on the chain rule, we have $v = \frac{ds}{d\lambda}\dot{\lambda}$ and, hence, $s' = \frac{ds}{d\lambda}$. Using the same analogy for Equations (7b) and (7c) and substituting them into the open-loop error states (Equations (5a) to (5d)) results in

$$\frac{ds(\lambda)}{d\lambda} = s'(s, \boldsymbol{\mathcal{S}}) \tag{9a}$$
$$\frac{dx_e(\lambda)}{d\lambda} = x_e'(s, \boldsymbol{\mathcal{S}}) = s'(\kappa_d(s)y_e - 1) + \cos(\psi_t - \psi_v) \tag{9b}$$
$$\frac{dy_e(\lambda)}{d\lambda} = y_e'(s, \boldsymbol{\mathcal{S}}) = -s'\kappa_d(s)x_e - \sin(\psi_t - \psi_v) \tag{9c}$$
$$\frac{d\theta_e(\lambda)}{d\lambda} = \theta_e'(s, \boldsymbol{\mathcal{S}}) = \frac{d\theta_d}{ds}s' - \kappa_b(s, \boldsymbol{\mathcal{S}}) \tag{9d}$$
$$\frac{d\psi_e(\lambda)}{d\lambda} = \psi_e'(s, \boldsymbol{\mathcal{S}}) = \psi_d' - \kappa_v(s, \boldsymbol{\mathcal{S}}), \tag{9e}$$

where,

$$\psi_d' = \frac{d\psi_d}{d\lambda} = \frac{\partial \psi_d}{\partial x_e}x_e' + \frac{\partial \psi_d}{\partial y_e}y_e' + \frac{\partial \psi_d}{\partial \theta_e}\theta_e'. \tag{10}$$

The above equations provide a set of differential equations for $(s, \boldsymbol{\mathcal{S}})$ based on the independent variable λ. The algebraic Equations (4a) to (4c) can be used to track the geometric evolution of $\boldsymbol{q}$, θ_b, and ψ_v as functions of λ, which are $\boldsymbol{P}_a(\lambda)$, $\theta_a(\lambda)$, and $\psi_a(\lambda)$, respectively, and are governed by

$$\boldsymbol{P}_a(\lambda) = \boldsymbol{P}_d(s(\lambda)) + {}^{\mathcal{U}}\boldsymbol{R}_{\mathcal{T}}(s(\lambda)) \begin{bmatrix} x_e(\lambda) \\ y_e(\lambda) \end{bmatrix} \tag{11a}$$

$$\theta_a(\lambda) = \theta_d(s(\lambda)) - \theta_e(\lambda) \tag{11b}$$

$$\psi_a(\lambda) = \psi_d(s(\lambda), \boldsymbol{\mathcal{S}}^*(\lambda)) - \psi_e(\lambda). \tag{11c}$$

Therefore, the differential Equations (9a) to (9e) together with the algebraic Equations (11a) to (11c) result in a set of expressions for the asymptotic path $\boldsymbol{P}_a(\lambda)$ and heading $\theta_a(\lambda)$. Notice that λ does not explicitly appear in any of the above equations; therefore, $(s(t), \boldsymbol{\mathcal{S}}(t))$, the base path-following states at time t, can be associated with initial conditions ($\lambda = 0$) of the above differential equations. □

The merit of Proposition 1 is that a majority of path-following controllers in the literature (e.g., [16,19]) cannot be written in the generic form of Equations (7a) to (7c). Therefore, while, in those path-followers, v is an exogenous input and does not have a direct role in the stability (as long as $v \geq v_m > 0$), the asymptotic path $\boldsymbol{P}_a$ cannot be determined independently of $v(t)$ and it is only after the assignment of the speed profile that one can derive the asymptotic trajectory $\boldsymbol{P}_a(\lambda(t))$. However, Proposition 1 enables us to determine $\boldsymbol{P}_a(\lambda)$ without specifying the future velocity commands by directly integrating closed-loop equations. In other words, a path-following controller in the form of Equations (7a) to (7d) acts as a feedback path-planner that has a certain error function as its cost function and plans asymptotic paths from the base current posture $\boldsymbol{\mathcal{X}}$, toward the desired posture $\boldsymbol{\mathcal{X}}_d$. We provide an example for such a controller in Section 4.1.

In this paper, there is no need to solve the closed-loop differential equations. The differentials of $(s, \boldsymbol{\mathcal{S}})$, obtained in the form of the above proposition, will be used for WMRs' kinematic constraints in the next proposition. For that purpose, the higher time differentiations of control signals and error states may also be written as differentials based on λ, which are listed below for future reference.

$$\ddot{s} = s''(s, \boldsymbol{\mathcal{S}}, \boldsymbol{\mathcal{S}}')v^2 + s'(s, \boldsymbol{\mathcal{S}})\dot{v} \tag{12a}$$

$$\dot{\omega}_b = \kappa_b'(s, \boldsymbol{\mathcal{S}}, \boldsymbol{\mathcal{S}}')v^2 + \kappa_b(s, \boldsymbol{\mathcal{S}})\dot{v} \tag{12b}$$

$$\dot{\omega}_v = \kappa_v'(s, \boldsymbol{\mathcal{S}}, \boldsymbol{\mathcal{S}}')v^2 + \kappa_v(s, \boldsymbol{\mathcal{S}})\dot{v}, \tag{12c}$$

where, $s'' = \frac{ds'}{d\lambda} = \frac{d^2s}{d\lambda^2}$, $\kappa_b' = \frac{d\kappa_b}{d\lambda}$, $\kappa_v' = \frac{d\kappa_v}{d\lambda}$, and

$$\boldsymbol{\mathcal{S}}' = \left(x_e', y_e', \theta_e', \psi_e'\right). \tag{13}$$

Finally, for the closed-loop system, we may transform ${}^{\mathcal{B}}\dot{\boldsymbol{\mathcal{X}}}$ and ${}^{\mathcal{B}}\ddot{\boldsymbol{\mathcal{X}}}$, the velocity and acceleration of the base posture (Equation (1)) expressed in the body frame $\boldsymbol{\mathcal{B}}$, into

$${}^{\mathcal{B}}\dot{\boldsymbol{\mathcal{X}}} = {}^{\mathcal{B}}\boldsymbol{\mathcal{X}}'v, \qquad {}^{\mathcal{B}}\dot{\boldsymbol{\mathcal{X}}}' = {}^{\mathcal{B}}\boldsymbol{\mathcal{X}}''v, \tag{14a}$$

$${}^{\mathcal{B}}\ddot{\boldsymbol{\mathcal{X}}} = {}^{\mathcal{B}}\boldsymbol{\mathcal{X}}''v^2 + {}^{\mathcal{B}}\boldsymbol{\mathcal{X}}'\dot{v}, \tag{14b}$$

$${}^{\mathcal{B}}\boldsymbol{\mathcal{X}}' = \begin{bmatrix}\cos\psi_{vb} & \sin\psi_{vb} & \kappa_b)\end{bmatrix}^{\mathrm{T}}, \tag{14c}$$

$${}^{\mathcal{B}}\boldsymbol{\mathcal{X}}'' = \begin{bmatrix}-\kappa_{vb}\sin\psi_{vb} & \kappa_{vb}\cos\psi_{vb} & \kappa_b'\end{bmatrix}^{\mathrm{T}}, \tag{14d}$$

in which, $\psi_{vb} \triangleq \psi_v - \theta_b$, and $\kappa_{vb} \triangleq \kappa_v - \kappa_b$. The above transformation facilitates the treatment of kinematic constraints in Section 3.2.

3.2. Wheels' Kinematic Constraints

Figure 2 depicts a schematic view of an abstract Generalized Wheel (GW) as the ith wheel of the WMR and its corresponding parameters. The GW represents both Swedish wheels and normal wheels. In this sense, r_{sr} and γ define the radius and the direction of the small rollers' axis, respectively; hence, for a functional Swedish wheel: $\gamma \neq \frac{\pi}{2}$ and $r_{sr} \neq 0$. For a normal, non-Swedish wheel, we simply set $\gamma = \frac{\pi}{2}$ and $r_{sr} = 0$. The wheel is mounted on an L-shaped rod, parametrized by off-center values: (d_i, c_i), at the point L'_i, with ${}^{\mathcal{B}}\boldsymbol{\ell}'_i = c_i\, {}^{\mathcal{B}}\hat{\boldsymbol{u}}_i + d_i\, {}^{\mathcal{B}}\hat{\boldsymbol{v}}_i$. The rod is connected to the base at the attachment point $\boldsymbol{L}_i$ by a revolute joint. As shown in the figure, ϕ_i represents the steering angle of the wheel, and $v_i\hat{\boldsymbol{v}}_i$ represents its *driving velocity vector*, generated by the wheel's actuator. We define the absolute steering angle Φ_i and the projection matrix $\boldsymbol{J}_i(\hat{\boldsymbol{x}}_i)$ as

$$\Phi_i = \theta_b + \phi_i, \tag{15a}$$

$$\boldsymbol{J}_i(\hat{\boldsymbol{x}}_i) = \begin{bmatrix} \hat{\boldsymbol{x}}_i^{\mathrm{T}} & \hat{\boldsymbol{x}}_i.(\hat{\boldsymbol{z}} \times {}^{\mathcal{B}}\boldsymbol{\ell}_i) \end{bmatrix}, \tag{15b}$$

in which $\hat{\boldsymbol{x}}_i$ is an arbitrary unit vector.

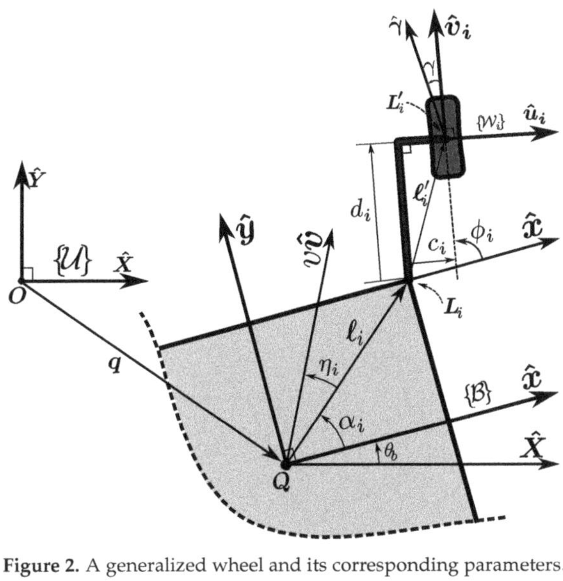

Figure 2. A generalized wheel and its corresponding parameters.

Moreover, for fixed and Swedish types, the steering direction ${}^{\mathcal{B}}\hat{\boldsymbol{v}}_i(\phi_i)$, is a mechanical design variable, and is measured for steerable types, except when it is set as the control signal—a case that will be further explored. Table 2, for each type of wheel, lists the required values for the GW variables and parameters. To derive the kinematic relations for the GW, we may differentiately form the vector relation $\boldsymbol{L}'_i = \boldsymbol{q} + \boldsymbol{\ell}_i + d_i\hat{\boldsymbol{v}}_i + c_i\hat{\boldsymbol{u}}_i$ and express it in the body frame to arrive at a velocity constraint between the wheel and the base, which is

$$\begin{aligned} v\, {}^{\mathcal{B}}\hat{\boldsymbol{v}} + \omega_b(\hat{\boldsymbol{z}} \times {}^{\mathcal{B}}\boldsymbol{\ell}_i) =& (v_i - c_i\dot{\Phi}_i)\, {}^{\mathcal{B}}\hat{\boldsymbol{v}}_i + d_i\dot{\Phi}_i\, {}^{\mathcal{B}}\hat{\boldsymbol{u}}_i \\ & - r_{sr}\dot{\phi}_{sr}(\hat{\boldsymbol{z}} \times {}^{\mathcal{B}}\hat{\boldsymbol{\gamma}}), \end{aligned} \tag{16}$$

in which $\dot{\phi}_{sr}$ is the angular velocity of the GW's small rollers. The above equation can be manipulated into scalar equations

$$v_i\, \hat{\boldsymbol{a}}_i.{}^{\mathcal{B}}\hat{\boldsymbol{v}}_i = \boldsymbol{J}_i(\hat{\boldsymbol{a}}_i)\, {}^{\mathcal{B}}\dot{\boldsymbol{\mathcal{X}}} + \dot{\Phi}_i\hat{\boldsymbol{a}}_i.(\hat{\boldsymbol{z}} \times {}^{\mathcal{B}}\boldsymbol{\ell}'_i) \tag{17}$$

where,

$$\hat{\boldsymbol{a}}_i = \begin{cases} {}^{\mathcal{B}}\hat{\boldsymbol{\gamma}}(\gamma) & \text{Swedish wheel}(\gamma \neq \frac{\pi}{2}, r_{sr} \neq 0) \\ {}^{\mathcal{B}}\hat{\boldsymbol{v}}_i(\phi_i) & \text{Other types}(\gamma = \frac{\pi}{2}, r_{sr} = 0) \end{cases}. \tag{18}$$

Equation (17) accompanied by the proper choice of $\hat{\boldsymbol{a}}_i$ (Equation (18)) determines the driving velocity v_i. By definition, for fixed and Swedish wheels $\dot{\phi}_i = 0$. For the steerable wheels, when $d_i \neq 0$, $\dot{\phi}_i$ is determined by setting $\hat{\boldsymbol{a}}_i = {}^{\mathcal{B}}\hat{\boldsymbol{u}}_i$ in Equation (17), which eliminates the left-hand side of the equation. When ${}^{\mathcal{B}}\hat{\boldsymbol{u}}_i.(\hat{\boldsymbol{z}} \times {}^{\mathcal{B}}\boldsymbol{\ell}'_i)$ is zero (equivalent to $d_i = 0$), setting $\hat{\boldsymbol{a}}_i = {}^{\mathcal{B}}\hat{\boldsymbol{u}}_i$ and $d_i = 0$ in Equation (17) and differentiating from this yields:

$$\dot{\Phi}_i \boldsymbol{J}_i({}^{\mathcal{B}}\hat{\boldsymbol{v}}_i)\, {}^{\mathcal{B}}\dot{\boldsymbol{\mathcal{X}}} + \boldsymbol{J}_i({}^{\mathcal{B}}\hat{\boldsymbol{u}}_i)\, {}^{\mathcal{B}}\ddot{\boldsymbol{\mathcal{X}}} = \omega_b^2\, {}^{\mathcal{B}}\hat{\boldsymbol{u}}_i.{}^{\mathcal{B}}\boldsymbol{\ell}_i. \tag{19}$$

Evidently, the above kinematic constraints cannot be analytically solved for a base speed that results in specified driving and steering velocities. However, as in the following proposition, we show that the incorporation of Proposition 1 into the kinematic constraints yields a set of direct relationships between the base speed and wheel velocities.

Table 2. GW Parameters for each Type of Wheel.

Variable \ Type	Fixed Wheel	Centered Steering Wheel	Caster Wheel	Swedish Wheel
${}^{\mathcal{B}}\hat{\boldsymbol{v}}_i(\phi_i)$	Fixed	Measurement or Equation (24)	Measurement	Fixed
d_i	$d_i \in \mathbb{R}$	$d_i = 0$	$d_i \in \mathbb{R}_{\neq 0}$	$d_i \in \mathbb{R}$
c_i	$c_i \in \mathbb{R}$	$c_i \in \mathbb{R}$	$c_i \in \mathbb{R}$	$c_i \in \mathbb{R}$
γ	$\gamma = \frac{\pi}{2}$	$\gamma = \frac{\pi}{2}$	$\gamma = \frac{\pi}{2}$	$\gamma \neq \frac{\pi}{2}$
r_{sr}	$r_{sr} = 0$	$r_{sr} = 0$	$r_{sr} = 0$	$r_{sr} \neq 0$

Proposition 2. *Consider a WMR defined in Definition 1. If the WMRs intermediary control signals $\boldsymbol{\mathcal{C}}$ conform to the generic form, as outlined in Proposition 1, then the driving and steering velocities of the ith wheel ($1 \leq i \leq n$) are in the form of*

$$v_i = v_i'(s, \boldsymbol{\mathcal{S}}, \boldsymbol{\mathcal{S}}', \boldsymbol{\mathcal{C}}')v \tag{20a}$$

$$\dot{\phi}_i = \phi_i'(s, \boldsymbol{\mathcal{S}}, \boldsymbol{\mathcal{S}}', \boldsymbol{\mathcal{C}}')v, \tag{20b}$$

in which, $v_i' = \frac{dv_i}{d\lambda}$, $\phi_i' = \frac{d\phi_i}{d\lambda}$ *, and*

$$\boldsymbol{\mathcal{C}}' = (s', \kappa_b, \kappa_v). \tag{21}$$

Proof. We prove this proposition by constructing v_i' and ϕ_i'. This is carried out by substituting the results of Proposition 1 into the kinematic constraints and performing some algebraic manipulations that can be simplified to the form of Equations (20a) and (20b). Based on Equations (7b) and (15a), we have $\dot{\Phi}_i = \omega_b + \dot{\phi}_i$ and $\omega_b = \kappa_b v$. Therefore, the proof for Equation (20b) is equivalent to finding an expression for $\dot{\Phi}_i = \Phi_i' v$ and setting $\phi_i' = \Phi_i' - \kappa_b$. Clearly, Equation (20b) automatically holds for zero steering wheels (fixed and Swedish wheels) with $\Phi_i' = \kappa_b$ and $\phi_i' = 0$. For the other types, setting $\hat{\boldsymbol{a}}_i = {}^{\mathcal{B}}\hat{\boldsymbol{u}}_i$ in Equation (17), and substituting ${}^{\mathcal{B}}\dot{\boldsymbol{\mathcal{X}}}$ by ${}^{\mathcal{B}}\boldsymbol{\mathcal{X}}'v$ from Equation (14c), and ${}^{\mathcal{B}}\ddot{\boldsymbol{\mathcal{X}}}$ from Equation (14b) in Equation (19), this can be rewritten in the form of $\dot{\Phi}_i = \Phi_i' v$ with

$$\Phi_i' = \begin{cases} d_i^{-1} \boldsymbol{J}_i({}^{\mathcal{B}}\hat{\boldsymbol{u}}_i){}^{\mathcal{B}}\boldsymbol{\mathcal{X}}' & d_i \neq 0 \\ \left(\boldsymbol{J}_i({}^{\mathcal{B}}\hat{\boldsymbol{v}}_i){}^{\mathcal{B}}\boldsymbol{\mathcal{X}}'\right)^{-1}\left(\kappa_b^2\, {}^{\mathcal{B}}\hat{\boldsymbol{u}}_i.{}^{\mathcal{B}}\boldsymbol{\ell}_i - \boldsymbol{J}_i({}^{\mathcal{B}}\hat{\boldsymbol{u}}_i){}^{\mathcal{B}}\boldsymbol{\mathcal{X}}''\right) & d_i = 0 \end{cases}, \tag{22}$$

which proves the second relation, Equation (20b). Notice that, for $d_i = 0$, the choice of control signals eliminates the acceleration terms from Equation (19) and simplifies it into the form of $\dot{\Phi}_i = \Phi_i' v$. In this case, the only caveat is that $\boldsymbol{J}_i({}^{\mathcal{B}}\hat{\boldsymbol{v}}_i){}^{\mathcal{B}}\boldsymbol{\mathcal{X}}'$ in the above

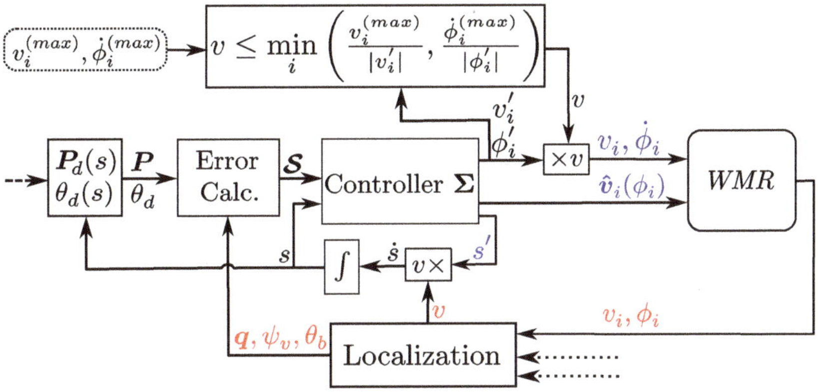

Figure 3. Schematic block diagram of the whole system.

equation may become very small, or even zero. This situation corresponds to the singularity configuration of wheels with $d_i = 0$ that are called centered steering wheels. This situation and its treatment is fully explained in Section 4.3. Finally, following the same paradigm, Equation (17) can be manipulated into the form of Equation (20a) with

$$v_i' = (\hat{\boldsymbol{a}}_i . {}^{\mathcal{B}}\hat{\boldsymbol{v}}_i)^{-1} \left(J_i(\hat{\boldsymbol{a}}_i)\ {}^{\mathcal{B}}\boldsymbol{\mathcal{X}}' + \Phi_i' \hat{\boldsymbol{a}}_i . (\hat{\boldsymbol{z}} \times {}^{\mathcal{B}}\boldsymbol{\ell}'_i) \right), \tag{23}$$

in which $\hat{\boldsymbol{a}}_i$ is determined by Equation (18). □

Remark 1. *For a WMR with centered steerable wheels ($d_i = 0$), under the reduced state model $\boldsymbol{\mathcal{S}}^*$, the wheels' steering angles ϕ_i can be set as control signals and are derived as follows. Based on Equation (16), the steering direction is ${}^{\mathcal{B}}\hat{\boldsymbol{v}}_i(\phi_i)$ is $v\ {}^{\mathcal{B}}\hat{\boldsymbol{v}} + \omega_b(\hat{\boldsymbol{z}} \times {}^{\mathcal{B}}\boldsymbol{\ell}_i)$. Therefore, for the closed-loop system, ω_b can be replaced with $\kappa_b v$ and the steering direction becomes*

$$ {}^{\mathcal{B}}\hat{\boldsymbol{v}}_i = \frac{{}^{\mathcal{B}}\hat{\boldsymbol{v}} + \kappa_b(\hat{\boldsymbol{z}} \times {}^{\mathcal{B}}\boldsymbol{\ell}_i)}{\sqrt{1 + \kappa_b^2 l_i^2 + 2 l_i k_b \sin \eta_i}}. \tag{24}$$

The major benefit of the above formulation is that it determines the proper steering direction of the wheels, even when the robot is stopped and the base speed is zero.

3.3. WMR Path-Following

The previous proposition most importantly shows that, for the closed-loop path-following of all WMR types, the kinematic constraints between the base and the wheels can be reduced to proportional functions of the base speed with proportions that are only functions of s and instant error states $\boldsymbol{\mathcal{S}}$. Consequently, a suitable solution to Subproblems (I) and (II) of Problem 1 is the selection of an arbitrary v, and a $\boldsymbol{C}$ that conforms to conditions in Proposition 1, and then using Equations (20a) and (20b) to evaluate the wheels' driving and steering velocities. Furthermore, to solve the Subproblem (III), instead of having an arbitrary profile for v, Proposition 2 can be used to find instant limits for v that bound the wheels' velocities. Such a solution exhaustively solves Problem 1 and is formally stated in the following Algorithm 1, with Figure 3 schematically depicting the corresponding block diagram.

Algorithm 1 WMR Bounded Velocity Path-Following.

Assume that the WMR possess n_d driving actuators and n_s steering actuators ($n_d + n_s \leq 2n$). The maximum driving velocity of the ith driving actuator is denoted as $v_i^{(\max)}$, and the maximum steering velocity of the ith steering actuator is denoted as $\dot{\phi}_i^{(\max)}$. At each time step, the control signals of wheels' actuators is evaluated by

1. **Desired Inputs:** Evaluate the internal state s, by integrating the internal feedback $\dot{s}$, and obtain the virtual target values $\boldsymbol{P}$ and θ_d by using $\boldsymbol{P}_d(s)$ and $\theta_d(s)$.
2. **Error Calculation:** Evaluate the error states $\boldsymbol{\mathcal{S}}$ by using the localization feedback (see Equations (4a) to (4c)).
3. **Controller $\boldsymbol{\Sigma}$:** Evaluate $\boldsymbol{\mathcal{S}}'$ (Equation (13)), and $\boldsymbol{\mathcal{C}}'$ (Equation (21)) and, by using the values of those signals, obtain $\hat{\boldsymbol{v}}_i$, v_i', and ϕ_i' of Equations (20a) and (20b) for all the wheels.
4. **Bounded Velocity:** Based on Equations (20a) and (20b), there are $n_d + n_s$ candidates for v, namely, $v^{(i)}$, which are

$$v^{(i)} = \frac{v_i^{(\max)}}{|v_i'|} \qquad \text{and} \qquad v^{(n_d+i)} = \frac{\dot{\phi}_i^{(\max)}}{|\phi_i'|} \tag{25}$$

Then, the maximum allowable base speed, denoted as $v^{(\max)}$, is

$$v^{(\max)} = \min_i v^{(i)}, \quad i \in \{1, 2, .., n_d + n_s\}. \tag{26}$$

5. **Wheels' Control Inputs:** Select a $v \leq v^{(\max)}$, and use $v_i = v_i' v$ and $\dot{\phi}_i = \phi_i' v$ to evaluate the actuators' velocity commands (Equations (20a) and (20b)).

Note that Equations (20a) and (20b), which are used in the forth step of the above algorithm to derive the velocity candidates, are strictly monotonic with respect to v_i, and $\dot{\phi}_i$ and so is their inverse with respect to v. Hence, applying the minimum of those m velocity candidates results in driving and steering velocities less than or equal to the given velocity bounds. Alternatively, at each instant, if $v = v^{(\max)}$ is selected, then at least one of the actuators is being driven at its maximum velocity, which renders the solution a bang-bang control [43] for the velocity v and, therefore, for a given desired path, heading, and control gains, the solution is time-optimal.

4. WMR Path-Following: Detailed Illustration

In this section, we provide the pertinent expressions for the parametrized controller presented in the previous section. In Section 4.1, we present an example controller for the base that complies with the conditions of Proposition 1, and thereby customize Algorithm 1 for WMRs based on their degree of maneuverability, δ_M.

4.1. Base Path-Following: The Controller

First, we define $\psi_d(s, \boldsymbol{\mathcal{S}}^*)$, the desired input for ψ_v as

$$\psi_d(s, \boldsymbol{\mathcal{S}}^*) = \psi_t - \sigma(y_e), \tag{27}$$

in which, $\sigma(y_e)$ is a function that generates a suitable approach angle from the base to $\boldsymbol{P}_d(s)$ and has the following features: $\sigma(0) = 0$ and $y_e\sigma(y_e) > 0 \quad \forall y_e \neq 0$. One candidate for $\sigma(y_e)$ is

$$\sigma(y_e) \triangleq \sin^{-1}\frac{k_2 y_e}{|y_e| + \epsilon}, \tag{28}$$

where $0 < k_2 \leq 1$ and $\epsilon > 0$.

Based on the above definition, it can be construed that, for large normal errors y_e, $\sigma(y_e)$ goes toward $\pi/2$ and, consequently, the base turns toward the virtual point $\boldsymbol{P}$ to decrease the error. As y_e decreases, $\sigma(y_e)$ moves toward zero and the base velocity turns toward the path tangent at $\boldsymbol{P}$; therefore, the robot motion becomes more aligned with the path. Evidently, larger values for k_2 result in sharper turns for the robot to reach the path.

Proposition 3. *The feedback control laws for signals $\boldsymbol{C}$ that are given by*

$$\dot{s} = s'(s, \boldsymbol{\mathcal{S}})v \qquad \omega_b = \theta_b'(s, \boldsymbol{\mathcal{S}})v \qquad \omega_v = \psi_v'(s, \boldsymbol{\mathcal{S}})v, \tag{29}$$

where,

$$s'(s, \boldsymbol{\mathcal{S}}) = k_1 x_e + \cos(\psi_t - \psi_v) \tag{30a}$$

$$\theta_b'(s, \boldsymbol{\mathcal{S}}) = k_3\theta_e + \frac{d\theta_d}{ds}s' \tag{30b}$$

$$\psi_v'(s, \boldsymbol{\mathcal{S}}) = \psi_d' \tag{30c}$$

$$\psi_d' = \kappa_d s' - y_e' \frac{d\sigma(y_e)}{dy_e}. \tag{30d}$$

and $k_1, k_3 > 0$, lead the reduced error states, $\boldsymbol{\mathcal{S}}^*$, to asymptotically converge to zero. Moreover, replacing Equation (30c) with

$$\psi_v'(s, \boldsymbol{\mathcal{S}}) = \kappa_d s' - \frac{d\sigma(y_e)}{dy_e} y_e' - \kappa_e^2 y_e \Delta + k_4 \psi_e \tag{31a}$$

$$\Delta = \begin{cases} \frac{\sin(\psi_t - \psi_v) - \sin\sigma(y_e)}{\psi_e} & \psi_e \neq 0 \\ \cos\sigma(y_e) & \psi_e = 0 \end{cases}, \tag{31b}$$

in which $\kappa_e \neq 0$ and $k_4 > 0$ result in error states $\boldsymbol{\mathcal{S}}$ asymptotically converging to zero. Consequently, the origin of the error space is stable and semi-globally exponentially stable by setting $v(t) \geq v_m > 0 \;\; \forall t$.

Proof. Here, we use standard quadratic form of error signals as the Lyapunov function but with modified control laws to make curvatures independent of speed, v. First, we provide a proof for the case where error states are $\boldsymbol{\mathcal{S}}^*$, and then we extend it for the full state $\boldsymbol{\mathcal{S}}$. Consider the following Lyapunov function:

$$V_1 = \frac{1}{2}x_e^2 + \frac{1}{2}y_e^2 + \frac{1}{2}\theta_e^2, \tag{32}$$

which is positive, definite and radially unbounded. The time differentiation of V_1, along with the solution of Equations (5a) to (5c), results in:

$$\dot{V}_1 = -(k_1 x_e^2 + k_2 \frac{y_e^2}{|y_e| + \epsilon} + k_3\theta_e^2)\, v(t), \tag{33}$$

which is negative; thus, the origin is stable. For a given $d_1 > 0$, if $v(t) \geq v_m > 0$ and, initially, $|y_e(t_0)| < d_1$, it is easy to show that $\dot{V}_1 < -\lambda V_1$. Thus, the origin is semi-globally exponentially stable.

To complete the proof, consider the following Lyapunov function:

$$V_2 = V_1 + \frac{1}{2\kappa_e^2}\psi_e^2. \tag{34}$$

The time differentiation of V_2 along the solution of Equations (5a) to (5d) results in:

$$\dot{V}_2 = \dot{V}_1 - \frac{k_4}{\kappa_e^2}\psi_e^2\, v(t), \tag{35}$$

which, again, is negative; therefore, the origin of the error state is stable. □

The above control laws clearly follow the generic format of Proposition 1. As mentioned in the above proposition, with the given control laws, the origin of $\boldsymbol{\mathcal{S}}$ is semi-globally exponentially stable for non-zero base speeds. The practical implication of this feature is that the WMR may stop ($v = 0$) for some time during the path-following operation, during which the states remain bounded and the path-following is naturally resumed once the WMR starts to move and the speed is not zero anymore. Note that there are several other functions for $\sigma(y_e)$ in the literature. However, while all of them result in negative $\dot{V}$, the one presented here is the one that results in a quadratic form for y_e in $\dot{V}$, and hence provides exponential stability.

4.2. Customization of the Path-Following Algorithm

WMRs are classified based into five different categories on the ordered pair $\delta = (\delta_m, \delta_s)$. Three of these categories possess the degree of maneuverability $\delta_M = \delta_m + \delta_s = 3$, and, for the other two, $\delta_M = 2$. In the following, we customize Algorithm 1 based on the WMR's degree of maneuverability and explain the accompanying details. The results of this section are listed in Table 3.

Table 3. Customization of the Path-Following.

Param. / δ_M	Error States	Desired Inputs	Selection of Q	κ_b	κ_v	s'
$\delta_M = 3$	$\boldsymbol{\mathcal{S}}^*$	$\boldsymbol{P}_d(s), \theta_d(s)$	Arbitrary	θ'_b (30b)	ψ'_d (30d)	s' (30a)
$\delta_M = 3$	$\boldsymbol{\mathcal{S}}$	$\boldsymbol{P}_d(s), \theta_d(s)$	Arbitrary	θ'_b (30b)	ψ'_v (31a)	s' (30a)
$\delta_M = 2$	$\boldsymbol{\mathcal{S}}^*$	$\boldsymbol{P}_d(s)$	Not on $\boldsymbol{a}_c$ [1]	ψ'_d (30d)	ψ'_d (30d)	s' (30a)
$\delta_M = 2$	$\boldsymbol{\mathcal{S}}$	$\boldsymbol{P}_d(s)$	Not on $\boldsymbol{a}_c$	κ_b (37)	ψ'_v (31a)	s' (30a)
$\delta_M = 2$	$\boldsymbol{\mathcal{S}}$	$\boldsymbol{P}_d(s)$	On $\boldsymbol{a}_c$	ψ'_v (31a)	ψ'_v (31a)	s' (30a)

[1] $\boldsymbol{a}_c$ is the common axis of the fixed wheels.

4.2.1. WMRs with $\delta_M = 3$

These types of WMRs are omnidirectional in nature, which means that they have independent heading and linear movements. However, the holonomic type with $\delta = (3,0)$ provides full mobility and, hence, instantaneous velocity in any direction. The other two categories ($\delta = (2,1)$ and $\delta = (1,2)$) are steerable and nonholonomic. They are capable of providing movement in any arbitrary direction, but only after they have steered their wheels to the corresponding configuration. For the problem at hand, the difference between holonomic and nonholonomic types only occurs at the beginning of the path, in which the holonomic type may start path-following instantly, but the nonholonomic types have to rearrange their steerable wheels. Other than this, on a smooth path and heading profile, both types provide the same functionality.

These types of WMRs are capable of changing the direction of their base linear velocity while the base is stationary. This can be instantly performed in the case of holonomic types or in the case of nonholonomic types by changing the direction of the steering wheels over time. This fact allows us to directly control ψ_v and have $\boldsymbol{\mathcal{S}}^*$ as the error states instead of $\boldsymbol{\mathcal{S}}$. Consequently, for this type of WMR, the heading and linear movements are independent. Hence, the controller's inputs are both $\boldsymbol{P}_d$ and θ_d, and the body frame $\boldsymbol{\mathcal{B}}$ is chosen arbitrarily. For the base path-following (Definition 2) of such WMRs, s', and κ_b of Proposition 1 are s', and θ'_b given by Equations (30a) and (30b) in Section 4.1, respectively. There are two viable options for κ_v. If the target states is set to be full states $(s, \boldsymbol{\mathcal{S}}^*)$, then, as mentioned earlier, κ_v should be set as ψ'_d given by Equation (30d). The second option is to set $(s, \boldsymbol{\mathcal{S}})$ as the target states and, therefore, κ_v should be evaluated using ψ'_v given by Equation (31a). Table 3 summarize these results.

4.2.2. WMRs with $\delta_M = 2$

These types of WMRs have limited mobility in their working plane, and the heading and linear movements are dependent. They are either *differential* with $\delta = (2, 0)$ or *carlike* with $\delta = (1, 1)$. Both categories have a set of coaxial fixed wheels. The only difference between these categories occurs at the beginning of the path-following. The *differential* type starts the path-following instantly, but the *carlike* type has to steer its steerable wheel according to the start of the path. Aside from this difference, both types provide the same functionality on a smooth path.

For such WMRs, there is no independent heading control and the progress of the base heading, as shown in Figure 4, is linked to the position of body origin Q. Therefore, in what follows, we strive to derive the control laws for ω_b, in the form of $\kappa_b v$ of Equation (7b), which observes the kinematic constraints. For a kinematically feasible WMR, all the fixed wheels are coaxial and, therefore, all $\hat{\boldsymbol{u}}_i$ are on the same line, which we call this common axis $\boldsymbol{a}_c$. To derive the constraint equations, set $\ddot{\phi}_i = \dot{\phi}_i = 0$ in Equation (16), and its time derivative, which can be rearranged into

$$\begin{aligned} \omega_b(d_i - {}^{\mathcal{B}}\hat{\boldsymbol{u}}_i.(\hat{\boldsymbol{z}} \times {}^{\mathcal{B}}\boldsymbol{\ell}_i)) &= v\, {}^{\mathcal{B}}\hat{\boldsymbol{u}}_i.{}^{\mathcal{B}}\hat{\boldsymbol{v}} && \text{(36a)} \\ v(\omega_b - \omega_v)(v + l_i \omega_b \sin \eta_i) &= (d_i v_i + v l_i \cos \eta_i)\dot{\omega}_b \\ &\quad - (d_i \dot{v}_i + \dot{v} l_i \cos \eta_i)\omega_b. && \text{(36b)} \end{aligned}$$

From the above equations, it follows that if Q, the origin of the body frame, is not on the common axis $\boldsymbol{a}_c$, a kinematically consistent expression for ω_b may be derived based on Equation (36a). However, when Q is placed on $\boldsymbol{a}_c$, Equation (36a) degenerates, since both ${}^{\mathcal{B}}\hat{\boldsymbol{u}}_i.{}^{\mathcal{B}}\hat{\boldsymbol{v}}$ and $d_i - {}^{\mathcal{B}}\hat{\boldsymbol{u}}_i.(\hat{\boldsymbol{z}} \times {}^{\mathcal{B}}\boldsymbol{\ell}_i)$ become zero and Equation (36b) should be employed. If Q has been placed on the common axis, then the right-hand side of Equation (36b) is zero and, in order for the constraint to be valid, ω_b should be set equal to ω_v at all times. In both cases, ω_b is in the form of $\kappa_b(s, \boldsymbol{S})v$, which are

$$\kappa_b = \begin{cases} \left(d_i - {}^{\mathcal{B}}\hat{\boldsymbol{u}}_i.(\hat{\boldsymbol{z}} \times {}^{\mathcal{B}}\boldsymbol{\ell}_i)\right)^{-1} {}^{\mathcal{B}}\hat{\boldsymbol{u}}_i.{}^{\mathcal{B}}\hat{\boldsymbol{v}} & Q \text{ is not on } \boldsymbol{a}_c \\ \kappa_v & Q \text{ is on } \boldsymbol{a}_c. \end{cases} \quad (37)$$

Finally, based on Equation (36b), if Q is on the common axis of the fixed wheels $\boldsymbol{a}_c$, then the heading is tangent to (or has a constant misalignment with) the footprint of Q, which, in the case of path-following, is eventually ψ_t. In this case, the WMR is not able to instantly provide any arbitrary ψ_v; hence, only the full states path-following $(s, \boldsymbol{S})$ are possible. Consequently, for the base path-following, s' is evaluated using Equation (30a) and κ_b, and κ_v are both set to be ψ_v' given by Equation (31a). On the other hand, if Q is not placed on $\boldsymbol{a}_c$ then, based on Equation (37), ω_b can be used to achieve any arbitrary ψ_v. Consequently, both the full states' path-following $(s, \boldsymbol{S})$, and the reduced states' path-following $(s, \boldsymbol{S}^*)$ are possible. For the base path-following, s' and κ_b are evaluated using Equations (30a) and (37), respectively. κ_v is set to ψ_v' given by Equation (31a) in the full states' case or ψ_d' given by Equation (30d) in the case of reduced states. These results are also listed in order in Table 3.

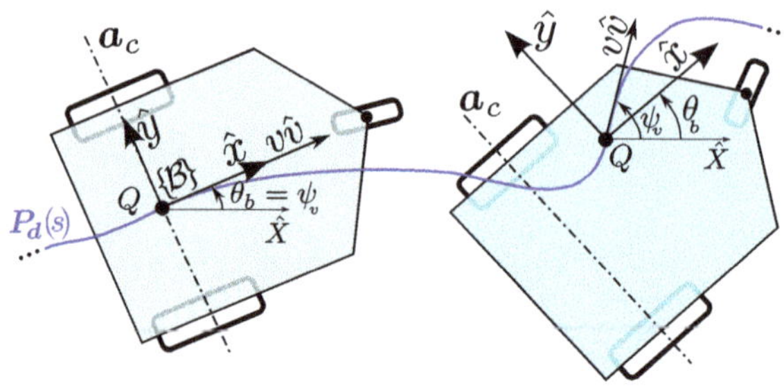

Figure 4. Two WMRs with $\delta_M = 2$. For the one on the left, the origin of $\boldsymbol{\mathcal{B}}$ is on the $\boldsymbol{a}_c$ and for the one on the right, the origin is outside of $\boldsymbol{a}_c$.

4.3. Analysis of Steering Wheels Singularities

In Section 4.1, we provided a stable path-following controller for the base path-following system defined in Definition 2. Next, we incorporated the constraints of Section 3.2 to map the base control signals to the wheels' velocities. The path-following of a WMR with a stable base controller is stable if there is no singularity in the mapping from the base onto the wheel. Hence, in this section, we elaborate on the singularities of the mapping and how the bounded velocity path-following algorithm treats and resolves those singularities, which specifically occurs with centered steering wheels.

As mentioned in the introduction, one way of examining the singularity of the steering wheels is by studying the base ICR. As the ICR moves closer to a wheel axis, the driving velocity of that wheel decreases and its steering velocity unboundedly increases. When the ICR coincides with the wheel's steering axis, the steering angle becomes undefined and singular. Here, we study this situation both geometrically and analytically based on the WMR's path-following.

Figure 5 shows the path-following of a WMR with four steering wheels that follow a straight line $\boldsymbol{P}_d$, during which it rotates around itself for 2π. As shown in the magnified area of the figure, during the operation, as the body ICR moves close to the first wheel, the curvature of $\boldsymbol{P}_1$, which is the wheel's footprint, increases; therefore, for the wheel to *keep up* with the rest of the WMR, it has to increase its steering velocity to pass the tight curvature. At the singularity, the ICR coincides with the steering axis, the curvature of the wheel's path becomes infinite and the steering direction becomes undefined. Theretofore, geometrically, the kinematic mapping between the base and a wheel becomes singular when a smooth path and movement toward a base result in an infinite curvature for the wheel's path.

In Section 3.2, for the closed-loop system, we derived the driving and steering velocities, v_i and ϕ_i, in the form of $v_i'v$ and $\phi_i'v$, respectively. The curvature of $\boldsymbol{P}_i$, the ith steering wheel's path, denoted as κ_i, becomes $\kappa_i = \phi_i'/v_i'$ and can be written as

$$\kappa_i = \frac{\kappa_b' l_i \cos\eta_i + (\kappa_v - \kappa_b)(1+l_i k_b \sin\eta_i)}{(1+\kappa_b^2 l_i^2 + 2l_i\kappa_b \sin\eta_i)^{\frac{3}{2}}}. \tag{38}$$

The singularity occurs when the denominator of the above equation becomes zero; that is, when $\sin\eta_i = -\operatorname{sgn}\kappa_b$ and $|\kappa_b| = 1/l_i$.

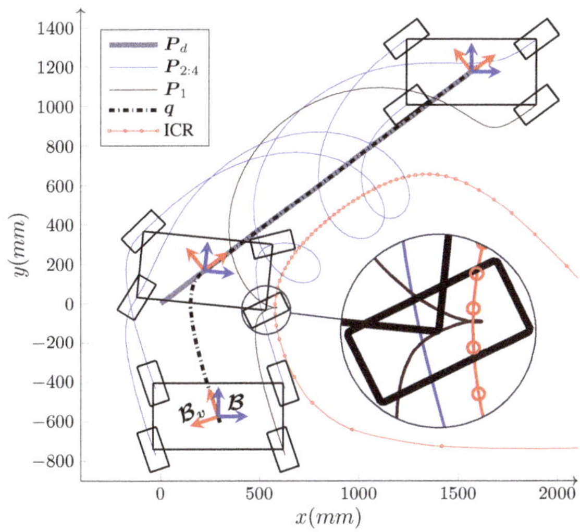

Figure 5. A nonholonomic omnidirectional WMR with four steering wheels following the desired path $\boldsymbol{P}(d)$, which is a straight line, while turning around itself.

One of the significant benefits of the bounded velocity path-following presented in Algorithm 1 is that it properly handles the steering wheels' singularities. As the wheel moves close to its singular configuration ϕ_i' unboundedly increases; therefore, based on Equation (25), that is, $v^{(n_d+i)} = \dot{\phi}_i^{(\max)}/|\phi_i'|$, the velocity candidate $v^{(n_d+i)}$, is reduced to limit $\dot{\phi}_i$ to its maximum. This, in turn, leads to the reduction in the base speed v. Hence, as the WMR moves close to its singular configuration, it reduces its speed and allows the steering wheel to make a tight turn. Figure 6 demonstrates this procedure for the path-following scenario depicted in Figure 5. If the WMR falls right into the singular configuration, ϕ_i' becomes infinity and the robot stops. At this configuration, the curvature of $\boldsymbol{P}_i$ is infinity and the path has inward and outward tangents at the singular point. This configuration can be seen as the start of a new path-following; the WMR reorientates its singular wheel from the inward tangent to the outward tangent and starts a new path-following.

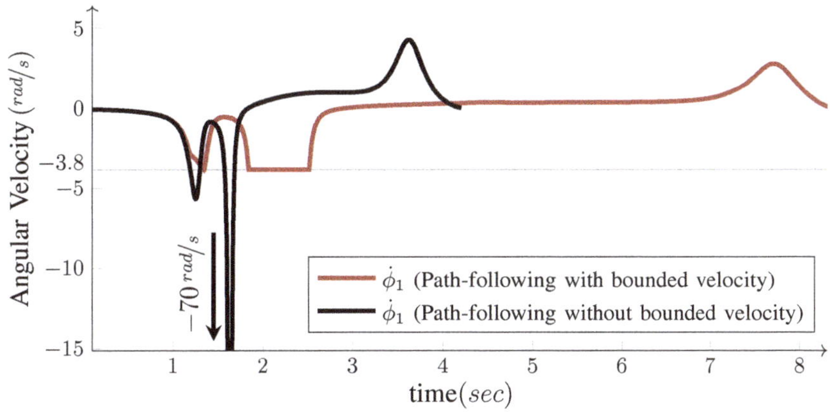

Figure 6. First wheel angular velocity $\dot{\phi}_1$, with and without bounded velocity.

5. Experimental and Simulation Results

In this section, we demonstrate some simulation and experimental data of the presented path-following controller in action. The results are for four categoiesy of WMR.

The experimental setup consists of two WMRs, shown in Figure 7. The robot in the left is a four-wheel, independently steered mobile manipulator called iMoro that is a non-holonomic omnidirectional WMR ($\delta = (1,2)$), also known as a two-steer. By manually fixing the steering of the rear wheels (setting $\dot{\phi}_i = 0$), it can also emulate the car-like type ($\delta = (1,1)$). In this case, the steering of the front wheels is naturally governed by the path-following based on the Ackerman principal. This feature was incorporated to test the algorithm for the car-like WMRs. For the last type, the WMR to the right of Figure 7, called LabRat, was employed, which is a differential drive mobile robot ($\delta = (2,0)$) and represents the unicycle kinematics. The path-following controller, namely, the Algorithm 1, was implemented on these WMRs in a real-time Linux environment based on [44]. This section is divided into three case studies; each focuses on different types of WMR and different scenarios (Videos of some of the experiments are provided in the multimedia attachment.). Finally, we finish this section by a brief discussion on some of the restrictions of the presented framework.

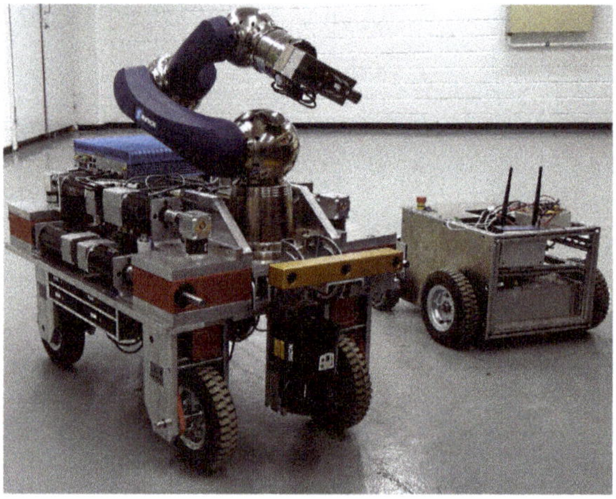

Figure 7. *Experimental setups*: iMoro (**left**): a four-wheel independently steering WMR ($\delta = (1,2)$), and LabRat (**right**): a differential drive WMR ($\delta = (2,0)$) with active fixed wheels at the rear.

5.1. Case Study I: $\delta = (3,0)$ and $\delta = (2,0)$

The simulation was performed on a holonomic omnidirectional WMR ($\delta = (3,0)$). The WMR has the same architecture as iMoro, but the steering wheels are virtually replaced by Swedish wheels of the same radii. Figures 8 and 9 portray the path-following scenario, the base footprint $\boldsymbol{q}$, and the two of the wheels' paths for the holonomic type (simulation) and the unicycle type (experiment using Labrat), respectively. As shown in the figures, the desired path is smooth, but has sharp turns and, hence, large curvatures at some of its points, which challenges the agility of the controller.

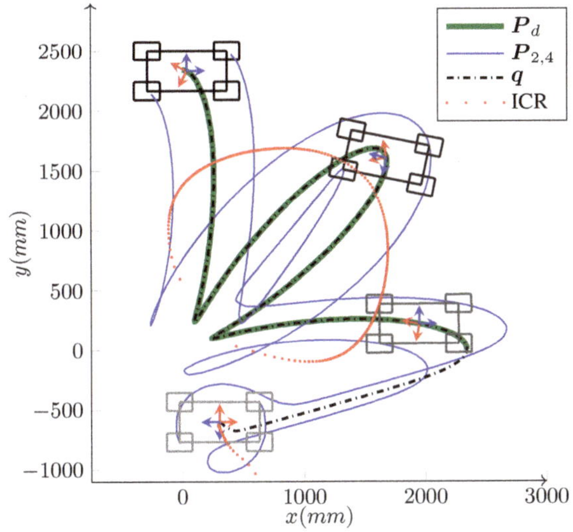

Figure 8. *Simulation*: Path-following with large initial errors of a WMR with four Swedish wheels ($\delta = (3,0)$). It seeks and follows the path $\boldsymbol{P}_d$, while correcting its heading from the initial error of $-180°$ to the desired heading of $360°$ at the end of the path.

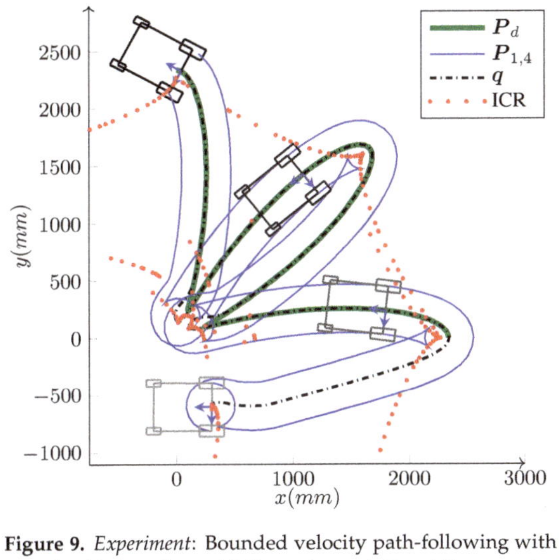

Figure 9. *Experiment*: Bounded velocity path-following with large initial errors for LabRat WMR ($\delta = (2,0)$). It seeks and follows the path $\boldsymbol{P}_d$ while correcting its heading from the initial heading error of $-180°$ toward the path tangent angle $\psi_t(s)$.

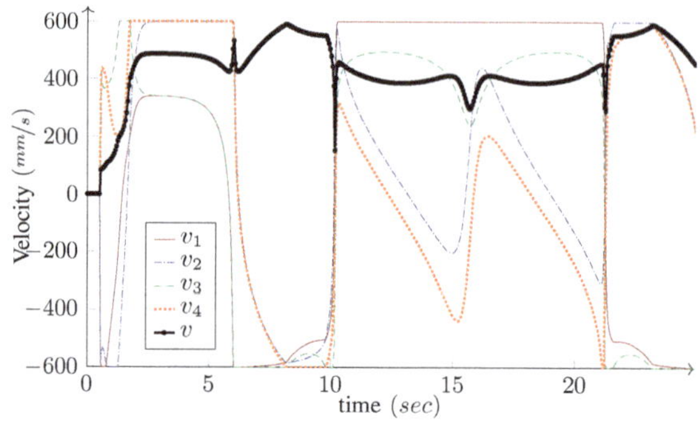

Figure 10. *Simulation*: Driving velocities v_i, and the base speed v, for path-following of a WMR with four Swedish wheels depicted in Figure 8.

We have intentionally set large initial errors to demonstrate the performance of the controller. The WMR is two meters off the starting point of the path and faces away from it (initial heading error is -180°). For the holonomic type, independent heading control is possible and the desired heading is 360° by the end of the path. Conversely, for the unicycle type ($\delta_M = 2$), independent heading control is not possible. Since the origin of the body frame is on the common axis of the fixed wheels, the heading and the linear velocity angle are the same. Hence, along with the base path-following, the algorithm corrects the initial heading error and the base heading follows the tangent of the path. As shown in the figures, the controller can maneuver the robot toward, and asymptotically onto, the path. The maximum driving velocity was set to 600 MMmm/s and the base speed was selected as $v = v^{(max)}$ given by Equation (26) in Algorithm 1. Figures 10 and 11 present the wheels' driving velocities v_i and the base speed v, for the holonomic, and unicycle type, respectively. As shown in the figures, the bounded velocity algorithm duly scales the base speed so that none of the wheels exceed their maximum driving velocity.

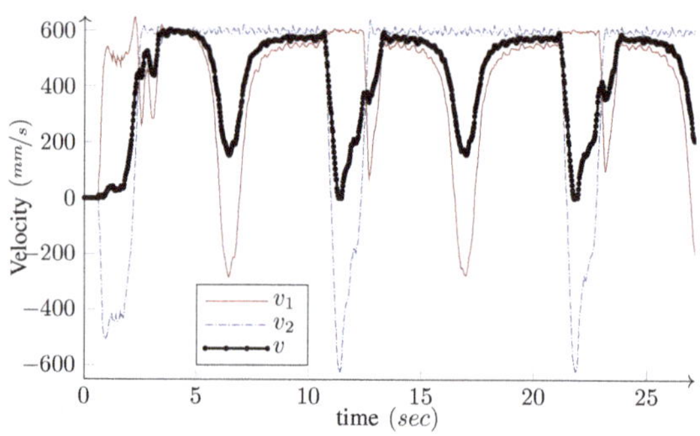

Figure 11. *Experiment*: Driving velocities v_1 and v_2, and the base speed v, for the bounded velocity path-following of LabRat ($\delta = (2, 0)$).

As shown in Figure 11, LabRat makes sharp turns by setting the velocities of both wheels to their maximum values but with different signs, which implies that one is driving forward and another is running backward. Hence, the base speed becomes almost zero and the velocity difference leads to the angular velocity that is needed for the turn. Collectively, these experiments demonstrate the agility of the proposed path-following algorithm to steadily realize sharp maneuvers without relying on switching procedures.

5.2. Case Study II: $\delta = (1, 2)$

This case study focuses on the emulation of a manipulation task with iMoro in the two-steer mode. As shown in Figure 12, the manipulation task consists of grasping the tip of a T-Slot aluminum profile, which is mounted on and extended from a wheeled table. A marker attached on the table is detected by a camera mounted on-board at the front of iMoro. The inertial frame is set on the marker and the pose of the tip is known with respect to the frame, while the wheeled table is placed randomly in the room. At the start, the WMRs' fingers are about to grasp the tip. It is desired for the WMR to start from this configuration, follow a desired tear shaped path around the room (Figure 13) and creturn to the exact same initial grasping configuration, which provides a means of investigating the repeatability of the system. In this experiment, the localization module consists of sensor fusion between vision data and wheel odometry (more information is given in [45]).

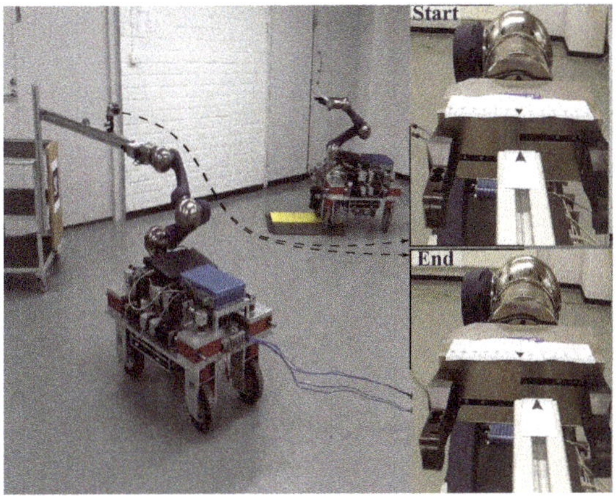

Figure 12. *Experiment*: The repeatability of the path-following controller. The robot starts from the grasping position marked by "Start" follows the path shown in Figure 13 and returns close to the initial pose.

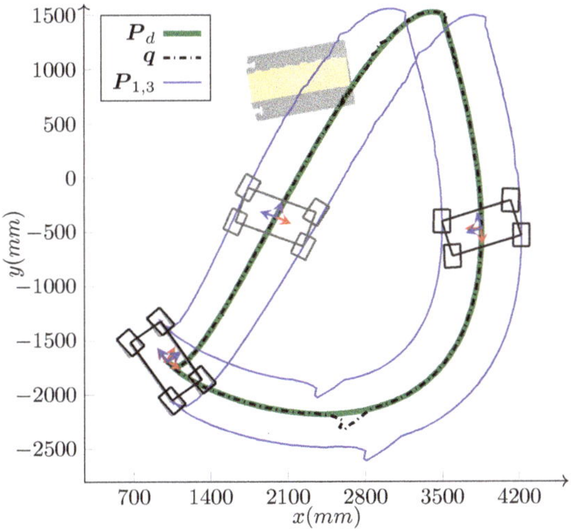

Figure 13. *Experiment*: The desired path and the localization feedback of the WMR, performing the task shown in Figure 12 (The ramp image is shown for the purpose of clarity and does not represent the exact position of the ramp).

To investigate the effects of uncertainties and disturbances, a cable protector ramp is placed somewhere on the path. The WMR driving up and down the ramp precipitates temporary but heavy localization disturbances that are evident in Figure 14 between 20 s and 30 s. Moreover, the independent desired heading is designed such that, for some time, the marker is out of the camera's field of view and the localization relies solely on wheel odometry. Once the marker returns into the camera's view, a sudden jump appears in the localization due to the accumulated drifting of the wheels' dead reckoning, which is also apparent in the figure before 50 s. Since this happens close to the end of the path, the controller has little time to correct the absolute error and bring the robot back to the initial configuration.

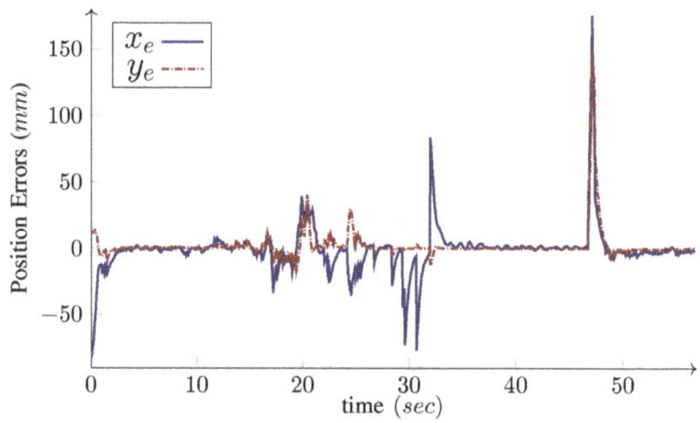

Figure 14. *Experiment*: Position errors in x and y directions of the inertial frame for the scenario depicted in Figure 13. It shows the disturbances due to the robot moving on a ramp and the localization jump due accumulated error of wheels' dead reckoning.

The two images on the right side of Figure 12 are from a separate camera mounted on the aluminum profile that show the position of the gripper's fingers at the start and end of the path-following, which attests to the successful return of the WMR to the initial configuration with around 15 mm of final error. This is also evident in Figures 13 and 14, which show the appearance of the errors and disturbances, and the controller's pertinent compensation. This experiment was repeated nine times with the ramp placed on several different locations on the path. The WMR successfully returned to the initial grasping configuration with the maximum error of 25 mm. Therefore, while the localization relying on vision and wheel odometry is very noisy, the control system shows sufficient robustness against this noise.

Another scenario using the same setup was also implemented to assess the controller's practical ability to bound the velocities and alleviate the singularities. In this scenario, the robot has to follow a given path, as shown in Figure 15 that ends in the same grasping configuration as before. However, during the path-following, the WMR is required to make a turn around itself, which not only pushes the platform near its singular configuration but also moves the marker out of the camera's field of view for some time. As illustrated in Figure 15, the turn happens somewhere near the start of the path, with the base ICR moving closer to one of the wheels steering axis. During the turn, the vision is lost and the accumulated error during this phase results in a localization jump once the camera can see the marker again. However, as shown in the figure, the controller manages to compensate for the errors and usher the robot toward the final grasping pose.

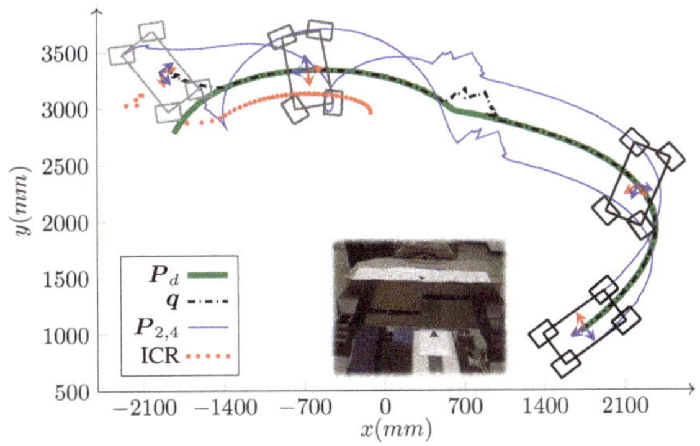

Figure 15. *Experiment*: Bounded velocity path-following with independent heading control of iMoro WMR ($\delta = (1,2)$). It seeks and follows the path P_d while correcting its heading from its initial heading to the desired heading of 360° at the end of the path.

The base speed is selected to be $v = v^{(max)}$ given by Equation (26) in Algorithm 1; therefore, at least one of the wheels runs with its maximum driving or steering velocity. Comparing Figures 16 and 17, it is clear that, most of the time, at least one of the wheels drives with its maximum driving velocity. However, when a tight turn is needed, the maximum velocity steadily changes from driving to steering, as shown in Figure 17. Notice that, near the beginning of the path, iMoro moves close to its singular configuration a couple of times, which corresponds to some of the peaks in the steering velocities. Figure 15 shows the high curvature of one of the wheels' footprint and the closeness of the body ICR to the wheel's steering near the singular configuration.

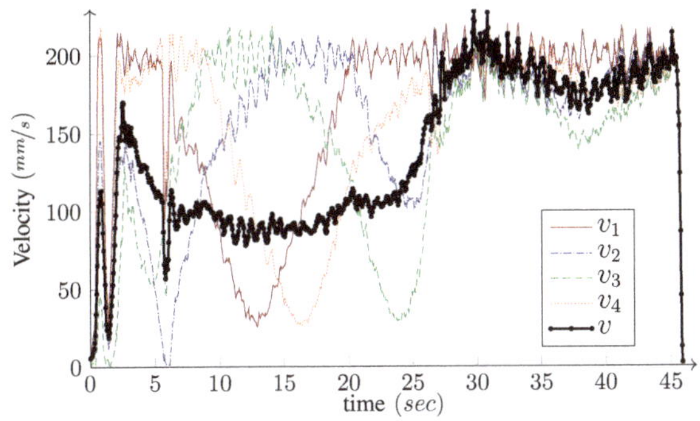

Figure 16. *Experiment*: Driving velocities v_i, and the base speed v, for the bounded-velocity path-following of iMoro depicted in Figure 15. The maximum driving velocity for all the wheels ($v_i^{(\max)}$) are set as 200 mm/s. The base speed is selected to be $v = v^{(\max)}$, given by Equation (26) in Algorithm 1.

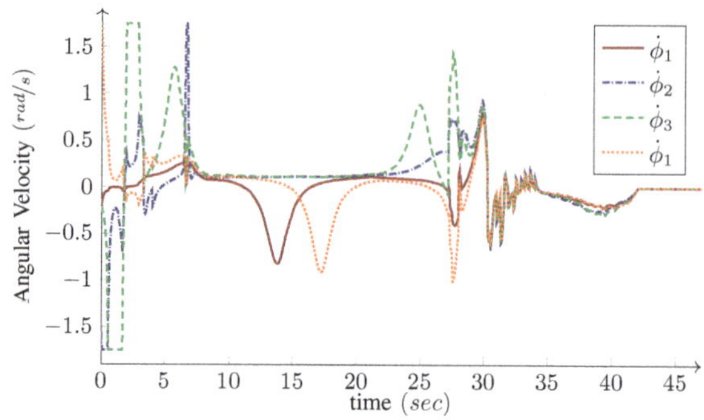

Figure 17. *Experiment*: Steering velocities $\dot{\phi}_i$ for the bounded velocity path-following of iMoro depicted in Figure 15. The maximum steering velocity for all the wheels ($\dot{\phi}_i^{(\max)}$) are set as 1.9 rad/s (110 deg/s).

5.3. Case Study III: $\delta = (1, 1)$

The last case study focuses on the path-following of car-like WMRs ($\delta = (1, 1)$) and, specifically, their bounded steering control. In this paper, we assumed that the steering wheels are free-turn. This assumption can easily be alleviated for some types of WMRs. Generally speaking, limited steering for WMRs with a degree of steerability greater than one ($\delta_s = 2$) is not favorable. This limitation greatly decreases the maneuverability of the platform to the point that it questions the benefits of allocating extra resources to obtain a WMR with $\delta_s > 1$. For example, the independence of heading and linear motion is greatly compromised and most of the results in the previous case study would not be possible. However, for configurations with $\delta_s = 1$, such as car-like WMRs, a limited steering range for steering wheels is common and widely used. Therefore, in this section we present a straightforward approach to account for bounded steering in car-like mode.

This experiment has been performed on iMoro. While the steering wheels on iMoro are free-turn, virtual limits are set to emulate bounded steering. Figure 18 shows the path-following of iMoro in car-like mode with three steering limits: $\phi_i^{\max} = \{45°, 65°, 90°\}$. The desired path is smooth but has a very high curvature at its turning point. The derivation of virtual bounds on control signals to achieve bounded steering is as follows. As shown in the figure, the body frame is on the common axis of the fixed wheels. Therefore, based on Equation (37) and the kinematic constraints of Section 3.2, the velocity constraint $v_i {}^{\mathcal{B}}\hat{\boldsymbol{v}}_i = v {}^{\mathcal{B}}\hat{\boldsymbol{v}} + \omega_b(\hat{\boldsymbol{z}} \times {}^{\mathcal{B}}\boldsymbol{\ell}_i)$ can be simplified to

$$ {}^{\mathcal{B}}\hat{\boldsymbol{v}}_i(\phi_i) = \frac{{}^{\mathcal{B}}\hat{\boldsymbol{v}} + \kappa_v(\hat{\boldsymbol{z}} \times {}^{\mathcal{B}}\boldsymbol{\ell}_i)}{||{}^{\mathcal{B}}\hat{\boldsymbol{v}} + \kappa_v(\hat{\boldsymbol{z}} \times {}^{\mathcal{B}}\boldsymbol{\ell}_i)||_2}. \tag{39} $$

In the above equation, ${}^{\mathcal{B}}\hat{\boldsymbol{v}}$ is known and constant (in case of Figure 18 it is $[1 \quad 0 \quad 0]^{\mathrm{T}}$). Replacing ϕ_i with the front wheels' steering limit $\phi_i^{\max}$, the above equation can be solved for the maximum κ_v, namely $\kappa_v^{\max} > 0$. Therefore, the bounded control signal $\bar{\kappa}_v$ that is used to derive actuator commands can be be found by saturating the output of the controller for κ_v, using

$$ \bar{\kappa}_v = \begin{cases} \kappa_v & |\kappa_v| \leq \kappa_v^{\max} \\ \mathrm{sgn}(\kappa_v)\kappa_v^{\max} & |\kappa_v| > \kappa_v^{\max} \end{cases}, \tag{40} $$

in which $\mathrm{sgn}(x)$ is the sign function. Note that, for the case where the base frame is not on the common axis of the fixed wheels, based on the first case of Equation (37), a similar procedure can be followed to derive the corresponding bound for the desired velocity direction ${}^{\mathcal{B}}\hat{\boldsymbol{v}}(\psi_d)$. In this case, the virtual steering bounds are achieved by using the reduced-state model $\boldsymbol{\mathcal{S}}^*$ and the saturated value of ψ_d.

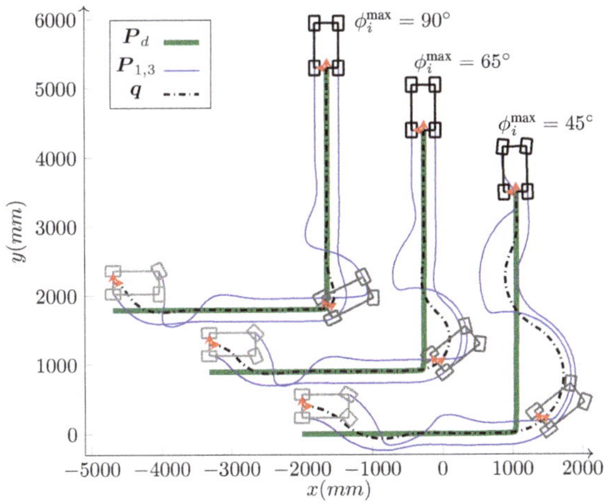

Figure 18. *Experiment*: Path-following of iMoro in car-like mode with three steering limits: $\phi_i^{\max} = \{45°, 65°, 90°\}$.

5.4. Restrictions

As presented in the results of this section, the proposed universal method navigates the WMRs while keeping the driving and steering actuators within their velocity boundaries. This approach is generally suitable in conjunction with a path-planner that generates obstacle-free paths for the WMR. While this approach is capable of navigating the robot toward the desired path, even when the errors are very large, this feature has to be used with care with respect to the obstacles that might be present on the corrective path. Moreover, in order to achieve higher velocity limits while performing tight maneuvers, bounding the velocities is not enough; the accelerations should be bounded too. We have presented a bounded acceleration solution for two-steer WMRs, such as iMoro, in [46], and are currently extending the results to cover all types of WMRs.

6. Conclusions

In this paper, we presented a universal bounded-velocity path-following algorithm for Wheeled Mobile Robots (WMRs) operating under the condition of pure rolling without skidding. The solution can be applied to various types of WMRs such as car-like, differential drive, and omnidirectional. The versatility of the framework is due to the generic representation of kinematic constraints. This representation accentuates the possibility of having universal controllers for kinematically different WMRs. We employed these results to derive a closed-form time-scaling solution for the base speed that keeps the velocities of the actuators within a set of pre-specified limits. Extending this establishment, we are currently working to enhance our solution to cover bounded accelerations, dynamic uncertainties and employ barrier functions for obstacle avoidance.

Supplementary Materials: The following are available online at https://www.mdpi.com/article/10.3390/s21227642/s1.

Author Contributions: The individual contributions of the authors are as follows. Investigation, formal analysis, software, writing—review and editing by R.O. Conceptualization, validation, writing—review, editing, and supervision by R.G. Supervision, project administration, and funding acquisition by J.M. All authors have read and agreed to the published version of the manuscript.

Funding: This work, supported by the European Union's Seventh Framework Program under the Marie Curie Initial Training Network, was carried out within the framework of the PURESAFE, Preventing hUman intervention for incREased SAfety in inFrastructures Emitting ionizing radiation, under REA grant agreement number 264336.

Institutional Review Board Statement: Not applicable.

Informed Consent Statement: Not applicable.

Data Availability Statement: Not applicable.

Conflicts of Interest: The authors declare no conflict of interest.

References

1. Morin, P.; Samson, C. Motion control of wheeled mobile robots. In *Springer Handbook of Robotics*; Springer: Berlin/Heidelberg, Germany, 2008; pp. 799–826.
2. Park, K.; Chung, H.; Lee, J.G. Point stabilization of mobile robots via state-space exact feedback linearization. *Robot. Comput.-Integr. Manuf.* **2000**, *16*, 353–363. [CrossRef]
3. Hwang, C.L.; Wu, H.M. Trajectory tracking of a mobile robot with frictions and uncertainties using hierarchical sliding-mode under-actuated control. *IET Control Theory Appl.* **2013**, *7*, 952–965. [CrossRef]
4. Fossen, T.; Pettersen, K.Y.; Galeazzi, R. Line-of-sight path following for Dubins paths with adaptive sideslip compensation of drift forces. *IEEE Trans. Control Syst. Technol.* **2015**, *23*, 820–827. [CrossRef]
5. Bullo, F.; Lynch, K.M. Kinematic controllability for decoupled trajectory planning in underactuated mechanical systems. *IEEE Trans. Robot. Autom.* **2001**, *17*, 402–412. [CrossRef]
6. Van Loock, W.; Pipeleers, G.; Diehl, M.; De Schutter, J.; Swevers, J. Optimal Path Following for Differentially Flat Robotic Systems Through a Geometric Problem Formulation. *IEEE Trans. Robot.* **2014**, *30*, 980–985. [CrossRef]
7. Debrouwere, F.; Van Loock, W.; Pipeleers, G.; Dinh, Q.T.; Diehl, M.; De Schutter, J.; Swevers, J. Time-optimal path following for robots with convex–concave constraints using sequential convex programming. *IEEE Trans. Robot.* **2013**, *29*, 1485–1495. [CrossRef]
8. Akhtar, A.; Nielsen, C.; Waslander, S.L. Path following using dynamic transverse feedback linearization for car-like robots. *IEEE Trans. Robot.* **2015**, *31*, 269–279. [CrossRef]
9. Ambrosino, G.; Ariola, M.; Ciniglio, U.; Corraro, F.; De Lellis, E.; Pironti, A. Path generation and tracking in 3-D for UAVs. *IEEE Trans. Control Syst. Technol.* **2009**, *17*, 980–988. [CrossRef]
10. Yamasaki, T.; Balakrishnan, S. Sliding mode-based pure pursuit guidance for unmanned aerial vehicle rendezvous and chase with a cooperative aircraft. *Proc. Inst. Mech. Eng. Part G J. Aerosp. Eng.* **2010**, *224*, 1057–1067. [CrossRef]
11. Nelson, D.R.; Barber, D.B.; McLain, T.W.; Beard, R.W. Vector field path following for miniature air vehicles. *IEEE Trans. Robot.* **2007**, *23*, 519–529. [CrossRef]
12. Sujit, P.; Saripalli, S.; Borges Sousa, J. Unmanned aerial vehicle path following: A survey and analysis of algorithms for fixed-wing unmanned aerial vehicless. *IEEE Control Syst.* **2014**, *34*, 42–59.
13. Samson, C. Time-varying feedback stabilization of car-like wheeled mobile robots. *Int. J. Robot. Res.* **1993**, *12*, 55–64. [CrossRef]
14. Aicardi, M.; Casalino, G.; Bicchi, A.; Balestrino, A. Closed loop steering of unicycle like vehicles via Lyapunov techniques. *IEEE Robot. Autom. Mag.* **1995**, *2*, 27–35. [CrossRef]
15. Micaelli, A.; Samson, C. *Trajectory Tracking for Unicycle-Type and Two-Steering-Wheels Mobile Robots*; Technical Report 2097; INRIA: Rocquencourt, France 1993.
16. Soetanto, D.; Lapierre, L.; Pascoal, A. Adaptive, non-singular path-following control of dynamic wheeled robots. In Proceedings of the 42nd IEEE Conference on Decision and Control, Maui, HI, USA, 9–12 December 2003; Volume 2, pp. 1765–1770.
17. Lapierre, L.; Soetanto, D.; Pascoal, A. Nonsingular path following control of a unicycle in the presence of parametric modelling uncertainties. *Int. J. Robust Nonlinear Control* **2006**, *16*, 485–503. [CrossRef]
18. Lapierre, L.; Indiverri, G. Path-Following Control of a Wheeled Robot under actuation saturation constraints. In Proceedings of the IAV07 Conference, Toulouse, France, 3–5 September 2007.
19. Lapierre, L.; Zapata, R.; Lepinay, P. Combined path-following and obstacle avoidance control of a wheeled robot. *Int. J. Robot. Res.* **2007**, *26*, 361–375. [CrossRef]
20. Kaminer, I.; Pascoal, A.; Xargay, E.; Hovakimyan, N.; Cao, C.; Dobrokhodov, V. Path following for small unmanned aerial vehicles using L1 adaptive augmentation of commercial autopilots. *J. Guid. Control. Dyn.* **2010**, *33*, 550–564. [CrossRef]
21. Encarnaçao, P.; Pascoal, A. 3D path following for autonomous underwater vehicle. In Proceedings of the 39 th IEEE Conference on Decision and Control, Sydney, NSW, Australia, 12–15 December 2000.
22. Ghabcheloo, R.; Hyvonen, M. Modeling and motion control of an articulated-frame-steering hydraulic mobile machine. In Proceedings of the 17th Mediterranean Conference on Control and Automation (MED'09), Thessaloniki, Greece, 24–26 June 2009; pp. 92–97.
23. Bibuli, M.; Bruzzone, G.; Caccia, M.; Lapierre, L. Path-following algorithms and experiments for an unmanned surface vehicle. *J. Field Robot.* **2009**, *26*, 669–688. [CrossRef]
24. Kanjanawanishkul, K.; Zell, A. Path following for an omnidirectional mobile robot based on model predictive control. In Proceedings of the IEEE International Conference on Robotics and Automation (ICRA'09), Kobe, Japan, 12–17 May 2009; pp. 3341–3346.
25. Thuilot, B.; d'Aandrea Novel, B.; Micaelli, A. Modeling and feedback control of mobile robots equipped with several steering wheels. *IEEE Trans. Robot. Autom.* **1996**, *12*, 375–390. [CrossRef]

26. Connette, C.P.; Pott, A.; Hagele, M.; Verl, A. Control of an pseudo-omnidirectional, non-holonomic, mobile robot based on an ICM representation in spherical coordinates. In Proceedings of the 47th IEEE Conference on Decision and Control (CDC 2008), Cancun, Mexico, 9–11 December 2008; pp. 4976–4983.
27. Song, J.B.; Byun, K.S. Steering control algorithm for efficient drive of a mobile robot with steerable omni-directional wheels. *J. Mech. Sci. Technol.* **2009**, *23*, 2747–2756. [CrossRef]
28. Moore, K.L.; Davidson, M.; Bahl, V.; Rich, S.; Jirgal, S. Modelling and control of a six-wheeled autonomous robot. In Proceedings of the IEEE 2000 American Control Conference, Chicago, IL, USA, 28–30 June 2000; Volume 3, pp. 1483–1490.
29. Connette, C.; Parlitz, C.; Hägele, M.; Verl, A. Singularity avoidance for over-actuated, pseudo-omnidirectional, wheeled mobile robots. In Proceedings of the IEEE International Conference on Robotics and Automation (ICRA), Kobe, Japan, 12–17 May 2009; pp. 4124–4130.
30. Schwesinger, U.; Pradalier, C.; Siegwart, R. A novel approach for steering wheel synchronization with velocity/acceleration limits and mechanical constraints. In Proceedings of the 2012 IEEE/RSJ International Conference on Intelligent Robots and Systems (IROS), Vilamoura-Algarve, Portugal, 7–12 October 2012; pp. 5360–5366.
31. Cousins, S. Ros on the pr2 [ros topics]. *IEEE Robot. Autom. Mag.* **2010**, *17*, 23–25. [CrossRef]
32. Graf, B.; Reiser, U.; Hagele, M.; Mauz, K.; Klein, P. Robotic home assistant Care-O-bot® 3-product vision and innovation platform. In Proceedings of the 2009 IEEE Workshop on Advanced Robotics and its Social Impacts (ARSO), Tokyo, Japan, 23–25 November 2009; pp. 139–144.
33. Dietrich, A.; Wimbock, T.; Albu-Schaffer, A.; Hirzinger, G. Reactive Whole-Body Control: Dynamic Mobile Manipulation Using a Large Number of Actuated Degrees of Freedom. *IEEE Robot. Autom. Mag.* **2012**, *19*, 20–33. [CrossRef]
34. Bak, T.; Jakobsen, H. Agricultural robotic platform with four wheel steering for weed detection. *Biosyst. Eng.* **2004**, *87*, 125–136. [CrossRef]
35. Cariou, C.; Lenain, R.; Thuilot, B.; Berducat, M. Automatic guidance of a four-wheel-steering mobile robot for accurate field operations. *J. Field Robot.* **2009**, *26*, 504–518. [CrossRef]
36. Frémy, J.; Ferland, F.; Lauria, M.; Michaud, F. Force-guidance of a compliant omnidirectional non-holonomic platform. *Robot. Auton. Syst.* **2014**, *62*, 579–590. [CrossRef]
37. Connette, C.; Hagele, M.; Verl, A. Singularity-free state-space representation for non-holonomic, omnidirectional undercarriages by means of coordinate switching. In Proceedings of the 2012 IEEE/RSJ International Conference on Intelligent Robots and Systems (IROS), Vilamoura-Algarve, Portugal, 7–12 October 2012; pp. 4959–4965.
38. Connette, C.; Pott, A.; Hägele, M.; Verl, A. Addressing input saturation and kinematic constraints of overactuated undercarriages by predictive potential fields. In Proceedings of the IEEE/RSJ International Conference on Intelligent Robots and Systems (IROS), Taipei, Taiwan, 18–22 October 2010; pp. 4775–4781.
39. Oftadeh, R.; Aref, M.M.; Ghabcheloo, R.; Mattila, J. Bounded-velocity motion control of four wheel steered mobile robots. In Proceedings of the 2013 IEEE/ASME International Conference on Advanced Intelligent Mechatronics (AIM), Wollongong, NSW, Australia, 9–12 July 2013; pp. 255–260.
40. Oftadeh, R.; Ghabcheloo, R.; Mattila, J. A novel time optimal path following controller with bounded velocities for mobile robots with independently steerable wheels. In Proceedings of the 2013 IEEE/RSJ International Conference on Intelligent Robots and Systems (IROS), Tokyo, Japan, 3–7 November 2013; pp. 4845–4851.
41. Oftadeh, R.; Ghabcheloo, R.; Mattila, J. A time-optimal bounded velocity path-following controller for generic Wheeled Mobile Robots. In Proceedings of the 2015 IEEE International Conference on Robotics and Automation (ICRA), Seattle, WA, USA, 26–30 May 2015; pp. 676–683.
42. Campion, G.; Bastin, G.; Dandrea-Novel, B. Structural properties and classification of kinematic and dynamic models of wheeled mobile robots. *IEEE Trans. Robot. Autom.* **1996**, *12*, 47–62. [CrossRef]
43. Osmolovskii, N.; Maurer, H. *Applications to Regular and Bang-Bang Control*; Advances in Design and Control, Society for Industrial and Applied Mathematics; SIAM: Philadelphia, PA, USA, 2012.
44. Oftadeh, R.; Aref, M.M.; Ghabcheloo, R.; Mattila, J. Unified framework for rapid prototyping of linux based real-time controllers with matlab and simulink. In Proceedings of the 2012 IEEE/ASME International Conference on Advanced Intelligent Mechatronics (AIM), Kaohsiung, Taiwan, 11–14 July 2012; pp. 274–279.
45. Aref, M.M.; Oftadeh, R.; Ghabcheloo, R.; Mattila, J. Real-time vision-based navigation for nonholonomic mobile robots. In Proceedings of the 2016 IEEE International Conference on Automation Science and Engineering (CASE), Fort Worth, TX, USA, 21–25 August 2016; pp. 515–522.
46. Oftadeh, R.; Ghabcheloo, R.; Mattila, J. Time Optimal Path Following with Bounded Velocities and Accelerations for Mobile Robots with Independently Steerable Wheels. In Proceedings of the 2014 IEEE International Conference on Robotics and Automation (ICRA), Hong Kong, China, 31 May–7 June 2014.

sensors

MDPI

Article

A Path-Following Controller for Marine Vehicles Using a Two-Scale Inner-Outer Loop Approach

Pramod Maurya [1,*], Helio Mitio Morishita [2], Antonio Pascoal [3] and A. Pedro Aguiar [4]

1 CSIR-National Institute of Oceanography, Dona Paula 403004, Goa, India
2 Department of Naval Architecture and Ocean Engineering, Polytechnic School, University of São Paulo, São Paulo Campus, São Paulo 05508-030, Brazil; hmmorish@usp.br
3 Institute for Systems and Robotics (ISR), IST, University of Lisbon, 1049-001 Lisbon, Portugal; antonio@isr.tecnico.ulisboa.pt
4 Research Center for Systems and Technologies and Department of Electrical and Computer Engineering, Faculty of Engineering, University of Porto, 4200-465 Porto, Portugal; pedro.aguiar@fe.up.pt
* Correspondence: maurya@nio.org

Abstract: This article addresses the problem of path following of marine vehicles along straight lines in the presence of currents by resorting to an inner-outer control loop strategy, with due account for the presence of currents. The inner-outer loop control structures exhibit a fast-slow temporal scale separation that yields simple "rules of thumb" for controller tuning. Stated intuitively, the inner-loop dynamics should be much faster than those of the outer loop. Conceptually, the procedure described has three key advantages: (i) it decouples the design of the inner and outer control loops, (ii) the structure of the outer-loop controller does not require exact knowledge of the vehicle dynamics, and (iii) it provides practitioners a very convenient method to effectively implement path-following controllers on a wide range of vehicles. The path-following controller discussed in this article is designed at the kinematic outer loop that commands the inner loop with the desired heading angles while the vehicle moves at an approximately constant speed. The key underlying idea is to provide a seamless implementation of path-following control algorithms on heterogeneous vehicles, which are often equipped with heading autopilots. To this end, we assume that the heading control system is characterized in terms of an IOS-like relationship without detailed knowledge of vehicle dynamics parameters. This paper quantitatively evaluates the combined inner-outer loop to obtain a relationship for assessing the combined system's stability. The methods used are based on nonlinear control theory, wherein the cascade and feedback systems of interest are characterized in terms of their IOS properties. We use the IOS small-gain theorem to obtain quantitative relationships for controller tuning that are applicable to a broad range of marine vehicles. Tests with AUVs and one ASV in real-life conditions have shown the efficacy of the path-following control structure developed.

Keywords: path following; inner-outer loop control; input-to-output stability; AUVs; ASVs

Citation: Maurya, P.; Morishita, H.M.; Pascoal, A.; Aguiar, A.P. A Path-Following Controller for Marine Vehicles Using a Two-Scale Inner-Outer Loop Approach. *Sensors* **2022**, *22*, 4293. https://doi.org/10.3390/s22114293

Academic Editor: Baochang Zhang

Received: 31 March 2022
Accepted: 29 May 2022
Published: 5 June 2022

1. Introduction

The use of autonomous marine vehicles, including surface and underwater robots, for various scientific and commercial applications at sea, has increased multi-fold in the last decade. Missions of interest include, among others, bathymetric surveys, seabed imaging, environmental monitoring, inspection of offshore critical infrastructures, and marine archaeology studies. In most of these missions, marine vehicles are required to follow spatial paths accurately. A representative example is the case where an AUV (autonomous underwater vehicle) or an ASV (autonomous surface vehicle) is requested to execute "lawn-mowing" maneuvers along desired paths in the presence of unknown ocean currents.

We recall the crucial difference between trajectory tracking and path following. In the latter, no explicit temporal constraints are imposed on the desired vehicle's motion and a path is planned using spatial coordinates only, along with a reference speed profile that may depend on where the vehicle is on the path. In contrast, in trajectory tracking, space and time explicitly define the reference coordinates for the desired vehicle's motion, that is, the vehicle is required to track a 3D curve parameterized in time. This strategy is only pursued in practice when simultaneous and temporal specifications play a decisive role. However, in the process of tracking a desired inertia trajectory, a vehicle may be required to reach a speed with respect to the water that may either be too small, leading to the loss of surface control authority or too high, exceeding the capability of the propulsion system installed onboard.

A properly designed path-following control systems can naturally lead to smoother vehicle trajectories without pushing the control signals into saturation, in contrast to what may happen when trajectory tracking controllers are used. The fundamental limitations of trajectory tracking can be found in [1,2]. In [2], Aguiar studied performance limitations of trajectory tracking strategies due to unstable zero-dynamics in terms of a lower bound on $\mathcal{L}_2$-norm of the tracking error even if the control effort is unlimited. It was also shown that path following is free from such a limitation. In [3], a hybrid solution to path following and trajectory tracking problem for underactuated vehicles was discussed for 2D and a more general 3D space. Aguiar and Hespanha in [3] followed a supervisory control architecture in which a switching logic is used to adapt an adequate estimator and control law from the family of estimators and candidate control laws. Supervisory control combined with a nonlinear Lyapunov-based tracking control was demonstrated and its robustness to parametric modeling uncertainties was shown with examples of a hovercraft and an AUV. Marine vehicles suffer from disturbances induced by ocean currents. The issue of disturbances was not addressed in [3]. The path following problem in 3D for an underwater vehicle was also described in [4], where the controller design builds on the Lyapunov theory and resorts to back-stepping techniques, demanding the knowledge of a complete hydrodynamic model of the vehicle.

In [5], Indiveri described a 3D kinematical solution to path following by recalling sliding-mode control techniques. Rather than designing the control input to drive a tracking error to zero, the sliding mode control uses the control input to drive and keep the state on a surface where the error has stable dynamics. However, the authors did not address the issues related to the vehicle dynamics and described the controller only at the kinematic level.

Pioneering work in solving the path following problem for wheeled robots has been addressed in [6,7]. Path-following problem for a car pulling several trailers is addressed in [8]. In [9], Altafini provided local asymptotic stability for a path of non-constant curvature for a trailer vehicle. More recently, in [10], a model predictive path-following control of a laboratory tower crane has been described to move a load along a predefined geometric path.

It is interesting to see in [11,12] that path following is at the core of cooperative motion control for multiple vehicles where these vehicles are supposed to follow a set of fixed spatial paths while holding a desired formation pattern. Each vehicle is equipped with a path-following algorithm to maneuver along its assigned spatial path, whereas a distributed control law performs the formation control by adjusting the speeds of different vehicles. It is therefore important to emphasize the need for a reliable path-following method that is suitable for heterogeneous vehicles with little knowledge of their dynamics. Cooperation among multiple vehicles with a view to performing different tasks plays a critical role in executing a number of mission scenarios [13]. The *GREX* project is an example of the use of cooperative-motion control strategies involving a number of vehicles developed by different oceanographic institutions for their needs. Figure 1 illustrates the diversity in the marine vehicles used for cooperation during the sea trials of the GREX project.

Figure 1. Heterogeneous vehicle used to demonstrate cooperative motion control during GREX trails at Sesimbra, Portugal.

The EU-funded project MORPH (FP7-ICT-2011-7 GA 288704, 2012–2016) [14] advanced the novel concept of an underwater robotic system composed of a number of spatially separated mobile robot-modules, carrying distinct and yet complementary resources with path-following algorithms implemented on every vehicle. MORPH provided the foundation for efficient methods to survey the underwater environment with great accuracy, especially in situations that defy classical technology. Namely, underwater surveys over rugged terrain and near vertical cliffs.

The WiMUST project (H2020-ICT-2014-1, 2015–2018) [15] witnessed the development of advanced cooperative and networked control/navigation systems to enable a group of marine robots (both on the surface and submerged) equipped with acoustic sources and towed acoustic streamers to perform geotechnical seismic surveys in a fully automatic manner. For the first time worldwide, a mission was performed in 2018 at sea in Sines, Portugal, with a fleet of seven autonomous marine robots performing high-resolution 3D sub-bottom mapping in cooperation. Every individual vehicle was required to be equipped with a path-following algorithm.

Path-following algorithms are fundamental for these vehicles to cooperate effectively. The straight-line path following problem for formations of multiple under-actuated marine surface vessels is addressed in [16]. The controller used is a combination of an LOS-based path-following controller and a nonlinear synchronization controller for the along-path synchronization of the vessels. The synchronization controller takes into account the loss of controllability at velocities close to zero for under-actuated vehicles. A unified analysis of stability properties of both the cross-track error dynamics and the synchronization error dynamics are discussed by using the tools from the theory of nonlinear cascaded systems. For the path-following controller of each vehicle, the authors used a line-of-sight guidance law in combination with a stabilizing heading controller. The guidance law is a function of cross-track error and look-ahead distance [17], which is an along-track distance between the nearest point on the track and a point that lies ahead of the vehicle. The look-ahead distance is used as design parameter. Depending on the damping on sway motion, the look-ahead distance can be increased or reduced to impose restrictions on the commanded yaw rate. However, the bound on the design parameter (look-ahead distance) requires the knowledge of the mass matrix and damping coefficients for surge and sway dynamics of the vehicle and the proposed controller does not take ocean currents into account.

Similar work that takes ocean currents into account is reported in [18]. The control law proposed to drive the cross-track error to zero is the same as that reported in [16] with an extra term which is a function of the ocean currents. The ocean currents for an individual vehicle are estimated using an adaptation law, which solves a differential Ricatti equation that is again a function of the mass matrix and damping coefficients.

In [19], instead of estimating the currents, the authors used an integrator similar to the work reported in this paper. Burger et al. introduced a conditional integrator to avoid large overshoots during the saturation of the control signal. The conditional integrator combines the benefits of integral action and sliding-mode control. It behaves either like a PI controller or like a sliding-mode controller, depending on the magnitude of the control signal to avoid the chattering caused by the sliding-mode control. However, the problem of integrator windup can also be addressed by using a smooth anti-windup scheme.

The tools used for stability analysis in [16,18,20] are similar to the one reported in this paper, which relies on the theory of interconnected and cascaded systems. In [20], path following problems for more general spatial paths (with constraints on their curvature) in the presence of constant ocean currents were addressed. The authors introduced a virtual Serret–Frenet reference frame that is anchored on and propagates along the desired path. When the vehicle reaches the vicinity of that point, the reference is updated, requesting the vehicle to converge to another point further on the reference path. A Luenberger-type observer is designed to estimate the currents by measuring the relative velocity of the vehicle w.r.t the water. The proposed guidance law is a nonlinear function of surge and sway velocities and involves solving a quadratic function of currents, cross-track error and look-ahead distance. The estimation of ocean currents requires the measurement of relative velocity from an Acoustic Doppler Current Profiler (ADCP). Most surface vehicles are equipped only with GPS and cannot measure the relative velocity of the vehicle. In the simplified case of [20], to follow straight lines while cruising at a constant speed, the measurement of the relative velocity was critical to estimate currents. Again, the design of the controller requires knowledge of surge and sway dynamics.

We now shift our attention to the importance of inner-outer loop control structures. In the field of aircraft control, path following has been addressed as dynamic and kinematic loop control structures similar to inner and outer loop in marine vehicles. In [21], path-following control in 3D was built on a nonlinear control strategy that is first derived at the kinematic level, followed by the design of a L_1-adaptive output-feedback control law that effectively augments an existing autopilot and yields an inner-outer loop control structure with guaranteed performance whereas, multiple vehicle coordination is achieved by enforcing temporal constraints on the speed profiles of the vehicles along their paths. A survey and analysis of algorithms for path following reported in [22] showed how inner-outer loop-based guidance schemes are implemented in most Unmanned Air Vehicles (UAV) where practitioners used inexpensive open-source autopilots.

There is extensive literature on the path following, displaying a vast choice of available control laws based on linear and nonlinear techniques. Representative examples of path-following controllers for marine vehicles can be found in [23–26]. Furthermore, in [27], the authors described a nonlinear path-following guidance method in inner-outer loop form, where the outer loop plays a role of a guidance scheme, generating lateral acceleration commands, and the inner loop follows. This paper does not address the stability of the outer loop in the presence of the inner loop. This topic was addressed in [28] with the assumption that there is complete knowledge of the vehicle model parameters.

No reference in the literature, to the best of our knowledge, addresses the problem of path following without prior knowledge of the inner loop dynamics. Path following is either designed at a kinematic level only or demands complete knowledge of horizontal plane dynamics.

The scarcity of publications on nonlinear path following for ocean vehicles without complete knowledge of vehicle dynamics somehow reflects the hardness of the problem mainly due to the presence of a nonzero lateral velocity and shows the relevance of the research topic here discussed. Motivated by the above considerations, this article addresses the problem of path following for marine vehicles by resorting to inner-outer control loops, with due account for the vehicle dynamics and currents. The inner-outer loop control structures exhibit a fast-slow temporal scale separation that yields simple "rules of thumb" for controller tuning. Stated intuitively, the inner loop dynamics should be much faster than those of the outer loop. This qualitative result is well rooted in singular perturbation theory [29]. Conceptually, the procedure described has three key advantages: (i) it decouples the design of the inner and outer control loops, (ii) the structure of the outer loop controller does not depend on the dynamics of the vehicle, and (iii) it provides practitioners a very convenient method to effectively implement path-following controllers on a wide range of vehicles.

The path-following controller discussed in this article is designed at the kinematic outer loop that commands the inner loop with the desired heading angles while the vehicle moves at an approximately constant speed. The idea is to provide a seamless implementation of path-following control algorithms on heterogeneous vehicles that may be pre-equipped with heading autopilots. To address this issue, we developed a novel methodology for the design of path-following controllers for marine vehicles which uses a simple characterization of the marine vehicle's dynamics, in the form of input-output gains or bandwidth-like characterization, without having to know the detailed dynamics of a marine vehicle. This is the key contribution of this article,which is rooted in and extends substantially the methodology described in [30]. The focus of the presentation is on AUVs; however, the techniques can be easily extended to autonomous surface vehicles (ASVs). The paper is organized as follows. We first discuss the nonlinear dynamics of two marine vehicles used in experiments, followed by the formal proof of the stability of a simple inner-loop PD controller applied to a nonlinear three-degree-of-freedom model. We then tackle the problem of path following without considering the dynamics of the vehicle. In Section 6.5, we consider path following for straight lines in 2D, propose an inner-outer loop control structure for its solution, and provide the proof of the stability of the resulting feedback control system. We describe the results of simulations and field tests performed with real marine vehicles, summarize the main conclusions, and discuss problems that warrant further research.

2. Notation and AUV Modeling

Depth and heading controllers are the core systems of autonomous marine. Depth control is used to maintain the depth of an AUV at a desired value, whereas heading control is used to steer both AUVs and ASVs along desired directions with respect to the magnetic north. The design of such controllers varies from simple proportional-integral-derivative (PID) and linear quadratic methods based on linearized dynamic models [31] to more complex Lyapunov-based nonlinear control. Modeling the dynamics of a vehicle is critical for its maneuvering, stabilization, and motion control. However, accurate modeling of the dynamics of such vehicles is oftentimes painstaking, time consuming, and quite costly. In [32], the hydrodynamic data required to model the Marius AUV have been determined by full-scale tests, using a towing tank equipped with a *Planar Motion Mechanism*. There are only a few test facilities of this kind which many researchers developing AUVs for scientific needs cannot afford. To avoid such expensive and time-consuming methods of determining the hydrodynamic coefficients, most of the users rely on semi-empirical and analytical methods [33], together with CFD analysis. Later, the parameters of importance in simplified models can be derived/verified by performing certain open loop maneuvers in the water. One such example is the *circular maneuver* for horizontal plane models [34]. Vehicle models obtained using such techniques are necessarily simplified but, if properly exploited, may be extremely useful in characterizing the system to be controlled in a form that is suitable for input-output stability analysis. A compelling example is the case where, using nonlinear system analysis, the dynamics of a system may be characterized in terms of parameters that play a role equivalent to static gain and bandwidth for first-order linear systems. Such models can be used to design the controllers with a simple structure. In what follows, the structure of a generic vehicle model that we adapt borrows from the work of Fossen [34].

2.1. Vehicle Modeling

Following usual practice, we define two reference frames: a body-fixed reference frame $\{\mathcal{B}\}$ in which the dynamics of the vehicle are naturally described and an earth-fixed reference frame $\{\mathcal{I}\}$ in which the position and orientation of the vehicle are expressed (see Figure 2). The following notation is required.

- $\nu_1 = [u \;\; v \;\; w]^T$ is the linear velocity of the origin of $\{\mathcal{B}\}$ with respect to $\{\mathcal{I}\}$ expressed in $\{\mathcal{B}\}$ (i.e., body-fixed linear velocity).
- $\nu_2 = [p \;\; q \;\; r]^T$ is the angular velocity of $\{\mathcal{B}\}$ with respect to $\{\mathcal{I}\}$ expressed in $\{\mathcal{B}\}$ (i.e., body-fixed angular velocity).
- $\eta_1 = [x \;\; y \;\; z]^T$ is the position of the origin of $\{\mathcal{B}\}$ measured in $\{\mathcal{I}\}$.
- $\eta_2 = [\phi \;\; \theta \;\; \psi]^T$ parameterizes locally the orientation of $\{\mathcal{B}\}$ with respect to $\{\mathcal{I}\}$.

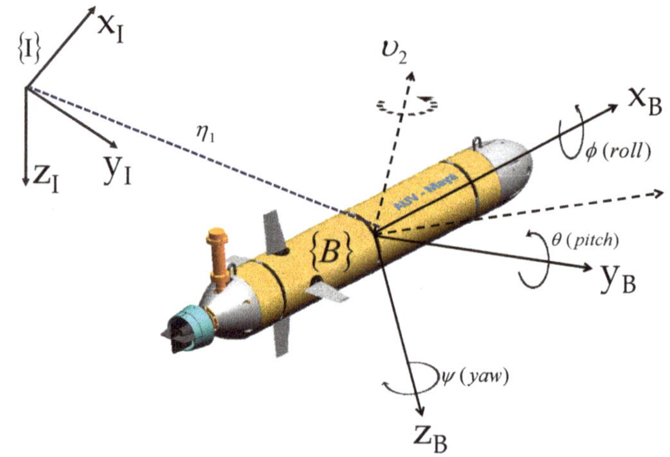

Figure 2. Notations and reference frames for an AUV.

An arbitrary vector ${}^{\mathcal{B}}V \in \mathbb{R}^3$ expressed in $\{\mathcal{B}\}$ can be expressed in $\{\mathcal{I}\}$ as ${}^{\mathcal{I}}V = {}^{\mathcal{I}}_{\mathcal{B}}R\,{}^{\mathcal{B}}V$. An important relation for ${}^{\mathcal{I}}_{\mathcal{B}}R$ (abbreviated as R) [35] is $R^TR = I$, implying that $R^T = R^{-1}$ and $\det(R) = 1$. The matrix $R \in \mathbb{R}^{3\times 3}$ can be described locally in terms of a sequence of 3 transformations that take $\{\mathcal{B}\}$ to $\{\mathcal{I}\}$ by rotating it sequentially about its current $z \to y \to x$ axis through the Euler angles ψ—yaw (rotation about z-axis), θ—pitch (rotation about y-axis) and ϕ—roll (rotation about x-axis) [34]. The final rotation matrix from $\{\mathcal{B}\}$ to $\{\mathcal{I}\}$ parameterized by the Euler angles $\eta_2 = [\phi \;\; \theta \;\; \psi]^T$ is given by

$$R(\eta_2) = \begin{pmatrix} c\psi c\theta & -s\psi c\phi + c\psi s\theta s\phi & s\psi s\phi + c\psi c\phi s\theta \\ s\psi c\theta & c\psi c\phi + s\phi s\theta s\psi & -c\psi s\phi + s\theta s\psi c\phi \\ -s\theta & c\theta s\phi & c\theta c\phi \end{pmatrix}, \tag{1}$$

where $s(.) = sin(.)$ and $c(.) = cos(.)$.

2.2. Kinematics

The kinematic equations that relate body-fixed with inertial linear and angular velocities are given by

$$\dot{\eta}_1 = R(\eta_2)\nu_1, \tag{2}$$
$$\dot{\eta}_2 = Q(\eta_2)\nu_2. \tag{3}$$

The transformation of body-fixed angular velocities is performed using the transformation matrix $Q(\eta_2)$ given by

$$Q(\eta_2) = \begin{pmatrix} 1 & s\phi t\theta & c\phi t\theta \\ 0 & c\phi & -s\phi \\ 0 & s\phi/c\theta & c\phi/c\theta \end{pmatrix}, \tag{4}$$

where $t(.) = tan(.)$. Note that Q is singular for $\theta = \pm\pi/2$ when using the above sequence of Euler angles and can be avoided by using quaternions (see [34] for details). However, for an AUV, an angle close to $\pi/2$ is practically not desirable and is avoided by design. Thus, it is reasonable to assume that most AUVs operate with small pitch angle. Finally, the combined 6-DOF kinematic equation can be written as

$$\begin{bmatrix} \dot{\eta}_1 \\ \dot{\eta}_2 \end{bmatrix} = \begin{bmatrix} R(\eta_2) & 0 \\ 0 & Q(\eta_2) \end{bmatrix} \begin{bmatrix} \nu_1 \\ \nu_2 \end{bmatrix} \Longleftrightarrow \dot{\eta} = J(\eta)\nu. \tag{5}$$

An important relation for the derivative of a rotation matrix is given by

$$\dot{R}(\eta_2) = R(\eta_2)S(\nu_2), \tag{6}$$

where S is a skew-symmetric matrix, for an arbitrary vector $u = [u_x \; u_y \; u_z]^T \in \mathbb{R}^3$, it takes the form

$$S(u) = \begin{bmatrix} 0 & -u_z & u_y \\ u_z & 0 & -u_x \\ -u_y & u_x & 0 \end{bmatrix}. \tag{7}$$

Furthermore,

$$S^T = -S; \;\; S(u)v = -S(v)u; \;\; S(u)v = u \times v. \tag{8}$$

2.3. Dynamics

With the assumptions that the center of mass of the rigid body is coincident with the origin of $\{\mathcal{B}\}$ and $\{\mathcal{I}\}$ is an inertial frame, Newton–Euler's laws apply in the latter frame and the dynamic equations for translation can be written as

$$\sum{}^{\mathcal{I}} F_{RB} = m\frac{d}{dt}\left({}^{\mathcal{I}}\nu_1\right) = m\frac{d}{dt}(R\nu_1) \tag{9}$$
$$= m\frac{dR}{dt}\nu_1 + mR\frac{d}{dt}\nu_1 \tag{10}$$
$$= mRS(\nu_2)\nu_1 + mR\frac{d}{dt}\nu_1 \tag{11}$$

where m is the mass matrix and $\sum^{\mathcal{I}} F_{RB}$ is the sum of external forces expressed in the inertial reference frame. The dynamic equations for an AUV are usually expressed in the body-fixed frame for convenience, where the inertia tensor is constant and the external forces are more easily expressed. The sum of external forces expressed in $\{\mathcal{B}\}$ is given by

$$\sum R^{-1\mathcal{I}}F_{RB} = mR^{-1}RS(\nu_2)\nu_1 + mR^{-1}R\frac{d}{dt}\nu_1 \tag{12}$$

$$\sum F_{RB} = m[\nu_2 \times \nu_1 + \dot{\nu}_1], \tag{13}$$

where F_{RB} are the external forces measured in $\{\mathcal{B}\}$. Similarly, applying Newton–Euler's laws in $\{\mathcal{B}\}$, the dynamic equations for rotational motion can be written as

$$\sum {}^{\mathcal{I}}N_{RB} = \frac{d}{dt}\left({}^{\mathcal{I}}L\right) = \frac{d}{dt}(RI_{RB}\nu_2) \tag{14}$$

$$= RS(\nu_2)I_{RB}\nu_2 + RI_{RB}\frac{d}{dt}\nu_2, \tag{15}$$

where I_{RB} is the moment of inertia matrix and ${}^{\mathcal{I}}L$ is the angular momentum measured in $\{\mathcal{I}\}$. Now, expressing the above dynamic equations in body-fixed reference frame $\{\mathcal{B}\}$ yields

$$\sum R^{-1\mathcal{I}}N_{RB} = R^{-1}RS(\nu_2)I_{RB}\nu_2 + R^{-1}RI_{RB}\frac{d}{dt}\nu_2 \tag{16}$$

$$\sum N_{RB} = I_{RB}\dot{\nu}_2 + \nu_2 \times I_{RB}\nu_2, \tag{17}$$

where N_{RB} includes the external torques measured in $\{\mathcal{B}\}$.

Combining the equations for translation and rotational motion, a simplified vectorial representation can be written as

$$M_{RB}\dot{\nu} + C_{RB}(\nu)\nu = \tau_{RB}, \tag{18}$$

where M_{RB} is the rigid-body inertia matrix, which in the general case, is given by

$$M_{RB} = \begin{bmatrix} mI_{3\times3} & -mS\left(r_g^b\right) \\ mS\left(r_g^b\right) & I_o \end{bmatrix} \tag{19}$$

$$= \begin{bmatrix} m & 0 & 0 & 0 & mz_g & -my_g \\ 0 & m & 0 & -mz_g & 0 & -mx_g \\ 0 & 0 & m & my_g & -mx_g & 0 \\ 0 & -mz_g & my_g & I_x & -I_{xy} & -I_{xz} \\ mz_g & 0 & -mx_g & -I_{xy} & I_y & -I_{yz} \\ -my_g & mx_g & 0 & -I_{zx} & -I_{zy} & I_z \end{bmatrix}, \tag{20}$$

and satisfies the properties

- $M_{RB} = M_{RB}^T > 0$,
- $\dot{M}_{RB} = 0_{6\times6}$,

where $I_{3\times3}$ is the identity matrix, $I_o = I_o^T > 0$ is the inertia matrix about O, and $S\left(r_g^b\right)$ in (19) is the matrix cross-product operator.

The rigid body Coriolis and centripetal matrix $C_{RB}(\nu)$ can always be represented in symmetric form, i.e., $C_{RB}(\nu) = C_{RB}^T(\nu)\ \forall \nu \in \mathbb{R}^6$. For the given inertia matrix,

$$M_{RB} = M_{RB}^T = \begin{bmatrix} M_{11} & M_{12} \\ M_{21} & M_{22} \end{bmatrix} > 0, \tag{21}$$

where $M_{21} = M_{12}^T$, $C_{RB}(\nu)$ can be written as

$$C_{RB}(\nu) = \begin{bmatrix} 0_{3\times3} & -mS(\nu_1) - mS\left(S(\nu_2)r_g^b\right) \\ -mS(\nu_1) - mS\left(S(\nu_2)r_g^b\right) & mS\left(S(\nu_1)r_g^b\right) - S(I_o\nu_2) \end{bmatrix} \tag{22}$$

with $S(\nu_1)\nu_1 = 0$. All external forces and torques are represented as a generalized vector $\tau_{RB} = \left[\sum F_{RB}^T \;\; \sum N^T RB\right]^T = [X\; Y\; Z\; K\; M\; N]^T$ where $[X\; Y\; Z]^T$ are the external forces and $[K\; M\; N]^T$ are the external torques both expressed in body $\{\mathcal{B}\}$. To explicitly take into account different types of external forces and torques, this vector can be decomposed as

$$\tau_{RB} = \tau + \tau_A + \tau_D + \tau_R + \tau_{dist}, \tag{23}$$

where

- τ—control inputs (forces and torques due to thrusters/surfaces);
- $\tau_A = -M_A\dot{\nu} - C_A(\nu)\nu$—terms due to added masses;
- $\tau_D = -D(\nu)\nu$—hydrodynamics terms due to lift, drag, skin friction, etc.;
- $\tau_R = -g(\eta)$—restoring forces and torques due to the interplay between gravity and buoyancy forces;
- τ_{dist}—terms due to external disturbances, e.g., waves, winds, etc.

Neglecting the term τ_{dist}, the final dynamic model of an AUV can be written as

$$[M_{RB} + M_A]\dot{\nu} + [C_{RB}(\nu) + C_A(\nu)]\nu + D(\nu)\nu + g(\eta) = \tau \tag{24}$$

$$\Leftrightarrow M\dot{\nu} + C(\nu)\nu + D(\nu)\nu + g(\eta) = \tau. \tag{25}$$

3. Examples of Horizontal Plane Dynamics

3.1. 3-DOF Nonlinear Model

In what follows, we consider the horizontal plane dynamics of an underwater vehicle with 3 degrees of freedom (surge, sway, and yaw rate). Assuming that the roll of the vehicle is negligible and vertical and the horizontal plane dynamics are decoupled, and the $\{\mathcal{B}\}$ frame coincides with the principal axes of inertia of the body, the corresponding nonlinear equations of motion can be written as

$$m_u\dot{u} - m_v vr + d_u u = \tau_u \tag{26}$$

$$m_v\dot{v} + m_u ur + d_v v = 0 \tag{27}$$

$$m_r\dot{r} - m_{uv} uv + d_r r = \tau_r \tag{28}$$

with

$$m_u = m - X_{\dot{u}} \qquad m_r = I_z - N_{\dot{r}} \qquad d_u = -X_u - X_{|u|u}|u| \tag{29}$$

$$m_v = m - Y_{\dot{v}} \qquad m_{uv} = m_u - m_v \qquad d_v = -Y_v - Y_{|v|v}|v| \tag{30}$$

$$d_r = -N_r - N_{|r|v}|r|, \tag{31}$$

where m is the mass, I_{zz} is the moment of inertia about the vertical axis, $X_{\dot{u}}$, $Y_{\dot{v}}$, and $N_{\dot{r}}$ are added mass coefficients, X_u, Y_v, and N_r are linear damping coefficients, and $X_{|u|u}$, $Y_{|v|v}$, and $N_{|r|r}$ are the nonlinear damping coefficients of the AUV model. The compact form of these equations describing the motion of a marine vehicle which is three-plane symmetric can be written as [36]

$$M\dot{\nu} + C(\nu)\nu + D(\nu)\nu = \tau, \tag{32}$$

where

$$\nu = [u \;\; v \;\; r]^{\mathrm{T}}, \qquad \tau = [\tau_u \;\; 0 \;\; \tau_r]^{\mathrm{T}}, \tag{33}$$

$$M = \begin{pmatrix} m_u & 0 & 0 \\ 0 & m_v & 0 \\ 0 & 0 & m_r \end{pmatrix} > 0, \tag{34}$$

$$C(\nu) = \begin{pmatrix} 0 & 0 & -m_v v \\ 0 & 0 & m_u u \\ m_v v & -m_u u & 0 \end{pmatrix}, \tag{35}$$

$$D(\nu) = \begin{pmatrix} d_u & 0 & 0 \\ 0 & d_v & 0 \\ 0 & 0 & d_r \end{pmatrix} > 0, \tag{36}$$

We shall now discuss the modeling of an underwater vehicle referring to two examples of actual vehicles: (1) the MEDUSA vehicle in which yaw and heave motions are controlled only by thrusters (see Figure 3), (2) Maya, a torpedo-shaped AUV where the yaw and pitch motions are controlled using conventional rudders and fins while the vehicle cruises propelled by a single thruster aligned with the x-axis of the body (see Figure 4). In this model, the surge equations are not considered, and the vehicle is assumed to be cruising at a constant speed u_o [31].

Figure 3. Medusa Autonomous Marine Vehicles, developed at DSOR, IST, Lisbon.

Figure 4. The MAYA Autonomous underwater vehicle developed at NIO, Goa.

3.2. MEDUSA-Class Vehicle as an Example

The MEDUSA-class vehicles are autonomous robotic marine vehicles, capable of working both as surface and underwater robots, developed at the Dynamical Systems and Ocean Robotics group (DSOR) of Instituto de Sistemas e Robótica, Instituto Superior Técnico (ISR-IST) [37]. The MEDUSA-class AUV is a twin hull vehicle separated by 150 mm (see Figure 3). It weighs around 17 kg, is 1 m long, and with a hull diameter of 150 mm. The two hulls contain the batteries, onboard electronics, sensors, and the main computer running the Robot Operating System (ROS) .

The surface operating vehicles have two thrusters placed on each side of the vehicle at 150 mm from the center line. The forward force $\tau_u = F_s + F_p$ is generated as a sum of two forces generated by starboard (F_s) and portside (F_p) thrusters and a yaw moment $\tau_r = 0.15(F_s - F_p)$. The restoring moments for the vehicle is large enough for the roll and pitch motions to be neglected due to large separation between center of gravity and center of buoyancy. The vehicles are not actuated in the sway axis (i.e., $\tau_v = 0$) [38]. The hydrodynamic parameters for Medusa are derived using the combination of semi-empirical and analytical methods and experimental data in calm waters. The parameters are tabulated in Table A1.

3.3. *Maya AUV: An Example*

The Maya AUV is an axis-symmetric underwater vehicle developed at the National Institute of Oceanography (NIO), Goa, India. The Maya AUV [39] follows a low-drag hull with a removable nose cone which carries scientific sensors (see Figures 2 and 4). It has a single propeller for propulsion and two pairs of stern planes to control depth and heading. The nose section can accommodate different sensors for specific missions at sea. The AUV is equipped with an attitude and heading reference system (AHRS), a Doppler velocity log (DVL) for navigation underwater, and GPS for surface navigation.

Based on the assumption that the complete six-degrees-of-freedom model for the AUV can be split into two non-interacting models for the vertical and horizontal planes (see [40]), the simplified sway and yaw dynamics at constant speed u_0 are given by [36],

$$m\dot{v} + mu_0 r = Y \tag{37}$$

$$I_z\dot{r} = N. \tag{38}$$

For small roll and pitch angles,

$$\dot{\psi} = \frac{sin\theta}{cos\theta}q + \frac{cos\phi}{cos\theta}r \approx r. \tag{39}$$

The linear modeling of hydrodynamic damping, added mass, and rudder angle gives

$$Y = Y_v\dot{v} + Y_R\dot{r} + Y_v v + Y_r r + Y_\delta \delta_r \tag{40}$$

$$N = N_v\dot{v} + N_R\dot{r} + N_v v + N_r r + N_\delta \delta_r. \tag{41}$$

The quadratic terms on damping are neglected because of limited magnitude of v and r. The model is linearized about the nominal cruising speed of $u_0 = 1.2$ m/s and the linearized dynamic equations of motion for the horizontal plane represented in matrix form is given by

$$\begin{bmatrix} m - Y_{\dot{v}} & -Y_{\dot{r}} & 0 \\ -N_{\dot{v}} & I_z - N_{\dot{r}} & 0 \\ 0 & -1 & 1 \end{bmatrix} \begin{bmatrix} \dot{v} \\ \dot{r} \\ \dot{\psi} \end{bmatrix} + \begin{bmatrix} -Y_v & -Y_r + mu_0 & 0 \\ -N_v & -N_r & 0 \\ 0 & -1 & 0 \end{bmatrix} \begin{bmatrix} v \\ r \\ \psi \end{bmatrix} = \begin{bmatrix} Y_\delta \\ N_\delta \\ 0 \end{bmatrix} \delta_r. \tag{42}$$

The vehicle parameters were estimated by resorting to analytic and semi-empirical methods for hydrodynamic parameter estimation (see [33]), and the details of the parameters for the MAYA AUV are given in Table A2.

4. Heading Control for a 3-DOF Nonlinear Model

In preparation for the analysis of combined guidance and control systems for a marine vehicle in 2D, we start by obtaining a compact description of the dynamics of a closed-loop yaw control system. This will serve as an important step to tackle the problem of path following. Most surface and underwater vehicles use simple Proportional-Derivative (PD) yaw controllers, oftentimes designed using a 3-DOF model linearized about a trimming condition with a view to steering the vehicle along a straight line at a fixed forward speed. However, there is no formal proof of stability for the convergence of the error (between desired and true heading) to zero when the controller is applied to a nonlinear model. This section provides proof of convergence for a PD controller coupled with the 3-DOF nonlinear dynamics of a Medusa Class marine vehicle, using concepts from the Lyapunov theory. We will be later using a simple characterization of the combination of the AUV dynamics with a heading controller, in the form input-output gains or bandwidth-like characterization and this is a key contribution of this article. The motion of an AUV in the horizontal plane expressed in the body-fixed frame is given in (27), and is repeated here for the reader's convenience:

$$M\dot{\nu} + C(\nu)\nu + D(\nu)\nu = \tau, \tag{43}$$

where $\nu = [u \quad v \quad r]^T$ is the state vector; $\tau = [\tau_u \quad 0 \quad \tau_r]^T$ is the control vector and

$$M = \begin{bmatrix} m_u & 0 & 0 \\ 0 & m_v & 0 \\ 0 & 0 & m_r \end{bmatrix} > 0,$$

$$C(\nu) = \begin{bmatrix} 0 & 0 & -m_v v \\ 0 & 0 & m_u u \\ m_v v & -m_u u & 0 \end{bmatrix},$$

$$D(\nu) = \begin{bmatrix} d_u & 0 & 0 \\ 0 & d_v & 0 \\ 0 & 0 & d_r \end{bmatrix} > 0,$$

$m_u = m - X_{\dot{u}}$; $m_v = m - Y_{\dot{v}}$; $m_r = I_{zz} - N_{\dot{r}}$,
$d_u = -X_u - X_{|u|u}|u| = d_{u1} + d_{u2}|u|$,
$d_v = -Y_v - Y_{|v|v}|v| = d_{v1} + d_{v2}|v|$,
$d_r = -N_r - N_{|r|r}|r| = d_{r1} + d_{r2}|r|$.

Considering practical issues, the following assumptions are made:

(a) The surge velocity u is much larger than the sway v so that $u \approx U$ is constant, where U is the total speed of the vehicle w.r.t the fluid;
(b) The yaw rate and therefore yaw are controlled using a proportional-derivative control law given by, i.e., $\tau_r = -K\tilde{\psi} - K_d(r - \dot{\psi}_d)$, where $\tilde{\psi} = \psi - \psi_d$ is the negative of the yaw heading error, and ψ_d is the heading command;
(c) The dynamics of the heading is replaced by the dynamics of the heading error, i.e., $\dot{\tilde{\psi}} = \dot{\psi} - \dot{\psi}_d$.

Thus, taking into account assumptions (a) to (c), the mathematical model of the system that describes, for a fixed U, the evolution of the closed-loop system under PD control, is given by

$$\begin{bmatrix} \dot{v} \\ \dot{r} \\ \dot{\tilde{\psi}} \end{bmatrix} = \underbrace{\begin{bmatrix} -\frac{d_{v1}}{m_v} & -\frac{m_u}{m_v}U & 0 \\ -\frac{m_v - m_u}{m_r}U & -\frac{d_{r1}+K_d}{m_r} & -\frac{K}{m_r} \\ 0 & 1 & 0 \end{bmatrix}}_{A} \underbrace{\begin{bmatrix} v \\ r \\ \tilde{\psi} \end{bmatrix}}_{x} + \underbrace{\begin{bmatrix} -\frac{d_{v2}|v|}{m_v} & 0 & 0 \\ 0 & -\frac{d_{r2}|r|}{m_r} & 0 \\ 0 & 0 & 0 \end{bmatrix}}_{G(x)} \begin{bmatrix} v \\ r \\ \tilde{\psi} \end{bmatrix} + \underbrace{\begin{bmatrix} 0 \\ \frac{K_d}{m_r} \\ -1 \end{bmatrix}}_{B} \dot{\psi}_d \tag{44}$$

Equation (44) can be rewritten as:

$$\dot{x} = Ax + G(x)x + Bu, \; u \in \mathbb{R} \tag{45}$$

for $x \in D_x = \{v, r, \tilde{\psi} \subset \mathbb{R}^3 : |v| \leq v_{max} \quad |r| \leq r_{max}, \quad |\tilde{\psi}| \leq \tilde{\psi}_{max}\}$ and $u \in D_u = \{u \subset \mathbb{R} : |u| \leq u_{max}\}$. Notice that Equation (45) contains linear and nonlinear terms.

5. Analysis of the Stability of the Origin

The stability analysis of the origin of the mathematical model of the system (45) is performed considering the influence of the vehicle speed U on the eigenvalues of matrix A, once the values for K and K_d have been defined. To analyze the stability of the origin, we use the properties of *input-to-state* (ISS) and *input-output stability* (IOS) of the control system. Furthermore, this analysis is important because ISS and IOS properties of the system are based on a Lyapunov equation that involves the matrix A. In what follows, we provide definitions of ISS and IOS and essential theorems from [29] that will be used to perform the stability analysis of the systems considered in this paper.

5.1. Input-to-State Stability (ISS)

Consider the system

$$\dot{x} = f(x, t, u) \quad x(t_0) = x_0 \tag{46}$$

where $x(t) \in \mathbb{R}^n$ and $u(t) \in \mathbb{R}^m$ denotes states and the input at time $t \geq 0$. The function $f : [0, \infty] \times \mathbb{R}^n \times \mathbb{R}^m \to \mathbb{R}^n$ is piecewise continuous in t and locally Lipschitz in x and u. The input $u(t)$ is a piecewise continuous, bounded function of t for all $t \geq 0$.

Definition 1. *The system (46) is said to be* input-to-state stable *(ISS) if there exists a class $\mathcal{KL}$-function β and α class $\mathcal{K}$-function, such that for any initial slate $x(t_0)$ and any bounded input $u(t)$, the solution $x(t)$ exists for all $t \geq t_0$ and satisfies*

$$\|x(t)\| \leq \beta(\|x_0\|, t - t_0) + \gamma\left(\sup_{t_0 \leq \tau \leq t} \|u(\tau)\|\right) \tag{47}$$

The function $\gamma \in \mathcal{K}$ describes the influence of the input on the solution of the system. The function $\beta \in \mathcal{KL}$ describes the transient behavior of the system [29,41].

The Lyapunov-based technique for the ISS verification gives a sufficient condition for input-to-state stability.

Theorem 1. *Let $V : [0, \infty) \times \mathbb{R}^n \to \mathbb{R}$ be a continuously differentiable function such that*

$$\alpha_1(\|x\|) \leq V(t, x) \leq \alpha_2(\|x\|) \tag{48}$$

$$\frac{\partial v}{\partial t} + \frac{\partial v}{\partial x} f(x, t, u) \leq -W_3(x), \;\; \forall \;\; \|x\| \geq \rho(\|u\|) > 0 \tag{49}$$

$\forall (t, x, u) \epsilon [0, \infty) \times \mathbb{R}^m \times \mathbb{R}^n$, where α_1, α_2 are class $\mathcal{K}_\infty$-functions, ρ is class $\mathcal{K}$-function and $W_3(x)$ is continuous definite function on $\mathbb{R}^n$. Then, the system (46) is input-to-state stable with $\gamma = \alpha_1^{-1} \; o \; \alpha_2 \; o \; \rho$.

5.2. Input-Output Stability (IOS)

To understand the notion of input-output stability, we start by considering a system H with input u and output y. The system H is a map between two signals spaces $y = H(u)$. The gain γ of H measures the largest amplification from u to y, where the magnitude of the latter is computed using appropriate function norms, see [29]. H can be a constant, a matrix, a linear system or a nonlinear system. The gain γ of H is defined as

$$\gamma(H) = \sup_{u \in \mathcal{L}_p} \frac{\|y\|_p}{\|u\|_p} = \sup_{u \in \mathcal{L}_p} \frac{\|H(u)\|_p}{\|u\|_p}. \tag{50}$$

The system H is finite-gain *bounded-input bounded-output* (BIBO) if $\gamma(H) < \infty$.

The above stability concept has been extended to capture the effect of initial conditions and possible biases in the input-output operator, yielding the concept of input-to-output (IOS) stability. The resulting concept and main stability result, taken from [29], are briefly summarized next. In this context, Lyapunov stability tools can be used to establish the IOS stability of nonlinear systems represented by state models. Consider a state model presented in (46), together with output function

$$y = h(x, t, u) \tag{51}$$

where $h : [0, \infty] \times D \times D_u \to \mathbb{R}^q$ is piecewise continuous in t and continuous in (x, u), where $D \subset \mathbb{R}^n$ is a domain that contains $x = 0$, and $D_u \subset \mathbb{R}^m$ is a domain that contains $u = 0$. The following theorem states conditions under which, following the terminology in Khalil, a system is $\mathcal{L}$-stable or a small signal is $\mathcal{L}$-stable for a given choice of signal space $\mathcal{L}$. Suppose $x = 0$ is an equilibrium point of the unforced system

$$\dot{x} = f(t, x, 0). \tag{52}$$

Theorem 2. *Consider the system* (46) *and* (51) *and take $r > 0$ and $r_u > 0$ such that $\{\|x\| \le r\} \subset D$ and $\{\|u\| \le r_u\} \subset D_u$. Suppose that*

- *$x = 0$ is an equilibrium point of* (52)*, and there is Lyapunov function $V(t, x)$ that satisfies*

$$c_1\|x\|^2 \le V(t, x) \le c_2\|x\|^2 \tag{53}$$

$$\frac{\partial v}{\partial t} + \frac{\partial v}{\partial x} f(t, x, 0) \le -c_3\|x\|^2 \tag{54}$$

$$\left\|\frac{\partial v}{\partial x}\right\| \le c_4\|x\| \tag{55}$$

for all $(t, x) \in [0, \infty) \times D$ for some positive constants c_1, c_2, c_3, and c_4.

- *f and h satisfy the inequalities*

$$\|f(t, x, u) - f(t, x, 0)\| \le L\|u\| \tag{56}$$

$$\|h(t, x, u)\| \le \eta_1\|x\| + \eta_2\|u\| \tag{57}$$

for all $(t, x) \in [0, \infty) \times D \times D_u$ for some non negative constants L, η_1, and η_2.

Then, for each x_0 with $\|x\| \le r\sqrt{c_1/c_2}$, the system (46) *and* (51) *is small-signal finite gain $\mathcal{L}_p$-stable for each $p \in [1, \infty]$. In particular, for each $u \in \mathcal{L}_{pe}$ with $\sup_{t_0 \le \tau \le t}\|u(t)\| \le min\{r_u, c_1c_3r/(c_2c_4L)\}$, the output $y(t)$ satisfies*

$$\|y_\tau\| \le \gamma\|u_\tau\|_{\mathcal{L}_p} + \beta \tag{58}$$

for all $\tau \in [0, \infty)$, with

$$\gamma = \eta_2 + \frac{\eta_1 c_2 c_4 L}{c_1 c_3}, \tag{59}$$

$$\beta = \eta_1\|x_0\|\sqrt{\frac{c_2}{c_1}}\rho, \quad where \quad \rho = \begin{cases} 1 & if \quad p = \infty \\ \left(\frac{2c_2}{c_3 p}\right)^{(1/p)}, & if \quad p \in [1, \infty) \end{cases} \tag{60}$$

Furthermore, if the origin is globally exponentially stable and the assumptions hold globally (with $D = \mathbb{R}^n$ and $D_u = \mathbb{R}^m$), then, for each $x_0 \in \mathbb{R}^n$, the system (46) *and* (51) *is finite gain $\mathcal{L}_p$-stable for each $p \in [1, \infty]$.*

We refer the reader to explore [29] for details of the proof. ISS and IOS are the key Lyapunov stability tools used further for the analysis of our system.

5.3. ISS Analysis

For ISS analysis, the system model should satisfy the following equations for a suitably defined Lyapunov function candidate $V(t,x)$:

$$\alpha_1(\|x\|) \le V(t,x) \le \alpha_2(\|x\|) \tag{61}$$

$$\frac{\partial V}{\partial t} + \frac{\partial V}{\partial x} f(t,x,u) \le -W_3(x) \quad \forall \|x\| \ge \rho(\|x\|) > 0. \tag{62}$$

where in this particular case, $f(t,x,u)$ is the right-hand side of (45). Then, the system is ISS with, $\gamma = \alpha_1^{-1} \circ \alpha_2 \circ \rho$ (please refer to [29] for definitions and theorems mentioned from now on).

Consider the following Lyapunov function candidate to check the ISS property of the system, given by

$$V = x^T P x \tag{63}$$

where P is a nonsingular symmetric matrix.

Then,

$$\lambda_{min}(P)\|x\|^2 \le x^T P x \le \lambda_{max}(P)\|x\|^2 \tag{64}$$

Thus, comparing (61) with Equation (64), we conclude that

$$\alpha_1(\|x\|) = \lambda_{min}(P)\|x\|^2 \tag{65}$$

$$\alpha_2(\|x\|) = \lambda_{max}(P)\|x\|^2 \tag{66}$$

The derivative of V along the trajectories defined by Equation (45) is given by

$$\dot{V} = x^T\left[A + G(x)\right]^T P + P[A + G(x)]x + 2uB^T P x \tag{67}$$

Equation (67) can be rewritten as

$$\dot{V} = x^T(A^T P + PA)x + 2x^T P G(x)x + 2uB^T P x. \tag{68}$$

The matrix product $PG(x)$ results in

$$F = PG = \begin{bmatrix} -p_{1,1}d_1|x_1| & -p_{1,2}d_2|x_2| & 0 \\ -p_{1,2}d_1|x_1| & -p_{2,2}d_2|x_2| & 0 \\ -p_{1,3}d_1|x_1| & -p_{2,3}d_2|x_2| & 0 \end{bmatrix} = \begin{bmatrix} \bar{f}_{1,1}|x_1| & \bar{f}_{1,2}|x_2| & 0 \\ \bar{f}_{2,1}|x_1| & \bar{f}_{2,2}|x_2| & 0 \\ \bar{f}_{3,1}|x_1| & \bar{f}_{3,2}|x_2| & 0 \end{bmatrix} \tag{69}$$

where $d_1 = \frac{d_{v2}}{m_v}$ and $d_2 = \frac{d_{r2}}{m_r}$, and p_{ij} is the entry i,j of P. Inserting both the Lyapunov equation $A^T P + PA = -Q$ and $x^T F x$ into Equation (68) yields

$$\begin{aligned} \dot{V} = &- x^T Q x + 2\left[\bar{f}_{1,1}|x_1|x_1^2 + \bar{f}_{2,2}|x_2|x_2^2 + (\bar{f}_{2,1}|x_1| + \bar{f}_{1,2}|x_2|)x_1x_2 + \bar{f}_{3,1}|x_1|x_3x_1 \right. \\ &\left. + \bar{f}_{3,2}|x_2|x_3x_2\right] + 2uB^T P x \end{aligned} \tag{70}$$

The following inequalities can be used in Equation (70):

(I) $\bar{f}_{1,1}|x_1|x_1^2 \le 0$ and $\bar{f}_{2,2}|x_2|x_2^2 \le 0$ because $p_{1,1} \ge 0$ and $p_{2,2} \ge 0$;

(II) $x_1x_2 + x_1x_3 + x_2x_3 \le x_1^2 + x_2^2 + x_3^2 = \|x\|^2$.
This follows from the inequality $(x_1 - x_2)^2 + (x_1 - x_3)^2 + (x_2 - x_3)^2 \ge 0$.

Taking into account inequality I in Equation (70) yields

$$\dot{V} \leq -x^T Qx + 2\left[|(\bar{f}_{2,1}|x_1| + \bar{f}_{1,2}|x_2|)||x_1 x_2| + |\bar{f}_{3,1}||x_1||x_3 x_1| + |\bar{f}_{3,2}||x_2||x_3 x_2| \right] + 2\|u\|\|B^T\|\|P\|\|x\| \tag{71}$$

Now, inserting inequality II into Equation (71) leads to

$$\dot{V} \leq -\lambda_{min}(Q)\|x\|^2 + 2\left[(|\bar{f}_{2,1}| + |\bar{f}_{3,1}|)|x_1| + (|\bar{f}_{1,2}| + |\bar{f}_{3,2}|)|x_2| \right] \|x\|^2 + 2\|u\|\|B^T\|\|P\|\|x\|. \tag{72}$$

Concisely, Equation (72) can be expressed as:

$$\dot{V} \leq -\lambda_{min}(Q)\|x\|^2 + 2|\bar{F}(x)|\|x\|^2 + 2\|u\|\|B^T\|\|P\|\|x\| \tag{73}$$

where

$$\begin{aligned} \bar{F}(x) &= (|\bar{f}_{2,1}| + |\bar{f}_{3,1}|)|x_1| + (|\bar{f}_{1,2}| + |\bar{f}_{3,2}|)|x_2| \\ &= \underbrace{\begin{bmatrix} |\bar{f}_{2,1}| + |\bar{f}_{3,1}| & |\bar{f}_{1,2}| + |\bar{f}_{3,2}| & 0 \end{bmatrix}}_{F'} \underbrace{\begin{bmatrix} |x_1| \\ |x_2| \\ |x_3| \end{bmatrix}}_{x'} = F'x' \end{aligned} \tag{74}$$

Then, using the fact that $\|x'\| = \|x\|$, yields

$$\begin{aligned} \dot{V} &\leq -(\lambda_{min}(Q) - 2|\bar{F}(x)|)\|x\|^2 + 2\|u\|\|B^T\|\|P\|\|x\| \\ &\leq -(\lambda_{min}(Q) - 2\|F'\|\|x'\|)\|x\|^2 + 2\|u\|\|B^T\|\|P\|\|x\| \\ &\leq -(\lambda_{min}(Q) - 2\|F'\|\|x\|)\|x\|^2 + 2\|u\|\|B^T\|\|P\|\|x\| \end{aligned} \tag{75}$$

Notice that for a given δ, there exists some $r > 0$ such that

$$|\bar{F}(x)| < \delta, \quad \forall \|x\| < r \tag{76}$$

Equation (75) can be rewritten as

$$\begin{aligned} \dot{V} \leq -(1-\theta)\left[\lambda_{min}(Q) - 2\delta\right]\|x\|^2 - \big[\theta(\lambda_{min}(Q) - 2\delta)\|x\|^2 \\ - 2|u|\|B\|\|P\|\|x\|\big] \quad \forall\, 0 < \theta < 1 \end{aligned} \tag{77}$$

Then,

$$\dot{V} \leq -(1-\theta)\left[\lambda_{min}(Q) - 2\delta\right]\|x\|^2 \tag{78}$$

for

$$\delta < \frac{\lambda_{min}(Q)}{2} \tag{79}$$

and

$$\|x\| > \frac{2|u|\|B^T\|\|P\|}{\theta\left[\lambda_{min}(Q) - 2\delta\right]}.$$

As a consequence, the system with closed-loop dynamics defined by Equation (44) is ISS with

$$W_3 = (1-\theta)\left[\lambda_{min}(Q) - 2\delta\right],$$

$$\rho = \frac{2|u|\|B^T\|\|P\|}{\theta\left[\lambda_{min}(Q) - 2\delta\right]},$$

and

$$\gamma = \frac{\lambda_{max}(P)}{\lambda_{min}(P)} \frac{2|u|\|B^T\|\|P\|}{\theta\left[\lambda_{min}(Q) - 2\delta\right]}.$$

5.4. IOS Analysis

The proof of the IOS property of the same system (represented by (44)) is based on Theorem 2 [29]. Consider the system

$$\dot{x} = f(x,u), \quad x(0) = x_0 \tag{80}$$

$$y = h(x) \tag{81}$$

According to Theorem 2, if the system (80) and (81) satisfies the conditions shown below, then the system is small-signal finite-gain $\mathcal{L}_p$-stable for each $p \in [1,\infty]$. In particular, for each $u \in \mathcal{L}_{pe}$ with $sup_{0 \le t \le \tau}\|u(t)\| \le min\{r_u, c_1c_3r/(c_2c_4\mathbb{L})\}$, the output satisfies

$$\|y_\tau\|_{\mathcal{L}_p} \le \gamma\|u_\tau\|_{\mathcal{L}_p} + \beta \tag{82}$$

where the parameters γ and β are given by

$$\gamma = \eta_2 + \frac{\eta_1 c_2 c_4}{c_1 c_3}, \tag{83}$$

and

$$\beta = \eta_1\|x_0\|\sqrt{\frac{c_2}{c_1}}\rho, \quad where \quad \rho = \begin{cases} 1 & if \quad p = \infty \\ \left(\frac{2c_2}{c_3 p}\right)^{(1/p)} & if \quad p \in [1,\infty) \end{cases}. \tag{84}$$

The parameters c_1, c_2, c_3, c_4, η_1, and η_2 are determined from inequalities in Theorem 2.

For the particular case of the system under consideration, Equations (80) and (81) are given as

$$\dot{x} = f(x) = Ax + G(x)x + Bu; \quad x(0) = x_0 \tag{85}$$

$$y = h(x) = Hx \tag{86}$$

where

$$H = \begin{bmatrix} 0 & 0 & 1 \end{bmatrix} \tag{87}$$

Thus, considering the Lyapunov function defined by Equation (63) yields the following sequence of partial results.

- From inequality (53)

$$\lambda_{min}(P)\|x\|^2 \le x^T P x \le \lambda_{max}(P)\|x\|^2 \tag{88}$$

  Thus,
  $c_1 = \lambda_{min}(P)$ and $c_2 = \lambda_{max}(P)$

- From inequality (54)

$$\frac{\partial V}{\partial x} f(x,0) \le -\left[\lambda_{min}(Q) - 2\delta\right]\|x\|^2 \tag{89}$$

  and therefore
  $c_3 = \left[\lambda_{min}(Q) - 2\delta\right]$

- From inequality (55)

$$\|\frac{\partial V}{\partial x}\| = 2\|x^T P\| \le 2\lambda_{max}(P)\|x\| \tag{90}$$

  Thus, $c_4 = 2\lambda_{max}(P)$

- From inequality (56)

$$\|f(t,x,u) - f(t,x,0)\| = \|Bu\| \le \|B\|\|u\| \tag{91}$$

and therefore $\mathbb{L} = \|B\| = \sqrt{\lambda_{max}(B^T B)}$.

- From inequality (57)

$$\|h(x)\| \leq \|H\| \|x\|, \tag{92}$$

we obtain that
$\eta_1 = \|H\| = \sqrt{\lambda_{max}(H^T H)}$ and $\eta_2 = 0$

From the above, the parameters $\gamma_f = \gamma$ and $\beta_f = \beta$ for the Medusa model are given by

$$\gamma_f = \frac{\eta_1 c_2 c_4}{c_1 c_3} = \frac{\|H\| \lambda^2_{max}(P) 2\sqrt{\lambda_{max}(B^T B)}}{\lambda_{min}(P)\left[\lambda_{min}(Q) - 2\delta\right]} \tag{93}$$

$$\beta_f = \eta_1 \|x_0\| \sqrt{\frac{c_2}{c_1}}\rho = \|H\| \|x_0\| \sqrt{\frac{\lambda_{max(P)}}{\lambda_{min}(P)}} \quad for \quad p = \infty \tag{94}$$

Remark 1. *From a mathematical standpoint, the machinery adopted for the description of the systems under study, that is, inner (dynamic) and outer (kinematic) dynamics, is rooted in their characterization as ISS (input-to-state stable) or IOS (input-to-output stable) systems. This allows us to use powerful tools of nonlinear stability analysis. The core mathematical characterization of IOS and ISS hinges on the assumption that all functions involved in the description of the systems of interest are piecewise continuous in time and locally Lipschitz in the state and input variables. This is clearly indicated in the results on ISS described in Sections 5.1 and 5.2, as applied to the system described by Equation (44). Clearly, all functions involved (which capture the physical description of the vehicle) satisfy the conditions stated above. Identical comments apply to the results on IOS described in Sections 5.3 and 5.4, as applied to the system described by Equations (85)–(87), consisting of Equation (44) together with the trivial output function $h(x) = Hx$, $H = [0\ 0\ 1]$ described in Equations (86) and (87). Again, all functions involved satisfy the conditions stated above (see Equation (56). In addition, $h(x)$ satisfies the extra output-related conditions in Equation (57).*

6. Path Following Problem

Equipped with the above mathematical definitions and results, we now tackle the problem of path following. In the current setup, we design the kinematic (outer) loop without considering the dynamic (inner) loop or by assuming the inner loop is infinitely fast, which is not valid in practice. Moreover, the characteristics of inner loop controllers (such as heading and speed controllers) for many vehicles are provided in very general terms by their vendors. An example of these characteristics in a linear case is the approximate bandwidth and input-to-output stability gain (IOS) in the case of a nonlinear system [29]. Therefore, it is required for the system engineers to design or tune the outer-loop controller by considering these characteristics, such that the overall combined system is stable with the desired performance. However, this step necessitates going beyond qualitative assertions about the fast-slow temporal scale separation and quantitatively evaluating the combined inner-outer loop to obtain a relationship for assessing the combined system's stability. The methods used are based on nonlinear control theory, wherein the cascade and feedback systems of interest are characterized in terms of their IOS properties. We use the IOS small-gain theorem to obtain quantitative relationships for best controller tuning applicable to a broad range of marine vehicles.

The path-following controller discussed in this article is designed at the kinematic outer-loop that commands the inner-loop with the desired heading angles while the vehicle moves at an approximately constant speed. The idea is to provide a seamless implementation of the path-following control algorithms on the heterogeneous vehicles, which are pre-equipped with heading autopilots. To this effect, we assume that the heading control system is characterized only in terms of an IOS-like relationship without knowing detailed vehicle dynamic parameters.

6.1. Path Following: Straight Lines Problem

Figure 5 shows the path following problem for straight lines. In the figure, $\{\mathcal{I}\} = \{x_I, y_I\}$ represents the inertial reference frame, and $\{\mathcal{B}\} = \{x_B, y_B\}$ denotes a body reference frame fixed to the vehicle. Let us denote the position of the vehicle as vector P expressed in $\{\mathcal{I}\}$. We assume that the ocean current velocity represented by V_c expressed in $\{\mathcal{I}\}$ is constant. The velocity of the vehicle expressed in $\{\mathcal{I}\}$ is given by

$$\dot{P} = R(\psi)V_w + V_c,$$

where ψ is the yaw angle, V_w denotes the velocity of the vehicle with respect to the water expressed in $\{\mathcal{B}\}$, and $R(.)$ is the rotation matrix from $\{\mathcal{B}\}$ to $\{\mathcal{I}\}$, parameterized by ψ. Equivalently,

$$\dot{P} = R(\psi + \beta)[||V_w||\ 0]^T + V_c,$$

where β is the sideslip angle. Without any loss of generality, the straight-line path to be followed can be assumed to be along the x-axis of the inertial reference frame $\{\mathcal{I}\}$. The evolution of the cross-track error e is given by

$$\dot{e} = \sin(\psi + \beta)||V_w|| + v_{cy},$$

where v_{cy} denotes the component of V_c along the unit vector y_I. The total speed of the vehicle is set by an equivalent speed of rotation of the stern propeller(s) and the heading of the vehicle is controlled either by differential mode of two stern propellers or by the stern rudders operated in common mode.

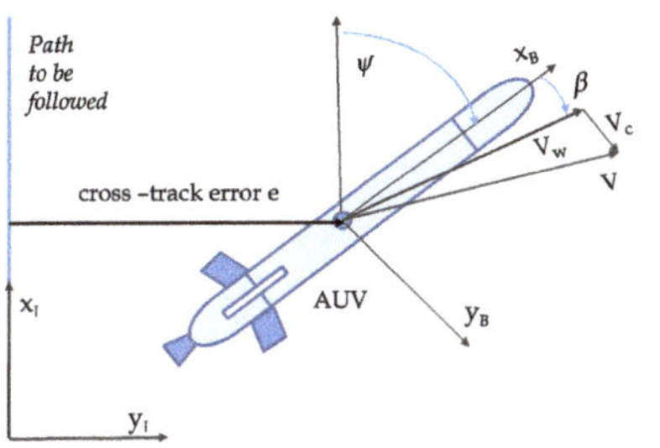

Figure 5. Marine vehicle body reference frame showing the cross-track error.

We assume that the total speed $||V_w|| = U > ||V_c||$ is constant. The objective is to command the heading angle which the vehicle can follow to drive the e to zero. In the following section, as a first step, we design an outer-loop controller at the kinematic level and show the convergence of the cross-track error to zero. In the second step, we include the yaw control dynamics (inner-loop) and determine the conditions and outer-loop tuning rules such that the complete inner-outer loop system is stable.

6.2. Path-Following Algorithm

To explain the rationale for the control law, we simplify the case by considering zero sideslip angle (this assumption will be lifted afterwards). In this case, the error dynamics are given by

$$\dot{e} = U\sin(\psi) + v_{cy}. \tag{95}$$

If we consider v_{cy} to be zero, then (95) can be re-written as

$$\dot{e} = Uu,$$

with $u = \sin(\psi)$. The choice of the control law $u = -(K_1/U)e$ would now ensure that e converges asymptotically and exponentially to the origin. In order to compensate for a fixed ocean current (bias) v_{cy}, an integral term is introduced in the virtual input u, which is now re-rewritten as

$$u = -\frac{1}{U}\left(K_1 e + K_2 \int_0^t e(\tau)\,d\tau\right).$$

As a consequence, the dynamics of e become

$$\dot{e} + K_1 e + K_2 \int_0^t e(\tau)\,d\tau = 0$$

Let

$$\varsigma = \int_0^t e(\tau)\,d\tau.$$

Then,

$$\ddot{\varsigma} + K_1\dot{\varsigma} + K_2\varsigma = 0 \tag{96}$$

The gains K_1 and K_2 can now be chosen so as to obtain a desired natural frequency and a desired damping factor in the above second-order system. The desired heading command obtained from the above virtual control is written as

$$\psi_d = \sin^{-1}(\sigma_e(u)),$$

where σ_e is a differentiable saturation function [42] bounded between $\pm e_s$ with $0 < e_s < 1$, defined as

$$\sigma_e(\varrho) = \begin{cases} \varrho & \text{if } |\varrho| < e_s - \epsilon \\ +e_s & \text{if } \varrho > e_s + \epsilon \\ -e_s & \text{if } \varrho < -e_s - \epsilon \\ p_1(\varrho) = -c_1\varrho^2 + c_2\varrho - c_3 & \text{if } \varrho \in \,]e_s - \epsilon, e_s + \epsilon] \\ p_2(\varrho) = c_1\varrho^2 + c_2\varrho + c_3 & \text{if } \varrho \in [-e_s - \epsilon, -e_s + \epsilon[\end{cases} \tag{97}$$

where $0 < \epsilon < e_s$ can be arbitrarily small, with $c_1 = \frac{1}{4\epsilon}$, $c_2 = \frac{1}{2} + \frac{e_s}{2\epsilon}$, and $c_3 = \frac{\epsilon^2 - 2\epsilon e_s + e_s^2}{4\epsilon}$. The saturation function is introduced to guarantee that the argument of $sin^{-1}(.)$ lies in the interval $[-1, +1]$, see Figure 6.

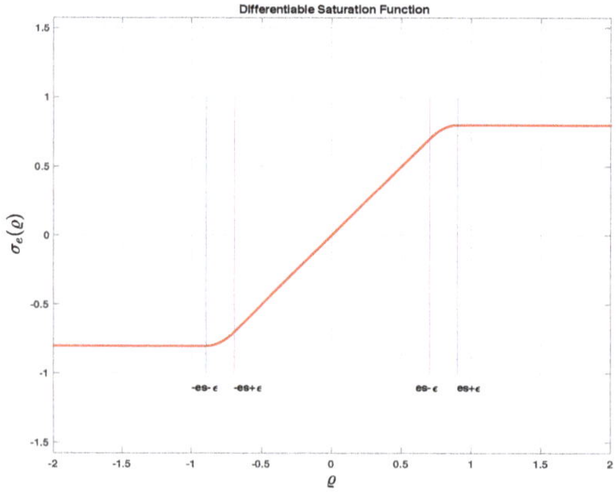

Figure 6. Differentiable saturation function.

With the introduction of integrator in the control law, it is important to have an anti-windup mechanism in the integral term of u. Thus, the final form of the control law for ψ_d involves a new definition of u and is given in terms of the operator $f : e \rightarrow \psi_d$ defined by

$$\psi_d = \sin^{-1}(\sigma_e(u)); u = \left(-\frac{K_1 e}{U} - \frac{K_2}{U}\varsigma\right) \tag{98}$$

where ς is the output of the dynamical system $f_{aw}(e) : e \rightarrow \varsigma$ with realization

$$\dot{\varsigma} = e + K_a\left[-\frac{K_1 e}{U} - \frac{K_2\varsigma}{U} - \sigma_e\left(-\frac{K_1 e}{U} - \frac{K_2\varsigma}{U}\right)\right], \tag{99}$$

and K_a is an anti-windup gain to control the integrator's charge and discharge rate. In what follows, we show with the help of Lyapunov-based analysis tools that using the above control law, the cross-track error converges to zero if the actual vehicle heading ψ equals ψ_d.

6.3. Convergence of Cross-Track Error without the Inner Loop Dynamics

Using the control law mentioned in (98) and (99), and $K_a = \frac{U}{K_1}$, the closed-loop kinematic equations can now be written as,

$$\dot{e} = U\sigma_e\left(-\frac{K_1 e}{U} - \frac{K_2\varsigma}{U}\right) + V_{yc} \tag{100}$$

$$\dot{\varsigma} = -\frac{K_2}{K_1}\varsigma - \frac{U}{K_1}\sigma_e\left(-\frac{K_1 e}{U} - \frac{K_2\varsigma}{U}\right), \tag{101}$$

Define the new set of variables

$$\begin{aligned} x_1 &= e \\ x_2 &= \varsigma - K, \end{aligned} \tag{102}$$

where K is a constant.

The equations of motion can then be written as

$$\dot{x}_1 = U\sigma_e\left(-\frac{K_1 x_1}{U} - \frac{K_2 x_2}{U} - \frac{KK_2}{U}\right) + V_{yc} \tag{103}$$

$$\dot{x}_2 = -\frac{K_2}{K_1}x_2 - \frac{K_2}{K_1}K - \frac{U}{K_1}\sigma_e\left(-\frac{K_1 x_1}{U} - \frac{K_2 x_2}{U} - \frac{KK_2}{U}\right). \tag{104}$$

We define another set of variables as

$$\begin{aligned} rly_1 &= \frac{K_1 x_1}{U} + \frac{K_2 x_2}{U}r \\ y_2 &= \frac{K_2 x_2}{U}, \end{aligned} \tag{105}$$

In terms of the new variables above,

$$\begin{aligned} \dot{y}_1 &= -K_1\sigma_e\left(y_1 + \frac{KK_2}{U}\right) + K_1\frac{V_{yc}}{U} + \dot{y}_2 \\ \dot{y}_2 &= -\frac{K_2}{K_1}y_2 - \frac{K_2}{K_1}\frac{KK_2}{U} + \frac{K_2}{K_1}\sigma_e\left(y_1 + \frac{KK_2}{U}\right), \end{aligned} \tag{106}$$

At this point, we explore an important property of the σ_e function defined in (97).

Property 1.

$$\sigma_e(Z + x) - Z = \sigma_{l_s}^{u_s}(x) \;\; \forall \;\; |Z| < e_s,$$

where $l_s = -e_s - |Z|$ *and* $u_s = e_s - |Z|$.

To simplify the notation, we use σ instead of $\sigma_{l_s}^{u_s}$ from here on. Using this property, we can write

$$\sigma_e\left(y_1 + \frac{KK_2}{U}\right) - \frac{KK_2}{U} = \sigma(y_1), \ \forall \ \left|\frac{KK_2}{U}\right| < e_s, \tag{107}$$

and later in the proof, it will be evident that $\left|\frac{KK_2}{U}\right| < e_s$. Thus, by simplifying further, we get

$$\begin{aligned} \dot{y}_1 &= -K_1\sigma(y_1) - K_1\frac{KK_2}{U} + K_1\frac{V_{yc}}{U} + \dot{y}_2 \\ \dot{y}_2 &= \frac{K_2}{K_1}\sigma(y_1) - \frac{K_2}{K_1}y_2 \end{aligned} \tag{108}$$

Choosing $V(y_1, y_2) = \int\limits_0^{y_1} \sigma(\eta)d\eta + \frac{1}{2}y_2^2$ as a Lyapunov candidate function yields

$$\begin{aligned} \dot{V} &= \sigma(y_1)\dot{y}_1 + y_2\dot{y}_2 \\ &= \sigma(y_1)\left[-K_1\sigma(y_1) - K_1\frac{KK_2}{U} + \right. \\ &\quad \left. K_1\frac{V_{yc}}{U} + \frac{K_2}{K_1}\sigma(y_1) - \frac{K_2}{K_1}y_2\right] \\ &= -\left(K_1 - \frac{K_2}{K_1}\right)\sigma^2(y_1) - \frac{K_2}{K_1}y_2^2 + \\ &\quad \sigma(y_1)\left[K_1\frac{V_{yc}}{U} - K_1\frac{KK_2}{U}\right]. \end{aligned}$$

Making

$$K = \frac{V_{yc}}{K_2} \tag{109}$$

yields

$$\dot{V} = -\left(K_1 - \frac{K_2}{K_1}\right)\sigma^2(y_1) - \frac{K_2}{K_1}y_2^2 \tag{110}$$

At this point, it is reasonable to assume that the vehicle speed with respect to water is larger than the intensity of the ocean current, that is,

$$U > \frac{1}{e_s}||V_c||. \tag{111}$$

Using the above assumption, it is now straightforward to show that $\left|\frac{KK_2}{U}\right| < e_s$. Thus,

$$\dot{V} < 0 \quad \forall \, K_1 > \frac{K_2}{K_1}. \tag{112}$$

We therefore conclude that the origin $y_1 = y_2 = 0$ is asymptotically stable. It is now trivial to show that the cross-track error e will tend to zero and the integrator ς will charge up to $\frac{V_{yc}}{K_2}$, in order to "learn" the currents as time increases.

6.4. Inner-Loop Dynamics

The key goal of this paper is to show that "identical behavior" is obtained when the dynamics of the heading autopilot (inner loop) and the sideslip of the vehicle are taken into account. In particular, we show that *the basic structure and the simplicity of the outer-loop control law are preserved*. The theoretical machinery used to prove stability borrows from

IOS concepts and a related small-gain theorem. See [29] for a fast-paced introduction to the subject and [43,44] for interesting applications of control techniques that bear affinity with inner-outer loop control structures. Here, we indicate briefly how the existence of the heading autopilot is taken into account without having to change the structure of the outer-loop described before. The resulting control scheme is depicted in Figure 7, where the heading autopilot plays the role of an inner loop.

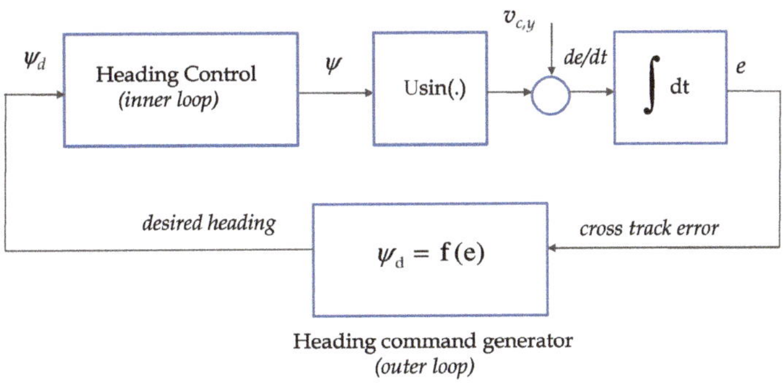

Figure 7. Path-following controller with two-scale inner-outer loop approach.

This section addresses explicitly the inclusion of the inner-loop dynamics, thus lifting the assumption that the actual heading ψ equals the desired heading ψ_d. Let

$$\tilde{\psi} = \psi - \psi_d$$

be the mismatch between actual and desired heading angles. We assume that the autopilot characteristics can be described in very general terms as an IOS system, see [29]. In order to understand the rationale for this characterization, notice that if the inner-loop dynamics are linear with static gain equal to 1, then its dynamics admit a realization of the form

$$\begin{aligned} \dot{x} &= Ax + B\psi_d \\ \psi &= Cx \end{aligned}$$

with $CA^{-1}B = 1$. In this case, the coordinate transformation $\eta = x + A^{-1}B\psi_d$ yields the realization

$$\begin{aligned} \dot{\eta} &= A\eta + A^{-1}B\dot{\psi}_d \\ \tilde{y} &= C\eta \end{aligned}$$

for the operator from $\dot{\psi}$ to $\tilde{y}$, with $\tilde{y} = \tilde{\psi} + \beta$ that characterizes the inner-loop dynamics, where β is sideslip angle and the output $\tilde{y}$ is the sum of the heading angle and sideslip angle. An IOS characterization of the loop can be easily derived from the above system matrices [29]. Notice, however, that this type of description applies also to general nonlinear systems of the form

$$\begin{aligned} \dot{\eta} &= g(\eta, \dot{\psi}_d) \\ \tilde{y} &= h(\eta, \dot{\psi}_d) \end{aligned}$$

and allows for a somewhat loose, yet quantifiable description of the inner-loop dynamics. This justifies the IOS characterization of the inner loop dynamics as

$$\|\tilde{y}(t)\| \leq \gamma_f \|\dot{\psi}_d(t)\| + \beta_f, \tag{113}$$

where γ_f and β_f are nonnegative constants. The above characterization captures in a rigorous mathematical framework simple physical facts about the inner-loop control system. Namely, (i) if the time-derivative of the heading reference ψ_d is bounded, then the heading-tracking error is bounded and (ii) the dynamics of the inner-loop system can be characterized in terms of bandwidth-like characteristics that are reflected in β_f and γ_f, see [29]. A simple exercise with a first-order system will convince the reader that as the bandwidth of the system increases, γ_f will decrease. For practical purposes, the latter can be viewed as a "tuning knob" during the path-following controller design phase. For analysis purposes, it is also required to ensure that not only $\tilde{y}$ but also the remaining variables in the inner loop be bounded in response to $\dot{\psi}_d$. This fact can be easily captured with an ISS condition of the type

$$\|\eta(t)\| \leq \beta_g(\|\eta(0)\|, t) + \gamma_g\left(\sup_{t_0 \leq \tau \leq t} \|\dot{\psi}_d(\tau)\|\right), \tag{114}$$

for some $\beta_g \in \mathcal{KL}$ and $\gamma_g \in \mathcal{K}$. We have shown before that such a condition holds. At this point, we make the key observation that the complete path-following control system can be represented as the interconnected structure depicted in Figure 8. The latter can be further abstracted to the scheme in Figure 9 consisting of blocks $H_1 : \tilde{y} \to \dot{\psi}_d$ and $H_2 : \dot{\psi}_d \to \tilde{y}$, a description of which is given next. To this effect, using the control law mentioned in (98) and (99), the system H_1 clearly admits the following representation

$$\begin{aligned} \dot{e} &= U\sin(\tilde{y} + \psi_d) + v_{cy}, \\ \dot{\varsigma} &= -\frac{K_2}{K_1}\varsigma - \frac{U}{K_1}\sigma_e\left(-\frac{K_1 e}{U} - \frac{K_2\varsigma}{U}\right) \\ \psi_d &= \sin^{-1}\left(\sigma_e\left(-\frac{K_1 e}{U} - \frac{K_2}{U}\varsigma\right)\right), \end{aligned} \tag{115}$$

and H_2 satisfies the IOS stability condition in (113).

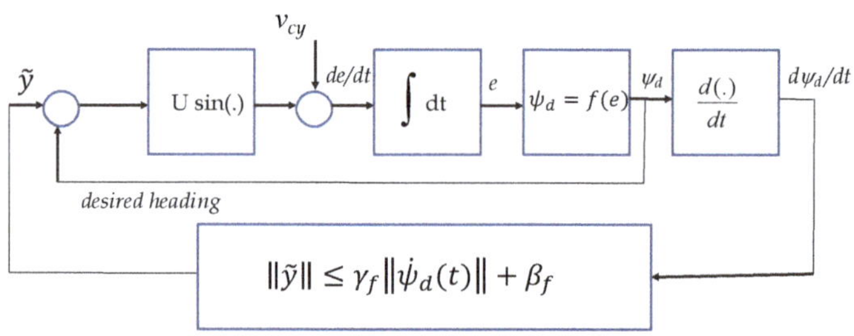

Figure 8. IOS characterization of inner-outer loop.

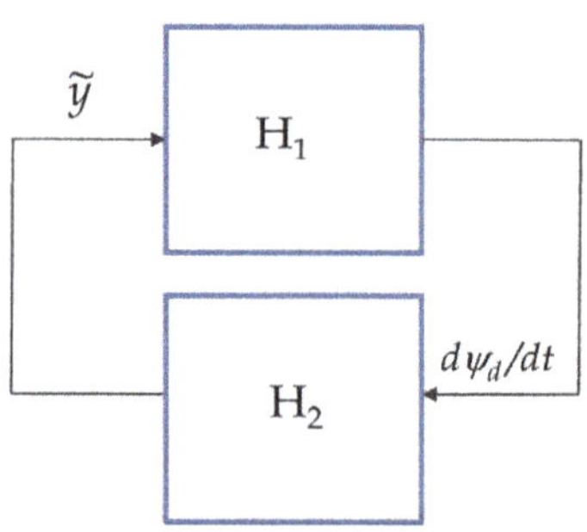

Figure 9. General feedback interconnection.

6.5. Convergence: Realistic Inner-Loop Dynamics

The proof that H_1 is IOS hinges on the facts that H_1 is the composition of two auxiliary systems $H_{a1} : \tilde{y} \to e$ and $H_{a2} : e \to \tilde{\psi}$ and that both are IOS. This is done next. Expanding Equation (115) and following the transformation of variables as mentioned in (105), the equation of motion can be rewritten as

$$\begin{aligned} \dot{y}_1 &= -K_1 \cos\tilde{y}\sigma_e\left(y_1 + \frac{KK_2}{U}\right) + K_1 \sin\tilde{y}\cos\psi_d + \\ &\frac{K_1}{U}V_{yc} + \dot{y}_2 \\ \dot{y}_2 &= \frac{K_2}{K_1}\left[\sigma_e\left(y_1 + \frac{KK_2}{U}\right) - \frac{KK_2}{U} - \frac{K_2}{K_1}y_2\right], \end{aligned} \tag{116}$$

By adding and subtracting the term $K_1 \cos\tilde{y}\frac{KK_2}{U}$, and by using the special property of the function σ_e in (107), we can write

$$\begin{aligned} \dot{y}_1 &= -K_1 \cos\tilde{y}\sigma(y_1) + K_1 \sin\tilde{y}\cos\psi_d + \frac{K_1}{U}V_{yc} \\ &\quad - K_1 \cos\tilde{y}\frac{KK_2}{U} + \frac{K_2}{K_1}\sigma(y_1) - \frac{K_2}{K_1}y_2 \\ \dot{y}_2 &= \frac{K_2}{K_1}\sigma(y_1) - \frac{K_2}{K_1}y_2 \end{aligned} \tag{117}$$

Choosing the same Lyapunov function as in Section 6.3 yields

$$\begin{aligned} \dot{V} &= \sigma(y_1)\dot{y}_1 + y_2\dot{y}_2 \\ &= \sigma(y_1)[-K_1 \cos\tilde{y}\sigma(y_1) + K_1 \sin\tilde{y}\cos\psi_d \\ &\quad + \frac{K_1}{U}V_{yc} - K_1\cos\tilde{y}\frac{KK_2}{U} + \frac{K_2}{K_1}\sigma(y_1) - \frac{K_2}{K_1}y_2] \\ &\quad + y_2\left[\frac{K_2}{K_1}\sigma(y_1) - \frac{K_2}{K_1}y_2\right] \\ &= -K_1\cos\tilde{y}\sigma^2(y_1) - \frac{K_2}{K_1}y_2^2 \\ &\quad + \sigma(y_1)\left[K_1 \sin\tilde{y}\cos\psi_d + \frac{K_1}{U}V_{yc} - K_1\cos\tilde{y}\frac{KK_2}{U}\right] \\ &\quad + \frac{K_2}{K_1}\sigma^2(y_1) \\ &= -\left(K_1\cos\tilde{y} - \frac{K_2}{K_1}\right)\sigma^2(y_1) - \frac{K_2}{K_1}y_2^2 \\ &\quad + \sigma(y_1)\left[K_1 \sin\tilde{y}\cos\psi_d + \frac{K_1}{U}V_{yc} - K_1\cos\tilde{y}\frac{KK_2}{U}\right], \end{aligned} \tag{118}$$

Now, substituting K from (109), we obtain

$$\begin{aligned} \dot{V} &= -\left(K_1\cos\tilde{y} - \frac{K_2}{K_1}\right)\sigma^2(y_1) - \frac{K_2}{K_1}y_2^2 \\ &\quad + \sigma(y_1)\left[K_1 \sin\tilde{y}\cos\psi_d + \frac{K_1}{U}V_{yc}(1 - \cos\tilde{y})\right]. \end{aligned}$$

For

$$K_1\cos\tilde{y} \geq \frac{K_2}{K_1} + \delta, \quad \text{with } 0 < \delta \leq K_1 - \frac{K_2}{K_1}, \tag{119}$$

we can further simplify the equations as

$$
\begin{aligned}
\dot{V} \leq & -\delta\sigma^2(y_1) - \frac{K_2}{K_1}y_2^2 \\
& + \sigma(y_1)\left[K_1 \sin\tilde{y}\cos\psi_d + \frac{K_1}{U}V_{yc}(1-\cos\tilde{y})\right], \\
= & -\delta\sigma^2(y_1) - \frac{K_2}{K_1}y_2^2 \\
& + |\sigma(y_1)|\left[|K_1 \sin\tilde{y}\cos\psi_d| + \left|\frac{K_1}{U}V_{yc}\right||1-\cos\tilde{y}|\right], \\
= & -\delta\sigma^2(y_1) - \frac{K_2}{K_1}y_2^2 \\
& + |\sigma(y_1)|K_1\left[\left|\frac{V_{yc}}{U}\right||1-\cos\tilde{y}| + |\sin\tilde{y}|\right], \\
= & -\delta(1-\theta)\sigma^2(y_1) - \delta\theta\sigma^2(y_1) - \frac{K_2}{K_1}y_2^2 \\
& + |\sigma(y_1)|K_1\left[\left|\frac{V_{yc}}{U}\right||1-\cos\tilde{y}| + |\sin\tilde{y}|\right],
\end{aligned}
$$

where $0 < \theta < 1$.

This implies that

$$
\begin{aligned}
\dot{V} & < -\delta(1-\theta)\sigma^2(y_1) - \frac{K_2}{K_1}y_2^2 \\
& \forall|\sigma(y_1)| > \frac{K_1}{\delta\theta}\left(\left|\frac{V_{yc}}{U}\right||1-\cos\tilde{y}| + |\sin\tilde{y}|\right).
\end{aligned} \tag{120}
$$

Since $\left(\left|\frac{V_{yc}}{U}\right||1-\cos\tilde{y}| + |\sin\tilde{y}|\right)$ is bounded by $\left(\left|\frac{V_{yc}}{U}\right| + 1\right)(|\tilde{y}|)$, it follows that

$$
\dot{V} < 0 \quad \forall|\sigma(y_1)| > \quad \frac{K_1}{\delta\theta}\left(\left|\frac{V_{yc}}{U}\right| + 1\right)|\tilde{y}| \tag{121}
$$

thus showing that $y = [y_1 \quad y_2]^T$ is ISS with restriction given by (119) on the input. Thus, from the definition of ISS, we obtain

$$
\|y(t)\| \leq \beta_l(\|y(0)\|, t) + \gamma_l\left(\sup_{0\leq\tau\leqslant t}|\tilde{y}(\tau)|\right) \quad \forall\, t \geq 0, \tag{122}
$$

where β_l is a class $\mathcal{KL}$-function, and γ_l is a class $\mathcal{K}$-function. To show that $H_1 : \tilde{y} \to \dot{\psi}_d$ is IOS, we start by computing $\dot{\psi}_d$. The control law ψ_d is given by

$$
\psi_d = \sin^{-1}\left\{\sigma_e\left(-\frac{K_1 e}{U} - \frac{K_2}{U}\varsigma\right)\right\}. \tag{123}
$$

Using (105), the above can be written as

$$
\psi_d = \sin^{-1}\left\{\sigma_e\left(-y_1 - \frac{KK_2}{U}\right)\right\}. \tag{124}
$$

Defining $\xi = -y_1 - \frac{KK_2}{U}$, the time derivative of ψ_d is given by

$$
\begin{aligned}
\frac{d\psi_d}{dt} & = \frac{d\psi_d}{d\sigma_e(\xi)} \quad \frac{d\sigma_e(\xi)}{d\xi} \quad \frac{d\xi}{dt} \\
& = -\dot{y}_1\frac{d}{d\xi}\sigma_e(\xi)\left[1 - \left\{\sigma_e\left(-y_1 - \frac{KK_2}{U}\right)\right\}^2\right]^{-\frac{1}{2}}
\end{aligned} \tag{125}
$$

with

$$\left[1-\left\{\sigma_e\left(-y_1-\frac{KK_2}{U}\right)\right\}^2\right]^{-\frac{1}{2}}\leq\eta \tag{126}$$

where $\eta=\frac{1}{\left(1-e_s^2\right)^{\frac{1}{2}}}$.

From the definition of $\sigma_e(.)$, $\frac{d}{d\xi}\sigma_e(\xi)$ is bounded by 1. Thus, $\dot{\psi}_d$ in (125) is bounded by

$$|\dot{\psi}_d|\leq\eta|\dot{y}_1|. \tag{127}$$

Equipped with the above result and the one in (122), we will now show that the system H_1 is IOS.

From Equation (117), we have

$$\begin{aligned}\dot{y}_1=&-K_1\cos\tilde{y}\sigma(y_1)+K_1\sin\tilde{y}\cos\psi_d+\frac{K_1}{U}V_{yc}-\\&K_1\cos\tilde{y}\frac{KK_2}{U}+\frac{K_2}{K_1}\sigma(y_1)-\frac{K_2}{K_1}y_2.\end{aligned}$$

With $K=\frac{V_{yc}}{K_2}$,

$$\begin{aligned}\dot{y}_1=&-K_1\cos\tilde{y}\sigma(y_1)+\frac{K_2}{K_1}\sigma(y_1)-\frac{K_2}{K_1}y_2+\\&K_1\left[\sin\tilde{y}\cos\psi_d+\frac{V_{yc}}{U}(1-\cos\tilde{y})\right].\end{aligned}$$

Taking the absolute value of both sides yields

$$\begin{aligned}|\dot{y}_1|\leq&\left(K_1+\frac{K_2}{K_1}\right)|\sigma(y_1)|+\frac{K_2}{K_1}|y_2|+\\&K_1\left[|\sin\tilde{y}|+\frac{V_{yc}}{U}|(1-\cos\tilde{y})|\right]\\|\dot{y}_1|\leq&\left(K_1+\frac{K_2}{K_1}\right)|\sigma(y_1)|+\frac{K_2}{K_1}|y_2|+K_1\left(1+\frac{V_{yc}}{U}\right)|\tilde{y}|\end{aligned}$$

Thus,

$$|\dot{\psi}_d|\leq C_1|\sigma(y_1)|+C_2|y_2|+C_3|\tilde{y}|$$

where $C_1=\eta\left(K_1+\frac{K_2}{K_1}\right), C_2=\eta\frac{K_2}{K_1}$, and $C_3=\eta K_1\left(1+\frac{V_{yc}}{U}\right)$. Using the fact $|y_1|+|y_2|=\|y\|_1$ and $|\sigma(y_1)|\leq|y_1|$, we can state that

$$C_1|\sigma(y_1)|+C_2|y_2|\leq max(C_1,C_2)\|y\|_1\leq C_1\|y\|_1.$$

Thus,

$$|\dot{\psi}_d|\leq\beta_1+\gamma_1|\tilde{y}| \tag{128}$$

where $\beta_1 = C_1\beta_l$ and $\gamma_1 = C_1\gamma_l + C_3$ (using the conditions in Theorem 2 and Equation (121)) is given by

$$\gamma_1=\eta K_1\left(\frac{V_{yc}}{U}+1\right)\left[\frac{1}{\delta\theta}\left(K_1+\frac{K_2}{K_1}\right)+1\right], \tag{129}$$

with γ_1 showing explicit dependence on K_1, K_2. In conclusion, the systems H_1 and H_2 are both IOS. It can now be shown, using the small gain theorem in [29], that the interconnected system is stable if $\gamma_1\gamma_f < 1$. This result yields a rule for the choice of gains K_1, K_2 (as functions of the inner-loop dynamic parameters) so that stability is obtained. Hence, we show using a small gain theorem that the above interconnected system is closed-loop-stable and all signals are bounded.

Notice that for restriction (119) to be feasible, it is important that $K_1 > \frac{K_2}{K_1}$. *In other words, if we choose* $K_1 = 2\xi\omega_n$ *and* $K_2 = \omega_n^2$, *where* ξ *and* ω_n *are damping factor and natural frequency, respectively, then* $\xi > 0.5$ *must be used as design parameter.*

6.6. An Example

Let us take a simple illustrative example, considering that the sideslip angle β is zero. In this situation, the inner Loop (Heading control) is characterized in terms of an IOS relationship given in (113) with $\tilde{y} = \tilde{\psi}$, where $\tilde{\psi} = \psi - \psi_d$, see (44). For such a system, it is straightforward to show that the system is finite-gain L_∞-stable, that is,

$$\|\tilde{\psi}\|_\infty \leq \gamma_f \|\dot{\psi}_d\|_\infty + \beta_f \tag{130}$$

with the gain γ_f given by

$$\gamma_f = \frac{2\lambda_{max}^2(Q)\|A^{-1}B\|_2\|C\|_2}{\lambda_{min}(Q)}, \tag{131}$$

where Q is the solution of the Lyapunov equation $QA + A^TQ = -I$. Approximating the inner loop as a first-order system with dynamics given by

$$\dot{\psi} = -a\psi + a\psi_d. \tag{132}$$

$$y_2 = \psi \tag{133}$$

it follows that

$$\gamma_f = \frac{2(\frac{1}{2a})^2}{(\frac{1}{2a})} \tag{134}$$

$$\gamma_f = \frac{1}{a}, \tag{135}$$

yielding the stability condition

$$\gamma_1 < a. \tag{136}$$

From (129), we have

$$\eta K_1\left(\frac{V_{yc}}{U} + 1\right)\left[\frac{1}{\delta\theta}\left(K_1 + \frac{K_2}{K_1}\right) + 1\right] < a. \tag{137}$$

For an inner-loop bandwidth $a = 1$ rad/s, using the parameters mentioned in Table 1 and equating γ_1 to a, the natural frequency ω_n for the outer loop should not be more than 0.095 rad/s.

Table 1. Parameters to design outer loop.

Parameters	Value
Speed U	1 m/s
y-component of current V_{yc}	0.1 m/s
saturation e_s	0.8
Parameters used for outer loop design	**with $K_2 = \omega_n^2$ and $K_1 = 2\xi\omega_n$**
damping factor ξ	0.8
δ	0.083
θ	0.99

Thus, in terms of bandwidth-like characterization, the inner loop bandwidth should be approximately 10 times higher than outer-loop bandwidth. The parameters δ and θ chosen in Table 1 impose a restriction of $\tilde{\psi} < 20.7$ degrees.

6.7. Relation between Outer-Loop Path Following and Using a Variable Look-Ahead Visibility Distance Line-of-Sight Guidance

Consider a case of simple path-following controller without the integral term, such that the control law can be written as

$$\psi_d = sin^{-1}\left(\frac{-K_1 e}{U}\right), \tag{138}$$

whereas, in the case of *look-ahead distance* line-of-sight guidance, the control law is given by

$$\psi_d = tan^{-1}\left(-\frac{e}{\Delta_d}\right), \tag{139}$$

where Δ_d is the *look-ahead distance* as shown in Figure 10.

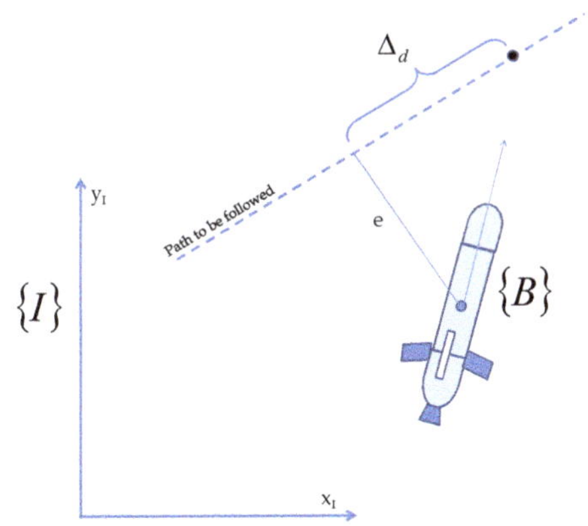

Figure 10. Line-of-sight guidance using *look-ahead distance*.

By equating Equations (138) and (139), we get a relationship between gain K_1 and the *look-ahead distance* Δ_d as follows:

$$-\frac{K_1 e}{U} = sin\left(tan^{-1}\left(-\frac{e}{\Delta_d}\right)\right) \tag{140}$$

$$K_1 = \frac{U}{(\Delta_d^2 + e^2)^{\frac{1}{2}}} \tag{141}$$

$$\Rightarrow \Delta_d = \pm\frac{1}{K_1}\left(U^2 - K_1^2 e^2\right)^{\frac{1}{2}} \tag{142}$$

Figure 11 shows the variation of look-ahead distance with cross-track error at different bandwidth in the above example. Notice that the look-ahead distance is a function of cross-track error (not fixed in this setup), and increases as the gain K_1 is reduced.

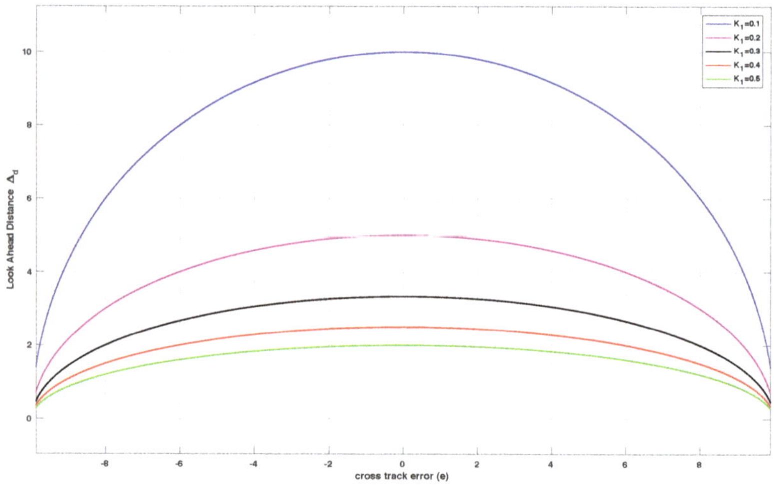

Figure 11. Look-ahead distance plotted against the cross-track error with different gains.

7. Path Following Problem: Arcs

Let P_a be the position of the vehicle (see Figure 12) in an inertial reference frame $\{\mathcal{I}\}$ and the associated Seret–Frenet frame T defined on the curve such that its origin (P_s) is the orthogonal projection of the point P_a onto curve; thus, e is the cross-track error with the coordinates of vehicle $(0, e)$ in $\{T\}$. With U as the speed of the vehicle, the kinematic equation for the evolution of e is given by

$$\dot{e} = U \sin(\psi - \theta_c) \tag{143}$$

where ψ is the vehicle orientation with respect to the $\{\mathcal{I}\}$, θ_c is the angle of tangent at point P_s measured from abscissa of $\{\mathcal{I}\}$, and $R_c(P_s)$ is the radius of curvature at $\{T\}$ (see [6]). Note that $\dot{\theta}_c$ becomes infinite when the $R_c(P_s) = e$, which in turn means that the vehicle is positioned exactly at the center of the circle with radius R_c and a tangent at $\{T\}$. However, for most of the practical applications, the reference paths to follow are curves with slowly varying curvature, this problem is unlikely to occur.

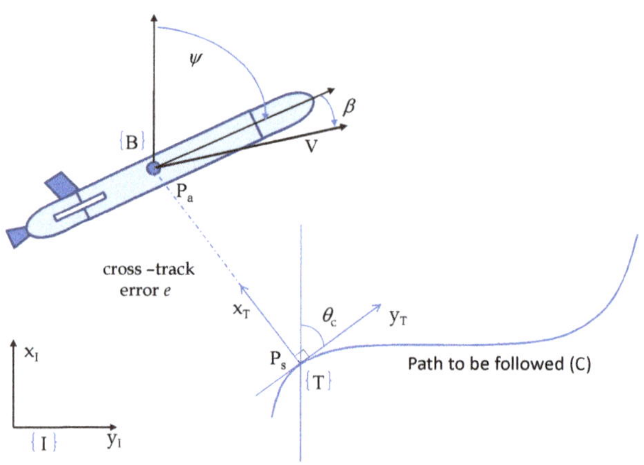

Figure 12. Cross-track error for straight-line following.

Following an approach similar to that in Section 6.2, the most suitable choice for the desired heading ψ_d will be

$$\psi_d = \sin^{-1}\left\{\sigma_e\left(-\frac{K_1 e}{U}\right)\right\} + \theta_c. \tag{144}$$

In order to follow a circumference, we compute at each instant the tangent to the path and act as if we were following a straight line. The algorithm applies to the case of straight lines and yields automatic compensation of the effect of unknown but constant currents. The methodology does not go through for the case of general paths even if we restrict ourselves to constant currents. Interestingly enough, as far as we know, this combined problem has not been solved yet. In the paper, we proposed a simple extension of the method to follow arcs of circumference that showed acceptable performance (small steady-state track error) in simulations and real test with the Medusa vehicle (figures are explained in next section); however, a general theoretical result is not available at this point [45].

8. Implementation and Field Test Results

The previous sections described the rationale and provided the mathematical machinery required for the study of a path-following controller for marine vehicles that relies on a two-scale inner-outer loop architecture. This approach effectively decouples the design of the inner and outer control loops, the combination of the two being studied at a later stage. The tools derived borrow extensively from nonlinear control theory and make use of the ISS and IOS characterization of dynamical systems. In this context, the analysis of the combined inner-outer loop structure is done using an appropriate small gain theorem [29]. For inner-loop controller design, the technique described in Section 4 was used, yielding a simple proportional and derivative control law that is pervasive in heading autopilots. In what concerns outer-loop design and stability analysis, despite the apparent complexity of the methodology adopted, the resulting outer-loop controller lends itself to the simple implementation structure shown in Equations (98) and (99) and depicted in Figure 13, where an anti-windup scheme is implemented using the so-called *D-methodology* introduced in [46]. Clearly, the implementation of the outer-loop controller does not require intensive computational power. We recall that the gains K_1 and K_2 can easily be computed by solving the characteristic Equation (96) for a choice of natural frequency ω_n, with $K_1 = 2\xi\omega_n$ and $K_2 = \omega_n^2$, where ξ is the damping factor.

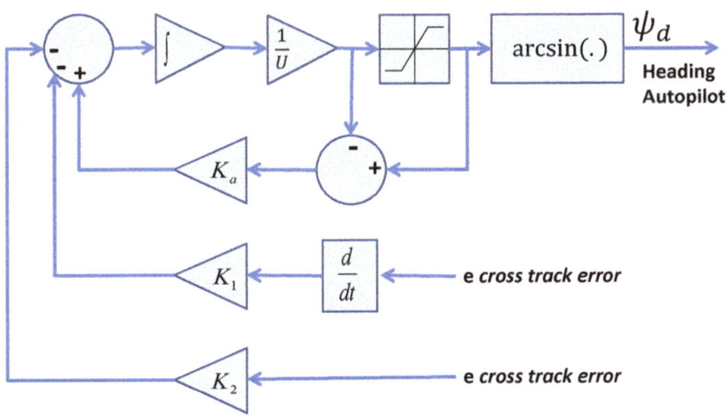

Figure 13. Implementation of the path-following algorithm using an anti-windup technique scheme that includes the so-called D-methodology in [46].

The algorithm for path following described was implemented and fully tested with success in three types of vehicles: the $DELFIM_x$ ASV, the $MAYA$ AUV, and several vehicles of the $MEDUSA$ class. The first is an autonomous surface vehicle that is the property of the Instituto Superior Tecnico, Lisbon, Portugal (see Figure 14). The second is an autonomous underwater vehicle (see Figure 15) described in Section 3.3. Implementation issues and results of tests carried out with the MAYA AUV are briefly discussed in [45]. The algorithm is also an part of the several $MEDUSA$ class vehicles developed by IST, Lisbon. The results are shown for one of the $MEDUSA$ class marine vehicle described in Section 3.2.

Figure 14. The DELFIM$_x$ ASV.

Figure 15. The MAYA AUV.

Prior to testing the path-following algorithm on the $DELFIM_x$ ASV, simulations were done with a full nonlinear model of the vessel. The outer-loop controller parameters were tuned based on the bandwidth of the linearized equations of motion of the vessel about 1.6 m/s. We call attention to the fact that we did not measure the ocean current during the sea trials of $DELFIM_x$. However, an estimate was obtained using the difference between the heading and course angles of the vehicle along the straight line components of the path. The estimated current of 0.2 m/s with direction from southwest to northeast was introduced in the simulation to allow for a fair comparison of real and simulated data.

We include both the results of simulations and actual tests at sea. Namely, Figures 16 and 17 show the results of simulations of a lawn-mowing maneuver for the ASV. Figure 16 illustrates the complete maneuver, whereas Figure 17 shows the cross error observed. The corresponding plots for real tests are shown in Figures 18 and 19, respectively. Clearly, the results of simulations and the real data are very similar, thus confirming the adequacy of the new method developed for path following. Notice in particular how both in simulated and related data the cross-track error converges to approximately zero over similar portions of the path (straight line segments), in the presence of a constant current. The variation in the cross-track error after the convergence at 300 s and 700 s are due to the transition from straight line to the arc and vice versa which is reflected in both simulations and the real tests. Notice that the heading of the vehicle (represented by a the symbol of ASV with the triangular-shaped head (see legend in Figure 18)) is different from course of the vehicle. This shows that the algorithm has learned the ocean currents and adjusted the heading accordingly.

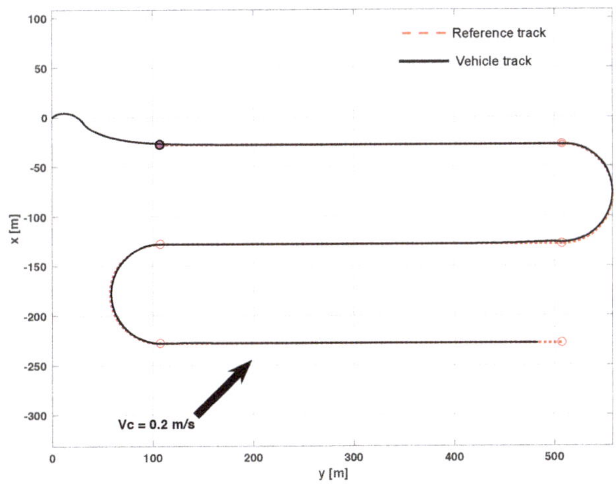

Figure 16. Simulated lawnmowing maneuver of the $DELFIM_x$ vehicle in the presence of ocean currents.

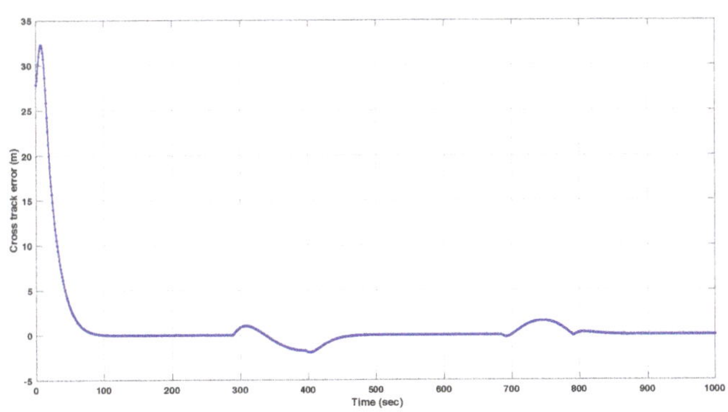

Figure 17. Cross-track error for the simulated track.

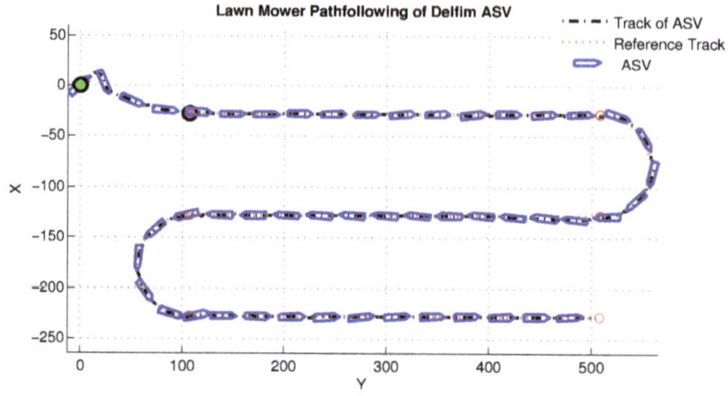

Figure 18. Delfim$_x$ performing a lawn-mowing maneuver in the Azores, PT.

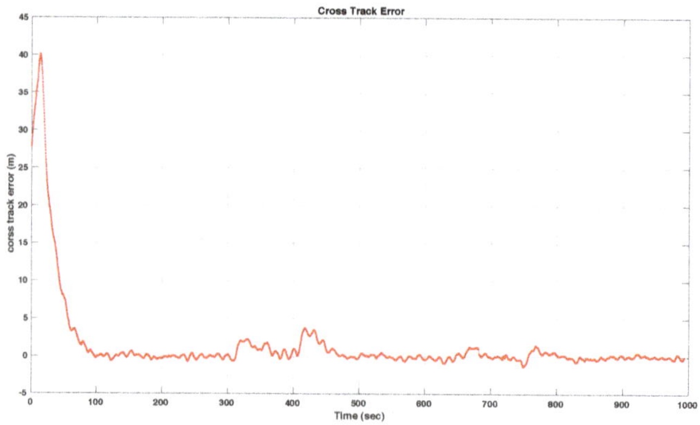

Figure 19. Delfim$_x$ cross-track error during the real mission.

Another test was conducted using the MAYA AUV at a lake to map chlorophyll at three different depths. The path-following algorithm used for these tests shows how the AUV was able to follow a path in a real environment independent of the depth control. The results are shown in Figure 20.

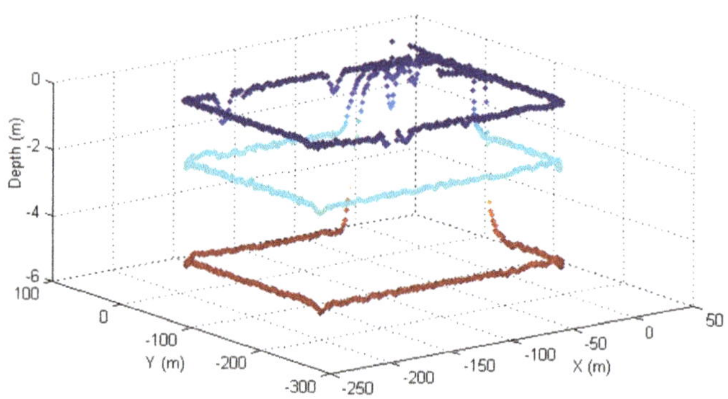

Figure 20. Square mission of MAYA at surface, 3 m and 5 m depth at Supa Dam, India.

The efficacy of the path following for straight lines was also shown during a real test with a Medusa class vehicle. The vehicle's GPS track and the reference path for the test performed are shown in Figure 21. It can be inferred from the fact that the course angle available from GPS and the vehicle angles were not the same that the vehicle was under the influence of an ocean current, see Figure 22. This figure shows clearly the role of the integrator to "learn" the constant ocean current and offset the heading angle accordingly.

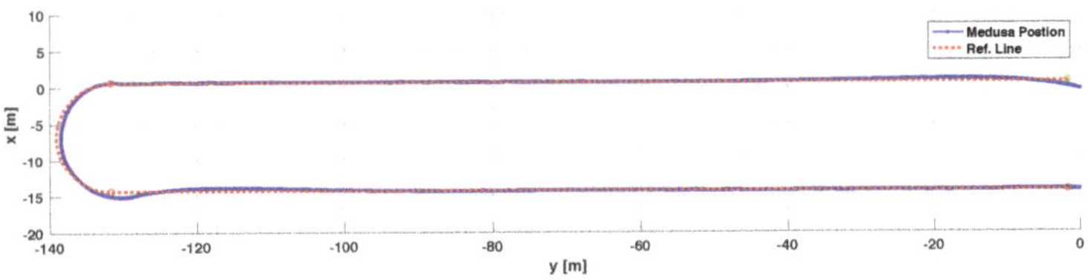

Figure 21. Medusa Vehicle performing lawnmower at Expo Site, Lisbon, Portugal.

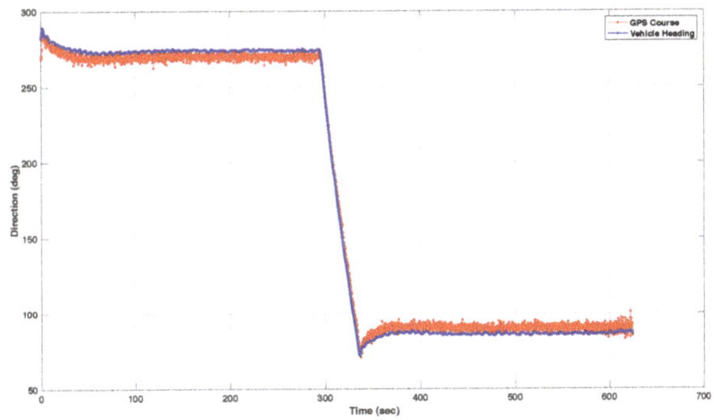

Figure 22. Heading and Course of the Medusa Vehicle Showing the effect of ocean currents.

The simulation results in Figures 23 and 24 illustrate the case where the $MAYA$ vehicle is requested to follow a segment of a circular path. The results show that in the presence of currents, the vehicle follows the arc with an error (i.e., the cross-track error will not converge to zero but to a neighborhood of zero). The convergence of the cross-track error to a neighborhood of zero along segments of arc is also captured in the sea tests performed by $DELFIM_x$ and $MEDUSA$ vehicles, as shown in Figures 18 and 21.

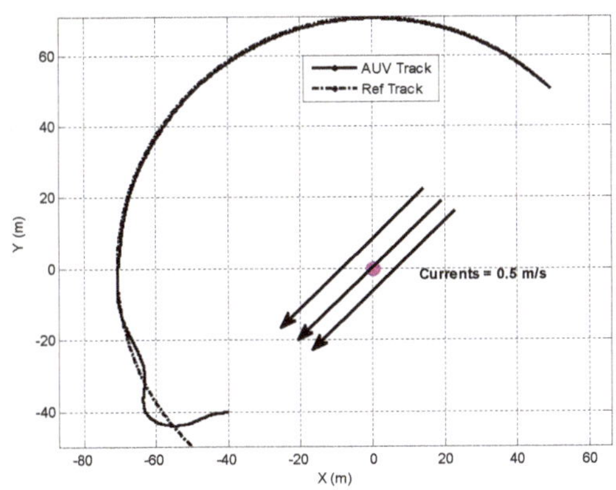

Figure 23. Simulation result of arc following.

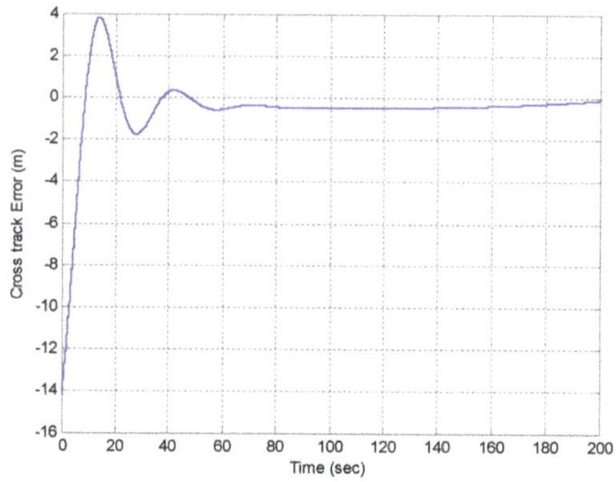

Figure 24. Evolution of cross-track error during arc following.

9. Conclusions and Future Work

This paper introduced an inner-outer control structure for marine vehicle path following in 2D, with due account for the vehicle dynamics and ocean currents. The structure is simple to implement and provides system designers a convenient way of tuning the outer-loop control law parameters as functions of a "bandwidth-like" characterization of the inner loop. Stability of the complete path-following control system was proven for straight paths, by resorting to nonlinear control theoretical tools that borrow from input-to-output stability concepts and a related small gain theorem. The efficacy of the inner-outer control structure adopted was shown during the rigorous tests with AUVs and ASVs at sea. These algorithms are now integral part of many autonomous marine vehicles used at NIO and IST. Moreover, the problem of cooperative control and navigation works on the assumption that the single vehicle is able to follow a desired path (without any temporal constraints).

The applications of this strategy include: (i) single-vehicle path following for a number of missions that include environmental surveying, seabed habitat mapping, and critical infrastructures security, and (ii) cooperative path following, which aims to steer a number of vehicles along pre-determined paths in a synchronized manner, with a view to overcoming the limitations imposed by the use of a single vehicle, effectively allowing for ocean exploration at unprecedented temporal and spatial scales. The method is easily extended to fully actuated or overly actuated vehicles where, besides having the center of mass of the vehicle follow a desired path, it is also required for the vehicle to track an arbitrary heading reference (that is, complete pose control). An obvious desired extension (pointed out before) is to derive a path-following controller capable of ensuring precise path following of general paths in the presence of constant currents. We conjecture that some form of an internal model principle should be used, which provides a good ground for future extension of the present work, see, for example, Refs. [47–49] and the references therein.

We also remark that we have addressed explicitly the effect of unknown but constant currents and showed how the path-following control law adopted allows for the rejection of this type of disturbance. We did not address the impact of waves, which cause first-order (oscillating) and second-order (drift) effects. We conjecture that the influence of waves may be studied by modeling (as is customary in the literature) their effect as a bounded output disturbance d_w, and characterizing the closed-loop operator with input d_w and output e (cross-track error) in terms of its input-output characteristics (IOS analysis).

Author Contributions: Conceptualization, P.M. and A.P.; methodology, P.M., A.P., H.M.M. and A.P.A.; software, P.M.; validation, P.M., A.P. and H.M.M.; formal analysis, P.M. and H.M.M.; investigation, P.M.; resources, P.M.; data curation, P.M.; writing—original draft preparation, P.M.; writing—review and editing, P.M., A.P. and H.M.M. All authors have read and agreed to the published version of the manuscript.

Funding: This work has been supported by RAMONES, funded by the European Union's Horizon 2020 research and innovation programme, under grant agreement No. 101017808 and by Fundação para a Ciência e a Tecnologia (FCT) through LARSyS—FCT Project UIDB/50009/2020. The design and development of "Maya" AUV was funded by Ministry of Electronics & Information Technology (MeitY), Government of India.

Institutional Review Board Statement: Not applicable.

Informed Consent Statement: Not applicable.

Data Availability Statement: Not applicable.

Acknowledgments: The authors are thankful to the Director CSIR—National Institute of Oceanography, Goa for encouragement and administrative support. Colleagues from the marine instrumentation division are acknowledged for their help during the field tests with MAYA AUV and preparation of the manuscript. We are also grateful to Elgar Desa from NIO for his constant encouragement, guidance, and support during the phases of design and testing of the MAYA AUV. This manuscript is NIO's contribution number 6928. A special word of thanks goes also to Rita Cunha, Carlos Silvestre, Luis Sebastiao, Bruno Cardeira, and Manuel Rufino for their support in conducting test at sea using $DELFIM_x$ and Medusa vehicles.

Conflicts of Interest: The authors declare no conflict of interest. The funders had no role in the design of the study; in the collection, analyses, or interpretation of data; in the writing of the manuscript, or in the decision to publish the results.

Appendix A. Vehicle Parameters

Table A1. Mass, inertia and hydrodynamics coefficients for a MEDUSA vehicle.

m = 17 kg	$I_z = 1$ kg·m^2	
$X_{\dot{u}} = -20$ kg	$\mathbf{Y}_{\dot{v}} = -30$ kg	$N_{\dot{r}} = -8.69$ kg·m^2
$X_u = -0.2$ kg/s	$\mathbf{Y}_v = -50$ kg/s	$N_r = -4.14$ kg·m^2/s
$X_{\lvert u \rvert u} = -25$ kg/m	$\mathbf{Y}_{\lvert v \rvert v} = -0.01$ kg/m	$N_{\lvert r \rvert r} = -6.23$ kg·m^2

Table A2. Vehicle parameters of MAYA AUV.

Physical Parameters	
Vehicle Speed (u_0)	: 1.2 m/s
Reynolds Number	: 1.6692×10^6
Sea Water Density(ρ)	: 1025 kg/m^3
Vehicle Parameters	
Length (L_{pp})	: 1.8 m
Center of mass (z_g)	: 0.52×10^{-2} m (w.r.t body geometric axis)
Center of Buoyancy (z_b)	: -0.172×10^{-2} m (w.r.t body geometric axis)
Weight (W)	: 53×9.8 kgf
Buoyancy (B)	: 53.4×9.8 kgf
Hydrodynamic Parameters	
Moments Coeff.	*Force Coeff.*
$N_r = -41.4397$ kg·m^2/s	$Y_r = 70.4195$ kg·m/s
$N_\delta = -33.8563$ kg·m^2/s^2	$Y_\delta = 96.3191$ kg·m/s^2
$N_v = -29.8727$ kg·m/s	$Y_v = -204.698$ kg/s
$N_{\dot{r}} = -41.4397$ kg·m^2	$Y_{\dot{r}} = 70.4195$ kg·m
$N_{\dot{v}} = -29.8727$ kg·m	$Y_{\dot{v}} = -204.698$ kg

References

1. Aguiar, A.P.; Dacic, D.B.; Hespanha, J.P.; Kokotovic, P. Path-Following or Reference-Tracking? An Answer relaxing the limits to performance. In Proceedings of the 5th IFAC Symposium on Intelligent Autonomous Vehicles (IAV), Lisbon, Portugal, 5–7 July 2004.
2. Aguiar, A.P.; Hespanha J.; Kokotovic, P. Path-Following for Non-Minimum Phase Systems Removes Performance Limitations. *IEEE Trans. Autom. Control* **2005**, *50*, 234–239. [CrossRef]
3. Aguiar, A.P.; Hespanha, J. Trajectory-Tracking and Path-Following of Underactuated Autonomous Vehicles with Parametric Modeling Uncertainty. *IEEE Trans. Autom. Control* **2007**, *52*, 1362–1379. [CrossRef]
4. Encarnacao, P.; Pascoal, A. 3D path following for autonomous underwater vehicle. In Proceedings of the 39th IEEE Conference on Decision and Control (Cat. No.00CH37187), Sydney, Australia, 12–15 December 2000; Volume 3, pp. 2977–2982.
5. Indiveri, G.; Zizzari, A.A. Kinematics Motion Control of an Underactuated Vehicle: A 3D Solution with Bounded Control Effort. In Proceedings of the 2nd IFAC Workshop Navigation, Guidance and Control of Underwater Vehicles (2008), Killaloe, Ireland, 8–10 April 2008.
6. Micaelli, A.; Samson, C . *Trajectory Tracking for Unicycle-Type and Two-Steering-Wheels Mobile Robots*; Roberts, G.; Sutton, R., Eds.; Technical Report 2097; INRIA: Sophia-Antipolis, France; 1993; pp. 353–386.
7. Samson, C. Path-following and time-varying feedback stabilization of a wheeled mobile robot. In Proceedings of the International Conference on Control, Automation, Robotics and Vision (ICARCV 92), Singapore, 16–18 September 1992; pp RO-13.1.1–RO-13.1.5.
8. Samson, C. Control of Chained Systems Application to Path Following and Time-Varying Point-Stabilization of Mobile Robots. *IEEE Trans. Autom. Control* **1995**, *40*, 64–76. [CrossRef]
9. Altafini. C.. Following a path of varying curvature as an output regulation problem. *IEEE Trans. Autom. Control* **2002**, *47*, 1551–1556. [CrossRef]
10. Böck, M.; Kugi, A. Real-time Nonlinear Model Predictive Path-Following Control of a Laboratory Tower Crane. *IEEE Trans. Control Syst. Technol.* **2014**, *22*, 1461–1473. [CrossRef]
11. Pascoal, A.; Silvestre, C.; Oliveira, P. Vehicle and Mission Control of Single Multiple Autonomous Marine Robots. In *Advances in Unmanned Marine Vehicles*; Roberts, G.; Sutton, R., Eds.; IEE Contol Engineering Series; IET: London, UK, 2006; pp. 353–386.
12. Ghabcheloo, R.; Aguiar, A.P.; Pascoal, A.C.S.; Kaminer I..; Hespanha, J. Coordinated Path-Following in the presence of Communication Losses and Time Delays. *SIAM J. Control Optim.* **2009**, *48*, 234–265. [CrossRef]
13. Pascoal, A.; Oliveira, P.; Silvestre, C.; Sebastiao, L.; Rufino, M.; Barroso, V.; Gomes, J.; Ayela, G.; Coince, P.; Cardew, M.; et al. Robotic ocean vehicles for marine science applications: The European ASIMOV project. In Proceedings of the OCEANS 2000 MTS/IEEE Conference and Exhibition. Conference Proceedings (Cat. No.00CH37158), Providence, RI, USA, 11–14 September 2000; Volume1, pp. 409–415. [CrossRef]
14. Kalwa, J.; Pascoal, A.; Ridao, P.; Birk, A.; Glotzbach, T.; Brignone, L.; Bibuli, M.; Alves, J.; Silva, M. EU project MORPH: Current Status After 3 Years of Cooperation Under and Above Water. *IFAC-PapersOnLine* **2015**, *48*, 119–124. [CrossRef]
15. Al-Khatib, H.; Antonelli, G.; Caffaz, A.; Caiti, A.; Casalino, G.; de Jong, I.B.; Duarte, H.; Indiveri, G.; Jesus, S.; Kebkal, K.; et al. The widely scalable Mobile Underwater Sonar Technology (WiMUST) project: An overview. In Proceedings of the OCEANS 2015—Genova, Genova, Italy, 18–21 May 2015; pp. 1–5. [CrossRef]
16. Børhaug, E.; Pavlov, A.; Panteley, E.; Pettersen, K. Straight Line Path Following for Formations of Underactuated Marine Surface Vessels. *IEEE Trans. Control Syst. Technol.* **2011**, *19*, 493–506. [CrossRef]
17. Papoulias, F.A. Bifurcation Analysis of Line of Sight Vehicle Guidance using Sliding Modes. *Int. J. Bifurc. Chaos* **1991**, *1*, 849–865. [CrossRef]
18. Burger, M.; Pavlov, A.; Bø rhaug, E.; Pettersen, K.Y. Straight Line Path Following for Formations of Underactuated Surface Vessels under Influence of Constant Ocean Currents. In Proceedings of the 2009 American Control Conference, St. Louis, MO, USA, 10–12 June 2009; pp. 3065–3070.
19. Burger, M.; Pavlov, A.; Pettersen, K.Y. Conditional Integrators for Path Following and Formation Control of Marine Vessels under Constant Disturbances. In Proceedings of the 8th IFAC Conference on Manoeuvring and Control of Marine Craft, Guaruja, Brazil, 16–18 September 2009.
20. Moe, S.; Caharija, W.; Pettersen, K.; Schjolberg, I. Path following of underactuated marine surface vessels in the presence of unknown ocean currents. In Proceedings of the 2014 American Control Conference, Portland, OR, USA, 4–6 June 2014; pp. 3856–3861. [CrossRef]
21. Aguiar, A.P.; Pascoal, A.M.; Kaminer, I.; Dobrokhodov, V.; Hovakimyan, N.; Xargay, E.; Cao, C.; Ghabcheloo, R. Time-Coordinated Path Following of Multiple UAVs over Time-Varying Networks using L1 Adaptation. In Proceedings of the AIAA Guidance, Navigation and Control Conference and Exhibit, Honolulu, HI, USA, 18–21 August 2008;
22. Sujit, P.; Saripalli, S.; Borges Sousa, J. Unmanned Aerial Vehicle Path Following: A Survey and Analysis of Algorithms for Fixed-Wing Unmanned Aerial Vehicles. *IEEE Control Syst. Mag.* **2014**, *34*, 42–59. [CrossRef]
23. Breivik, M.; Fossen, T.I. Principles of Guidance-Based Path Following in 2D and 3D. In Proceedings of the 44th IEEE Conference on Decision and Control, Seville, Spain, 15 December 2005; pp. 627–634.
24. Breivik, M.; Subbotin, M.V.; Fossen, T.I. Guided Formation Control for Wheeled Mobile Robots. In Proceedings of the 2006 9th International Conference on Control, Automation, Robotics and Vision, Singapore, 5–8 December 2006.

25. Carona, R.; Aguiar, A.P. Control of Unicycle Type Robots: Tracking , Path Following and Point Stabilization. In Proceedings of the IV Jornadas de Engineharia Electronica e Telecomunicacoes e de Computadores, Lisbon, Portugal, 20–21 November 2008; pp. 180–185.
26. Tayebi, A.; Rachid, A. Path Following Control Law for an Industrial Mobile Robot. In Proceedings of the 1996 IEEE International Conference on Control Applications, Dearborn, MI, USA, 15–18 September 1996; pp. 703–707.
27. Park, S.; Deyst, J.; How, J. Performance and Lyapunov Stability of a Nonlinear Path Following Guidance Method. *J. Guid. Control Dyn.* **2007**, *30*, 1718–1728. [CrossRef]
28. Breivik, M.; Fossen, T.I. Path Following for Marine Surface Vessels. In Proceedings of the Oceans '04 MTS/IEEE Techno-Ocean '04 (IEEE Cat. No.04CH37600), Kobe, Japan, 9–12 November 2004; pp. 2282–2289.
29. Khalil, H.K. *Nonlinear Systems*, 3rd ed.; Prentice Hall: Hoboken, NJ, USA, 2001.
30. Maurya, P.; Aguiar, A.P.; Pascoal, A. Marine Vehicle Path Following Using Inner-Outer Loop Control. In Proceedings of the 8th IFAC International Conference on Manoeuvring and Control of Marine Craft, Guaruja, Brazil, 16–18 September 2009. [CrossRef]
31. Maurya, P.; Desa, E.; Pascoal, A.; Barros, E.; Navelkar, G.; Madhan, R.; Mascarenhas, A.; Prabudesai, S.; Afzalpurkar, S.; Gouveia, A.; et al. Control of the Maya AUV in the Vertical and Horizontal Planes: Theory and Practical Results. In Proceedings of the 7th Conference on Manoeuvring and Control of Marine Craft (MCMC2006), Lisbon, Portugal, 20–22 September 2006.
32. Aage, C.; Smitt, L. Hydrodynamic manoeuvrability data of a flatfish type AUV. In Proceedings of the Oceans Engineering for Today's Technology and Tomorrow's Preservation (OCEANS '94), Brest, France, 13–16 September 1994; Volume 3, pp. III/425–III/430. [CrossRef]
33. De barros, E.; Dantas, J.; Pascoal, A.; de Sa, E. Investigation of Normal Force and Moment Coefficients for an AUV at Nonlinear Angle of Attack and Sideslip Range. *IEEE J. Ocean. Eng.* **2008**, *33*, 538–549. [CrossRef]
34. Fossen, T.I. *Marine Control System: Guidance, Navigation and Control of Ships, Rigs and Underwater Vehicles*; Marine Cybernetics AS: Trondheim, Norway, 2002.
35. Craig, J.J. *Introduction to Robotics: Mechanics and Control*; Includes bibliographies and index; Addison-Wesley Pub. Co.: Reading, MA, USA, 1986.
36. Fossen, T.I. Nonlinear Modeling and Control of Underwater Vehicles. Ph.D. Thesis, Department of Engineering Cybernetics, Norwegian University of Science and Technology, Trondheim, Norway, 1991.
37. Técnico Lisboa. MEDUSA. Available online: http://dsor.isr.ist.utl.pt/vehicles/medusa/ (accessed on 30 March 2022).
38. Abreu, P.C. Positioning and Navigation Systems for Robotic Underwater Vehicles. Master's Thesis, Instituto Superior Tecnico, University of Lisbon, Lisbon, Portugal, 2014,
39. Madhan, R.; Desa, E.; Prabhudesai, S.; Sebastiao, L.; Pascoal, A.; Desa, E.; Mascarenhas, A.; Maurya, P.; Navelkar, G.; Afzulpurkar, S.; et al. Mechanical Design and Development Aspects of a Small AUV MAYA. In *7th IFAC Conference on Manoeuvring and Control of Marine Craft*; IFAC: Lisbon, Portugal, 2006; pp. 1–6.
40. Jalving, B. The NDRE-AUV Flight Control System. *IEEE J. Ocean. Eng.* **1994**, *19*. [CrossRef]
41. Sontag, E. Smooth stabilization implies coprime factorization. *IEEE Trans. Autom. Control* **1989**, *34*, 435–443. [CrossRef]
42. Lopez-Araujo, D.; Zavala-Rio, A.; Fantoni, I.; Salazar, S.; Lozano, R. Global stabilization of the PVTOL aircraft with lateral force coupling and bounded inputs. *Int. J. Control* **2010**, *7*, 1427–1441. [CrossRef]
43. Panteleyt, E.; Ortega, R. Cascaded Control of Feedback Interconnected Nonlinear Systems: Application to Robots with AC Drives. *Automatica* **1997**, *33*, 1935–1947. [CrossRef]
44. Vanni, F. Coordinated Motion Control of Multiple Autonomous Underwater Vehicles. Master's Thesis, Instituto Superior {Técnico}, Lisbon, Portugal, 2007; pp. 46–50.
45. Maurya, P.; Navelkar, G.; Madhan, R.; Afzalpurkar, S.; Prabhudesai, S.; Desa, E.; Pascoal, A. Navigation and path following guidance of the Maya AUV: From concept to practice. In Proceedings of the 2nd International Conference on Underwater System Technology: Theory and Applications, Bali, Indonesia, 4–5 November 2008.
46. Kaminer, I.; Pascoal, A.; Khargonekar, P.; Coleman, E. A Velocity Algorithm for the Implementation of Gain-Scheduled Controllers. *Automatica* **1995**, *31*, 1185–1191. [CrossRef]
47. Bengtsson, G. Output regulation and internal models: A frequency domain approach. *Automatica* **1977**, *13*, 333–345. [CrossRef]
48. Dorf, R.; Bishop, R. *Modern Control Systems*; Pearson Prentice Hall: Hoboken, NJ, USA, 2008.
49. Kawato, M. Internal models for motor control and trajectory planning. *Curr. Opin. Neurobiol.* **1999**, *9*, 718–727. [CrossRef]

sensors

MDPI

Article

Chemical Spill Encircling Using a Quadrotor and Autonomous Surface Vehicles: A Distributed Cooperative Approach

Marcelo Jacinto *, Rita Cunha and António Pascoal

Laboratory of Robotics and Systems in Engineering and Science (LARSyS), Instituto Superior Técnico, University of Lisbon, 1049-001 Lisboa, Portugal; rita@isr.tecnico.ulisboa.pt (R.C.); antonio@isr.tecnico.ulisboa.pt (A.P.)

* Correspondence: marcelo.jacinto@tecnico.ulisboa.pt

Abstract: This article addresses the problem of formation control of a quadrotor and one (or more) marine vehicles operating at the surface of the water with the end goal of encircling the boundary of a chemical spill, enabling such vehicles to carry and release chemical dispersants used during ocean cleanup missions to break up oil molecules. Firstly, the mathematical models of the Medusa class of marine robots and quadrotor aircrafts are introduced, followed by the design of single vehicle motion controllers that allow these vehicles to follow a parameterised path individually using Lyapunov-based techniques. At a second stage, a distributed controller using event-triggered communications is introduced, enabling the vehicles to perform cooperative path following missions according to a pre-defined geometric formation. In the next step, a real-time path planning algorithm is developed that makes use of a camera sensor, installed on-board the quadrotor. This sensor enables the detection in the image of which pixels encode parts of a chemical spill boundary and use them to generate and update, in real time, a set of smooth B-spline-based paths for all the vehicles to follow cooperatively. The performance of the complete system is evaluated by resorting to 3-D simulation software, making it possible to visually simulate a chemical spill. Results from real water trials are also provided for parts of the system, where two Medusa vehicles are required to perform a static lawn-mowing path following mission cooperatively at the surface of the water.

Keywords: quadrotor control; autonomous surface vehicle control; cooperative path following; online path planning; chemical spill boundary encircling

Citation: Jacinto, M.; Cunha, R.; Pascoal, A. Chemical Spill Encircling Using a Quadrotor and Autonomous Surface Vehicles: A Distributed Cooperative Approach. *Sensors* **2022**, *22*, 2178. https://doi.org/10.3390/s22062178

Academic Editor: Maria Gabriella Xibilia

Received: 4 February 2022
Accepted: 7 March 2022
Published: 10 March 2022

1. Introduction

The problems of perimeter detection, boundary searching, and encircling have been widely researched topics with a variety of practical applications, ranging from the monitoring of wildfire spread in forests [1], to the control and encircling of oil spills [2] and harmful invasive algae blooms [3] at the surface of the ocean. In this paper, we will focus on the problem of chemical spill encircling.

The two main phenomena that contribute to the transportation and spread of hazardous chemicals over water, such as oil, are advection and diffusion. In the first, the chemical is transported due to the flow of water, while the second refers to the motion of the fluid caused by the existence of concentration gradients. One way of modelling the flow field of the incompressible fluid is by solving iteratively the convection–diffusion equations [4]. In the literature, many works address the problem of dynamic boundary tracking at the surface of the ocean by proposing control schemes which require (at least one) surface vessel to measure the concentration gradient of a hazardous contaminant. These measurements of the chemical plume are used by potential field controllers with the end goal of steering the robots to the boundary of the plume [5,6]. A completely different approach, adopted by Saldaña et al. [7], is to consider that a general environmental boundary can be approximated by a closed curve that is slowly varying over time and that can be described by a general parametric equation. In his research, the author proposes a model for the curve described spatially by a truncated Fourier series that changes its shape smoothly over time. To achieve this, it is assumed that a team of Autonomous Surface Vehicles (ASVs) are distributed equally around the chemical spill, and every vehicle is capable of taking local measurements of the boundary as it moves around it. These local measurements are then used to update the shape of the closed curve using recursive least squares. Although this is a very general solution to the problem, it can be argued that the use of a truncated Fourier series to represent a path for underactuated vehicles to follow is a rather poor choice of function, as the resulting curve can self-intersect and exhibit substantial oscillations. Moreover, it does not take into consideration the physical constraints imposed by the vehicles. In order to lift the limitations imposed by this method, more stable parametric curves could be considered, such as Bernstein polynomials or B-splines [8].

In recent years, there has been a massive development of and demand for Autonomous Underwater Vehicles (AUVs), due not only to their agility when it comes to the execution of scientific and comercial missions, but also to their low cost when compared to traditional ships, which require an on-board crew to be operated. Additionally, there has also been an exponential growth in demand for Unmanned Aerial Vehicles (UAVs), with a special emphasis on multirotor systems, which usually offer high-quality camera sensors at low market prices. Aerial vehicles can have a top-down view of the environment, making them the tool par excellence for surveillance and maintenance missions. On the other hand, AUVs and ASVs can be used to carry and release chemical dispersants used in cleanup missions to break oil molecules [9]. Together, these unmanned vehicles have huge potential to automate and reduce the cost of ocean cleanup operations.

In this paper, we address the problem of chemical spill encircling and focus on the development of a set of control and path planning tools that allow a team of robots constituted of a quadrotor (equipped with an onboard camera) and ASVs to detect and encircle the dynamical boundary of a chemical spill closely, as depicted in Figure 1. In our proposal, the quadrotor is responsible for detecting in real time the boundary of a chemical spill in the image stream produced by its onboard camera, and producing a path that itself and one or more ASVs are required to follow cooperatively. To achieve this, we start by proposing a set of single-vehicle motion control laws based on non-linear Lyapunov techniques that allow individual ASVs to follow a pre-defined parametric curve, based on previous works by Aguiar et al. [10–12]. These control techniques are then extended to the case of quadrotor vehicles. Borrowing from the work of N. Hung and F. Rego [13], a distributed controller using event-triggered communications is presented, allowing the vehicles to perform Cooperative Path Following (CPF) missions, according to a pre-defined geometric formation. Finally, a new real-time path planning framework that uses growing unclamped (and uniform) cubic B-splines is proposed, which fits a 2-D point cloud generated from the drone's image stream.

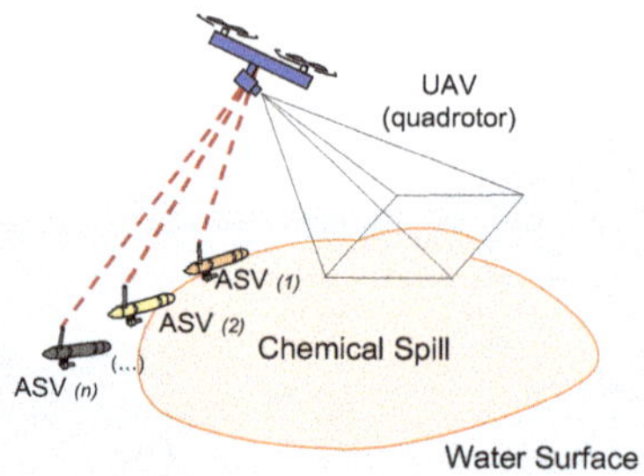

Figure 1. Cooperative path following along an environmental boundary.

A set of real experiments are performed with the Medusa class of marine vehicles [14] (property of ISR-DSOR) to access the real-life performance of the proposed path following and CPF algorithms. Additionally, the complete path planning solution is evaluated by resorting to the Gazebo 3-D simulator, PX4-SITL [15], and UUVSimulator [16], using a dynamic model of a Medusa vehicle and an Iris quadrotor equipped with a virtual RGB camera.

2. Preliminaries

2.1. Notation

The unit vector $\mathbf{e}_3$ is defined as $\mathbf{e}_3 = [0,0,1]^T$. For a vector $\mathbf{x} \in \mathbb{R}^n$, the symbol x_i denotes the ith element of the vector. We shall use $\|\mathbf{x}\| = \sqrt{\mathbf{x}^T\mathbf{x}}$ to denote the Euclidean norm of a vector. The notation $K \succeq 0$ is used to denote a matrix $K \in \mathbb{R}^{n\times n}$ that is positive semi-definite. The symbol I is used to denote the identity matrix and $\mathbf{1}$ is a vector with all elements equal to one. The symbols $\lfloor x \rceil, x \in \mathbb{R}$ denote the x nearest integer, $\lfloor x \rfloor$ denotes the floor of x, and $\lceil x \rceil$ denotes the ceiling of x. The symbol $R(.)$ is used to denote a rotation matrix with properties $R^T = R^{-1}$ and $det(R) = 1$. The map $S(\cdot) : \mathbb{R}^n \rightarrow \mathbb{R}^{n\times n}, n = 2,3$ yields a skew-symmetric matrix $S(\mathbf{x})\mathbf{y} = \mathbf{x} \times \mathbf{y}, \forall \mathbf{x}, \mathbf{y} \in \mathbb{R}^n$. When considering an estimator for an unknown variable x, we use the hat nomenclature $\hat{x}$ to denote its estimate and $\tilde{x}$ when referring to the estimation error.

2.2. Graph Theory

A weighted digraph $\mathcal{G} = \mathcal{G}(\mathcal{V},\mathcal{E},\mathcal{A})$ consists of a set of N vertices $\mathcal{V} = [V_1, ..., V_N]^T$, a set of directed edges $\mathcal{E} \subseteq \mathcal{V} \times \mathcal{V}$, and a weighted adjacency matrix $\mathcal{A} = [a_{ij}] \in \mathbb{R}^{N\times N}$, such that $a_{ij} > 0$ if the edge that connects vertex i to j belongs to the graph, and 0 otherwise. The set of in-neighbours of a vertex i is given by $\mathcal{N}_i^{in} = \{j \in \mathcal{V} : (j,i) \in \mathcal{E}\}$, and the set of out-neighbours by $\mathcal{N}_i^{out} = \{j \in \mathcal{V} : (i,j) \in \mathcal{E}\}$. The in- and out-degree matrices D^{in} and D^{out} are a set of diagonal matrices defined by:

$$D^{in/out} = diag(d_i^{in/out}), \text{ with } d_i^{in} = \sum_{j\in\mathcal{N}_i^{in}} a_{ij} \text{ and } d_i^{out} = \sum_{j\in\mathcal{N}_i^{out}} a_{ji}. \tag{1}$$

A graph $\mathcal{G}$ is undirected if communication links are unidirectional. If $\mathcal{G}$ is an undirected graph, then $\mathcal{G}$ is also balanced, i.e., $D^{in} = D^{out} := D$, and its Laplacian matrix L is symmetric, positive semi-definite, and defined according to $L := (D - \mathcal{A})$. In these conditions, it is well known that L has a simple eigenvalue at zero associated with eigen vector $\mathbf{1}$, with the remaining eigen values being positive. Moreover, $L\mathbf{1} = \mathbf{0}$.

Remark: With the graph definition given above, we adopt the convention that an agent i can receive information from its neighbors in $\mathcal{N}_i^{in}$ and send information to its neighbors in $\mathcal{N}_i^{out}$.

2.3. Uniform B-Spline Curves

A 2-D B-spline curve of degree $k + 1$ in $\mathbb{R}^2$ is a piecewise polynomial function formed by several components of degree k, defined as:

$$C(\gamma) = \sum_{i=0}^{n} B_{i,k}(\gamma)P_i, \tag{2}$$

where $\mathcal{P} = \{P_i \in \mathbb{R}^2, i = 0, ..., n\}$ are a set of control points and $B_{i,k}(\gamma)$ are the B-spline basis functions. It follows from the Cox–De Boor's recursive algorithm, according to L. Piegl and W. Tiller ([8], Chapter 2.2), that:

$$B_{i,0}(\gamma) = \begin{cases} 1, \text{if } \gamma_i \leq \gamma \leq \gamma_{i+1} \\ 0, \text{otherwise} \end{cases}, \tag{3}$$

$$B_{i,j}(\gamma) = \frac{\gamma - \gamma_i}{\gamma_{i+j} - \gamma_i} B_{i,j-1}(\gamma) + \frac{\gamma_{i+j+1} - \gamma}{\gamma_{i+j+1} - \gamma_{i+1}} B_{i+1,j-1}(\gamma), \tag{4}$$

where the index $j = 0, ..., k$ and the values γ_i belong to the m-dimensional knot vector $\mathbf{U} = \{\gamma_i\}_{i=0}^{m}$, with the number of knots related to the degree of the curve and the number of control points by $m = k + 1 + n$.

For the particular case of 2-D, uniform, non-clamped cubic B-splines with $n - k + 1$ segments, each segment's x- and y-coordinates of the parametric curve can be described according to the vectorial notation [17] as follows:

$$C_i(\gamma) := \left[C_i^x(\gamma) \quad C_i^y(\gamma)\right]^T, \tag{5}$$

with $C_i^x(\gamma)$ and $C_i^y(\gamma)$ computed according to:

$$C_i^{x/y}(\gamma) := \underbrace{\frac{1}{6}\left[(\gamma-i)^3 \quad (\gamma-i)^2 \quad (\gamma-i) \quad 1\right]\begin{bmatrix}-1 & 3 & -3 & 1\\ 3 & -6 & 3 & 0\\ -3 & 0 & 3 & 0\\ 1 & 4 & 1 & 0\end{bmatrix}}_{\left[B_{i,3}(\gamma) \quad B_{i+1,3}(\gamma) \quad B_{i+2,3}(\gamma) \quad B_{i+3,3}(\gamma)\right]}\begin{bmatrix}P_i^{x/y}\\ P_{i+1}^{x/y}\\ P_{i+2}^{x/y}\\ P_{i+3}^{x/y}\end{bmatrix}, \tag{6}$$

where $\gamma \in [0, n - k + 1]$ and $i := \lfloor\gamma\rfloor$, such that $\gamma - i \in [0, 1]$ and each curve segment is only defined by four distinct control points. Defining a unidimensional vector with all control points $\mathbf{P} = [P_0^x, ..., P_n^x, P_0^y, ..., P_n^y]^T \in \mathbb{R}^{2(n+1)}$, where both x- and y-coordinates are concatenated, and a vector of distinct curve parameters $\boldsymbol{\gamma} = [\gamma_0, ..., \gamma_q] \in \mathbb{R}^{q+1}$ at which we wish to evaluate our curve, $\mathbf{C}(\boldsymbol{\gamma}) \in \mathbb{R}^{2(q+1)}$ is given by:

$$\mathbf{C}(\boldsymbol{\gamma}) = B(\boldsymbol{\gamma}) \cdot \mathbf{P}, \tag{7}$$

where $B(\boldsymbol{\gamma}) \in \mathbb{R}^{2(q+1) \times 2(n+1)}$ is a diagonal by blocks matrix, and for each line of B, only four basis functions are different then zero and computed according to (6).

3. Vehicle Modelling

Let $\{U\}$ denote an inertial reference frame and $\{B\}$ a body-fixed reference frame attached to the geometric center of mass of each vehicle, according to Figure 2.

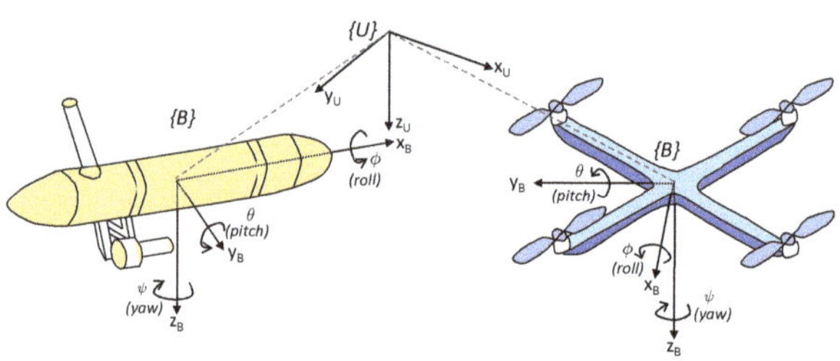

Figure 2. Adopted reference frames for a surface vehicle (**left**) and a quadrotor (**right**).

3.1. ASV Model

The ASV vehicle is modeled as a rigid body whose motion is restricted to a 2-D plane at the surface of the water, such that the roll and pitch angles are zero, i.e., $\phi = \theta = 0$. Let the kinematic equations of the vehicle be given by:

$$\underbrace{\begin{bmatrix}\dot{x}\\ \dot{y}\end{bmatrix}}_{\dot{\mathbf{p}}} = \underbrace{\begin{bmatrix}cos(\psi) & -sin(\psi)\\ sin(\psi) & cos(\psi)\end{bmatrix}}_{{}^U_B R(\psi)}\underbrace{\begin{bmatrix}u\\ v\end{bmatrix}}_{\mathbf{v}} + \underbrace{\begin{bmatrix}v_{cx}\\ v_{cy}\end{bmatrix}}_{\mathbf{v}_c}, \tag{8}$$

$$\dot{\psi} = r, \tag{9}$$

where $\mathbf{p} := [x, y]^T$ denotes the ASV position expressed in $\{U\}$, $\mathbf{v} := [u, v]^T$ denotes the body-velocity vector, ${}^U_B R(\psi) \in \mathbb{R}^{2\times 2}$ denotes the rotation matrix, and $\mathbf{v} := [v_{cx}, v_{cy}]^T$ denotes the ocean current, expressed in $\{U\}$ and assumed to be constant, irrotational, and bounded. The ASV model is considered to be underactuated, with the input of the system being given by $\mathbf{u} = [u, r]^T \in \mathbb{R}^2$.

3.2. Quadrotor Model

The kinematic equations that describe the motion of a rigid body in 3-D space can be described by a double integrator model, according to:

$$\ddot{\mathbf{p}} := \underbrace{g\mathbf{e}_3 - \frac{1}{m}{}^U_B R(\boldsymbol{\theta}) T \mathbf{e}_3}_{\mathbf{u}} + \mathbf{d}, \tag{10}$$

where $\mathbf{p} := [x, y, z]^T$ denotes the quadrotor's position expressed in $\{U\}$, $\boldsymbol{\theta} := [\phi, \theta, \psi]^T$ denotes the orientation vector of $\{B\}$ expressed in $\{U\}$, and $\mathbf{u} \in \mathbb{R}^3$ can be regarded as the input of the system, comprising both the attitude of the vehicle and the total thrust T. The vector $\mathbf{d} \in \mathbb{R}^3$ represents unmeasured external disturbances, such as wind, acting on the vehicle, assumed to be constant and bounded such that $\|\mathbf{d}\| \leq d_{max}$. The 3-D rotation matrix adopted for the quadrotor model follows the Z-Y-X convention, and is given by:

$${}^U_B R(\boldsymbol{\theta}) = R_z(\psi) R_y(\theta) R_x(\phi). \tag{11}$$

4. Path Following

The path following (PF) problem concerns the problem of making a vehicle move along a desired path $\mathbf{p}_d(\gamma)$ parameterised by a variable γ (for example, the arc-length of the curve). The key idea is that each vehicle must approach a virtual target that moves along the path with a desired speed profile $v_d(\gamma)$, according to Figure 3. Since the end goal is to have more than one vehicle performing path following with a pre-defined inter-vehicle formation, this speed profile is given as the sum of another single-vehicle speed profile and an inter-vehicle coordination term, according to:

$$v_d(\gamma) := v_L(\gamma) + v^{coord}, \text{ with } |v_L(\gamma)| \leq v_L^{max}, \tag{12}$$

where $v_L(\gamma)$ is a desired speed profile defined only as a function of the path, v_L^{max} is a pre-defined speed upper-bound, and v^{coord} is the speed coordination term that will be used in Section 5 to enable the CPF behaviour. It is important to notice that the desired speed profile $v_L(\gamma)$ should be the same for all the vehicles, enabling them to follow a given path at the same rate. On the other hand, the speed coordination term v^{coord} will not be the same for all vehicles and will be used to adjust the progression speed of each individual robot based on how aligned they are with each other.

Remark 1. *Speed profile $v_d(\gamma)$ might not correspond directly to an inertial speed, especially if the curve is not parameterised in terms of the arc-length. Nonetheless, a relation between the inertial speed and the desired speed profile is addressed in detail in Section 6.4.*

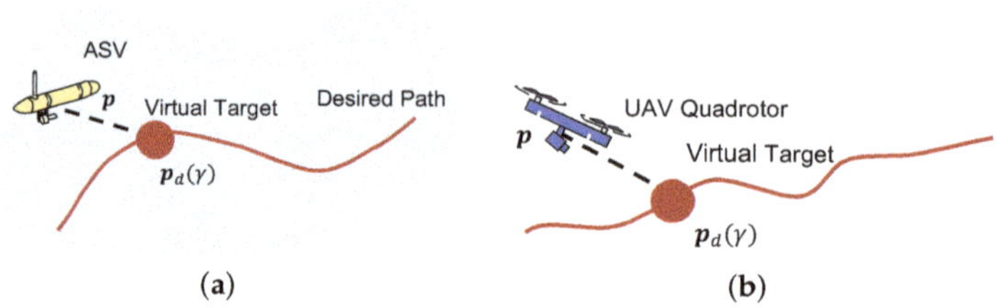

Figure 3. Path following schematic: (**a**) ASV path following. (**b**) Quadrotor path following.

Problem 1. *Given a generic vehicle (ASV or quadrotor), consider the geometric path* $\mathbf{p}_d(\gamma) : [0,\infty] \rightarrow \mathbb{R}^2/\mathbb{R}^3$ *for the ASV/quadrotor respectively, parameterised by a continuous variable* $\gamma \in \mathbb{R}$ *and* $v_d(\gamma,t) \in \mathbb{R}$ *a desired speed profile for a virtual target moving along the desired path. Furthermore, consider* $\mathbf{p}_d(\gamma)$ *to be* C^2 *and have its first and second derivatives with respect to* γ *bounded. Assume the vehicle is equipped with inner-loop controllers allowing it to track a desired control reference* $\mathbf{u}_d \in \mathbb{R}^2/\mathbb{R}^3$*, assumed to be bounded, by recruiting the appropriate forces and torques to apply to the vehicle. Design a feedback control law for the system input* $\mathbf{u}_d$ *and virtual target* $\ddot{\gamma}$ *such that:*

- *the vehicle's position converges to a tube around the desired position that can be made arbitrarily small, i.e.,* $\|\mathbf{p}(t) - \mathbf{p}_d(\gamma)\|$ *converges to a neighbourhood of the origin;*
- *the speed of the virtual target moving along the path converges to the desired speed profile, i.e.,* $|\dot{\gamma} - v_d(\gamma,t)| \rightarrow 0$ *as* $t \rightarrow \infty$.

4.1. ASV Path Following

Following the approach proposed by Aguiar et al. [10–12], consider the global diffeomorphic coordinate transformation which expresses the position error defined in the body-frame of the vehicle $\{B\}$ as:

$$\mathbf{e}_p(t) := {}^B_U R(\psi)(\mathbf{p}(t) - \mathbf{p}_d(\gamma)), \tag{13}$$

and let the speed-tracking error be defined as:

$$e_\gamma := \dot{\gamma} - v_d(\gamma,t). \tag{14}$$

With these definitions, the body-fixed position error dynamics are given by:

$$\dot{\mathbf{e}}_p(t) = {}^B_U \dot{R}(\psi)(\mathbf{p}(t) - \mathbf{p}_d(\gamma)) + {}^B_U R(\psi)(\dot{\mathbf{p}}(t) - \dot{\mathbf{p}}_d(\gamma)). \tag{15}$$

We recall that the derivative of a rotation matrix can be expressed as the product of a skew-symmetric matrix with the transposed rotation matrix, that is:

$${}^B_U \dot{R}(\psi) = -S(r) {}^B_U R(\psi). \tag{16}$$

Replacing (16) in (15) yields the position error dynamics expressed in the body-fixed frame as:

$$\dot{\mathbf{e}}_p(t) = -S(r) \underbrace{{}^B_U R(\psi)(\mathbf{p}(t) - \mathbf{p}_d(\gamma))}_{\mathbf{e}_p(t)} + \mathbf{v} + \underbrace{{}^B_U R(\psi)\mathbf{v}_c}_{v_c} - {}^B_U R(\psi) \frac{\partial \mathbf{p}_d(\gamma)}{\partial \gamma} \dot{\gamma}. \tag{17}$$

Since there is no direct control in the sway motion, the goal is to generate surge speed and heading rate control references. Therefore, we must make these references appear explicitly in the error expression. By introducing an offset $\boldsymbol{\delta} = [0, \delta]^T \in \mathbb{R}^2$ (with $\delta < 0$) in the standard position error, it is possible to re-write (17) as:

$$\dot{\mathbf{e}}_p(t) = -S(r)(\mathbf{e}_p - \boldsymbol{\delta}) + \underbrace{\begin{bmatrix} 1 & 0 \\ 0 & -\delta \end{bmatrix}}_{\Delta} \underbrace{\begin{bmatrix} u \\ r \end{bmatrix}}_{\mathbf{u}} + \begin{bmatrix} 0 \\ v \end{bmatrix} + \boldsymbol{v}_c - {}^B_U R(\psi) \frac{\partial \mathbf{p}_d(\gamma)}{\partial \gamma} \dot{\gamma}. \tag{18}$$

Consider that each ASV is equipped with a Doppler Velocity Logger (DVL) capable of providing the vehicle's relative velocity with respect to the water $\mathbf{v}$, expressed in $\{B\}$, and a Global Positioning System (GPS) unit which provides measurements of the position of the vehicle $\mathbf{p}$, expressed in $\{U\}$. To estimate the ocean current, Pascoal et al. [18] and Sanches et al. [19] propose the use of a complementary filter. Consider the process model given by (8) and the candidate complementary filter model described by:

$$\mathcal{F} := \begin{cases} \dot{\hat{\mathbf{p}}} = k_1(\mathbf{p} - \hat{\mathbf{p}}) + {}^U_B R(\psi)\mathbf{v} + \hat{\mathbf{v}}_c \\ \dot{\hat{\mathbf{v}}}_c = k_2(\mathbf{p} - \hat{\mathbf{p}}), \end{cases} \tag{19}$$

with k_1 and k_2 positive constants. The proposed complementary filter is asymptotically stable. For a formal stability analysis of this complementary filter, refer to Pascoal et al. [18].

At this point, it is important to notice that the current velocity $\boldsymbol{v}_c$ and the requested input $\mathbf{u}_d$ that are to be applied to a vehicle's kinematic model cannot be estimated and tracked, respectively, with infinite precision. For this reason, we define the current estimation error and the inner-loop tracking error given by:

$$\begin{aligned} \tilde{\boldsymbol{v}}_c &:= \boldsymbol{v}_c - \hat{\boldsymbol{v}}_c, \\ \tilde{\mathbf{u}} &:= \mathbf{u} - \mathbf{u}_d. \end{aligned} \tag{20}$$

Consider the Proposition 1 introduced below, in which a solution to Problem 1, applied to an ASV, is provided along with convergence guarantees in the presence of bounded estimation and tracking errors.

Proposition 1. *Consider the system described by the kinematics in (8), with the outer-loop control laws given by:*

$$\mathbf{u}_d := \Delta^{-1}\Bigg(- K_p \boldsymbol{\sigma}(\mathbf{e}_p - \boldsymbol{\delta}) - \begin{bmatrix} 0 \\ v \end{bmatrix} - \hat{\boldsymbol{v}}_c + {}^B_U R(\psi) \frac{\partial \mathbf{p}_d(\gamma)}{\partial \gamma} v_d(\gamma, t) \Bigg), \tag{21}$$

$$\ddot{\gamma} := -k_\gamma e_\gamma + \dot{v}_d(\gamma, t) + (\mathbf{e}_p - \boldsymbol{\delta})^T {}^B_U R(\psi) \frac{\partial \mathbf{p}_d(\gamma)}{\partial \gamma}, \tag{22}$$

where $K_p \succeq 0$, $k_\gamma > 0$, and $\boldsymbol{\sigma}(\mathbf{e}_p) = tanh(\|\mathbf{e}_p\|) \frac{\mathbf{e}_p}{\|\mathbf{e}_p\|}$ is a saturation function. The closed-loop system is input-to-state stable (ISS) with respect to $\Delta\tilde{\mathbf{u}} + \tilde{\boldsymbol{v}}_\mathbf{c}$, and the proposed control law solves Problem 1 for the ASV vehicle.

Proof. Appendix A. □

4.2. Quadrotor Path Following

Given that the quadrotor system is modelled by a double integrator in the inertial frame $\{U\}$, as stated in (10), consider the position and velocity errors defined in $\{U\}$ as:

$$\mathbf{e}_p := \mathbf{p}(t) - \mathbf{p}_d(\gamma), \tag{23}$$

$$\mathbf{e}_v := \dot{\mathbf{p}} - \frac{\partial \mathbf{p}_d}{\partial \gamma} v_d(\gamma, t), \tag{24}$$

and a virtual target speed tracking error defined by (14). Consider also a new auxiliary error $\mathbf{z}$, defined as:

$$\mathbf{z} := \mathbf{e}_v + K_1 \mathbf{e}_p, \tag{25}$$

where $K_1 \succeq 0$ is a gain matrix. The position and velocity error dynamics can be written as:

$$\dot{\mathbf{e}}_p = \dot{\mathbf{p}} - \frac{\partial \mathbf{p}_d}{\partial \gamma}\dot{\gamma}, \tag{26}$$

$$\dot{\mathbf{e}}_v = \ddot{\mathbf{p}} - \frac{d}{dt}\left(\frac{\partial \mathbf{p}_d}{\partial \gamma} v_d(\gamma, t)\right). \tag{27}$$

Furthermore, consider the time derivative introduced in (27), the desired virtual target speed function (12), and the virtual target speed tracking error function (14). Then, the time derivative term introduced in (27) can be expanded as:

$$\frac{d}{dt}\left(\frac{\partial \mathbf{p}_d}{\partial \gamma} v_d(\gamma, t)\right) = \underbrace{\left[\frac{\partial^2 \mathbf{p}_d}{\partial \gamma^2} v_d(\gamma, t) + \frac{\partial \mathbf{p}_d}{\partial \gamma}\frac{\partial v_L(\gamma)}{\partial \gamma}\right]}_{\mathbf{h}(\gamma)}(e_\gamma + v_d(\gamma, t)) + \frac{\partial \mathbf{p}_d}{\partial \gamma}\dot{v}^{coord}(t). \tag{28}$$

Replacing (10) and (28) in (27) yields:

$$\dot{\mathbf{e}}_v = \mathbf{u} + \mathbf{d} - \mathbf{h}(\gamma)(e_\gamma + v_d(\gamma, t)) - \frac{\partial \mathbf{p}_d}{\partial \gamma}\dot{v}^{coord}(t). \tag{29}$$

Unlike the case of the ASVs where current estimates are given by a complementary filter, in the case of a quadrotor, a different direction is taken towards estimating disturbances such as wind. According to Xie and Cabecinhas et al. [20,21], straightforward implementations of estimators can lead to windup and result in unbounded growth of an external disturbance estimate. To avoid such problems, Xie and Cabecinhas propose the use of a sufficiently smooth projection operator in the estimator design. Consider the disturbance observer given by:

$$\dot{\hat{\mathbf{d}}} := K_d \text{Proj}(\mathbf{z}, \hat{\mathbf{d}}) = \mathbf{z} - \frac{\eta_1 \eta_2}{2(\beta^2 + 2\beta d_{max})^{n+1} d_{max}^2}\hat{\mathbf{d}}, \tag{30}$$

where K_d denotes a diagonal gain matrix and:

$$\eta_1 = \begin{cases} (\hat{\mathbf{d}}^T\hat{\mathbf{d}} - d_{max}^2)^{n+1}, \text{ if } (\hat{\mathbf{d}}^T\hat{\mathbf{d}} - d_{max}^2) > 0 \\ 0, \text{ otherwise,} \end{cases} \tag{31}$$

and:

$$\eta_2 = \hat{\mathbf{d}}^T\mathbf{z} + \sqrt{(\hat{\mathbf{d}}^T\mathbf{z})^2 + \varsigma^2}, \tag{32}$$

where $\varsigma, \beta > 0$ are arbitrary constants. This projection operator, first proposed in Cai et al. [22], enjoys the useful properties:

$$\tilde{\mathbf{d}}^T Proj(\mathbf{z}, \hat{\mathbf{d}}) \geq \tilde{\mathbf{d}}^T\mathbf{z}, \tag{33}$$

and:

$$\left\|\hat{\mathbf{d}}\right\| \leq d_{max} + \beta, \forall t \geq 0. \tag{34}$$

Once again, consider the inner-loop tracking error and disturbance estimation error given by:

$$\tilde{\mathbf{u}} := \mathbf{u} - \mathbf{u}_d, \tag{35}$$

$$\tilde{\mathbf{d}} := \mathbf{d} - \hat{\mathbf{d}}. \tag{36}$$

Proposition 2. *Consider the system described by (10), the disturbance estimator dynamics given by (30), and the inner-loop tracking error given by (35). Furthermore, consider the control law given by:*

$$\mathbf{u}_d := -\hat{\mathbf{d}} + \mathbf{h}(\gamma)v_d(\gamma,t) + \frac{\partial \mathbf{p}_d}{\partial \gamma}\dot{v}^{coord}(t) - \mathbf{e}_v K_v - \mathbf{e}_p K_p, \tag{37}$$

$$\ddot{\gamma} := -k_\gamma e_\gamma + \dot{v}_d(\gamma,t) + \mathbf{e}_p^T \frac{\partial \mathbf{p}_d}{\partial \gamma} + \mathbf{z}^T\left(\mathbf{h}(\gamma) + K_1 \frac{\partial \mathbf{p}_d}{\partial \gamma}\right), \tag{38}$$

where $K_p, K_v \succeq 0$, and k_γ is a positive gain. For sufficiently small initial position and velocity errors ($\mathbf{e}_p$, $\mathbf{e}_v$), and a sufficiently large separation between the time-scales of the inner and outer loop systems, it can be guaranteed that the system error converges to a neighbourhood of zero. The proposed control law solves Problem 1 for the quadrotor vehicle.

Remark 2. *In-depth and quantitative overall stability analysis can be conducted for the inner–outer loop control system, but this will be dependent directly on the type of inner loop adopted. This results from the fact that the desired accelerations $\mathbf{u_d}$ must be decoupled in a set of desired thrust and attitude for the quadrotor to track. Given that this analysis is out of the scope of this work, we assume that the quadrotor is equipped with a generic inner loop that is capable of keeping the tracking error $\tilde{\mathbf{u}}$ small and bounded.*

Proof. Appendix B. □

5. Cooperative Path Following

In this section, the problem of CPF is addressed. The end goal is to have an algorithm that allows one quadrotor and multiple ASVs to perform a path following mission cooperatively, using a distributed architecture. The vehicles are required to execute their mission according to a fixed geometric configuration. To cope with limitations imposed by real environments where inter-vehicle communications are discrete, an Event-Triggered Communications (ETC) mechanism is adopted, based on previous work developed by A. Aguiar and A. Pascoal [23] and N. Hung and F. Rego [13].

Synchronisation Problem with Event-Triggered Communications

Consider a group of $N \in \mathbb{R}^+ \setminus \{1\}$ autonomous vehicles/agents in a network that can be described mathematically by a digraph $\mathcal{G}(\mathcal{V},\mathcal{E},\mathcal{A})$, consisting of N vertices, a set of directed edges $\mathcal{E} \subseteq \mathcal{V} \times \mathcal{V}$, where the edge ε_{ij} represents the flow of information from agent i to agent j, and a weighted adjacency matrix $\mathcal{A} = [a_{ij}] \in \mathbb{R}^{N\times N}$. Furthermore, each vehicle i is able to receive information from its neighbours in $\mathcal{N}_i^{in}$ and send information to its neighbours in $\mathcal{N}_i^{out}$, i.e., $\mathcal{G}$ is undirected. Moreover, consider that the communication topology of the vehicles is fixed; hence, the Laplacian L associated to $\mathcal{G}$ is constant. Let the state vector of the system be composed by the path parameter of each individual vehicle $\gamma = [\gamma_1, ..., \gamma_N]^T$. In addition, each vehicle is equipped with the PF controllers proposed in Section 4, and has an assigned path to follow, appropriately parameterised in order to ensure that a given desired formation between the vehicles is met. The CPF problem consists in designing a distributed control scheme that adjusts the speed of the vehicles such that all path parameters γ reach a consensus. Consider the problem formulation below.

Problem 2. *For each agent i, with $i = 1, ..., N$, derive a consensus protocol for the speed correction term $\mathbf{v^{coord}} = [v_1^{coord}, ..., v_N^{coord}]^T$, such that $\lim_{t\to\infty} |\gamma_i - \gamma_j| = 0, \forall j \in N_i^{in}$, and the formation of vehicles achieves the desired speed assignment $\mathbf{v_L}(\gamma) = [v_{L1}, ..., v_{LN}]^T$ as $t \to \infty$.*

Note that, according to the previously developed (PF) controllers, for each vehicle i, $|\dot{\gamma}_i - v_d(\gamma,t)| = 0$ is only guaranteed as $t \to \infty$, as the controlled variable is $\ddot{\gamma}_i$ and not $\dot{\gamma}_i$. Having this fact in mind, and assuming that the vehicles have already converged to their desired paths, i.e., $e_p \approx 0$ (and $e_v \approx 0$ in the case of the quadrotor), then the following simplifying assumption can be made:

Assumption 1. *The speed progression of all the virtual targets along the desired path is always assumed to be modelled by a single integrator system, which can be expressed in vectorial form as:*

$$\dot{\gamma} = \mathbf{v_d}(\gamma, t) = \mathbf{v_L}(\gamma) + \mathbf{v^{coord}}. \tag{39}$$

Let the synchronisation error vector be defined as $\boldsymbol{\varepsilon} = [\varepsilon_1, ..., \varepsilon_N]^T$ where, for each i:

$$\varepsilon_i := \sum_{j \in \mathcal{N}_i^{in}} a_{ij}(\gamma_i - \gamma_j), \tag{40}$$

with a_{ij} elements of the weighted adjacency matrix that describes the vehicle network. This error can also be expressed in vectorial form as:

$$\boldsymbol{\varepsilon} := L\gamma, \tag{41}$$

where ε_i denotes the coordination error between vehicle i and its neighbours. With the above notation, the coordination error dynamics of the multi-vehicle system are given by:

$$\dot{\boldsymbol{\varepsilon}} := L\dot{\gamma}. \tag{42}$$

In the work of N. Hung and F. Rego [13], the authors propose a scheme where each agent i has a set of estimators $\hat{\gamma}_j, j \in \mathcal{N}_i^{in}$ for the true state of each in-neighbour virtual target γ_j. In addition, each agent i has an estimator for its own state $\hat{\gamma}_i$, which is reset whenever vehicle i broadcasts its true state γ_i. The other estimators are reset whenever agent i receives the true state from its in-neighbours $j \in \mathcal{N}_i^{in}$. In this work, a time-dependent broadcast condition is adopted.

Proposition 3. *Consider the distributed control law given by:*

$$v_i^{coord} := -k_\varepsilon \sum_{j \in \mathcal{N}_i^{in}} a_{ij}(\gamma_i - \hat{\gamma}_j), \tag{43}$$

where $k_\varepsilon > 0$ and $\hat{\gamma}_j$ is vehicle i's estimate of vehicle j's real virtual target value. Consider also that the bank of estimators that each vehicle i is running is described by the dynamics equation:

$$\dot{\hat{\gamma}}_i := v_L(\hat{\gamma}_i). \tag{44}$$

At any time instant t, under negligible transmission delays, the vehicle j's self-state estimate $\hat{\gamma}_j$ is equal to vehicle i's estimate of $\hat{\gamma}_j$, which allows us to express the estimator dynamics using vectorial notation as:

$$\dot{\hat{\gamma}} := \mathbf{v_L}(\hat{\gamma}), \tag{45}$$

where $\hat{\gamma} = [\hat{\gamma}_1, ..., \hat{\gamma}_N]^T$ is the self-estimate of the virtual target of each vehicle. Let $\tilde{\gamma} = [\tilde{\gamma}_1, ..., \tilde{\gamma}_N]^T$ denote the local estimation errors of each vehicle, such that $\tilde{\gamma} = \gamma - \hat{\gamma}$. Then, $\mathbf{v^{coord}}$ can also be given in vectorial notation, according to:

$$\mathbf{v^{coord}} := -k_\varepsilon[D\gamma - \mathcal{A}\hat{\gamma}] = -k_\varepsilon(\boldsymbol{\varepsilon} + \mathcal{A}\tilde{\gamma}), \tag{46}$$

where D is a diagonal matrix and A the graph adjacency matrix. Consider also a triggering function used to define when to broadcast the along-path position of the virtual target of each vehicle, defined as:

$$\begin{cases} \delta_i(t) := |\tilde{\gamma}_i(t)| - g_i(t) \\ \tilde{\gamma}_i(t) = \hat{\gamma}_i(t) - \gamma_i(t), \end{cases} \tag{47}$$

where $\tilde{\gamma}_i(t)$ is the local estimation error of agent i and $g_i(t)$ is a time-dependent threshold function, such that if the estimation error exceeds this threshold, i.e., $\delta_i(t) \geq 0$, vehicle i broadcasts its state to the out-neighbours $\mathcal{N}_i^{out}$ and resets its local estimator. Furthermore, consider $g_i(t)$ to belong to a class of non-negative functions, given by:

$$g_i(t) = c_i + b_i e^{-\alpha_i t}, \tag{48}$$

with c_i, b_i and α_i being positive constants and $\mathbf{g}(t) = [g_1, ..., g_N]^T$ being the collection of functions g_i for each individual vehicle i. Consider also that $\mathbf{v_L}(\gamma) = v_L\mathbf{1} + \tilde{\mathbf{v}}_\mathbf{L}$, where $\tilde{\mathbf{v}}_\mathbf{L}$ is a bounded and arbitrarily small term that accounts for a transient period in which the vehicles are on different sections of the path, with slightly different desired speed profiles. Then, under Assumption 1, the system is ISS with respect to the error vector $\boldsymbol{\varepsilon}$ and the inputs $\tilde{\gamma}$ and $\tilde{\mathbf{v}}_\mathbf{L}$.

Proof. Appendix C. □

The proposed control scheme used for achieving CPF using ETC is summarised in Algorithm 1.

Algorithm 1 Event-Triggered Communication for vehicle i (adapted from [24]).

At every time instant t, each vehicle i follows the procedure:
procedure COORDINATION AND COMMUNICATION
    if $\delta_i(t) \geq 0$ where δ_i is computed using (47) and (48) **then**
        Broadcast $\gamma_i(t)$;
        Reset the estimator $\hat{\gamma}_i$;
    if Receive a new message from agent $j \in \mathcal{N}_i^{in}$ **then**
        Reset $\hat{\gamma}_j(t)$;
    Run the estimators according to (44);
    Update the first order control protocol u_i using (43).

Given the general distributed control scheme, we now elaborate and address a specific formation, in the context of this work, in the sections that follow.

6. Path Planning

This section addresses the problem of generating a set of smooth and planar reference paths for each individual vehicle to follow, with the end goal of encircling the boundary of a chemical spill. In order to make the vehicles follow the dynamic boundary according to a pre-defined formation (such as a triangle) multiple paths should be generated from one reference path that encodes the boundary. Borrowing from the work of Saldaña et al. [7], we start by presenting a rigorous mathematical definition of a dynamic boundary below.

Definition 1. *A dynamic boundary is a set of planar points Ω_t, such that $\forall z \in \Omega_t$, and for any $\xi > 0$, the open disk centered at point z with radius ξ contains points of Ω_t and its complement set Ω_t^C. Moreover, the dynamic boundary can be approximated by a parametric closed curve (Jordan curve) $\mathbf{C}(\gamma, t) : [0, \infty] \times [0, \infty] \to \mathbb{R}^2$, mapped by a parameter $\gamma \in \mathbb{R}_0^+$ and time $t \in \mathbb{R}_0^+$. The curve is continuous with no self-intersecting points, and changes smoothly with respect to both time t and parameter γ, as depicted in Figure 4a.*

Since the chemical spill boundary is assumed to be dynamic, a path planning problem can be formulated in which a quadrotor is actively re-planning the path that the ASVs should follow at the water surface, as the group of vehicles moves along it and more up-to-date data is acquired by the quadrotor's vision system. Consider, therefore, Problem 3.

Problem 3. *Consider a quadrotor flying over a body of water at a pre-defined fixed altitude, equipped with a camera sensor pointing downwards with a fixed pitch angle relative to the vehicle's body reference frame $\{B\}$. Consider also that the vehicle is capable of detecting the boundary of a chemical* *spill in the 2-D image provided by the camera sensor. Furthermore, one or more ASVs at the surface of the water are required to follow a path dictated by the quadrotor, according to a pre-defined vehicle formation. As the quadrotor detects the dynamic boundary in the image:*

1. *use the data provided by its navigation system to convert the pixels to a 2-D point cloud expressed in the inertial frame* $\{U\}$;
2. *remove outliers and perform pre-processing on the 2-D point cloud;*
3. *generate a smooth and planar reference path by formulating an online optimisation problem that fits the data with open uniform B-splines;*
4. *send the updated path to the vehicle network;*
5. *make each vehicle generate an unique path for itself, capturing the pre-defined vehicle formation;*
6. *repeat the process.*

In order to solve Problem 3, a few simplifying assumptions are made:

Assumption 2. *The dynamic boundary is located at the ocean's surface, assumed to be a 2-D plane at* $Z_U = 0$ *in the inertial frame of reference* $\{U\}$.

Assumption 3. *The quadrotor has a navigation system that can track the vehicle's pose with good accuracy.*

Assumption 4. *The quadrotor has a limited field of view of the environment, i.e, the camera sensor might not be able to capture the entire chemical spill boundary, but rather sections of it, according to Figure 4b.*

Assumption 5. *The detection of the pixels that encode the boundary in the image frame is a sub-system that is assumed to be already available, such as the one proposed in [25].*

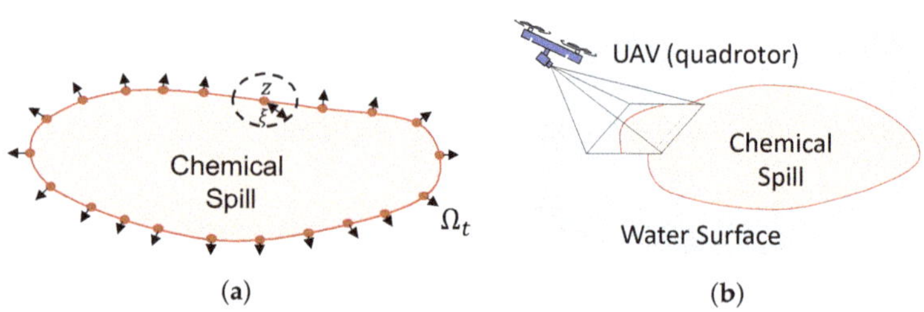

Figure 4. Dynamic Boundary schematic: (**a**) Boundary formal definition. (**b**) Drone's field of view.

6.1. Planar Point Cloud Generation

The camera model adopted is characterised by: (i) a set of extrinsic parameters, which model the conversion between coordinates expressed in the world/inertial reference frame $\{U\}$ and the camera reference frame $\{C\}$; (ii) intrinsic parameters which describe how a set of points in $\{C\}$ are represented in the image frame, according to Figure 5.

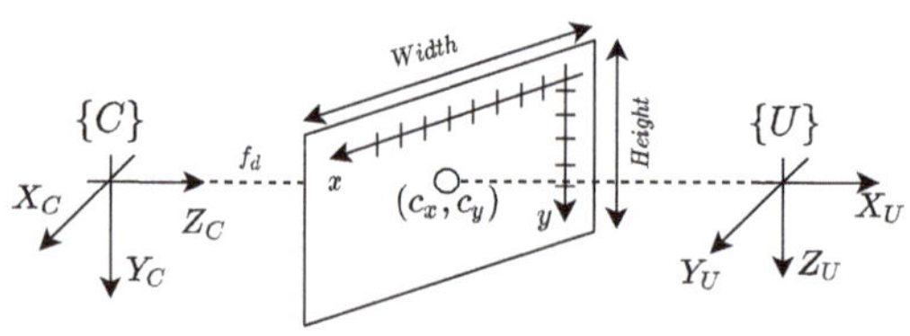

Figure 5. Camera model and reference frames.

The intrinsic parameters consist of the focal distance f_d, the scale factors (s_x, s_y) in the X- and Y-axis, respectively, and (c_x, c_y), which corresponds to the offset of the focal point in the image plane. These parameters can be obtained a priori by resorting to a camera calibration process, described in detail in [26]. Combining the matrices of intrinsic parameters K, also known as the full-rank calibration matrix, and the matrix of external parameters ${}^{C}_{U}[R|T]$, and expressing the inertial frame coordinates as homogeneous coordinates, the transformation between inertial frame and camera plane is described by:

$$\lambda \begin{bmatrix} x \\ y \\ 1 \end{bmatrix} = \underbrace{\begin{bmatrix} f_d s_x & 0 & c_x \\ 0 & f_d s_y & c_y \\ 0 & 0 & 1 \end{bmatrix} \begin{bmatrix} 1 & 0 & 0 & 0 \\ 0 & 1 & 0 & 0 \\ 0 & 0 & 1 & 0 \end{bmatrix}}_{K} \underbrace{\begin{bmatrix} {}^{C}_{U}R & {}^{C}_{U}\boldsymbol{T} \\ \mathbf{0}_{1\times 3} & 1 \end{bmatrix}}_{{}^{C}_{U}[R|T]} \begin{bmatrix} X_U \\ Y_U \\ Z_U \\ 1 \end{bmatrix}, \tag{49}$$

where x and y denote the coordinates in the image frame and λ is a scale factor. It is important to mention that ${}^{C}_{U}[R|T]$ results from a series of successive rigid-body transformations (rotations and translations) given by:

$${}^{C}_{U}[R|T] = {}^{C}_{B}[R|T]\,{}^{B}_{U}[R|T], \tag{50}$$

where ${}^{B}_{U}[R|T]$ denotes the conversion of coordinates expressed in the inertial frame $\{U\}$ to the quadrotor's body frame $\{B\}$, provided by its navigation system, and ${}^{C}_{B}[R|T]$ is a matrix known a priori, as the camera attached to the vehicle is assumed to be fixed. The intrinsic and extrinsic parameters can be aggregated in a matrix Ω according to:

$$\Omega = K \cdot {}^{C}_{U}[R|T]. \tag{51}$$

In order to convert a given set of pixels (x, y) that encode the chemical spill boundary in the image frame to a point cloud expressed in the inertial frame, depth information about the scene is required. Taking into consideration Assumption 2, all the points in the inertial frame will lie on the plane described by $Z_U = 0$, which solves the depth requirement. Moreover, from Assumption 3, it can be concluded that the linear system of Equation (49) is well defined and can be inverted such that for each pixel representing the boundary of the chemical spill, X_U and Y_U are extracted from:

$$\frac{1}{\lambda} \begin{bmatrix} X_U \\ Y_U \\ 1 \end{bmatrix} = \begin{bmatrix} \Omega_1 & \Omega_2 & \Omega_4 \\ \Omega_5 & \Omega_6 & \Omega_8 \\ \Omega_9 & \Omega_{10} & \Omega_{12} \end{bmatrix}^{-1} \begin{bmatrix} x \\ y \\ 1 \end{bmatrix}. \tag{52}$$

Remark 3. *This methodology relies heavily on the assumption that the quadrotor has a good navigation system, since small estimation errors in the altitude of the vehicle can lead to errors of several meters in the generated point cloud.*

6.2. Pre-Processing the Planar Point Cloud

Before using the 2-D point cloud to generate a path, it is important to pre-process the information provided in it. Consider, for instance, the example in Figure 6, where the quadrotor produces a 2-D point cloud, representing the boundary, at an arbitrary time-step t_k. In the point cloud, some points represent the chemical spill boundary in a region that is close to the vehicle—the region of interest, i.e., where the main cluster of points is expected to be located (in region B). The separation between regions A and B is defined by drawing a normal to the path at the point where the re-planning starts (defined formally in Section 6.3.1). Some points are outliers as a result of either noisy measurements or regions of the boundary that are not entirely captured by the field of view of the camera. The latter can be seen as disconnected from the main cluster and should be disregarded in the path planning process. According to Figure 6, the original path (in purple) obtained at time t_{k-1} should be re-planned in order to obtain a new one (in red) that better fits the main cluster of points.

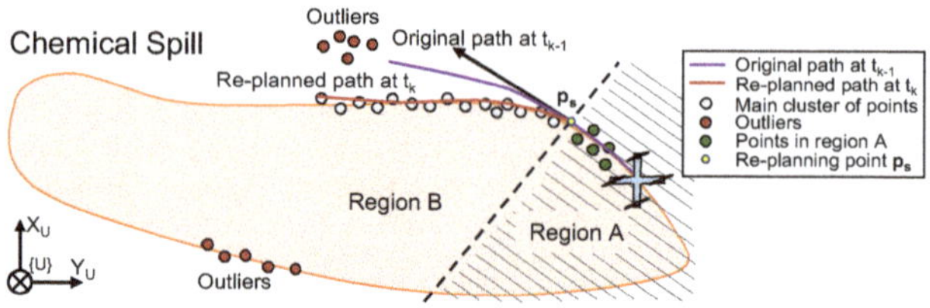

Figure 6. Pre-processing the point cloud and re-planning schematic.

Unlike conventional motion planning problems, the main cluster of points does not have an explicit ordering, yielding a sequence of waypoints that the vehicle should visit sequentially in time—this information must be inferred. On the other hand, it is possible to define explicitly where the path re-planning process starts—at a point $\mathbf{p}_s := C(\gamma_s)$ arbitrarily further ahead of the drone's position on the current curve $C(\gamma)$, such that $\gamma_{drone} \leq \gamma_s$. Motivated by this example, and inspired by the work of Liu Y. et al. [27], the following pre-processing steps are introduced:

- Remove unused points that are behind the point $\mathbf{p}_s$, i.e., points in region A;
- Order the remaining set of points and remove outliers in region B.

6.2.1. Removing Unused Points

Consider $\mathbf{p}_s \in \mathbb{R}^2$ to be the point at which the path re-planning starts. In order to remove the points that are in region A, consider that ψ_s is the tangent angle to the current path at $\mathbf{p}_s$. A coordinate transformation can be applied to the 2-D point cloud $\boldsymbol{X} := \{\boldsymbol{X}_l\}_{l=1}^{L} \in \mathbb{R}^2$, such that in a new reference frame, points that are behind $\mathbf{p}_s$ (in region A) have a negative X-coordinate. This coordinate transformation is given by:

$$\boldsymbol{X}_l^{\circ} = R(\psi_s) \cdot (\boldsymbol{X}_l - \mathbf{p}_s), \forall l = 1, ..., L, \tag{53}$$

where $\boldsymbol{X}_l^{\circ} = [\boldsymbol{X}_l^{\circ x}, \boldsymbol{X}_l^{\circ y}]^T$. Each point $\boldsymbol{X}_l$ is discarded if $\boldsymbol{X}_l^{\circ x} < 0$. The points that belong to set $\boldsymbol{X}$ and are not discarded, and should be saved in a new set $\boldsymbol{X}^{\star} := \{\boldsymbol{X}_j\}_{j=1}^{J} \in \mathbb{R}^2$ with $J \leq L$. The pseudo-code is shown in Algorithm 2.

Algorithm 2 Remove points "behind" the re-planning point.

```
Obtain a new 2-D point cloud X := {X_l}_{l=1}^{L} ∈ R^2;
Define p_s as the desired initial point for the re-planning to start;
Define ψ_s as the tangent angle to the current path at t_k at p_s;
Follow the procedure:
procedure REMOVE UNUSED POINTS(X, p_s, ψ_s)
    for l = 1, ..., L do
        Compute X_l^∘ according to (53);
        if X_l^∘x < 0 then
            Discard X_l;
    return the new set X* := {X_j}_{j=1}^{J} ∈ R^2 with J ≤ L.
```

6.2.2. Ordering a Set of Points and Removing Outliers

In order to avoid clustering outliers, reduce the point cloud to a curve-like shape, and extract some implicit ordering from the data, Lee I. [28] proposes an algorithm that seeks to extract a structure "as simple as possible" from the data, by resorting to an Euclidean Minimum Spanning Tree (EMST). Consider the unordered set of points $\boldsymbol{X}^{\star}$ obtained previously and a graph $\mathcal{G} = (\mathcal{V}, \mathcal{E})$, such that $\mathcal{V} = \{\boldsymbol{X}_j = (x_j, y_j) | j = 1, ..., J\}$ and $\mathcal{E} = \{(\boldsymbol{X}_i, \boldsymbol{X}_j) | i, j = 1, ..., J, i \neq j\}$. The EMST is a tree that connects all points in $\mathcal{G}$ with the weight of its edges corresponding to the Euclidean distance between each pair of points, that can be computed according to the very popular Kruskal's algorithm. In order to reduce the time complexity of this process, a threshold distance N_J can be used to define whether each pair of points is connected and a KDTree [29] can be used to compute a sparse graph $\mathcal{G}$ where each point has a limited set of neighbours, as shown in Figure 7.

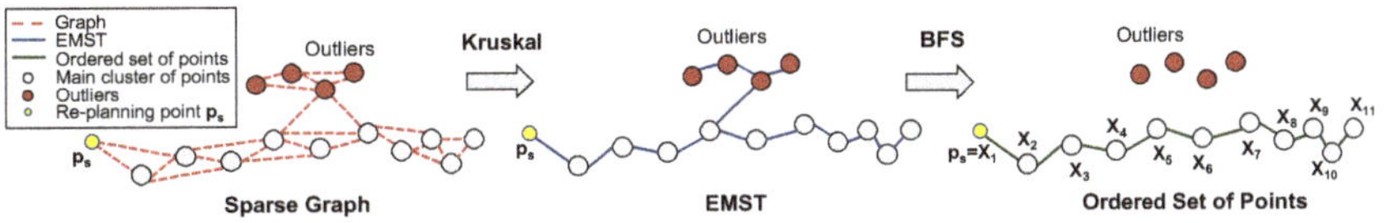

Figure 7. From sparse graph to an ordered list of points (example).

To remove outliers and define a coarse path to follow, Breadth First Search (BFS) can be applied to the EMST, starting from $\mathbf{p}_s$. This removal of outliers from the point cloud is crucial to avoid smaller clusters of points being considered later in the curve fitting problem. The resulting ordered list of points that forms the path with the highest number of points should be saved in a new ordered set $\mathbf{X}^{\dagger} := \{X_k\}_{k=1}^{K} \in \mathbb{R}^2$. The proposed steps are summarised in Algorithm 3.

Algorithm 3 Order a set of 2-D points.

Add the desired initial point for the path $\mathbf{p}_s$ to $\boldsymbol{X}^{\star}$;
Define a threshold distance for the neighbours N_J;
Follow the procedure:
procedure ORDER POINTS($\boldsymbol{X}^{\star}$, N_J)
  Construct a KDTree from $\boldsymbol{X}^{\star}$ and use N_J as a distance threshold;
  Create a graph $\mathcal{G}$ with J vertices and no edges;
  for $\boldsymbol{X}_j$, $j = 1, ..., J$ **do**
    Query the KDTree for the neighbours of $\boldsymbol{X}_j$ and their euclidean distances;
    Add the corresponding edges to the graph $\mathcal{G}$;
  Compute the MST of the graph $\mathcal{G}$ starting from vertex corresponding to $\mathbf{p}_s$;
  Compute the path with the highest number of points, starting at $\mathbf{p}_s$ using BFS;
  return the new ordered set of points $\mathbf{X}^{\dagger} := \{\mathbf{X}_k\}_{k=1}^{K} \in \mathbb{R}^2$.

6.3. Path Generation—Approximating the Point Cloud with a Parametric Curve

In order to have a suitable representation of a path that the proposed controllers can follow, it is a requirement to generate a curve that is smooth and at least $\mathcal{C}^2$. In order to fulfil this requirement, the ordered set of points produced previously can be approximated by non-clamped uniform cubic B-splines, composed of multiple spline segments, where each segment is paramaterised by $\gamma \in [0, 1)$.

6.3.1. Define the Number of Segments to Use

Consider now the ordered sequence of K points obtained via the application of Algorithms 2 and 3 to the original 2-D point cloud. In order to fit the points with a parametric curve, we are required to attribute to each point $\boldsymbol{X}_k \in \mathbb{R}^2$ a corresponding γ_k in the target parametric curve. This problem could be formulated as a nonlinear optimisation problem—which is computationally demanding to solve for real-time applications. A non-optimal, but more efficient solution, proposed by Liu M. et al. [30] for Simultaneous Localisation and Mapping (SLAM) applications, is to consider D_X to be the total distance between the points, given by:

$$D_X := \sum_{k=2}^{K} \|\boldsymbol{X}_k - \boldsymbol{X}_{k-1}\|, \tag{54}$$

and the corresponding vector of parametric values $\gamma = [\gamma_1, ..., \gamma_k]^T$ to be given by:

$$\begin{cases} \gamma_1 = 0, \\ \gamma_k = \gamma_{k-1} + \frac{\|\mathbf{X}_k - \mathbf{X}_{k-1}\|}{D_X}\gamma_{max}, k = 2, ..., K, \end{cases} \tag{55}$$

where γ_{max} is the maximum parameter value of the parametric curve. For cubic B-splines, this number depends directly on the number of control points N_C that the target curve will have, such that $\gamma_{max} = N_C - 3$. The number of control points also dictates how many spline segments are used for the fitting problem. The optimal number of control points can be obtained by solving yet another nonlinear optimisation problem, but due to the real-time nature of the problem, this option is disregarded. Given that a uniform cubic B-spline must have at least four control points to define one segment, and that a low number of sections can under-fit a long set of points whilst a high number leads to over-fitting issues, this number should not be a static constant either. A non-optimal yet dynamic way of defining the number of control points N_C is by taking:

$$N_C := max\left\{\left\lfloor \frac{D_X}{\rho} \right\rceil, 4\right\}, \tag{56}$$

with $(1/\rho) > 0$ being a control point's density (tunning parameter defined a priori). A smaller ρ leads to a higher N_C. Applying this method to the previous example, and considering $N_C = 7$, $\gamma_{max} = \gamma_{11} = 4$, the result in Figure 8 is obtained.

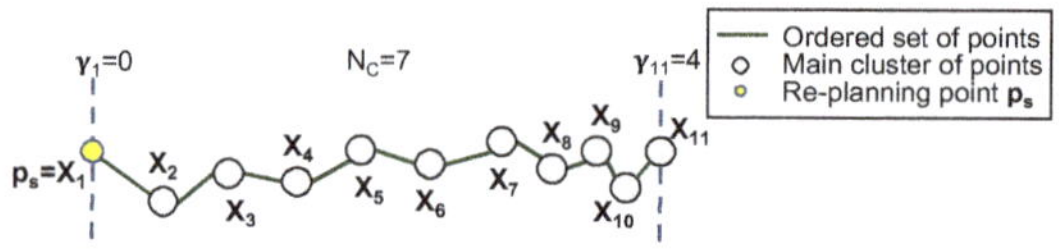

Figure 8. Ordered set of points with parametric values associated to them (example).

6.3.2. Fitting the Points with a Uniform Cubic B-Spline

For fitting the ordered set of points $\mathbf{X}^\dagger$ with a non-clamped uniform cubic B-spline $C(\gamma, \mathbf{P})$, an optimisation problem is formulated. Consider the objective function given by:

$$f(\mathbf{P}) := \underbrace{\sum_{k=1}^{K} \|\mathbf{C}(\gamma_k, \mathbf{P}) - \mathbf{X}_k\|^2}_{goal} + F_r, \tag{57}$$

with:

$$F_r = \underbrace{\lambda \int_0^{\gamma_{max}} \left\| \frac{\partial C(\gamma, \mathbf{P})}{\partial \gamma} \right\|^2 d\gamma + \beta \int_0^{\gamma_{max}} \left\| \frac{\partial^2 C(\gamma, \mathbf{P})}{\partial \gamma^2} \right\|^2 d\gamma}_{\text{regularisation term}}, \tag{58}$$

where $\mathbf{P} = [P_0^x, ..., P_{N_c-1}^x, P_0^y, ..., P_{N_c-1}^y]^T \in \mathbb{R}^{2N_c}$ is the vector of control points that defines the target curve. The first term minimises the distance between the target B-spline curve and the set of points, whilst F_r is a regularisation term and $\alpha, \gamma \geq 0$ are the regularisation variables. The integral of the L_2^2 norm of the first derivative penalises the total length of the curve, while the integral of the L_2^2 norm of the second derivative penalises bends in the path. This objective function can also be expressed using vector notation, according to:

$$f(\mathbf{P}) = \underbrace{\|B(\gamma)\mathbf{P} - \mathbf{X}\|^2}_{goal} + \underbrace{\lambda \mathbf{P}^T R_1 \mathbf{P} + \beta \mathbf{P}^T R_2 \mathbf{P}}_{\text{regularisation term}}, \tag{59}$$

where $\mathbf{X} = [X_1^x, ..., X_K^x, X_1^y, ..., X_K^y]$ denotes the points to fit, and R_1, R_2 are constant matrices that can be computed numerically (see Appendix D).

In order to define the new path, it would not suffice to discard the previously planned curve defined after γ_s and minimise the objective function with respect to the control points. To guarantee $\mathcal{C}^2$ continuity between the previous path and the newly planned one, linear equality constraints should be imposed on the values of $\mathbf{C}^{new}(0)$, $\mathbf{C}^{new'}(0)$, and $\mathbf{C}^{new''}(0)$ of the new curve. Moreover, it is a requirement to save the old curve up to γ_s, as it may still be in use by other vehicles in the network.

Consider the re-planning point $\mathbf{p}_s$ introduced previously, chosen such that it corresponds to the transition between the spline segment that the virtual target of the drone is "sitting on", and the next segment, according to:

$$\mathbf{p}_s = \mathbf{C}^{old}(\gamma_s) \text{ with } \gamma_s = \lceil \gamma_{drone} \rceil, \tag{60}$$

where γ_{drone} corresponds to the quadrotor's virtual target at time instant t_k. With this choice of γ_s, it is possible to take advantage of the local support property of B-splines and simplify the equality constraints of the problem, while at the same time simplifying the storage of the curves in memory.

Considering that $\mathbf{p}_s$ is dictated by (60), the old curve segments that are described by parametric values such as $\gamma \geq \gamma_s$ should be discarded and replaced by a newer curve. Since each curve segment is defined by only four control points, discarding those segments is equivalent to removing control points with indexes $i \geq \gamma_s + 3$ from the old control points vector. This operation results in a vector given by:

$$\mathbf{P}^{old} = [P_0^x, P_1^x, ..., P_{\gamma_s}^x, P_{\gamma_s+1}^x, P_{\gamma_s+2}^x, P_0^y, P_1^y, ..., P_{\gamma_s}^y, P_{\gamma_s+1}^y, P_{\gamma_s+2}^y]^T. \tag{61}$$

For the particular example in Figure 9, spline 1 (in green) should be discarded given that $\gamma_{drone} \in [0, 1)$; hence, $\gamma_s = 1$ and spline 0 are kept. To achieve this, all the control points with indexes $i \geq 1 + 3$ should be removed from the control points vector $\mathbf{P}^{old}$, i.e., $P_4 = (P_4^x, P_4^y)$.

Making use of the local support property once more, it is known that $\mathcal{C}^2$ continuity between two consecutive cubic spline segments is guaranteed, as long as the last three control points of the first segment coincide with the first three control points of the second segment. A trivial way of generating a new B-spline with guarantees of $\mathcal{C}^2$ continuity in the transition with the old curve, without explicitly defining equality constraints on the derivatives of the function, is to solve the following optimisation problem:

$$\begin{aligned} \mathbf{P}^{new} &= \underset{\mathbf{P}^{new}}{\operatorname{argmin}} f(\mathbf{P}^{new}) \\ \text{subject to} & \\ \begin{bmatrix} P_0^{x\,new} \\ P_1^{x\,new} \\ P_2^{x\,new} \\ P_0^{y\,new} \\ P_1^{y\,new} \\ P_2^{y\,new} \end{bmatrix} &= \begin{bmatrix} P_{\gamma_s}^x \\ P_{\gamma_s+1}^x \\ P_{\gamma_s+2}^x \\ P_{\gamma_s}^y \\ P_{\gamma_s+1}^y \\ P_{\gamma_s+2}^y \end{bmatrix}, \end{aligned} \tag{62}$$

where $\mathbf{P}^{new} = [P_0^{x\,new}, ..., P_{N_C-1}^{x\,new}, P_0^{y\,new}, ..., P_{N_C-1}^{y\,new}]^T$ is a new control points vector.

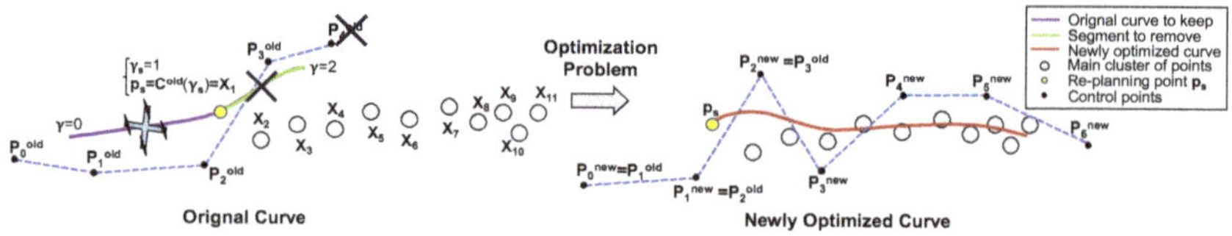

Figure 9. Solving the optimisation problem (example).

To keep track of old and new curves, it is possible to concatenate only the new control points vector $\mathbf{P}^{new}$ with the old control points vector $\mathbf{P}^{old}$, ignoring the first three control points, i.e., P_0^{new}, P_1^{new}, and P_2^{new}, which are repeated as a result of the equality constraints imposed by (62). Applying this methodology to the previous example, the final control points vector is given according to Figure 10.

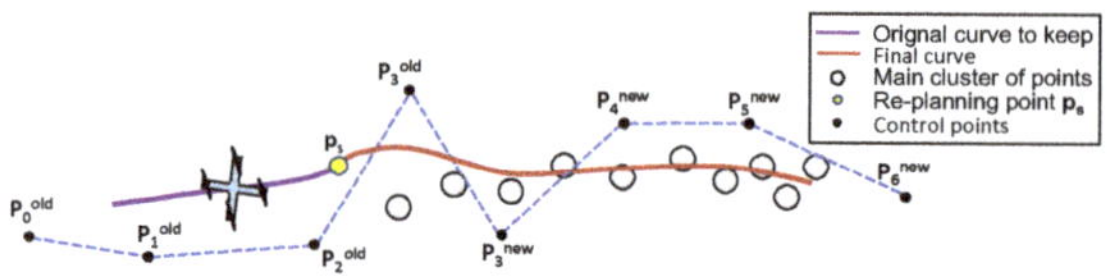

Figure 10. Final curve with control points concatenated (example).

These series of procedures are summarised in Algorithm 4. For the sake of simplicity, the separation between the X- and Y-coordinates of the control points was omitted.

Algorithm 4 Fitting the points—growing a uniform cubic B-spline

1: Compute D_X, γ and N_C according to Equations (54), (55), and (56), respectively;

2: Consider $\gamma_s = \lceil \gamma_{t_k} \rceil$ and the original control points vector:

$$\mathbf{P} = \left[P_0, P_1, ..., P_{\gamma_s}, P_{\gamma_s+1}, P_{\gamma_s+2}, P_{\gamma_s+3}, P_{\gamma_s+4} ..., P_n\right]^T; \tag{63}$$

3: Remove control points (corresponding to splines to be re-planned) from the original control points vector, such that:

$$\mathbf{P}^{old} = \left[P_0, P_1, ..., P_{\gamma_s}, P_{\gamma_s+1}, P_{\gamma_s+2}\right]^T; \tag{64}$$

4: Solve the optimisation problem in (62) and obtain a new vector with N_C control points:

$$\begin{aligned} \mathbf{P}^{new} &= \left[P_0^{new}, P_1^{new}, P_2^{new}, ..., P_{N_C-1}^{new}\right]^T, \\ \text{with } P_0^{new} &= P_{\gamma_s}, P_1^{new} = P_{\gamma_s+1}, P_2^{new} = P_{\gamma_s+2}; \end{aligned} \tag{65}$$

5: Concatenate the new vector with the old vector (ignoring the first three control points, which are repeated):

$$\mathbf{P}^{final} = \left[P_0, P_1, ..., P_{\gamma_s}, P_{\gamma_s+1}, P_{\gamma_s+2}, P_3^{new}, ..., P_{N_C-1}^{new}\right]^T. \tag{66}$$

6.4. From 2-D Path to Vehicle Formation

To generate individual paths for each vehicle to follow, we can consider a reference path (obtained via the application of the previous algorithms) and offset each point according to an expression that captures a desired vehicle formation. Start by considering a formation vector denominated $\boldsymbol{\mu}_i \in \mathbb{R}^3$ for each vehicle i, with each distance defined in the tangential reference frame $\{T\}$ to the virtual target's position in the original curve, according to Figure 11. According to Xie et al. [31], it is possible to define a desired path for each vehicle given by:

$$\mathbf{p}_{di}(\gamma_i) = \mathbf{C}(\gamma_i) + {}^{U}_{T}R(\gamma_i)\boldsymbol{\mu}_i, \tag{67}$$

where $\mathbf{p}_{di}$ is the desired path for the vehicle i, $\mathbf{C}(\gamma_i)$ is the planned curve, and ${}^{U}_{T}R(\gamma_i)$ is a rotation matrix computed according to:

$${}^{U}_{T}R(\gamma_i) = [\mathbf{r}_1(\gamma_i), \mathbf{r}_3(\gamma_i) \times \mathbf{r}_1(\gamma_i), \mathbf{r}_3(\gamma_i)], \tag{68}$$

with:

$$\mathbf{r}_1(\gamma_i) = \frac{\partial \mathbf{p}_d/\partial\gamma}{\|\partial \mathbf{p}_d/\partial\gamma\|}, \text{ with } \|\partial \mathbf{p}_d/\partial\gamma\| \neq 0 \qquad \mathbf{r}_3(\gamma_i) = \frac{\mathbf{r}_d - (\mathbf{r}_d \cdot \mathbf{r}_1(\gamma_i))\mathbf{r}_1(\gamma_i)}{\|\mathbf{r}_d - (\mathbf{r}_d \cdot \mathbf{r}_1(\gamma_i))\mathbf{r}_1(\gamma_i)\|}, \tag{69}$$

such that $\mathbf{r}_1$ is the tangent to the curve. Moreover, since all vehicles will only be required to operate in a 2-D plane, a trivial definition for one of the axis of the tangential frame $\{T\}$ is $\mathbf{r}_d = [0, 0, 1]^T$.

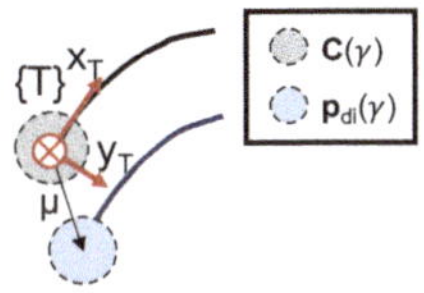

Figure 11. Formation vector.

A path might not be not parameterised according to the arc length and, for B-splines in particular, each spline segment is such that $\gamma \in [0, 1]$. Therefore, it is commonplace to define a constant required speed $V \leq V_{max}$ for the vehicle and let the desired speed profile for the virtual targets be given by:

$$v_L(\gamma) = \frac{V}{\|\mathbf{p}'_d(\gamma)\|}. \tag{70}$$

7. Implementation Details

To evaluate the performance of the proposed PF and CPF algorithms applied to marine ASVs, real water trials were conducted at Doca dos Olivais (Lisbon, Portugal) using the Medusa class of underactuated marine vehicles [14], shown in Figure 12. The vehicles used in the real trials were equipped with a GPS Astech MB100, a NavQuest600 Micro DVL, and a Vectornav VN-100T Attitude and Heading Reference System (AHRS). The operating system used during development was Ubuntu 18.04LTS along with ROS Melodic.

Figure 12. Real Medusa vehicles at Doca dos Olivais, Lisbon (Portugal).

To analyse the performance of the proposed online path planning algorithm, a realistic simulation environment that closely resembles the Doca dos Olivais site was developed and incorporated into the Gazebo simulator. Given the main goal of having a fleet of vehicles encircling a chemical spill, is was necessary to overlay a red stain mesh on top of the ocean's surface (Figure 13).

Figure 13. Simulated world of Doca dos Olivais with red chemical spill in Gazebo.

For simulating the Medusa ASVs, a CAD model of the vehicles was incorporated into the simulator (Figure 14a). The virtual vehicle was also equipped with DVL, AHRS, and GPS sensors provided by the UUVSimulator plugin [16]. To simulate the quadrotor, the Iris vehicle provided by the PX4 SITL Gazebo plugin [15] was used; see Figure 14b.

(**a**) (**b**)

Figure 14. Simulated vehicles in gazebo: (**a**) Medusa ASV. (**b**) Iris quadrotor with a fixed camera.

The simulated quadrotor was equipped with a virtual camera mounted 21 mm below the vehicle's center of mass and with a pitch angle of $-45°$, pointing downwards, and produced an image with a resolution of 640 × 480 px, according to Figure 15a. Its intrinsic parameters are given by:

$$\begin{cases} (c_x, c_y) = (320.5, 240.5) \\ (f_d s_x, f_d s_y) = (381.4, 381.4). \end{cases} \tag{71}$$

Given assumption 5, the detection of the boundary region between the spill and the ocean surface was out of the scope of this work. Therefore, we resorted to OpenCV library [32] to mask and threshold the red colours in the image feed. After this step, the Canny edge detection algorithm was applied to the binary image to retrieve the pixels corresponding to the boundary, according to Figure 15b. To solve the optimisation problem proposed in Section 6.3.2, we resorted to Scipy's SQP solver [33].

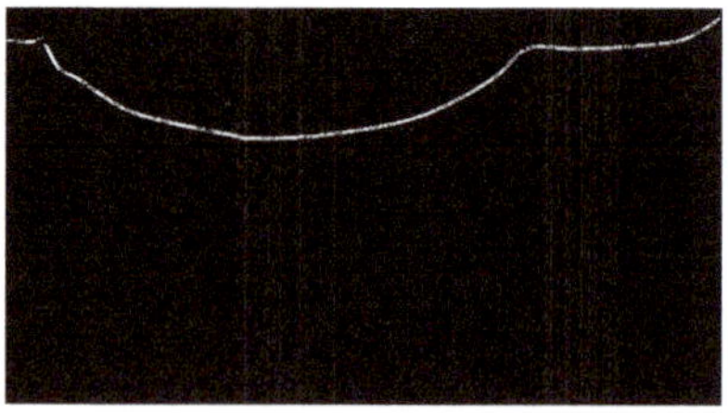

(**a**) (**b**)

Figure 15. Simulated camera feed: (**a**) Quadrotor's camera output. (**b**) Binary image.

The entire system architecture is shown in Figure 16. The inner-loop controls adopted for the quadrotor were the ones already provided by PX4, while for the ASV, we resorted to PID inner-loop controllers to steer the vehicles.

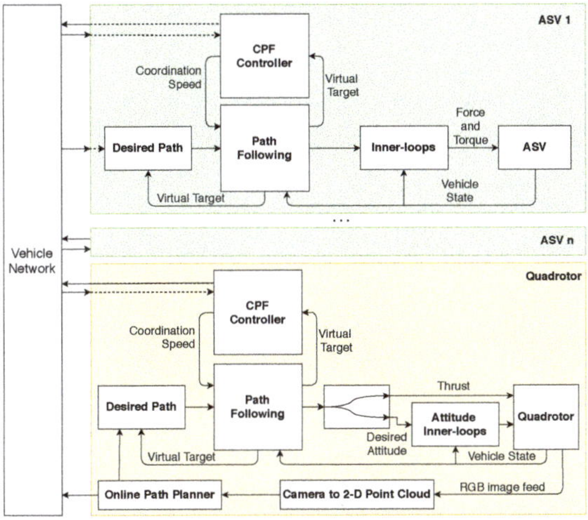

Figure 16. Planning and control architecture.

8. Experimental and Simulation Results

In this section, we present some real experimental results regarding the PF and CPF controllers applied to two Medusa ASV vehicles. In addition, realistic 3-D simulation results are also presented for the case study where two Medusa vehicles were required to perform a CPF mission on a pre-defined path with a quadrotor, in a leader–follower formation. Finally, a third case study is presented, where a simulated quadrotor had to detect the boundary of a chemical spill, and plan, in real-time, a path for both itself and a Medusa ASV to follow cooperatively. The control gains adopted are available in Appendix E.

8.1. Cpf with ETC between 2 Medusa Vehicles (Real)

For the real trial, performed at Doca dos Olivais (Lisbon, Portugal), two Medusa vehicles were required to perform a lawn-mowing mission cooperatively at the surface of the water, according to Figure 17. The black vehicle (Medusa 1) was required to follow the leader (Medusa 2) according to the formation vector $\boldsymbol{\mu} = [-5, -5, 0]^T$. Both vehicles were required to follow the path at $V = 0.5\text{m/s}$ and communications were bi-directional.

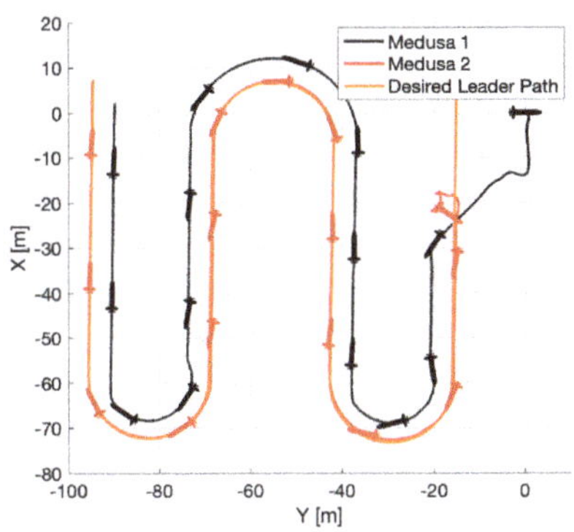

Figure 17. Real CPF mission with 2 Medusa vehicles.

According to the results in Figure 18a, the along-track error of Medusa 1 increases quickly as the virtual target tries not only to minimise the distance to the vehicle but also its distance to its neighbour's (Medusa 2) virtual target. As the vehicles start to move, this error starts to decrease, and according to Figure 18b, after approximately 50 s, the vehicles align themselves according to the desired formation, approach the desired speed profile and, as a consequence, the rate of information exchange decreases. This decrease in the rate of communication is due to the bank of estimators for the virtual targets running in each vehicle being able to better predict the evolution of the virtual targets.

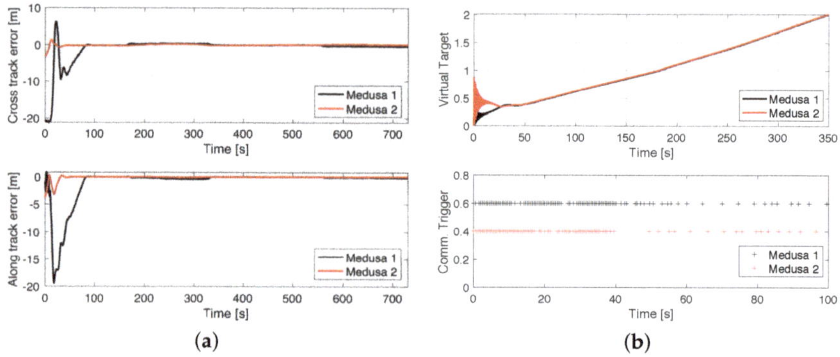

Figure 18. CPF with 2 real Medusa vehicles: (**a**) X–Y view. (**b**) Communication metrics.

8.2. Cpf with ETC between a Quadrotor and Medusa Vehicles (Simulation)

For this case study, a CPF mission was performed such that a simulated quadrotor and two Medusa vehicles were required to perform a lawn-mowing mission, according to Figure 19a. In this experiment, the aircraft was required to fly at a fixed altitude of 30 m; the formation vector for Medusa 1 was given by $\mu_1 = [-5, 5, 0]^T$m, and for Medusa 2, by $\mu_2 = [-5, -5, 0]^T$m, leading to a triangular formation with 2 ASVs side by side, behind the quadrotor. In this experiment, there was bi-directional communication between the pairs of vehicles: (quadrotor, Medusa 1) and (quadrotor, Medusa 2). From the results in Figure 19b, it is observable that the vehicles converge to their desired formation at around 25 s. After this period of time, the position error converges to a neighbourhood of zero and the virtual target speeds converge to their desired value. As a consequence, the number of communication events between the vehicles drops as the bank of observers in each vehicle can more accurately track the state of the virtual target of their peers.

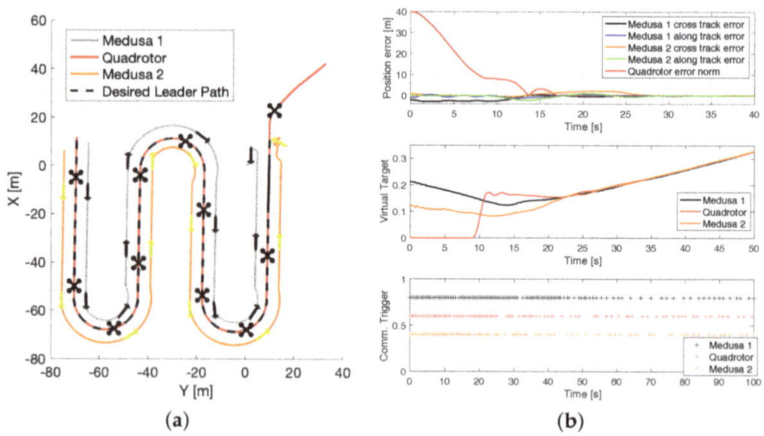

Figure 19. CPF with simulated Iris and Medusa vehicles: (**a**) X–Y view. (**b**). Performance metrics.

8.3. Boundary Encircling with a Quadrotor and a Medusa Vehicle (Simulation)

For the last simulated experiment, the quadrotor was required to start the same lawn-mowing that was adopted for the mission with one Medusa ASV. As soon as a chemical spill boundary was detected in the drone's image stream, the quadrotor was required to start the path planning algorithm at a rate of 1 Hz and send the most up-to-date path to the ASV, according to Figure 20. The drone was required to follow the path at 30 m of altitude with a desired constant speed of 0.5 m/s. Since the quadrotor was equipped with a fixed-mounted camera, it was also required to align its yaw angle with the tangent to the path in order not to lose sight of the boundary being followed.

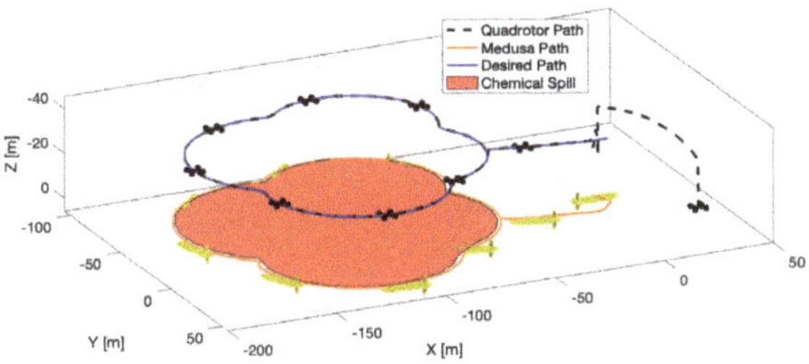

Figure 20. 3-D view of simulated boundary encircling mission with Iris and Medusa vehicles.

In order to guarantee that the path further ahead could be generated for the ASV to follow, it was desirable for the marine vehicle to follow the the quadrotor from behind, i.e., with a formation vector $\boldsymbol{\mu} = [-5, 5, 0]^T$m. In Figure 21a, a top-down view of the executed mission is shown. In Figure 21b, plots of the PF errors are provided along with the norm of the horizontal distance of each vehicle to the real boundary being followed. It is observable that the tracking error only increased in zones where the chemical spill had a crease. This is justified by the fact that the Medusa vehicle, when performing tight turns, was not able to cope with its virtual target speed and slowed down, leading to sudden spikes in the along-track error. These tracking errors were instantly compensated by the adaptive virtual target dynamics, which attempted to minimise the distance between itself and the vehicle. It is also observable that the norm of the distance between the marine vehicle and the chemical spill is much lower than its aerial counterpart, with the Medusa always following the boundary from its outskirts, due to the formation vector adopted.

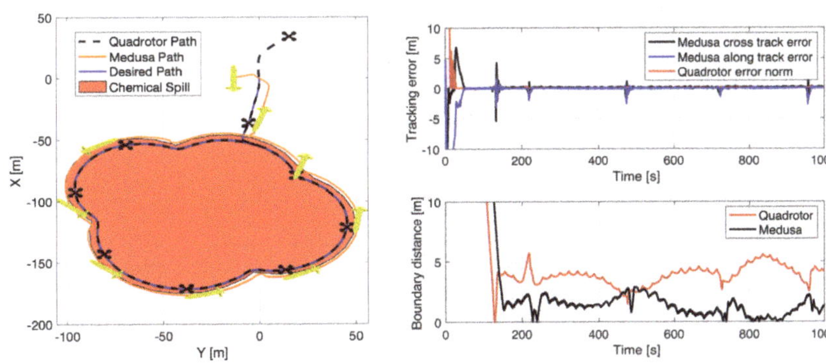

(a) (b)

Figure 21. Boundary encircling with simulated Iris and Medusa vehicles: (**a**) X–Y view. (**b**) Performance metrics.

From Figure 21b, it is also evident that the horizontal distance between the real drone's position and the boundary is bounded by 6m. This result is to be expected, as the altitude estimates are mainly provided by the simulated GPS system and small errors in the estimated attitude, especially yaw angle, will lead to errors of several meters in the generated 2-D point cloud. Due to the type of application at hand, and given that it is typical to have errors of several meters in underwater scenarios, these errors are considered within an acceptable range. In addition, the small oscillations in the boundary distance plot result from the simulated chemical spill boundary mesh being a composition of discrete lines which are picked up by the drone's camera.

In Figure 22, a plot of the point cloud generated by the algorithm is shown at two different time instants (in green), as well as the corresponding planned B-spline paths (in blue). Note that in Figure 22a, some of the green dots further away from the vehicle were discarded by the planning algorithm, as they were too far away from the main cluster of points.

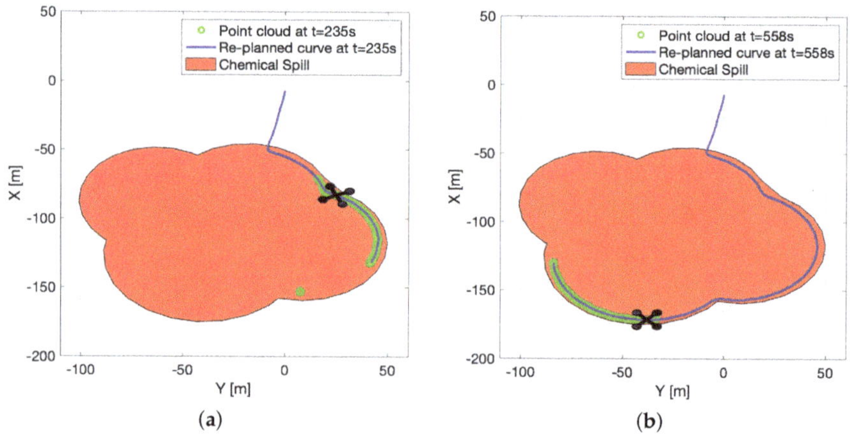

Figure 22. Path generation (**a**) Time = 235 s. (**b**) Time = 558 s.

9. Conclusions and Future Work

This paper addressed the problem of encircling an environmental boundary caused by a chemical spill using a team of robots composed of an aerial quadrotor and Medusa marine vehicles. The path following problem was introduced, and a non-linear control law derived for the ASV, exploiting the technique described in P. Aguiar and F. Vanni [10–12]. Inspired by this control law, a new one was derived for a quadrotor following the same methodology, with some key differences due to the nature of the aircraft. For the section that followed, the CPF problem was formulated and a proposal to solve the problem was presented, such that the synchronisation controller was distributed and the same for all vehicles (aerial and marine) using event-triggered communications based on previous work by N. Hung and F. Rego [13]. In addition, a new real-time path planning algorithm was developed that made use of the camera sensor onboard of the quadrotor to have a local view of the boundary and generate a point cloud expressed in the inertial frame. This data was then used to solve an optimisation problem which generates a B-spline-based path that grows dynamically as the drone moves along the boundary and acquires more data. The path is then shared with all ASV vehicles in the network in real time. The proposed algorithms were implemented in ROS, and a 3-D virtual scenario was generated, allowing for a mixture of real and simulated results. Future work includes making the height at which the quadrotor operates dynamic and introducing curvature limits as inequality constraints to the path planning problem, as well as obstacle avoidance before carrying out integrated experiments with real vehicles.

Author Contributions: The individual contributions of the authors are as follows: Conceptualisation, software, formal analysis, investigation, writing, review, and editing by M.J.; Conceptualisation, validation, review, editing, project administration, and funding acquisition by A.P. and R.C. All authors have read and agreed to the published version of the manuscript.

Funding: This work was partially funded by the H2020-INFRAIA-2017-1-two-stage EUMarineRobots-Marine Robotics Research Infrastructure Network (Grant agreement ID: 731103), the H2020-FETPROACT-2020-2 RAMONES-Radioactivity Monitoring in Ocean Ecosystems (Grant agreement ID: 101017808), H2020-MSCA-RISE-2018 ECOBOTICS.SEA-Bio-inspired Technologies for a Sustainable Marine Ecosystem (Grant agreement ID: 824043).

Institutional Review Board Statement: Not applicable.

Informed Consent Statement: Not applicable.

Data Availability Statement: The path planning library that implements the algorithms proposed in Section 6 is available at https://github.com/MarceloJacinto/BSplineFit (accessed on 8 March 2022).

Acknowledgments: The authors of this work would like to express a deep gratitude to Nguyen Hung, Francisco Rego, João Quintas, and João Cruz for all the support provided during the preparation of the results presented in this work.

Conflicts of Interest: The authors declare no conflict of interest.

Appendix A. Proof of Preposition 1

Proof. Consider the candidate Lyapunov Function given by:

$$V_1(\mathbf{e}_p) = \frac{1}{2}(\mathbf{e}_p - \delta)^T(\mathbf{e}_p - \delta). \tag{A1}$$

Taking the first time derivative of (A1) and replacing in (17) and (14) leads to:

$$\begin{aligned}\dot{V}_1(\mathbf{e}_p) = (\mathbf{e}_p - \delta)^T\Bigg(&-S(r)(\mathbf{e}_p - \delta) + \Delta\mathbf{u} + \begin{bmatrix}0\\v\end{bmatrix}\\ &+\mathbf{v}_c - {}^B_U R(\psi)\frac{\partial \mathbf{p}_d(\gamma)}{\partial\gamma}(e_\gamma + v_d(\gamma,t))\Bigg).\end{aligned} \tag{A2}$$

Taking into account the properties of the skew-symmetric matrix S:

$$(\mathbf{e}_p - \delta)^T S(r)(\mathbf{e}_p - \delta) = 0. \tag{A3}$$

Replacing (20) and (21) in $\dot{V}_1$ yields:

$$\begin{aligned}\dot{V}_1(\mathbf{e}_p) &= (\mathbf{e}_p - \delta)^T\left(\Delta(\tilde{\mathbf{u}} + \mathbf{u}_d) + \begin{bmatrix}0\\v\end{bmatrix} + \tilde{\mathbf{v}}_c + \hat{\mathbf{v}}_c - {}^B_U R(\psi)\frac{\partial \mathbf{p}_d(\gamma)}{\partial\gamma}(e_\gamma + v_d(\gamma,t))\right)\\ &= -(\mathbf{e}_p - \delta)^T K_p\sigma(\mathbf{e}_p - \delta) - (\mathbf{e}_p - \delta)^T{}^B_U R(\psi)\frac{\partial \mathbf{p}_d(\gamma)}{\partial\gamma}e_\gamma + \Delta\tilde{\mathbf{u}} + \tilde{\mathbf{v}}_c.\end{aligned} \tag{A4}$$

By taking a backstepping approach, consider a second candidate Lyapunov function:

$$V_2(\mathbf{e}_p, e_\gamma) = V_1(\mathbf{e}_p) + \frac{1}{2}e_\gamma^2. \tag{A5}$$

Taking the first derivative and replacing in the control law (22) for the virtual target, we obtain:

$$
\begin{aligned}
\dot{V}_2(\mathbf{e}_p, e_\gamma) =& \dot{V}_1(\mathbf{e}_p) + e_\gamma(\ddot{\gamma} - \dot{v}_d(\gamma, t)) \\
=& -(\mathbf{e}_p - \delta)^T K_p \sigma(\mathbf{e}_p - \delta) - k_\gamma e_\gamma^2 + \Delta\tilde{\mathbf{u}} + \tilde{v}_c \\
\leq& -(1-\theta)(\mathbf{e}_p - \delta)^T K_p \sigma(\mathbf{e}_p - \delta) - \theta(\mathbf{e}_p - \delta)^T K_p \sigma(\mathbf{e}_p - \delta) \\
&- k_\gamma |e_\gamma|^2 + \|\mathbf{e}_p - \delta\| \|\Delta\tilde{\mathbf{u}} + \tilde{v}_c\|,
\end{aligned}
\tag{A6}
$$

where $0 < \theta < 1$. The term:

$$
\begin{aligned}
&-\theta(\mathbf{e}_p - \delta)^T K_p \sigma(\mathbf{e}_p - \delta) + \|\mathbf{e}_p - \delta\| \|\Delta\tilde{\mathbf{u}} + \tilde{v}_c\| \\
&= -\theta(\mathbf{e}_p - \delta)^T K_p \frac{\mathbf{e}_p - \delta}{\|\mathbf{e}_p - \delta\|} \sigma(\|\mathbf{e}_p - \delta\|) + \|\mathbf{e}_p - \delta\| \|\Delta\tilde{\mathbf{u}} + \tilde{v}_c\|,
\end{aligned}
\tag{A7}
$$

will be ≤ 0 if:

$$\theta \lambda_{min}(K_p) \sigma(\|\mathbf{e}_p - \delta\|) \geq \|\Delta\tilde{\mathbf{u}} + \tilde{v}_c\|, \tag{A8}$$

which, in turn, implies that:

$$\|\mathbf{e}_p - \delta\| \geq \sigma^{-1}\left(\frac{1}{\theta \lambda_{min}(K_p)} \|\Delta\tilde{\mathbf{u}} + \tilde{v}_c\|\right), \tag{A9}$$

and:

$$\dot{V}_2 \leq -(1-\theta)(\mathbf{e}_p - \delta)^T K_p \sigma(\mathbf{e}_p - \delta) - k_\gamma |e_\gamma|^2, \tag{A10}$$

as the right side of inequality (A9) can be made arbitrarily small through the choice of the gain matrix K_p. It follows directly from H. Khalil ([34], Theorem 4.19) that the controlled system is ISS. □

Appendix B. Proof of Preposition 2

Proof. Consider the candidate Lyapunov function given by:

$$V_1(\mathbf{e}_p) := \frac{1}{2}\mathbf{e}_p^T \mathbf{e}_p. \tag{A11}$$

Taking the first derivative of (A11) and replacing in (24)–(26), yields:

$$\dot{V}_1(\mathbf{e}_p) = -\mathbf{e}_p^T K_1 \mathbf{e}_p + \mathbf{e}_p^T\left(\mathbf{z} - \frac{\partial \mathbf{p}_d}{\partial \gamma} e_\gamma\right). \tag{A12}$$

With a view to applying backstepping techniques, consider:

$$V_2(\mathbf{e}_p, \mathbf{e}_v) := V_1(\mathbf{e}_p) + \frac{1}{2}\mathbf{z}^T \mathbf{z}. \tag{A13}$$

Replacing (26), (27), and (29) in $\dot{V}_2$ yields:

$$
\begin{aligned}
\dot{V}_2 =& -\mathbf{e}_p^T K_1 \mathbf{e}_p + \mathbf{e}_p^T \mathbf{z} - \mathbf{e}_p^T \frac{\partial \mathbf{p}_d}{\partial \gamma} e_\gamma \\
&+ \mathbf{z}^T\left(\mathbf{u} + \mathbf{d} - \mathbf{h}(\gamma)(e_\gamma + v_d(\gamma, t)) - \frac{\partial \mathbf{p}_d}{\partial \gamma} \dot{v}^{coord}(t) + K_1 \mathbf{e}_v - K_1 \frac{\partial \mathbf{p}_d}{\partial \gamma} e_\gamma\right).
\end{aligned}
\tag{A14}
$$

By replacing (35)–(37) in the Lyapunov time derivative, it follows that:

$$\dot{V}_2 = -\mathbf{e}_p^T K_1 \mathbf{e}_p - \mathbf{z}^T K_2 \mathbf{z} - \mathbf{e}_p^T \frac{\partial \mathbf{p}_d}{\partial \gamma} e_\gamma - \mathbf{z}^T\left(\mathbf{h}(\gamma) + K_1 \frac{\partial \mathbf{p}_d}{\partial \gamma}\right) e_\gamma + \mathbf{z}^T(\tilde{\mathbf{u}} + \tilde{\mathbf{d}}). \tag{A15}$$

Consider a third candidate Lyapunov, obtained by backstepping, defined as:

$$V_3 := V_2 + \frac{1}{2}e_\gamma^2. \tag{A16}$$

Taking its derivative, with respect to time, and replacing in (38), we obtain:

$$\dot{V}_3 = -\mathbf{e}_p^T K_1 \mathbf{e}_p - \mathbf{z}^T K_2 \mathbf{z} - k_\gamma e_\gamma^2 + \mathbf{z}^T(\tilde{\mathbf{u}} + \tilde{\mathbf{d}}). \tag{A17}$$

Consider one last backstepping that involves the construction of:

$$V_4 = V_3 + \frac{1}{2}\tilde{\mathbf{d}}^T K_d^{-1}\tilde{\mathbf{d}}. \tag{A18}$$

Taking its time derivative, and taking into consideration (33):

$$\begin{aligned}\dot{V}_4 &= -\mathbf{e}_p^T K_1 \mathbf{e}_p - \mathbf{z}^T K_2 \mathbf{z} - k_\gamma e_\gamma^2 + \underbrace{\tilde{\mathbf{d}}^T(\mathbf{z} - \text{Proj}(\mathbf{z}, \hat{\mathbf{d}}))}_{\leq 0} + \mathbf{z}^T\tilde{\mathbf{u}} \\ &\leq -W(\mathbf{e}_p, \mathbf{e}_v, e_\gamma) + \mathbf{z}^T\tilde{\mathbf{u}}.\end{aligned} \tag{A19}$$

Assuming that the quadrotor is equipped with a generic inner loop that is capable of keeping the tracking error $\tilde{\mathbf{u}}$ small and bounded, the right side of inequality (A19) can be made small enough such that the controlled system is stable. A more in-depth stability analysis can be conducted for the inner–outer loop control system, but this will be dependent directly on the type of inner loop adopted. This results from the fact that the desired accelerations $\mathbf{u_d}$ must be decoupled in a set of desired thrusts and attitudes for the quadrotor to track.

In order to simplify the designed control law $\mathbf{u}_d$, consider the final algebraic manipulation:

$$\begin{aligned}\mathbf{u}_d^\diamond &= -\hat{\mathbf{d}} + \mathbf{h}(\gamma)v_d(\gamma, t) + \frac{\partial \mathbf{p}_d}{\partial \gamma}\dot{v}^{coord}(t) - K_1\mathbf{e}_v - \mathbf{e}_p - K_2\mathbf{z} \\ &= -\hat{\mathbf{d}} + \mathbf{h}(\gamma)v_d(\gamma, t) + \frac{\partial \mathbf{p}_d}{\partial \gamma}\dot{v}^{coord}(t) - \mathbf{e}_v\underbrace{(K_1 + K_2)}_{K_v} - \mathbf{e}_p\underbrace{(I + K_1K_2)}_{K_p}.\end{aligned} \tag{A20}$$

□

Appendix C. Proof of Preposition 3

Proof. Consider that:

$$\mathbf{v_L}(\gamma) = v_L\mathbf{1} + \tilde{\mathbf{v}}_\mathbf{L}, \tag{A21}$$

where $\tilde{\mathbf{v}}_\mathbf{L}$ is a bounded and arbitrarily small term that accounts for a transient period in which the vehicles are on different sections of the path, with slightly different desired speed profiles. Replacing (39), the speed correction term proposed in (46), and (A21) in (42) yields:

$$\begin{aligned}\dot{\boldsymbol{\varepsilon}} &= L(\mathbf{v_L}(\gamma) - k_\varepsilon(\boldsymbol{\varepsilon} + \mathcal{A}\tilde{\gamma})) \\ &= v_L \cancelto{0}{L\mathbf{1}} + L\tilde{\mathbf{v}}_\mathbf{L} - k_\varepsilon L(\boldsymbol{\varepsilon} + \mathcal{A}\tilde{\gamma}) \\ &= -k_\varepsilon L(\boldsymbol{\varepsilon} + \mathbf{d}) \text{ with } \mathbf{d} = \frac{\tilde{\mathbf{v}}_\mathbf{L}}{k_\varepsilon} + \mathcal{A}\tilde{\gamma},\end{aligned} \tag{A22}$$

where $\mathbf{d}$ is a disturbance that results from combining the terms dependent on $\tilde{\mathbf{v}}_\mathbf{L}$ and $\tilde{\gamma}$. Consider the Laplacian matrix L, expressed in canonical Jordan form as:

$$L = V\Lambda V^{-1}, \tag{A23}$$

and the change of variables:

$$\bar{\boldsymbol{\varepsilon}} = V^{-1}\boldsymbol{\varepsilon}. \tag{A24}$$

Applying (A24) to (A22) yields:

$$\dot{\bar{\boldsymbol{\varepsilon}}} = -k_\varepsilon \Lambda(\bar{\boldsymbol{\varepsilon}} + \bar{\mathbf{d}}), \text{ with } \bar{\mathbf{d}} = V^{-1}\mathbf{d}. \tag{A25}$$

It is possible to decompose the above equality according to the notation:

$$\begin{bmatrix} \dot{\bar{\varepsilon}}_1 \\ \hline \dot{\bar{\boldsymbol{\varepsilon}}}_2 \end{bmatrix} = \begin{bmatrix} 0 \\ \hline -k_\varepsilon \Lambda_2(\bar{\boldsymbol{\varepsilon}}_2 + \bar{\mathbf{d}}_2) \end{bmatrix}, \tag{A26}$$

where the first half of the vector denotes the term that depends on the null eigenvalue of the Laplacian, while the second term is a vector that depends only on the positive eigenvalues of the Laplacian. Consider now the candidate Lyapunov function:

$$V_{\bar{\varepsilon}_2} = \frac{1}{2}\bar{\boldsymbol{\varepsilon}}_2^T \bar{\boldsymbol{\varepsilon}}_2, \tag{A27}$$

and its time derivative, given by:

$$\begin{aligned} \dot{V}_{\bar{\varepsilon}_2} &= -k_\varepsilon \bar{\boldsymbol{\varepsilon}}_2^T \Lambda_2(\bar{\boldsymbol{\varepsilon}}_2 + \bar{\mathbf{d}}_2) \\ &= -(1-\theta)k_\varepsilon \bar{\boldsymbol{\varepsilon}}_2^T \Lambda_2 \bar{\boldsymbol{\varepsilon}}_2 - \theta k_\varepsilon \bar{\boldsymbol{\varepsilon}}_2^T \Lambda_2 \bar{\boldsymbol{\varepsilon}}_2 - k_\varepsilon \bar{\boldsymbol{\varepsilon}}_2^T \Lambda_2 \bar{\mathbf{d}}_2, \end{aligned} \tag{A28}$$

where $0 < \theta < 1$. The term:

$$-\theta k_\varepsilon \bar{\boldsymbol{\varepsilon}}_2^T \Lambda_2 \bar{\boldsymbol{\varepsilon}}_2 - k_\varepsilon \bar{\boldsymbol{\varepsilon}}_2^T \Lambda_2 \bar{\mathbf{d}}_2, \tag{A29}$$

will be ≤ 0 if:

$$||\bar{\boldsymbol{\varepsilon}}_2|| \geq \frac{1}{\theta}||\bar{\mathbf{d}}_2||, \tag{A30}$$

and, therefore:

$$\dot{V}_{\bar{\varepsilon}_2} \leq -(1-\theta)k_\varepsilon \bar{\boldsymbol{\varepsilon}}_2^T \Lambda_2 \bar{\boldsymbol{\varepsilon}}_2. \tag{A31}$$

The term $||\tilde{\boldsymbol{\gamma}}||$ can be made arbitrarily small by controlling the gains that dictate the broadcasting scheme. Moreover, the term $\tilde{\mathbf{v}}_\mathbf{L}$ can be dominated by a proper choice k_ε. Hence, $||\mathbf{d}||$ can be made arbitrarily small and, thus, $||\bar{\mathbf{d}}_2||$ can be as well. It follows directly from H. Khalil ([34], Theorem 4.19) that the controlled system is ISS with respect to the error vector $\boldsymbol{\varepsilon}$ and the inputs $\tilde{\boldsymbol{\gamma}}$ and $\tilde{\mathbf{v}}_\mathbf{L}$. □

Appendix D. Computing the Regularisation Term Using Vectorial Notation

Consider the simplest unclamped uniform cubic B-spline with only one segment, such that $\gamma \in [0,1]$ and is described by (6). Then, its first derivative $\mathbf{C}'(\gamma)$ is given by:

$$\frac{\partial C^{x/y}}{\partial \gamma}(\gamma) = \underbrace{\underbrace{\begin{bmatrix} \gamma^2 & \gamma & 1 & 0 \end{bmatrix}}_{\mathbf{T}(\gamma)} \underbrace{\frac{1}{6}\begin{bmatrix} 3 & 0 & 0 & 0 \\ 0 & 2 & 0 & 0 \\ 0 & 0 & 1 & 0 \\ 0 & 0 & 0 & 0 \end{bmatrix} \begin{bmatrix} -1 & 3 & -3 & 1 \\ 3 & -6 & 3 & 0 \\ -3 & 0 & 3 & 0 \\ 1 & 4 & 1 & 0 \end{bmatrix}}_{M}}_{B'(\gamma)} \underbrace{\begin{bmatrix} P_0^{x/y} \\ P_1^{x/y} \\ P_2^{x/y} \\ P_3^{x/y} \end{bmatrix}}_{\mathbf{P}}. \tag{A32}$$

Therefore, the term $\int_\gamma \|\mathbf{C}'(\gamma)\|^2 d\gamma$ is computed according to:

$$\begin{aligned}\int_\gamma \|\mathbf{C}'(\gamma)\|^2 d\gamma &= \int_\gamma (B'(\gamma)\mathbf{P})^T(B'(\gamma)\mathbf{P})d\gamma \\ &= \int_0^1 \mathbf{P}^T B'(\gamma)^T B'(\gamma)\mathbf{P} d\gamma \\ &= \mathbf{P}^T M^T \left[\int_0^1 \mathbf{T}(\gamma)^T \mathbf{T}(\gamma) d\gamma\right] M\mathbf{P}.\end{aligned} \tag{A33}$$

Further, note that:

$$\mathbf{T}(\gamma)^T\mathbf{T}(\gamma) = \begin{bmatrix}\gamma^4 & \gamma^3 & \gamma^2 & 0\\ \gamma^3 & \gamma^2 & \gamma & 0\\ \gamma^2 & \gamma & 1 & 0\\ 0 & 0 & 0 & 0\end{bmatrix} \tag{A34}$$

and, as a consequence:

$$\int_0^1 \mathbf{T}(\gamma)^T\mathbf{T}(\gamma)d\gamma = Q = \begin{bmatrix}1/5 & 1/4 & 1/3 & 0\\ 1/4 & 1/3 & 1/2 & 0\\ 1/3 & 1/2 & 1 & 0\\ 0 & 0 & 0 & 0.\end{bmatrix} \tag{A35}$$

Hence, for the simplest case of a single B-spline segment, it is known that:

$$\int_\gamma \|\mathbf{C}'(\gamma)\|^2 d\gamma = \mathbf{P}^T \underbrace{M^T Q M}_{R_1} \mathbf{P}. \tag{A36}$$

The easiest way to extend this technique to a B-spline with n segments is to consider the modified vector $\mathbf{T}(\gamma) = [(\gamma - i)^2, (\gamma - i), 1, 0]^T$, where $i = \lfloor \gamma \rfloor$, according to the notation introduced in Section 2.3. Then, since $(\gamma - i) \in [0, 1]$, one can compute individually, for each segment, intermediate matrices R_1^i, according to (A36). Due to the locality property of B-splines, it is possible to "stack" these intermediate matrices to form the final matrix R_1. An analogous rationale can be applied to compute R_2.

Appendix E. Controller Gains Adopted

The controller gains used to obtain the results in Section 8 are presented in Table A1.

Table A1. Controller and path planning gains.

	Currents Observer (ASV)		Projection Operator (Quadrotor)
k_1	2.0	K_d	$diag(0.5, 0.5, 0.2)$
k_2	0.2	ς and β	10.0
	Path Following (ASV)		**Path Following (Quadrotor)**
K_p	$diag(0.5, 0.5)$	K_p	$diag(5.5, 5.5, 5.5)$
δ	−1.0	K_d	$diag(4.5, 4.5, 4.0)$
k_γ	0.5	k_γ	0.5
	Cooperative Path Following		**Path Planning**
k_ε	1.0	N_J	0.6m
c	0.001	$1/\rho$	4.0
b	5.0	λ	0.05
α	1.0	β	0.01

References

1. Casbeer, D.W.; Kingston, D.B.; Beard, R.W.; McLain, T.W. Cooperative forest fire surveillance using a team of small unmanned air vehicles. *Int. J. Syst. Sci.* **2006**, *37*, 351–360. [CrossRef]
2. Fahad, M.; Saul, N.; Guo, Y.; Bingham, B. Robotic simulation of dynamic plume tracking by Unmanned Surface Vessels. In Proceedings of the 2015 IEEE International Conference on Robotics and Automation (ICRA), Seattle, WA, USA, 26–30 May 2015; pp. 2654–2659. [CrossRef]
3. Pettersson, L.; Pozdnyakov, D. *Monitoring of Harmful Algal Blooms*, 1st ed.; Springer Science & Business Media: Berlin, Germany, 2013; p. 309. [CrossRef]
4. Sukhinov, A.; Chistyakov, A.; Nikitina, A.; Semenyakina, A.; Korovin, I.; Schaefer, G. Modelling of oil spill spread. In Proceedings of the 2016 5th International Conference on Informatics, Electronics and Vision (ICIEV), Dhaka, Bangladesh, 13–14 May 2016; pp. 1134–1139. [CrossRef]
5. Li, S.; Guo, Y.; Bingham, B. Multi-robot cooperative control for monitoring and tracking dynamic plumes. In Proceedings of the 2014 IEEE International Conference on Robotics and Automation (ICRA), Hong Kong, China, 31 May–7 June 2014; pp. 67–73. [CrossRef]
6. Clark, J.; Fierro, R. Cooperative hybrid control of robotic sensors for perimeter detection and tracking. In Proceedings of the 2005, American Control Conference, Portland, OR, USA, 8–10 June 2005; Volume 5, pp. 3500–3505. [CrossRef]
7. Saldaña, D.; Assunção, R.; Hsieh, M.A.; Campos, M.F.M.; Kumar, V. Cooperative prediction of time-varying boundaries with a team of robots. In Proceedings of the 2017 International Symposium on Multi-Robot and Multi-Agent Systems (MRS), Los Angeles, CA, USA, 4–5 December 2017; pp. 9–16. [CrossRef]
8. Piegl, L.; Tiller, W. *The Nurbs Book*, 1 ed.; Springer Science & Business Media: Berlin, Germany, 1995; p. 646. [CrossRef]
9. International Petroleum Industry Environmental Conservation Association (IPIECA). Dispersants and Their Role in Oil Spill Response. Volume 5. Available online: https://www.amn.pt/DCPM/Documents/DispersantsII.pdf (accessed on 8 March 2022).
10. Aguiar, A.P.; Hespanha, J.P. Trajectory-Tracking and Path-Following of Underactuated Autonomous Vehicles with Parametric Modeling Uncertainty. *IEEE Trans. Autom. Control.* **2007**, *52*, 1362–1379. [CrossRef]
11. Vanni, F.; Aguiar, A.P.; Pascoal, A.M. Cooperative path-following of underactuated autonomous marine vehicles with logic-based communication. In Proceedings of the 2nd IFAC Workshop on Navigation, Guidance and Control of Underwater Vehicles, Killaloe, Ireland, 8–10 April 2008; Volume 41, pp. 107–112. [CrossRef]
12. Aguiar, A.P.; Ghabcheloo, R.; Pascoal, A.M.; Silvestre, C. Coordinated Path-Following Control of Multiple Autonomous Underwater Vehicles. In Proceedings of the Seventeenth International Offshore and Polar Engineering Conference, Lisbon, Portugal, 1–6 July 2007; ISOPE-I-07-006.
13. Hung, N.T.; Rego, F.C.; Pascoal, A.M. Event-Triggered Communications for the Synchronization of Nonlinear Multi Agent Systems on Weight-Balanced Digraphs. In Proceedings of the 2019 18th European Control Conference (ECC), Naples, Italy, 25–28 June 2019; pp. 2713–2718. [CrossRef]
14. Abreu, P.; Botelho, J.; Góis, P.; Pascoal, A.; Ribeiro, J.; Ribeiro, M.; Rufino, M.; Sebastião, L.; Silva, H. The MEDUSA class of autonomous marine vehicles and their role in EU projects. In Proceedings of the OCEANS Shanghai, Shanghai, China, 10–13 April 2016; pp. 1–10. [CrossRef]
15. Furrer, F.; Burri, M.; Achtelik, M.; Siegwart, R. *Robot Operating System (ROS): The Complete Reference (Volume 1)*; Springer International Publishing: Cham, Switzerland, 2016; pp. 595–625; chapter RotorS—A Modular Gazebo MAV Simulator Framework. [CrossRef]
16. Manhães, M.; Scherer, S.A.; Voss, M.; Douat, L.R.; Rauschenbach, T. UUV Simulator: A Gazebo-based package for underwater intervention and multi-robot simulation. In Proceedings of the OCEANS 2016 MTS/IEEE Monterey, Monterey, CA, USA, 19–23 September 2016; pp. 1–8. [CrossRef]
17. Dey, T.K. Course 784 Notes—The Ohio State University—Lecture 7: Matrix Form for B-Spline Curves. Available online: https://web.cse.ohio-state.edu/~dey.8/course/784/note7.pdf (accessed on 16 January 2022).
18. Pascoal, A.; Kaminer, I.; Oliveira, P. Navigation system design using time-varying complementary filters. *IEEE Trans. Aerosp. Electron. Syst.* **2000**, *36*, 1099–1114. [CrossRef]
19. Sanches, G. Sensor-Based Formation Control of Autonomous Marine Robots. Master's Thesis, Instituto Superior Técnico, Lisbon, Portugal, 2015.
20. Xie, W.; Cabecinhas, D.; Cunha, R.; Silvestre, C. Robust Motion Control of an Underactuated Hovercraft. *IEEE Trans. Control. Syst. Technol.* **2019**, *27*, 2195–2208. [CrossRef]
21. Cabecinhas, D.; Cunha, R.; Silvestre, C. A nonlinear quadrotor trajectory tracking controller with disturbance rejection. *Control. Eng. Pract.* **2014**, *26*, 1–10. [CrossRef]
22. Cai, Z.; de Queiroz, M.; Dawson, D. A sufficiently smooth projection operator. *IEEE Trans. Autom. Control.* **2006**, *51*, 135–139. [CrossRef]
23. Aguiar, A.P.; Pascoal, A.M. Coordinated path-following control for nonlinear systems with logic-based communication. In Proceedings of the 2007 46th IEEE Conference on Decision and Control, New Orleans, LA, USA, 12–14 December 2007; pp. 1473–1479. [CrossRef]
24. Hung, N.T.; Pascoal, A.M. Consensus/synchronisation of networked nonlinear multiple agent systems with event-triggered communications. *Int. J. Control.* **2020**, 1–10. [CrossRef]

25. Odonkor, P.; Ball, Z.; Chowdhury, S. Distributed operation of collaborating unmanned aerial vehicles for time-sensitive oil spill mapping. *Swarm Evol. Comput.* **2019**, *46*, 52–68. [CrossRef]
26. Szeliski, R. *Computer Vision: Algorithms and Applications*; Springer Science & Business Media: Berlin, Germany, 2011; Volume 5. [CrossRef]
27. Liu, Y.; Yang, H.; Wang, W. Reconstructing B-spline Curves from Point Clouds—A Tangential Flow Approach Using Least Squares Minimization. In Proceedings of the International Conference on Shape Modeling and Applications 2005 (SMI' 05), Cambridge, MA, USA, 13–17 June 2005; pp. 4–12. [CrossRef]
28. Lee, I.K. Curve reconstruction from unorganized points. *Comput. Aided Geom. Des.* **2000**, *17*, 161–177. [CrossRef]
29. Bentley, J.L. Multidimensional Binary Search Trees Used for Associative Searching. *Commun. ACM* **1975**, *18*, 509–517. [CrossRef]
30. Liu, M.; Huang, S.; Dissanayake, G.; Kodagoda, S. Towards a consistent SLAM algorithm using B-Splines to represent environments. In Proceedings of the 2010 IEEE/RSJ International Conference on Intelligent Robots and Systems, Taipei, Taiwan, 18–22 October 2010; pp. 2065–2070. [CrossRef]
31. Xie, W.; Cabecinhas, D.; Cunha, R.; Silvestre, C. Cooperative Path Following Control of Multiple Quadcopters with Unknown External Disturbances. *IEEE Trans. Syst. Man Cybern. Syst.* **2020**, *52*, 667–679. [CrossRef]
32. Bradski, G. The OpenCV library. *Dr. Dobb's J. Softw. Tools* **2000**, *120*, 122–125.
33. Virtanen, P.; Gommers, R.; Oliphant, T.E.; Haberland, M.; Reddy, T.; Cournapeau, D.; Burovski, E.; Peterson, P.; Weckesser, W.; Bright, J.; et al. SciPy 1.0: Fundamental algorithms for scientific computing in Python. *Nat. Methods* **2020**, *17*, 261–272. [CrossRef] [PubMed]
34. Khalil, H. *Nonlinear Systems*, 3rd ed.; Always Learning; Pearson Education Limited: New York, NY, USA, 2013.

sensors

MDPI

Article

Visual Servoed Autonomous Landing of an UAV on a Catamaran in a Marine Environment

Andrea Delbene [1,*], Marco Baglietto [1,2] and Enrico Simetti [1,2]

1 Department of Informatics, Bioengineering, Robotics and Systems Engineering (DIBRIS), Università degli Studi di Genova, Via all'Opera Pia 13, 16145 Genoa, Italy; marco.baglietto@unige.it (M.B.); enrico.simetti@unige.it (E.S.)
2 Interuniversity Research Center on Integrated Systems for the Marine Environment, Via all'Opera Pia 13, 16145 Genoa, Italy
* Correspondence: andrea.delbene@edu.unige.it

Abstract: This paper introduces a procedure for autonomous landing of a quadrotor on an unmanned surface vehicle in a marine environment. The relative pose and velocity of the vehicle with respect to the quadrotor are estimated using a combination of data coming from a vision system, which recognizes a set of AprilTags located on the vehicle itself, and an ultrasonic sensor, to achieve further robustness during the final landing phase. The considered software and hardware architecture is provided, and the details about the landing procedure are presented. Software-in-the-loop tests were performed as a validation step for the proposed algorithms; to recreate realistic conditions, the movements of the landing platform have been replicated from data of a test in a real marine environment. In order to provide further proof of the reliability of the vision system, a video sequence from a manual landing of a quadrotor on the surface vehicle in a real marine environment has been processed, and the results are presented.

Keywords: UAV; ASV; splashproof quadrotor; vision system; state machine; autonomous landing; marine robotics; aerial robotics

Citation: Delbene, A.; Baglietto, M.; Simetti, E. Visual Servoed Autonomous Landing of an UAV on a Catamaran in a Marine Environment. *Sensors* **2022**, *22*, 3544. https://doi.org/10.3390/s22093544

Academic Editors: Reza Ghabcheloo and Antonio M. Pascoal

Received: 31 March 2022
Accepted: 4 May 2022
Published: 6 May 2022

1. Introduction

During the last years, unmanned aerial vehicles (UAVs) have been used in a wide variety of applications, such as in agriculture [1], civil protection [2], infrastructure, inspection and maintenance [3], and light shows [4]. Their uses are not limited to the Earth's surface, but extend to space environments as well, where they could be used for ad hoc missions [5]. When considering applications in a marine environment, different types of robots can be involved. Generally, a set of underwater and surface vehicles allow the execution of complex tasks, such as monitoring wide areas or cooperative collaboration for a mutual goal. For instance, in [6], an autonomous robotic team composed of underwater and surface vehicles was considered for geotechnical survey purposes. By considering aerial agents also, a wider variety of missions can be designed, such as the protection and security of marine areas, or humanitarian search and rescue activities [7]. The landing of aerial agents requires ad hoc procedures [8], and when working in complex conditions, this is often performed by a human operator. In fact, in marine applications, the sea conditions could alter the pose of the landing target, and could determine the success or the failure of the landing procedure itself. Therefore, in fully autonomous missions, the landing maneuvers of an aerial agent must be robust to difficulties and reliable.

On the basis of the previous works [9,10], we aimed to provide an efficient, reliable, and modular solution to autonomously land a quadrotor on a catamaran in a marine environment.

1.1. Related Work

Performing an autonomous landing procedure on a platform is a complex task that requires several steps to be achieved. In outdoor scenarios, global navigation satellite system (GNSS) receivers usually provide the positions of the quadrotor and the catamaran. Still, the data coming from GNSS are not sufficient to perform an autonomous landing, due to their inaccuracy. Even if properly filtered [11], the accuracy and precision are not enough for such complex and precise maneuvers. Additionally, the catamaran is subjected to unpredictable oscillatory dynamics caused by sea behavior and weather conditions to which the quadrotor must be able to react. For these reasons, alternative approaches have to be taken into consideration. For instance, a video system would allow increasing the reliability and the performance of the landing procedure. The GNSS data are used by the quadrotor to move close to the position of the catamaran, and from there, a vision system gives the quadrotor the relative pose of the platform on the catamaran. Computer vision algorithms are extremely useful to close the loop when the landing platform is in an uncertain position or it is moving [12], since they allow directly estimating the horizontal and vertical tracking errors with respect to the target point, instead of providing its coordinates in an absolute frame. An interesting and efficient solution was proposed in [13], where an extended Kalman filter was developed to combine data coming from different sensors (inertial navigation, GNSS receiver, and visual sensor) to build a navigation system and perform a landing procedure. In [14], a solution composed of several LEDs and an "H" sign placed on the landing platform was proposed: the LEDs give the possibility to the unmanned aerial vehicle (UAV) of recognizing the platform from high altitudes by using an infrared camera, and the "H" sign helps the estimation of the center of the platform itself when the quadrotor is closer. Instead, in [15], helipads composed of different geometric shapes (a cross, a circle, and a square) were proposed, to test the designed vision-based autonomous landing algorithm. Experimental results have been achieved outside of the marine environment, with a mobile robot carrying a landing platform moving on the ground. Another solution for the estimation of the relative pose between the quadrotor and the landing platform is the one proposed in [16], where a specific marker composed of a series of concentric circles would allow the detection of the platform from close by.

A similar methodology is the one presented in [17], where a landing platform composed of several AprilTags [18,19] with different dimensions is introduced: the larger tags permit the detection from higher altitudes, and the smaller ones from lower altitudes. This allows the quadrotor to constantly track the landing platform while decreasing its altitude during the landing procedure. The choice of using AprilTags is mainly related to their versatility and robustness [20].

1.2. Contributions

The innovation of this paper with respect to the state of art is mainly the development of a set of software packages able to perform an autonomous landing procedure in a sea environment, where the catamaran is subject to wave-induced oscillations. The landing procedure is tackled by implementing a set of strategies, such as a preliminary positioning of the drone, platform searching, horizontal tracking to keep it aligned, and vertical compensation with respect to the landing platform. The behavior of the quadrotor during the whole landing procedure is handled by a finite state machine: a set of states and conditions that describe the actions the quadrotor has to perform, depending on the data coming from different sensors. An improved landing platform composed of more tags with respect to the past solution [9] has been designed. Simulations of autonomous landing have been performed in an environment composed by several tools, such as Gazebo (more info at: http://gazebosim.org/ last access: 30 March 2022), ROS2 (more info at: https://docs.ros.org/en/foxy/index.html, last access: 30 March 2022), and PX4 (more info at: https://px4.io/, last access: 30 March 2022), where the pose of the catamaran was replicated from data coming from sea tests involving only the catamaran itself so that the motion of the landing platform was realistic.

The proposed software architecture allows both the validation of the considered methodology via software-in-the-loop simulations and the integration of most of those components in the real hardware, as a preparation for tests in a real environment. To further validate the reliability and robustness of the onboard vision system, an onboard video captured from a manual flight landing of the quadrotor on the catamaran has been processed offline using the adopted vision system.

Therefore, with respect to the previous works [9,10], the contributions of the present manuscript are:

1. A new software architecture powered by the ROS2 middleware and designed specifically to be modular for both simulation tests and outdoor tests in a real marine environment;
2. Design and implementation of an improved landing state machine, with the addition of a new state and the introduction of a new procedure to synchronize the position of the quadrotor with the catamaran before the landing approach;
3. Simulations in a software-in-the-loop environment of a safe landing on a landing platform, whose movements were replicated from the telemetry of the catamaran recorded during outdoor tests in a marine environment;
4. Realization and integration on the catamaran of a new landing platform, with more tags to gain better robustness during the landing procedure;
5. Validation of the vision system using the recordings of a manual landing on the catamaran in a marine environment.

This paper is organized as follows: In Section 2, an overview of the whole system is proposed. In Section 3, the methodology of the proposed landing procedure is detailed, and in Section 4 the developed software/firmware architecture is presented. In Section 5, the main results obtained in flight emulation tests are shown. Finally, some conclusions are given in Section 6.

2. System Overview

The considered experimental system is composed mainly of two different agents, each one with unique characteristics and features.

2.1. Catamaran

The ULISSE autonomous surface vehicle (ASV), developed by the interuniversity research center for Integrated Systems for Marine Environment (ISME, University of Genova node), is a 3 m long and 1.8 m wide catamaran, constructed in fiberglass (see Figure 1). It was designed as a modular vehicle for various applications. When used for marine geotechnical surveys [21] or acting as an intelligent buoy for underwater vehicles, it carries a deck with an underwater mast with acoustic sensors. When used as a means to extend the action range of the aerial drones, the catamaran is equipped with a dedicated landing platform (as in Figure 1). In each hull of the catamaran, a compartment hosts batteries (around 3.2 kWh of energy each), the hardware architecture where the control software runs the *ROS2* middleware, and a wide range of sensors (GNSS receiver, gyroscopes, accelerometers, and a compass sensor) to collect ego-motion measurements. The catamaran is provided with a roll-bar where the GNSS antenna is located, along with a 5 GHz one for Wi-Fi communication. The catamaran is propelled by two Torqeedo Cruise 2R electric thrusters, with electrical power of 2 kW each, which offer high maneuverability of the vessel even at low speeds, making it very agile in cluttered areas.

Figure 1. The ULISSE catamaran, equipped with the landing platform and the splash-proof quadrotor, deployed in one of the tests at sea.

2.2. Quadrotor

The chosen quadrotor model is the SwellPro Splash Drone 3 (more info at: https://swellpro.com/, last access: 30 March 2022), a drone that provides an external waterproof structure specifically designed for marine applications, along with various internal hardware components that allow performing manual flights. For the purpose of the realization of the proposed strategy, these components were substituted to robotize the agent itself. In particular, a Raspberry Pi Model B+ (more info at: https://www.raspberrypi.com/, last access: 30 March 2022) was embedded, along with a Raspicam v2, to allow onboard computations, and a Pixracer (more info at: https://docs.px4.io/master/en/flight_controller/pixracer.html, last access: 30 March 2022) autopilot system containing several embedded sensors, such as an accelerometer, a magnetometer, a gyroscope, and a barometer. The autopilot receives the setpoints computed by the algorithm running on the Raspberry Pi, and translates them into pulse width modulation (PWM) signals for the single motors of the quadrotor. The quadrotor is also endowed with a GNSS receiver. An ultrasonic sensor was included, as it is essential during the landing procedure, and so was a payload release mechanism actuated by a servo motor, to enable the quadrotor to carry out delivery tasks in a marine environment.

3. Methodology

The proposed landing solution is composed of different modules. Each one is detailed hereafter.

3.1. Perception and Pose Estimation

The relative pose of the quadrotor with respect to the landing platform is estimated by an onboard vision system that processes the video stream coming from the Raspicam. The platform is equipped with a set of visually distinguishable tags, each one being different from the others and characterized by a unique ID. The adopted vision system, named AprilTag, is an open-source, robust, and well documented tool [18,19] that allows 3D position and orientation computation of the considered tags with respect to the camera [20]. The use of a single tag does not guarantee its identification during the whole landing procedure; hence, the landing platform was equipped with 13 unique AprilTags, following a similar configuration as [17].

The tags, as shown in Figure 2, were placed in such a way as to guarantee visibility from different distances and robustness in the landing phase. The AprilTag markers on the outer edges are large, and thus easily recognizable at higher altitudes. The smaller internal ones play a crucial role in the final instants of the landing maneuver when the quadrotor is closer to the platform, improving safety and reliability.

Figure 2. The set S_l of AprilTags of different sizes as printed on the landing platform.

The list of the detectable tags is represented by the set $S_l = \{1, \ldots, 13\}$. At each iteration of the vision system, the detected tags ID are stored in a subset $S_d \subseteq S_l$. For each detected tag ID $i \in S_d$, the vision system computes a transformation matrix ${}^c_iT$ that describes the position in the scene of the identified tag i with respect to the camera frame c. In order to compute the pose of the platform center with respect to the camera for each tag $i \in S_l$, a set of transformation matrices ${}^i_pT$—$i \in S_l$ describing the position of each tag with respect to the platform center—is calibrated and computed offline. Thus, a post-multiplication gives the needed transformation matrix:

$$ {}^c_pT_i = {}^c_iT \, {}^i_pT \, . \tag{1} $$

Theoretically, each detected tag gives equally correct information. However, to improve the quality of the estimation, these measures are merged and weighted by the areas (a_i) of each detected tag in the camera frame: $i \in S_d$. Thus, the weighted transformation matrix between the center of the platform and the camera on the quadrotor ${}^c_pT$ is obtained by:

$$ {}^c_pT = \frac{1}{w} \sum_{i \in S_d} {}^c_pT_i a_i \, , \tag{2} $$

where w is the normalization term defined as:

$$ w = \sum_{i \in S_d} a_i \, . \tag{3} $$

The reference error is then transformed in the inertial frame, taking into account the quadrotor's attitude (see Figure 3), and sent to the guidance controller, which generates the desired commands for the autopilot.

Still, depending on sea conditions, the vertical velocity of the landing platform could vary a lot, and the estimated vertical error using the vision system alone does not provide a reliable measure in the final instants of the landing phase. To increase robustness, an ultrasonic sensor pointing downward was installed on the quadrotor, providing distance information at a limited range. More precisely, the camera provides information at 20 fps (frames per second). The ultrasonic sensor provides information at 30 Hz, and at distances less than 0.75 m, provides more precise and reliable data. The distance data coming from the ultrasonic sensor are used to estimate the platform's vertical velocity via a basic

Kalman filter [9]. This information is merged with the estimated vertical error, as shown in Section 3.3.

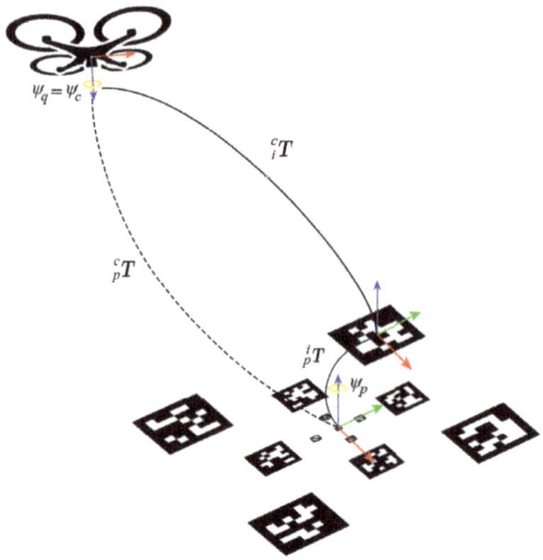

Figure 3. A representation of the different transformation matrices involved in the relative pose estimation between the UAV and the landing platform's center.

3.2. Horizontal Platform Tracking

One of the tasks the quadrotor has to perform during the landing procedure is horizontal tracking of the platform, reducing the estimated horizontal error provided by the camera. By defining the horizontal positions of the quadrotor and the platform as $p_{q,xy}$ and $p_{l,xy}$, respectively, the horizontal position error is $e_{p,xy} = p_{l,xy} - p_{q,xy}$. A measure of this error is taken from the onboard vision system. Thus, a PI regulator is designed to produce position setpoints $p^*_{q,xy}$:

$$p^*_{q,xy} = p_{q,xy} + K_P\, e_{p,xy} + K_I \int e_{p,xy}\, dt\,, \tag{4}$$

where K_P and K_I are the proportional and integral gains of the controller, respectively.

3.3. Vertical Platform Compensation

Once the quadrotor is at a certain distance from the landing platform, it needs to take care of the heave motions of the landing pad, induced by the waves. In this delicate phase, the altitude setpoints are generated to keep the relative velocity between quadrotor and catamaran to a specific value $v^{des}_{r,z}$. More in detail, a vertical target absolute velocity can be defined as:

$$v^{des}_{q,z} = v^{des}_{r,z} + v_{l,z}\,, \tag{5}$$

where $v_{l,z}$ is obtained by:

$$v_{l,z} = v_{q,z} - v_{r,z}\,, \tag{6}$$

and $v_{r,z}$ is the estimation of the relative vertical velocity obtained by the above mentioned Kalman filter. The vertical error velocity is defined as:

$$e_{v,z} = v^{des}_{q,z} - v_{q,z}\,. \tag{7}$$

Thus, the desired vertical velocity is a proportional scale of (7) by a gain K_1:

$$v^*_{q,z} = v_{q,z} + K_1\, e_{v,z}\,. \tag{8}$$

Finally, the altitude setpoints are generated as:

$$p^*_{q,z} = p_{q,z} + K_2\, v^*_{q,z}\,, \tag{9}$$

where $p_{q,z}$ is the current altitude of the quadrotor, and K_2 is a scale gain. Vision system data are not used at this stage, since the ultrasonic sensor gives information at a higher frequency.

3.4. Finite State Machine

The landing phase is described by a series of connected states, whose transitions are handled by a finite state machine. The behavior of the quadrotor is described by eight states: initialization, searching, tracking, hovering, descending, ascending, compensation, and landing; Figure 4 describes how the states are linked. The transitions among them are triggered by boolean algebra operations.

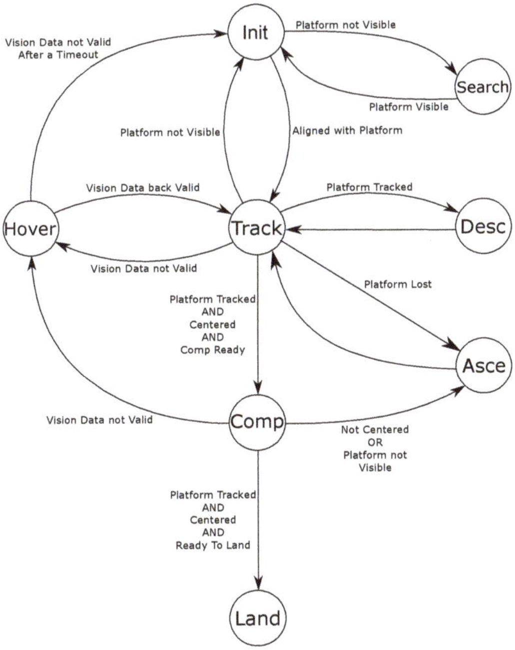

Figure 4. The diagram of the proposed finite state machine for the autonomous landing.

Initially, the quadrotor performs the rendezvous with the catamaran. The latter sends a stream of its GNSS position to the former, which flies to reach it. The catamaran's GNSS position is exploited only in the initial phase, as it can be imprecise, and would not guarantee a robust and reliable landing, especially in cases of signal loss. When the quadrotor reaches the area described by the received GNSS coordinates, the finite state machine starts, whose states are detailed in the following subsections.

3.4.1. Initialization

This is the entry point of the procedure. In this state, the quadrotor reaches the starting altitude and starts to look for the landing platform. If the landing pad is not detected, the finite state machine changes the state to *searching*. Otherwise, the quadrotor places itself in a specific position and orientation with respect to the catamaran. The basic idea is to prevent landing from a position where the quadrotor could hit the roll-bar located on the stern side of the catamaran (see Figure 1).

For this purpose, as seen in Figure 5, the quadrotor is placed in front of the landing platform, at a certain distance from the center. In detail, the desired positions $p^*_{q,x}$ and $p^*_{q,y}$ are computed directly by:

$$p^*_{q,x} = p_{q,x} + e_{p,x} + R\sin(\psi_l) \quad (10)$$
$$p^*_{q,y} = p_{q,y} + e_{p,y} + R\cos(\psi_l)\,, \quad (11)$$

where $e_{p,x}$ and $e_{p,y}$ are the estimated horizontal error components (see Section 3.2), R is the desired fixed distance the quadrotor has to keep from the platform, and ψ_l is the catamaran's yaw angle. To prevent further complications in the quadrotor's movements, its yaw is kept constant for the whole landing phase.

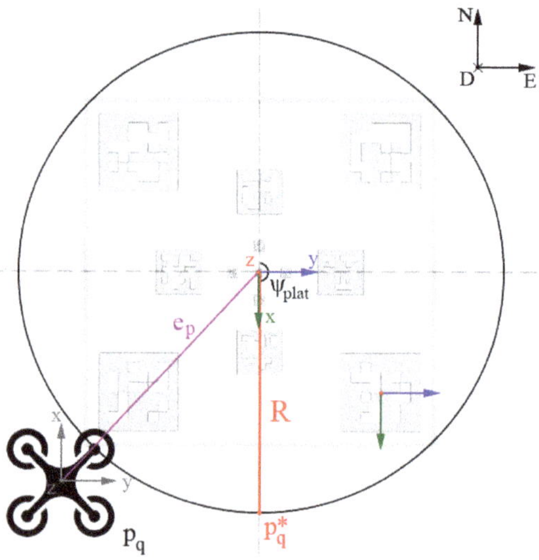

Figure 5. The generation of the initial relative position p^*_q during the initialization phase. The quadrotor needs to place itself in front of the catamaran, and it does that by moving around the platform and placing at a certain distance from it.

3.4.2. Searching

If the quadrotor has no visual information about the position of the landing platform, it enters a state where it searches for it. To this end, the quadrotor reaches a predefined altitude and flies in circles increasing large in radius. In particular, by taking as the center point the quadrotor's position at the initialization time of the searching phase $p_{q,x}(t_0)$, $p_{q,y}(t_0)$, the desired position of the quadrotor is defined by:

$$p^*_{q,x}(t) = p_{q,x}(t_0) + R\cos\left(\frac{v_s}{R}(t-t_0)\right) \quad (12)$$
$$p^*_{q,y}(t) = p_{q,y}(t_0) + R\sin\left(\frac{v_s}{R}(t-t_0)\right), \quad (13)$$

where R is the desired radius of the first circle and v_s is the desired linear velocity to be tracked during this phase. The searching continues until the following condition is verified:

$$(t-t_0) < \frac{2\pi R}{v_s}\,. \quad (14)$$

When this condition is no longer true, the parameters are updated: R is increased by 0.5 m so that the quadrotor inspects a new area while overlapping a part of the previous one, t_0 is set to the current value of t ($t_0 = t$). By doing that, the condition returns true, and at the next iteration, the quadrotor starts a new circle, but with an increased radius. The structure of the second term of (14) makes sure the quadrotor starts a new circle in the exact instant when it finishes the first one.

Once the quadrotor detects the platform, the landing procedure begins, and the quadrotor goes back to the initialization state. This process guarantees the success of the action even if only the approximate position of the platform is known. If the quadrotor loses the platform when already landing, the searching state takes also into account the last computed vision error, so that the quadrotor centers itself in the last known position of the catamaran to restart the search.

3.4.3. Tracking

When the quadrotor is correctly positioned with respect to the catamaran, the tracking state is triggered, handling the reduction of the vertical and horizontal error between the two agents. The descent of the quadrotor toward the catamaran becomes slanted; in particular, at the time instant t_h indicating the moment this state starts, a slope between its current altitude and an altitude point z_{max} (ideally, the maximum distance from the platform that allows the horizontal tracking of the smaller tags) is chosen. The z reference is computed using:

$$z(t_h) = m(t_h)e_{p,x}(t_h)\ , \tag{15}$$

where

$$m(t_h) = -\frac{p_{q,z}(t_h) - z_{max}}{R\sin(\psi_l(t_h))}\ . \tag{16}$$

3.4.4. Hovering

This state handles the case when the vision data coming from the camera have not been updated for more than a second. In that case, the quadrotor is asked to keep its position for a certain period until the vision system is back online, sending again the required data. Then, once the feedback is restored, the quadrotor will resume its mission.

3.4.5. Descending

This state gets triggered if the quadrotor is tracking the landing platform under a certain threshold and for a number of consecutive frames, but its current altitude is over the ideal horizontal maximum tracking altitude. The altitude waypoints are autonomously adjusted by being decreased by 0.1 m at each iteration.

3.4.6. Ascending

This state gets triggered if the quadrotor has no visual contact with the landing platform for a number of consecutive frames, and its current altitude is below the ideal horizontal minimum tracking altitude. The altitude waypoints are autonomously adjusted by being increased by 0.1 m at each iteration.

3.4.7. Compensation

When the quadrotor is under a certain vertical distance from the landing platform and it is centered with respect to it, the horizontal tracking (Section 3.2) and the vertical compensation (Section 3.3) tasks generate the position setpoints. In this state, to compensate for the catamaran's oscillations, the vertical position setpoints are generated using the estimations coming from the Kalman filter and the equations reported in Section 3.3. The measures coming from the ultrasonic sensor are the only ones used, due to their higher-frequency updating.

3.4.8. Landing

When the quadrotor is sufficiently close to the platform and the relative velocity between the two agents is under a certain threshold, the finite state machine enters the landing state, where the motors of the quadrotors are shut down, allowing it to land on the catamaran. Due to the criticality of this decision, the altitude and velocity thresholds have been set to very low values, to prevent crashes or mishaps.

4. Software Architecture

To process information coming from the various sensors and generate the setpoints necessary to fulfill the assigned tasks, the quadrotor needs to be equipped with a set of software tools communicating with each other, which were chosen strictly due to the hardware components that were installed on the quadrotor itself, presented in Section 2.2. An overview of the considered software architecture is presented in Figure 6.

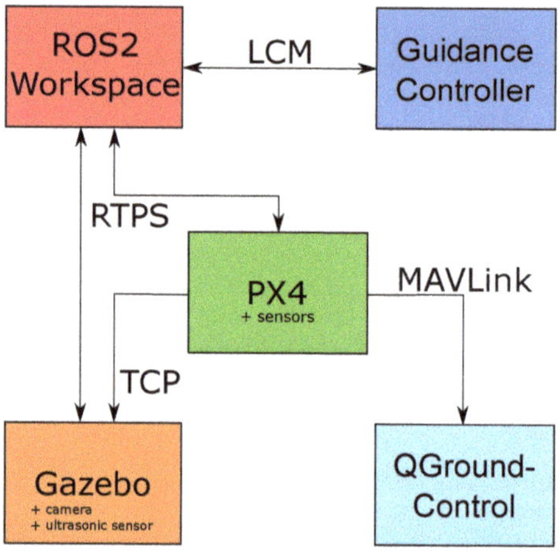

Figure 6. The main blocks composing the software architecture in the ROS2 simulation environment.

4.1. Gazebo

Gazebo is software that makes it possible to simulate accurately and efficiently the dynamic behavior of populations of robots in complex environments. It offers an environment where the dynamics of the quarotor are approximately simulated. This was a necessary tool to test the developed algorithms, as an intermediate preparation for outdoor tests on real hardware.

4.2. PX4

PX4 is open-source flight control software for quadrotors and other unmanned vehicles. In this project, it is mainly used as a means to translate the pose setpoints coming from ROS2 nodes into PWM signals that are directly injected into the motors.

4.3. ROS2

ROS2 is a real-time version of the more common ROS (Robot Operating System), a set of open-access software libraries and tools for building robot applications. Its tools allow structuring the software in different modules interacting with each other through well defined message-based interfaces. In our study, ROS2 was used to create a workspace composed by a set of different modules, each one implementing a different feature: from packages collecting and processing the data coming from the camera and the ultrasonic sensor, to the ones handling the information transfer between the PX4 and the guidance controller. When performing a simulation, the sensors were replaced by their software counterparts implemented in Gazebo, and ROS2 nodes retrieved the data via RTPS (real-time publish–subscribe; see Section 4.6). Other software packages included the Kalman filter and the vision system described in Section 3.1. The vision system node is asynchronous and processes the compressed images coming from the camera whenever they are available (20 fps), requiring an average computational time of ∼7.0 ms. The measured maximum computational time was 29.86 ms, and the minimum was 1.56 ms; it depends on the number of visible markers. The node implementing the Kalman filter instead is synchronous with the ultrasonic sensor's refresh rate of 30 ms, processing the coming data in an average of ∼3.0 ms; 4.7 ms maximally and 2.0 ms minimally.

4.4. Guidance Controller

The guidance controller is a software package that implements autonomous flight actions for the quadrotor. It is a modular open-source architecture written in C++, composed of several modules that communicate internally and externally via a communication protocol middleware named Lightweight Communications and Marshalling (LCM) [22]. The proposed architecture includes several macrotasks the quadrotor can perform—some of them simple, such as take-off from a point, navigation to point, and landing on a specific point; and others more complicated, designed for marine missions. The latter include searching and rescuing a shipwrecked person, landing on a platform, and criticality and failure handling. However, this paper mainly focuses on the autonomous landing on the catamaran.

4.5. QGroundControl

QGC (QGroundControl, more info at: http://qgroundcontrol.com/, last access: 30 March 2022) is software providing a graphical interface useful for monitoring a quadrotor's status, full flight control, mission planning, and tuning of an autopilot system's parameters.

4.6. RTPS

PX4-Fast RTPS Bridge (or more commonly RTPS, more info at: https://dev.px4.io/v1.11_noredirect/en/middleware/micrortps.html , last access: 30 March 2022) is a communication protocol that adds a real-time publish–subscribe (RTPS) interface to the PX4 Autopilot system, enabling the exchange messages between the various internal PX4 Autopilot components and ROS2 applications in real-time.

5. Emulation Results

5.1. Software-in-the-Loop Simulation

This section describes the results obtained with software-in-the-loop (SITL) tests performed with the help of the experimental architecture introduced in Section 4. To test the proposed methodology under realistic conditions, particularly as concerns the motion of the landing pad, we proceeded as follows. We first recorded to a log file of the telemetry (GNSS position and attitude) of the ULISSE catamaran executing a rendezvous with the quadrotor. The catamaran was directed toward a point and then instructed to hold its position. To do so, the catamaran positioned itself against the estimated direction of the current. However, note that the effects of waves and the fact that the catamaran is nonholonomic still induced a small lateral drift.

Then, we replicated the movement of the catamaran from the log file within the Gazebo environment and carried out several simulations of the quadrotor landing on it. Notice that, as the heave motion is not measurable in the real ULISSE ASV, we generated simulated motions using the Pierson–Moskowitz spectrum. Three different log files replicating the behavior of the catamaran were used. Over than 30 simulations have been performed; due to the similarity of the data among the logs, the catamaran's behavior was replicated from the same log file for the majority of the tests. In a few of them, it happened that the quadrotor lost visual contact with the tags on the platform; in these cases, the quadrotor restarted the algorithm by resetting the state machine and approaching the landing pad again. Despite these setbacks, the quadrotor was able to land successfully on the platform in each simulation. For the sake of brevity, only the results of one simulations are reported hereafter.

Let us begin by comparing the roll and pitch references and estimated values in Figure 7a and Figure 7b, respectively. From the figures, we can notice when the biggest adjustments in terms of position occurred from 82 to 98 s, when the procedure started with the searching state. Indeed, in this simulation, the UAV did not find the platform at the rendezvous point. Thus, the "searching" state was entered and the quadrotor started to move around in circles. From 98 to 118 s, the quadrotor placed itself in the front of the landing platform, to prepare for the landing. After then, before the quadrotor landed, the generated references varied slightly, because the effort needed to keep the alignment with the landing platform was minor. At around $t = 135$ s, the quadrotor successfully landed on the platform.

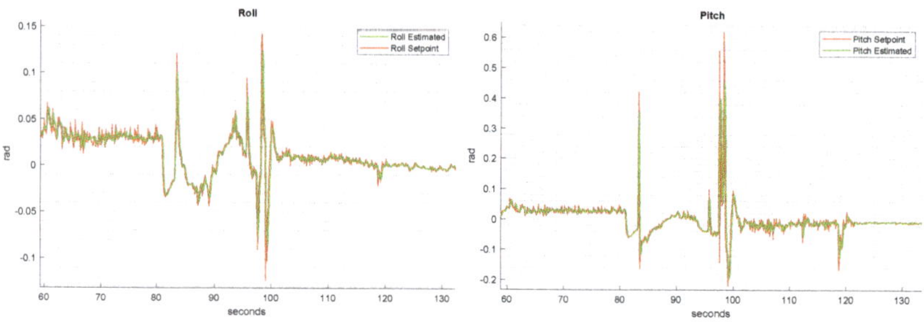

(**a**) (**b**)

Figure 7. Time-wise behavior of the desired and estimated roll (**a**) and pitch (**b**) of the UAV during the landing procedure.

The yaw reference was kept constant during the entire simulation, as it does not have a significant influence on the landing performance. Thus, its graph is omitted here.

The data about the positions of the quadrotor and the catamaran with respect to the starting point are depicted in Figure 8a,b. The first figure shows a bird's eye view on the x and y dimensions only, and the second includes all the axes. The circular movements generated by Equations (12) and (13) during the searching state can be seen in the interval ($x \in [24, 26]$ m, $y \in [8, 10]$ m). The instant where the radius changed was point 1 ($x = 25.9$ m, $y = 8.9$ m). In this test, the quadrotor had to perform just one complete circle before seeing the platform at the start of the second one. The initialization phase came afterwards, where the quadrotor had to perform some adjustments to place itself on the bow side of the catamaran. It can be seen how the references slightly changed in the following period: this happened because the catamaran was sliding a bit while trying to keep its position against the sea current. More precisely, in this test the catamaran drifted laterally at an average velocity of ~0.21 m/s. In order to find the maximum slide velocity the catamaran can have without compromising the landing of the quadrotor, several landing tests were performed in simulations with increasing drift velocity. The results of these simulations show that the quadrotor is able to successfully land at a drift velocity of up to 0.35 m/s, approximately. In Figure 8b, it can aksi be seen how the altitude of the catamaran had oscillating behavior in the final steps of the landing procedure. In fact, the catamaran's heave movement is more affected by waves when it is not moving forwards. Still, the quadrotor managed to follow the catamaran for the entire period and accomplish the land at point 2 ($x = 33.6$ m, $y = 10.1$ m, $z = 2.0$ m), showing the robustness of the proposed solution.

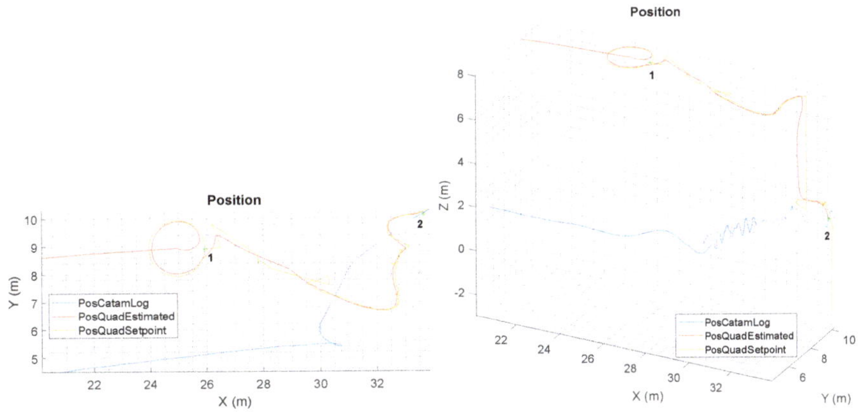

(a) (b)

Figure 8. (**a**) Bird's eye view of the trajectories of the UAV and of the ULISSE ASV during the simulation. 1 indicates the position at which the search radius changed, and 2 indicates where the final landing was accomplished. (**b**) A 3D representation.

Figure 9 shows how the x and y axis velocities are generated starting from the pose references. In detail, during the searching phase from 82 to 98 s, the sum of the velocity components is approximately constant, corresponding to the parameter v_s in Equations (12) and (13). This is not true when the quadrotor changes its circle's radius, at 84 and 96 s. After that, the quadrotor makes visual contact with the landing platform; thus, the generated velocities are the ones needed to place the quadrotor appropriately for the landing.

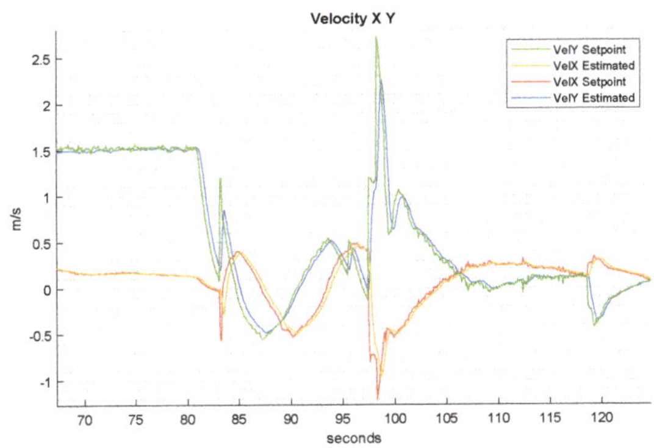

Figure 9. Time-wise behavior of the UAV's x and y velocities and their setpoints during the whole landing procedure.

Figure 10a shows the quadrotor's performances on the z-axis. As the inertial frame is a NED (north–east–down) frame, the sign of z is negative. Note the step-change reference at around $t = 118$ s, stating the transition between the initialization and tracking state (Section 3.4.3) of the landing state machine. This brings the quadrotor to a prefixed minimum altitude from the landing platform, to trigger as soon as possible the switch to the compensation state. When the quadrotor is centered with respect to the landing platform and the Kalman filter taking data from the ultrasonic sensor gives reliable outputs, the state is switched to the compensation one, where the quadrotor compensates for the platform's vertical motions (see Section 3.4.7).

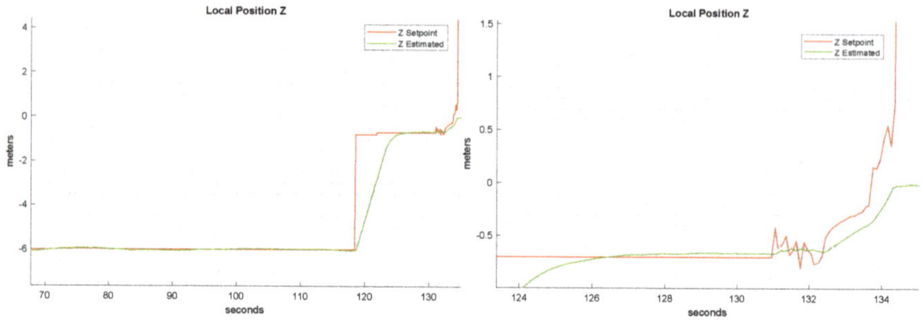

(a) (b)

Figure 10. Time-wise behavior of the absolute z position and setpoint: (**a**) overall simulation; (**b**) a zoom of the compensation state. Notice that at $t = 134$ s, a high setpoint was generated to force the UAV to freefall on top of the landing pad.

Figure 10b shows a zoom of Figure 10a. The generated references change step by step, allowing the quadrotor to be as reactive as possible with respect to the platform's oscillations. The peak at around $t = 134$ s in Figure 10b shows the transition from the compensation to the landing state. At that time, the z reference is set to a value (20 m) well below the platform height above the sea, to force the UAV to set its motors to the minimum, making it free-fall directly onto the platform. Once the Autopilot's firmware, PX4, has detected the landing, it automatically shuts off the motors in a few moments.

Figure 11 depicts the velocity on the z inertial axis. The generated references show the correlation among the z position setpoints. A full video of the simulation is available online and can be seen at: https://youtu.be/hILq4kUn9XY last access: 30 March 2022.

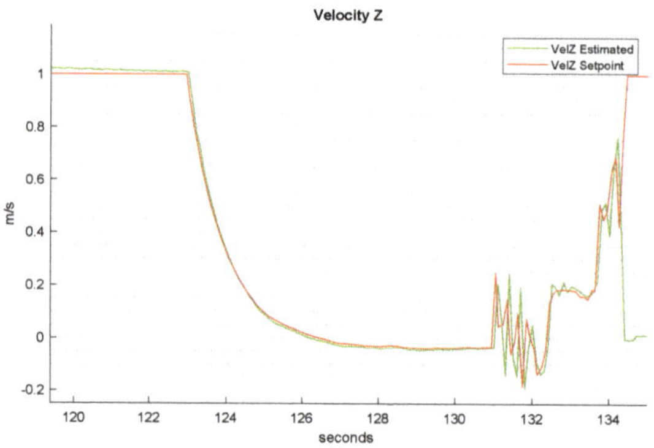

Figure 11. Time-wise behavior of the UAV z velocity and its setpoint during the whole landing procedure.

5.2. Vision System Validaton

To validate the pose estimation algorithm, an experimental trial where the quadrotor was manually controlled was conducted, and the output of the UAV's camera was recorded at a resolution of 480p. Then, the video was replayed offline. Each frame was sent to the pose estimation algorithm. This experiment was not conceived to test the system under different lightning conditions. We selected a day with clear sky, as we wanted to test the material of the landing platform, which was selected to ensure a high level of opacity, to prevent light reflections on the markers. Note that in poor light conditions, for instance, during the evening, at night or on a cloudy day, the landing platform should be backlit to ensure reliable detection of the tags.

In the following, different images taken at various altitudes are presented, showing the detection performances of the proposed algorithm. In particular, the outputs of the pose estimation node, i.e., the detected nodes, are shown in magenta together with the tag identifier, on top of the same input image, to highlight the detection results.

Figure 12a depicts a situation where the distance bewtween quadrotor and catamaran is around 6–7 m, a case similar to the initialization phase. At that distance, in the best case, the detected tags are eight in number; the bigger tags play a fundamental role, offering reliability and robustness in the positioning of the quadrotor. The smaller ones are instead not recognizable yet. In Figure 12b, the quadrotor is landing on the catamaran. In a medium-distance case like this one, the number of tags detected is higher. In fact, the smaller ones become visible, except for the smallest one at the center, and some of the bigger ones can be out of camera range. Tag 11, on the upper-left corner, is a little outside of the image plane, and tag 12, in the lower-left corner, is obscured by the roll-bar's shadow. Finally, Figure 12c shows the instants before the landing; there, the bigger tags are almost all not visible, whereas the smallest ones come into play to keep the quadrotor centered with respect to landing platform.

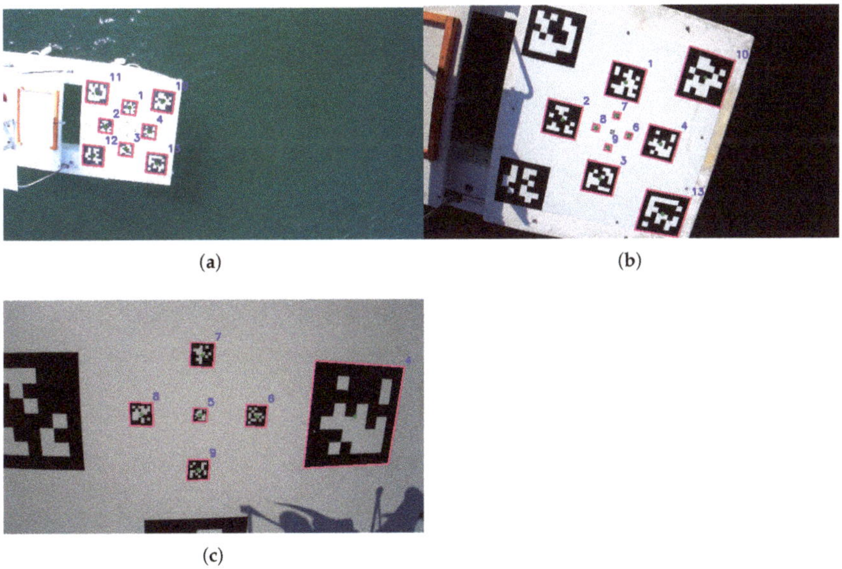

Figure 12. Detected tags at various altitudes: (**a**) initialization phase, high altitude; (**b**) tracking phase, medium altitude; (**c**) compensation phase, low altitude. Each detected tag is highlighted in magenta together with the tag identifier in blue.

These last three graphs depict additional data from the vision system. In Figure 13a, the number of detected tags is shown, varying between 1 and 10 tags (corresponding to Figure 12b). It can be seen that the number of detected tags was always equal to or greater than one, assuring the continuity of the platform detection during the whole landing procedure, confirming the performances achieved in the simulation environment. The estimated horizontal (Figure 13b) and vertical (Figure 13c) errors with respect to the catamaran instead are pretty consistent with the quadrotor's flight. A video of this emulation is available at https://youtu.be/iGNDCoQ2zaY last access: 30 March 2022. A further comparison of the vision system's performance in the real test and in simulations has been performed in terms of marker detectability: it has been noticed that the smaller markers (IDs 6, 7, 8, 9) became visible when the relative distance of quadrotor–landing platform was approximately under 1.5 m both in simulations and in the real tests. Instead, the bigger ones (IDs 10, 11, 12, 13) could be detected at distances up to 9.0 m in the simulations.

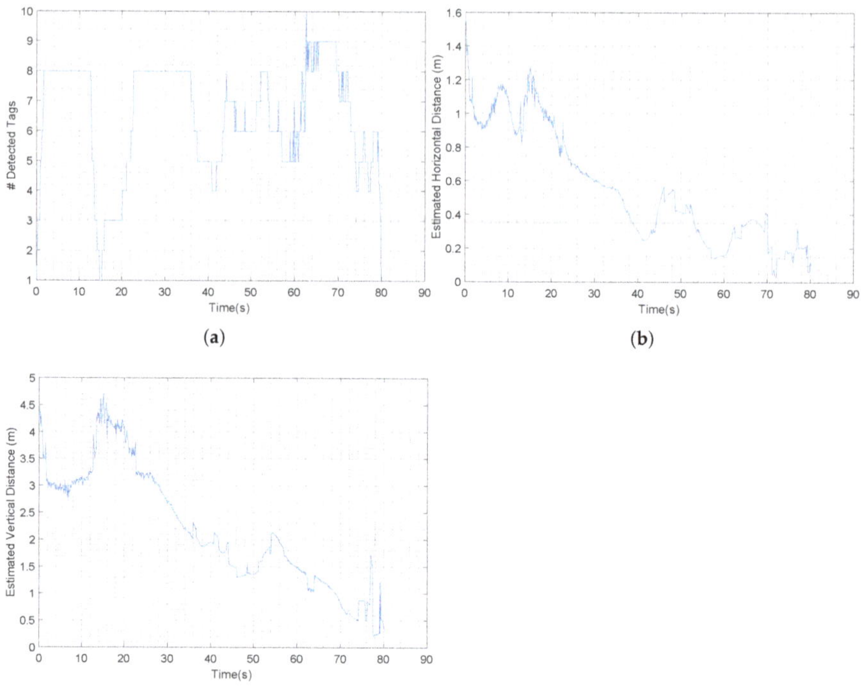

Figure 13. Evolution during manual landing experiment of: (**a**) number of tags detected; (**b**) horizontal error; (**c**) vertical error.

6. Conclusions

In this work, a procedure for the autonomous landing of a quadrotor on a catamaran has been presented. The objective was to propose a specific, reliable, and robust architecture composed of different modules, adaptable to both simulations and tests in a real environment. A vision system relying on AprilTags has been proposed to recognize the landing platform; several tags have been placed on the platform itself to always assure recognizability during the entire procedure.

A finite state machine handles the landing procedure. It is composed of different states, each one describing a specific behavior the quadrotor has to engage.

The validity of the proposed methodology has been shown in a simulation environment. To test the landing procedure with realistic motions of the landing pad, the motion of the ULISSE ASV was recorded at sea and then replayed within the Gazebo environment. The proposed vision system was further verified using a pre-recorded video of a landing performed under direct teleoperation of the quadrotor.

Author Contributions: Conceptualization, A.D., M.B. and E.S.; methodology, A.D., M.B. and E.S.; software, A.D.; data curation, A.D.; writing—original draft preparation, A.D.; writing—review and editing, A.D., M.B. and E.S.; visualization, A.D.; supervision, M.B. and E.S.; funding acquisition, E.S. All authors have read and agreed to the published version of the manuscript.

Funding: This publication was made possible by NPRP grant 10-0213-170458 from the Qatar National Research Fund (a member of Qatar Foundation).

Institutional Review Board Statement: Not applicable.

Informed Consent Statement: Not applicable.

Data Availability Statement: Not applicable.

Conflicts of Interest: The authors declare no conflict of interest.

References

1. Tsouros, D.C.; Bibi, S.; Sarigiannidis, P.G. A Review on UAV-Based Applications for Precision Agriculture. *Information* **2019**, *10*, 349. [CrossRef]
2. Boccardo, P.; Chiabrando, F.; Dutto, F.; Tonolo, F.G.; Lingua, A. UAV Deployment Exercise for Mapping Purposes: Evaluation of Emergency Response Applications. *Sensors* **2015**, *15*, 15717–15737. [CrossRef] [PubMed]
3. Jordan, S.; Moore, J.; Hovet, S.; Box, J.; Perry, J.; Kirsche, K.; Lewis, D.; Tse, Z.T.H. State-of-the-art technologies for UAV inspections. *IET Radar Sonar Navig.* **2018**, *12*, 151–164. [CrossRef]
4. Huang, J.; Tian, G.; Zhang, J.; Chen, Y. On Unmanned Aerial Vehicles Light Show Systems: Algorithms, Software and Hardware. *Appl. Sci.* **2021**, *11*, 7687. [CrossRef]
5. Sharma, M.; Gupta, A.; Gupta, S.K.; Alsamhi, S.H.; Shvetsov, A.V. Survey on Unmanned Aerial Vehicle for Mars Exploration: Deployment Use Case. *Drones* **2022**, *6*, 4. [CrossRef]
6. Simetti, E.; Indiveri, G.; Pascoal, A.M. WiMUST: A cooperative marine robotic system for autonomous geotechnical surveys. *J. Field Robot.* **2021**, *38*, 268–288. [CrossRef]
7. Casalino, G.; Allotta, B.; Antonelli, G.; Caiti, A.; Conte, G.; Indiveri, G.; Melchiorri, C.; Simetti, E. ISME research trends: Marine robotics for emergencies at sea. In Proceedings of the 2016 OCEANS, Shanghai, China, 10–13 April 2016; pp. 1–5.
8. Kong, W.; Zhou, D.; Zhang, D.; Zhang, J. Vision-based autonomous landing system for unmanned aerial vehicle: A survey. In Proceedings of the 2014 International Conference on Multisensor Fusion and Information Integration for Intelligent Systems (MFI), Beijing, China, 28–29 September 2014; pp. 1–8.
9. Bastianelli Naticchi, N.; Baglietto, M.; Sperindé, A.; Simetti, E.; Casalino, G. Visual Servoed Autonomous Landing on a Surface Vessel. In Proceedings of the OCEANS 2019 MTS/IEEE, Marseille, France, 17–20 June 2019.
10. Nisticó, A.; Baglietto, M.; Simetti, E.; Casalino, G.; Sperindé, A. Marea project: UAV landing procedure on a moving and floating platform. In Proceedings of the OCEANS 2017, Anchorage, AK, USA, 18–21 September 2017; pp. 1–10.
11. Abdelkrim, N.; Aouf, N.; Tsourdos, A.; White, B. Robust nonlinear filtering for INS/GPS UAV localization. In Proceedings of the 2008 16 th Mediterranean Conference on Control and Automation, Ajaccio, France, 25–27 June 2008; pp. 695–702.
12. Gautam, A.; Sujit, P.B.; Saripalli, S. A survey of autonomous landing techniques for UAVs. In Proceedings of the 2014 International Conference on Unmanned Aircraft Systems (ICUAS), Orlando, FL, USA, 27–30 May 2014; pp. 1210–1218.
13. Yang, X.; Mejias, L.; Garratt, M. Multi sensor data fusion for UAV navigation during landing operations. In Proceedings of the 2011 Australian Conference on Robotics and Automation (ACRA), Melbourne, Australia, 7–9 December 2011; pp. 1–10.
14. Wang, L.; Bai, X. Quadrotor Autonomous Approaching and Landing on a Vessel Deck. *J. Intell. Robot. Syst.* **2018**, *92*, 125–143. [CrossRef]
15. Falanga, D.; Zanchettin, A.; Simovic, A.; Delmerico, J.; Scaramuzza, D. Vision-based autonomous quadrotor landing on a moving platform. In Proceedings of the 2017 IEEE International Symposium on Safety, Security and Rescue Robotics (SSRR), Shanghai, China, 11–13 October 2017; pp. 200–207.
16. Verbandt, M.; Theys, B.; Schutter, J.D. Robust marker-tracking system for vision-based autonomous landing of VTOL UAVs. In Proceedings of the 2014 International Micro Air Vehicle Conference and Competition (IMAV), Delft, The Netherlands, 12–15 August 2014.
17. Araar, O.; Aouf, N.; Vitanov, I. Vision Based Autonomous Landing of Multirotor UAV on Moving Platform. *J. Intell. Robot. Syst.* **2017**, *85*, 369–384. [CrossRef]
18. Olson, E.; Strom, J.; Morton, R.; Richardson, A.; Ranganathan, P.; Goeddel, R.; Bulic, M.; Crossman, J.; Marinier, R. Progress toward multi-robot reconnaissance and the MAGIC 2010 competition. *J. Field Robot.* **2012**, *29*, 762–792. [CrossRef]
19. Mersch, D.; Crespi, A.; Keller, L. Tracking Individuals Shows Spatial Fidelity Is a Key Regulator of Ant Social Organization. *Science* **2013**, *340*, 1090–1093. [CrossRef] [PubMed]
20. Olson, E. AprilTag: A robust and flexible visual fiducial system. In Proceedings of the 2011 IEEE International Conference on Robotics and Automation, Shanghai, China, 9–13 May 2011; pp. 3400–3407.
21. Simetti, E.; Indiveri, G. Control oriented modeling of a twin thruster autonomous surface vehicle. *Ocean Eng.* **2022**, *243*, 110260. [CrossRef]
22. Huang, A.; Olson, E.; Moore, D.C. Lcm: Lightweight communications and marshalling. In Proceedings of the 2010 IEEE/RSJ International Conference on Intelligent Robots and Systems, Taipei, Taiwan, 18–22 October 2010; pp. 4057–4062.

sensors

MDPI

Article

Defense against Adversarial Swarms with Parameter Uncertainty

Claire Walton [1,2,*], Isaac Kaminer [3], Qi Gong [4], Abram H. Clark [5] and Theodoros Tsatsanifos [3]

1 Department of Electrical and Computer Engineering, University of Texas at San Antonio, San Antonio, TX 78249, USA
2 Department of Mathematics, University of Texas at San Antonio, San Antonio, TX 78249, USA
3 Department of Mechanical and Aerospace Engineering, Naval Postgraduate School, Monterey, CA 93943, USA; kaminer@nps.edu (I.K.); theodoros.tsatsanifos.gr@nps.edu (T.T.)
4 Department of Applied Mathematics, University of California Santa Cruz, Santa Cruz, CA 95064, USA; qigong@ucsc.edu
5 Department of Physics, Naval Postgraduate School, Monterey, CA 93943, USA; abe.clark@nps.edu
* Correspondence: claire.walton@utsa.edu

Abstract: This paper addresses the problem of optimal defense of a high-value unit (HVU) against a large-scale swarm attack. We discuss multiple models for intra-swarm cooperation strategies and provide a framework for combining these cooperative models with HVU tracking and adversarial interaction forces. We show that the problem of defending against a swarm attack can be cast in the framework of uncertain parameter optimal control. We discuss numerical solution methods, then derive a consistency result for the dual problem of this framework, providing a tool for verifying computational results. We also show that the dual conditions can be computed numerically, providing further computational utility. Finally, we apply these numerical results to derive optimal defender strategies against a 100-agent swarm attack.

Keywords: optimal control; parameter uncertainty; swarming

Citation: Walton, C.; Kaminer, I.; Gong, Q.; Clark, A.H.; Tsatsanifos, T. Defense against Adversarial Swarms with Parameter Uncertainty. *Sensors* **2022**, *22*, 4773. https://doi.org/10.3390/s22134773

Academic Editor: Carlo Alberto Avizzano

Received: 11 April 2022
Accepted: 3 June 2022
Published: 24 June 2022

1. Introduction

Swarms are characterized by large numbers of agents which act individually, yet produce collective, herd-like behaviors. Implementing cooperating swarm strategies for a large-scale swarm is a technical challenge which can be considered to be from the "insider's perspective". It assumes inside control over the swarm's operating algorithms. However, as large-scale 'swarm' systems of autonomous systems become achievable—such as those proposed by autonomous driving, UAV package delivery, and military applications—interactions with swarms outside our direct control become another challenge. This generates its own "outsider's perspective" issues.

In this paper, we look at the specific challenge of protecting an asset against an adversarial swarm. Autonomous defensive agents are tasked with protected a high-value unit (HVU) from an incoming swarm attack. The defenders do not fully know the cooperating strategy employed by the adversarial swarm. Nevertheless, the task of the defenders is to maximize the probability of survival of the HVU against an attack by such a swarm. This challenge raises many issues—for instance, how to search for the swarm [1], how to observe and infer swarm operating algorithms [2], and how to best defend against the swarm given algorithm unknowns, and only limited, indirect control through external means. In this paper, we restrict ourselves to the last issue. However, these problems share multiple technical challenges. The preliminary approach we apply in this paper demonstrates some basic methods which we hope will stimulate the development of more sophisticated tools.

For objectives achieved via external control of the swarm, several features of swarm behavior must be characterized: capturing the dynamic nature of the swarm, tracking the collective risk profile created by a swarm, and engaging with a swarm via dynamic inputs, such as autonomous defenders. The many modeling layers create a challenge for generating an effective response to the swarm, as model uncertainty and model error are almost certain. In this paper, we look at several dynamic systems where the network structure is determined by parameters. These parameters set neighborhood relations and interaction rules. Additional parameters establish defender input and swarm risk.

We consider the generation of optimal defense strategies given uncertainty in parameter values. We demonstrate that small deviances in parameter values can have catastrophic effects on defense trajectories optimized without taking error into account. We then demonstrate the contrasting robustness of applying an uncertain parameter optimal control framework instead of optimizing with nominal values. The robustness against these parameter values suggests that refined parameter knowledge may not be necessary given appropriate computational tools. These computational tools—and the modeling of the high-dimensional swarm itself—are expensive. To assist with this issue, we provide dual conditions for this problem in the form of a Pontryagin minimum principle and prove the consistency of these conditions for the numerical algorithm. These dual conditions can, thus, be computed from the numerical solution of the computational method and provide a tool for solution verification and parameter sensitivity analysis.

Although in this paper, optimal strategies against swarms motivate the framework of uncertain parameter optimal control, and the subsequent development of the dual conditions, both the framework and the dual conditions have many applications beyond swarm defense. Optimal control with parameter uncertainty is relevant to robotics—where parts, such as wheels, may have small size and calibration uncertainties; aerospace—where both components and exogeneous factors, such as wind, may be modeled using parameter uncertainty; and search and rescue—where the location of a target object can be considered a parameter uncertainty [3,4]. It is also an instance of mean-field optimal control (which includes this framework, but also more general probability distributions), which is finding application in the training of neural networks [5]. The dual conditions provided in this paper provide both a tool for verification of numerical solutions, as well as another potential route for generating numerical solutions.

The structure of this paper is as follows. Section 2 provides examples of dynamic swarming models and extensions for defensive interactions. Section 3 discusses optimization challenges and describes a general uncertain parameter optimal control framework that this problem could be addressed with. Section 4 provides a proof of the consistency of the dual problem for this control framework, which expands on the results initially presented in the conference paper [6]. Section 5 gives an example of numerical implementation that demonstrates optimal defense against a large-scale swarm of 100 agents. Section 6 discusses the results and future work.

2. Modeling Adverserial Swarms

2.1. Cooperative Swarm Models

The literature on the design of swarm strategies which produce coherent, stable collective behavior has become vast. A quick review of the literature points to two main trends/categories in swarm behavior design. The first relies on dynamic modeling of the agents and potential functions to control their behavior (see [7,8] and references therein). The second trend relates to the use of rules to describe agents' motion and local rule-based algorithms to control them [9,10].

We present two examples of dynamic swarming strategies from the literature. These examples are illustrative of the forces considered in many swarming models:

- collision avoidance between swarm members;
- alignment forces between neighboring swarm members;
- stabilizing forces.

These intra-swarm goals are aggregated to provide a swarm control law, which we will refer to as F_S, to each swarm agent. Both example models in this paper share the same double integrator form with respect to this control law. For n swarm agents, the dynamics are defined by

$$\ddot{x}_i = u_i. \; i = 1, \ldots, n, \tag{1}$$

$$u_i = F_S(x_i, \dot{x}_i, \forall j \neq i : x_j, \dot{x}_j | \theta). \tag{2}$$

2.1.1. Example Model 1: Virtual Body Artificial Potential

In this model [11,12], swarm agents track to a virtual body (or bodies) guiding their course, while also reacting to intra-swarm forces of collision avoidance and group cohesion. The input u_i is the sum of intra-swarm forces, virtual body tracking, and a velocity dampening term. In addition, in this adversarial scenario, swarm agents are influenced to avoid intruding defense agents. The intra-swarm force between two swarm agents has magnitude f_I and is a gradient of an artificial potential V_I. Let

$$x_{ij} = x_i - x_j. \tag{3}$$

The artificial potential V_I depends on the distance $||x_{ij}||$ between swarm agents i and j. The artificial potential V_I is defined as:

$$V_I = \begin{cases} \alpha\left(\ln(||x_{ij}||) + \dfrac{d_0}{||x_{ij}||}\right), & 0 < ||x_{ij}|| < d_1 \\ \alpha\left(\ln(d_1) + \dfrac{d_0}{d_1}\right), & ||x_{ij}|| \geq d_1 \end{cases} \tag{4}$$

where α is a scalar control gain, and d_0 and d_1 are scalar constants for distance ranges. Then the magnitude of interaction force is given by

$$f_I = \begin{cases} \nabla_{||x_{ij}||} V_I, & 0 < ||x_{ij}|| < d_1 \\ 0, & ||x_{ij}|| \geq d_1 \end{cases} \tag{5}$$

The swarm body is guided by 'virtual leaders', non-corporeal reference trajectories which lead the swarm. We assign a potential V_h on a given swarm agent i associated with the k-th virtual leader, defined with the distance $||h_{ik}||$ between the swarm agent i and leader k. Mirroring the parameters α, d_0, and d_1 defining V_I, we assign V_h the parameters α_h, h_0, and h_1. An additional dissipative force f_{v_i} is included for stability. The control law u_i for the vehicle i associated with m defenders is given by

$$\begin{aligned} u_i &= -\sum_{j\neq i}^{n} \nabla_{x_i} V_I(x_{ij}) - \sum_{k=1}^{m} \nabla_{x_i} V_h(h_{ik}) + f_{v_i} \\ &= -\sum_{j\neq i}^{n} \frac{f_I(x_{ij})}{||x_{ij}||} x_{ij} - \sum_{k=1}^{m} \frac{f_h(h_{ik})}{||h_{ik}||} h_{ik} + f_{v_i}. \end{aligned} \tag{6}$$

2.1.2. Example Model 2: Reynolds Boid Model

In this model [8,13], for radius r, $j = 1, \ldots, N$, define the neighbors of agent i at position $x_i \in \mathbb{R}^n$ by the set

$$\mathcal{N}_i = \{j | j \neq i \wedge \|x_i - x_j\| < r\} \tag{7}$$

Swarm control is designated by three forces.

Alignment of velocity vectors:

$$f_{al} = -w_{al}\left(\dot{x}_i - \frac{1}{|\mathcal{N}_i|} \sum_{j\in\mathcal{N}_i} \dot{x}_j\right) \tag{8}$$

Cohesion of swarm:

$$f_{coh} = -w_{coh}\left(x_i - \frac{1}{|\mathcal{N}_i|}\sum_{j\in\mathcal{N}_i} x_j\right) \tag{9}$$

Separation between agents:

$$f_{sep} = -w_{sep}\frac{1}{|\mathcal{N}_i|}\left(\sum_{j\in\mathcal{N}_i}\frac{x_j - x_i}{\|x_i - x_j\|}\right) \tag{10}$$

for positive constant parameters w_{al}, w_{coh}, w_{sep}.

$$u_i = f_{al} + f_{coh} + f_{sep} \tag{11}$$

2.2. Adversarial Swarm Models

The previous subsection provides several examples of inner swarm cooperative forces, F_S. In order to enable adversarial behavior and defense, these inner swarm cooperative forces need to be supplemented by additional forces of exogenous input into the collective. As written, the above cooperative swarming models neither respond to outside agents nor 'attack' (swarm towards) a specific target. We, thus, supplement the control laws above with two additional forces. The first, we refer to as F_{HVU}; the goal of the swarm, in this paper, is limited to tracking an HVU. An example of F_{HVU} is provided in the example of Section 5, in Equation (28).

We also supplement by an adversarial force, which we refer to as F_D. The review [7] discusses several approaches to adversarial control. Examples include containment strategies modeled after dolphins [14], sheep-dogs [15,16], and birds of prey [17]. In [18], the authors studied the interaction between two swarms, one of which could be considered adversarial. In these examples of adversarial swarm control, the mechanism of interaction and defense is provided through the swarm's own pursuit and evasion responses. This indirectly uses the swarm's own response strategy against it—an approach which can be termed 'herding'.

In addition to herding reactions, one can consider more direct additional forces of disruption, to model neutralizing swarm agents and/or physically remove them from the swarm. One form this can take, for example, is the removal of agents from the communications network, as considered in [19]. Another approach is taken in [20], which uses survival probabilities based on damage attrition. Defenders and the attacking swarm engage in mutual damage attrition while the swarm also damages the HVU when in proximity to it. Probable damage between agents is tracked as damage rates over time, where the rate of damage is based on features such as distance between agents and angle of attack. The damage rate at time t provides the probability of a successful 'hit' in time period $[t, t+\Delta t]$. The probability of agent survival can be modeled based on the aggregate number of hits it takes to incapacitate the agent. The authors of [20] provide derivations for multiple possibilities, such as single-shot destruction and N-shot destruction. These probabilities take the form of ODE equations. Tracking survival probabilities thus adds an additional state to the dynamics of each agent—a survival probability state.

We, thus, summarize a control scheme with HVU target-tracking and herding driven by the reactive forces of collision avoidance with the defenders as the following, for HVU states y_0 and defender states y_k, $k = 1, \ldots, K$:

$$\begin{aligned} u_i =& F_S(x_i, \dot{x}_i, \forall j \neq i : x_j, \dot{x}_j|\theta) && \leftarrow \textit{intra-swarm} \\ &+F_{HVU}(x_i, \dot{x}_i, y_0, \dot{y}_0|\theta) && \leftarrow \textit{target tracking} \\ &+F_D(x_i, \dot{x}_i, \forall k : y_k, \dot{y}_k|\theta) && \leftarrow \textit{herding and/or damage} \end{aligned} \tag{12}$$

Example Attrition Model: Single-Shot Destruction

From [20]: let $P_0(t)$ be the probability the HVU has survived up to time t, $P_k(t)$, $k = 1, \dots, K$, the probability defender k has survived, and $Q_j(t), j = 1, \dots, N$ the probability swarm attacker j has survived. Let $d_y^{j,k}(x_j(t), y_k(t))$ be the damage the defender y_k inflicts on swarm attacker x_j and let $d_x^{k,j}(y_k(t), x_j)$ be the damage the swarm attacker x_j inflicts on the defender y_k, with the HVU represented by $k = 0$.

Then the survival probabilities for attackers and defenders from single-shot destruction are given by the coupled ODEs:

$$\begin{cases} \dot{Q}_j(t) = -Q_j(t) \sum_{k=1}^{K} P_k(t) d_y^{j,k}(x_j(t), y_k(t)), & Q_j(0) = 1 \\ \dot{P}_k(t) = -P_k(t) \sum_{j=1}^{N} Q_j(t) d_x^{k,j}(y_k(t), x_j(t)), & P_k(0) = 1 \end{cases}$$

for $j = 1, \dots, N, k = 0 \dots, K$.

3. Problem Formulation

The above models depend on a large number of parameters. The dynamic swarming model coupled with attrition functions results in over a dozen key parameters, and many more would result from a non-homogeneous swarm. A concern would be that this adds too much model specificity, making optimal defense strategies lack robustness due to sensitivity to the specific set of model parameters. This concern turns out to be justified. When defense strategies are optimized for fixed, nominal parameter values, they display catastrophic failure for small perturbations of certain parameters, as can be seen in Figure 1. In fact, the plots included in Figure 1 clearly demonstrate that the sensitivity of the cost with respect to the uncertain parameters is highly non-linear. Thus, generating robust defense strategies requires a more sophisticated formalism introduced in the next Section 3.1.

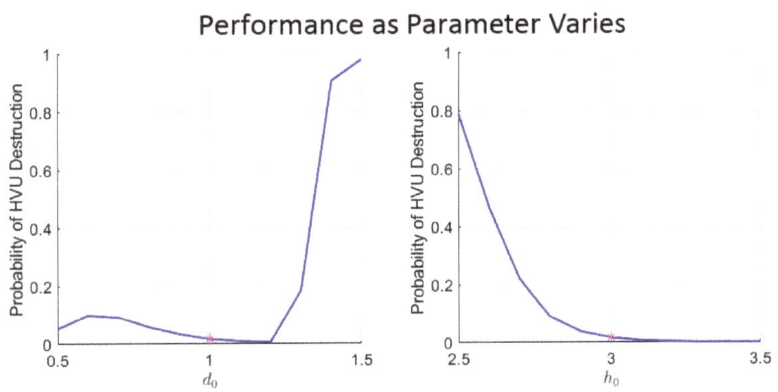

Figure 1. Example performance of solutions calculated using nominal values when parameter value is varied. Calculated using values in Section 5.1. Magenta dot marks the nominal value used in the optimization problem. In the left panel, d_0 is varied as the parameter; in the right panel h_0 is varied.

3.1. Uncertain Parameter Optimal Control

The class of problems addressed by the computational algorithm is defined as follows:

Problem P. Determine the function pair (x, u) with $x \in W_{1,\infty}([0, T] \times \Theta; \mathbb{R}^{n_x})$, $u \in L_\infty([0, T]; \mathbb{R}^{n_u})$ that minimizes the cost

$$J[x, u] = \int_\Theta \left[F(x(T, \theta), \theta) + \int_0^T r(x(t, \theta), u(t), t, \theta) dt \right] d\theta \tag{13}$$

subject to the dynamics

$$\frac{\partial x}{\partial t}(t,\theta) = f(x(t,\theta), u(t), \theta), \tag{14}$$

initial condition $x(0,\theta) = x_0(\theta)$, and the control constraint $g(u(t)) \leq 0$ for all $t \in [0,T]$. The set $L_\infty([0,T];\mathbb{R}^{n_u})$ is the set of all essentially bounded functions, $W_{1,\infty}([0,T] \times \Theta; \mathbb{R}^{n_x})$ the Sobolev space of all essentially bounded functions with essentially bounded distributional derivatives, and $F : \mathbb{R}^{n_x} \times \mathbb{R}^{n_\theta} \mapsto \mathbb{R}$, $r : \mathbb{R}^{n_x} \times \mathbb{R}^{n_u} \times \mathbb{R} \times \mathbb{R}^{n_\theta} \mapsto \mathbb{R}$, $g : \mathbb{R}^{n_u} \mapsto \mathbb{R}^{n_g}$. Additional conditions imposed on the state and control space and component functions are specified in Appendix A.

In Problem **P**, the set Θ is the domain of a parameter $\theta \in \mathbb{R}^{n_\theta}$. The format of the cost functional is that of the integral over Θ of a Mayer–Bolza type cost with parameter θ. This parameter can represent a range of values for a feature of the system, such as in ensemble control [21], or a stochastic parameter with a known probability density function.

For computation of numerical solutions, we introduce an approximation of Problem **P**, referred to as Problem $\mathbf{P^M}$. Problem $\mathbf{P^M}$ is created by approximating the parameter space, Θ, with a numerical integration scheme. This numerical integration scheme is defined in terms of a finite set of M nodes $\{\theta_i^M\}_{i=1}^M$ and an associated set of M weights $\{\alpha_i^M\}_{i=1}^M \subset \mathbb{R}$ such that

$$\int_\Theta h(\theta)d\theta = \lim_{M\to\infty} \sum_{i=1}^{M} h(\theta_i^M)\alpha_i^M. \tag{15}$$

given certain function smoothness assumptions. See Appendix A Assumption A1 for formal assumptions. Throughout the paper, M is used to denote the number of nodes used in this approximation of parameter space.

For a given set of nodes $\{\theta_i^M\}_{i=1}^M$, and control $u(t)$, let $\bar{x}_i^M(t)$, $i = 1, \ldots, M$, be defined as the solution to the ODE created by the state dynamics of Problem **P** evaluated at θ_i^M:

$$\begin{cases} \frac{d\bar{x}_i^M}{dt}(t) = f(\bar{x}_i^M(t), u(t), \theta_i^M) \\ \bar{x}_i^M(0) = x_0(\theta_i^M), \end{cases} \quad i = 1, \ldots, M. \tag{16}$$

Let $\bar{X}^M(t) = [\bar{x}_1^M(t), \ldots, \bar{x}_M^M(t)]$. The system of ODEs defining $\bar{X}^M$ has dimension $n_x \times M$, where n_x is the dimension of the original state space and M is the number of nodes. The numerical integration scheme for parameter space creates an approximate objective functional, defined by:

$$\bar{J}^M[\bar{X}^M, u] = \sum_{i=1}^{M} \left[F\left(\bar{x}_i^M(T), \theta_i^M\right) + \int_0^T r(\bar{x}_i^M(t), u(t), t, \theta_i^M)dt \right] \alpha_i^M. \tag{17}$$

In [4], the consistency of $\mathbf{P^M}$ is proved. This is the property that, if optimal solutions to Problem $\mathbf{P^M}$ converge as the number of nodes $M \to \infty$, they converge to feasible, optimal solutions of Problem **P**. See [4] for detailed proof and assumptions.

3.2. *Computational Efficiency*

The computation time of the numerical solution to the discretized problem defined in Equations (16) and (17) will depend on the value of M. Ideally, it should be sufficiently small so as to allow for a fast solution. On the other hand, a value of M that is too small will result in a solution that is not particularly useful, i.e., too far from the optimal. Naturally, the question arises: how far is a particular solution from the optimal? One tool for assessing this lies in computing the Hamiltonian and is addressed in Section 4.

4. Consistency of Dual Variables

The dual variables provide a method to determine the solution of an optimal control problem or a tool to validate a numerically computed solution. For numerical schemes based on direct discretization of the control problem, analyzing the properties of the dual variables and their resultant Hamiltonian may also lead to insight into the validity of an approximation scheme [22,23]. This could be especially helpful in high-dimensional problems, such as swarming, where parsimonious discretization is crucial to computational tractability.

Previous work shows the consistency of the primal variables in approximate Problem $\mathbf{P^M}$ to the original parameter uncertainty framework of Problem **P**. Here, we build on that and prove the consistency of the dual problem of Problem **P** as well. This theoretical contribution is diagrammed in Figure 2. The consistency of the dual problem in parameter space enables approximate computation of the Hamiltonian from numerical solutions.

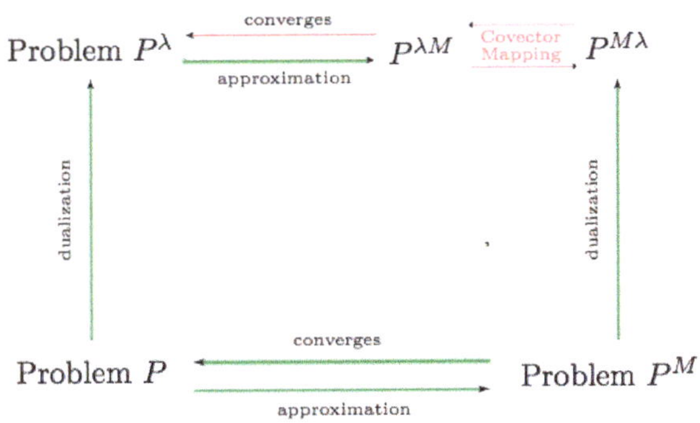

Figure 2. Diagram of primal and dual relations for parameter uncertainty control. Red lines designate the contribution of this paper.

In [24], necessary conditions for Problem **P** were established. These conditions are as follows:

Problem $\mathbf{P^\lambda}$ [([24], pp. 80–82)]. If (x^*, u^*) is an optimal solution to Problem **P**, then there exists an absolutely continuous costate vector $\lambda^*(t,\theta)$, such that for $\theta \in \Theta$:

$$\frac{\partial \lambda^*}{\partial t}(t,\theta) = -\frac{\partial H(x^*, \lambda^*, u^*, t, \theta)}{\partial x},$$

$$\lambda^*(T,\theta) = \frac{\partial F(x^*(T,\theta),\theta)}{\partial x} \tag{18}$$

where H is defined as:

$$H(x,\lambda,u,t,\theta) = \\ \lambda f(x(t,\theta),u(t),\theta) + r(x(t,\theta),u(t),t,\theta). \tag{19}$$

Furthermore, the optimal control u^* satisfies

$$u^*(t) = \arg\min_{u\in U} \mathbf{H}(x^*, \lambda^*, u, t),$$

where $\mathbf{H}$ is given by

$$\mathbf{H}(x,\lambda,u,t) = \int_\Theta H(x,\lambda,u,t,\theta)d\theta. \tag{20}$$

Because Problem $\mathbf{P^M}$ is a standard non-linear optimal control problem, it admits a dual problem as well. Problem $\mathbf{P^{M\lambda}}$, provided by the Pontryagin minimum principle (a survey of minimum principle conditions is given by [25]). Applied to $\mathbf{P^M}$ this generates:

Problem $\mathbf{P}^{M\lambda}$. For feasible solution $(\bar{X}^M, u)$ to Problem $\mathbf{P^M}$, find $\bar{\Lambda}(t) = [\bar{\lambda}_1^M(t), \ldots \bar{\lambda}_M^M(t)]$, $\bar{\lambda}_i^M : [0, T] \to \mathbb{R}^{N_x}$, that satisfies the following conditions:

$$\frac{d\bar{\lambda}_i^M}{dt}(t) = -\frac{\partial \bar{H}^M(\bar{x}_i^M, \bar{\lambda}_i^M, u, t)}{\partial x_i^M},$$

$$\bar{\lambda}_i^M(T) = \alpha_i^M \frac{\partial F(\bar{x}_i^M, \theta_i^M)}{\partial \bar{x}_i^M}, \tag{21}$$

where $\bar{H}^M$ is defined as:

$$\bar{H}^M(\bar{X}^M, \bar{\Lambda}_M, u, t) = \sum_{i=1}^{M} \Big[\bar{\lambda}_i^M f(\bar{x}_i^M(t), u(t), \theta_i^M) + \alpha_i^M r(\bar{x}_i^M(t), u(t), t, \theta_i^M) \Big]. \tag{22}$$

An alternate direction from which to approach solving Problem $\mathbf{P}$ overall is to approximate the necessary conditions of Problem $\mathbf{P}$, i.e., Problem $\mathbf{P}^{\lambda}$, directly rather than to approximate Problem $\mathbf{P}$. This creates the system of equations:

$$\frac{d\lambda}{dt}(t, \theta_i^M) = -\frac{\partial H(x, u, t, \theta_i^M)}{\partial x}$$

$$\lambda(T, \theta_i^M) = \frac{\partial F(x(T, \theta_i^M), \theta_i^M)}{\partial x} \tag{23}$$

for $i = 1, \ldots, M$, where H is defined as:

$$H(x, \lambda, u, t, \theta) = \lambda f(x(t, \theta), u(t), \theta) + r(x(t, \theta), u(t), t, \theta).$$

This system of equations can be re-written in terms of the quadrature approximation of the stationary Hamiltonian defined in Equation (20). Define

$$\tilde{H}^M(x, \lambda, u, t) := \sum_{i=1}^{M} \alpha_i^M H(x(t, \theta_i^M), \lambda(t, \theta_i^M), u(t), t, \theta_i^M).$$

Let

$$\tilde{\Lambda}(t) = [\tilde{\lambda}_1^M(t), \ldots \tilde{\lambda}_M^M(t)] = [\lambda(t, \theta_1^M), \ldots, \lambda(t, \theta_M^M)]$$

and let

$$\tilde{X}_M = [\tilde{x}_1^M(t), \ldots, \tilde{x}_M^M(t)]$$

denote the semi-discretized states from Equation (16). Equation (23) can then be written as:

$$\frac{d\tilde{\lambda}_i^M}{dt}(t) = -\frac{1}{\alpha_i^M} \cdot \frac{\partial \tilde{H}^M(\tilde{X}_M, \tilde{\Lambda}, u, t)}{\partial \tilde{x}_i^M}$$

$$\tilde{\lambda}_i^M(T) = \frac{\partial F(\tilde{x}_i^M(T), \theta_i^M)}{\partial \tilde{x}_i^M} \tag{24}$$

for $i = 1, \ldots, M$. Thus, we reach the following discretized dual problem:

Problem $\mathbf{P}^{\lambda M}$. For feasible controls u and solutions $\tilde{X}_M$ to Equation (16), find $\tilde{\Lambda}(t) = [\tilde{\lambda}_1^M(t), \ldots \tilde{\lambda}_M^M(t)]$, $\tilde{\lambda}_i^M : [0, T] \to \mathbb{R}^{n_x}$, that satisfies the following conditions:

$$\frac{d\tilde{\lambda}_i^M}{dt}(t) = -\frac{1}{\alpha_i^M} \cdot \frac{\partial \tilde{H}^M(\tilde{X}_M, \tilde{\Lambda}, u, t)}{\partial \tilde{x}_i^M},$$

$$\tilde{\lambda}_i^M(T) = \frac{\partial F(\tilde{x}_i^M, \theta_i^M)}{\partial \tilde{x}_i^M}, \tag{25}$$

where $\tilde{H}^M$ is defined as:

$$\tilde{H}^M(\tilde{X}_M, \tilde{\Lambda}_M, u, t) = \sum_{i=1}^{M} \left[\alpha_i^M \tilde{\lambda}_i^M f(\tilde{x}_i^M(t), u(t), \theta_i^M) + \alpha_i^M r(\tilde{x}_i^M(t), u(t), t, \theta_i^M) \right]. \tag{26}$$

Lemma 1. *The mapping:*

$$(\bar{x}_i^M, \bar{u}) \mapsto (\tilde{x}_i^M, \tilde{u}), \quad \frac{\bar{\lambda}_i^M}{\alpha_i^M} \mapsto \tilde{\lambda}_i^M,$$

for $i = 1, \ldots, M$ is a bijective mapping from solutions of Problem $\mathbf{P}^{M\lambda}$ to Problem $\mathbf{P}^{\lambda M}$.

Proof. In the case of this particular problem, unlike standard control, the collocation of the relevant dynamics involves no approximation of differentiation (since the discretization is in the parameter domain rather than the time domain), and, thus, the mapping of covectors between Problem $\mathbf{P}^{M\lambda}$ and Problem $\mathbf{P}^{\tilde{H}^M(\tilde{X}_M, \tilde{\Lambda}_M, u, t) = \lambda M}$ is straightforward and simply constructively provided by the lemma itself. The two mappings of the lemma, $(\bar{x}_i^M, \bar{u}) \mapsto (\tilde{x}_i^M, \tilde{u})$ (identity mapping) and $\frac{\bar{\lambda}_i^M}{\alpha_i^M} \mapsto \tilde{\lambda}_i^M$ (scaling by $\frac{1}{\alpha_i^M}$) are both bijections. □

Theorem 1. *Let $\{\tilde{X}_M, \tilde{\Lambda}_M, u_M\}_{M \in V}$ be a sequence of solutions for Problem $\mathbf{P}^{\lambda M}$ with an accumulation point $\{\tilde{X}^\infty, \tilde{\Lambda}^\infty, u^\infty\}$. Let $(x^\infty, \lambda^\infty, u^\infty)$ be the solutions to Problem $\mathbf{P}^{\lambda}$ for the control u^∞. Then*

$$\lim_{M \in V} \tilde{H}^M(\tilde{X}_M, \tilde{\Lambda}_M, u_M, t) = \mathbf{H}(x^\infty, \lambda^\infty, u^\infty, t)$$

where $\tilde{H}^M$ is the Hamiltonian of Problem $\mathbf{P}^{\lambda M}$ as defined by Equation (26) and $\mathbf{H}$ is the Hamiltonian of Problem $\mathbf{P}$ as defined by Equation (20). The proof of this theorem can be found in the Appendix B.

The convergence of the Hamiltonians of the approximate, standard control problems to the Hamiltonian of the general problem, $\mathbf{H}(x^\infty, \lambda^\infty, u^\infty, t)$, means that many of the useful features of the Hamiltonians of standard optimal control problems are preserved. For instance, it is straightforward to show that the satisfaction of Pontryagin's minimum principle by the approximate Hamiltonians implies minimization of $\mathbf{H}(x^\infty, \lambda^\infty, u^\infty, t)$ as well. That is, that

$$\mathbf{H}(x^\infty, \lambda^\infty, u^\infty, t) \leq \mathbf{H}(x^\infty, \lambda^\infty, u, t)$$

for all feasible u. Furthermore, when applicable, the stationarity properties of the standard control Hamiltonian, such as a constant-valued Hamiltonian in time-invariant problems, or stationarity with respect to $u(t)$ in problems with open control regions, are also preserved.

5. Numerical Example

In a slight refashioning of the notation in the Section 2.2, Equation (12), let the parameter vector θ be defined by all the *unknown* parameters defining the interaction functions. Assuming prior distribution $\phi(\theta)$ over these unknowns and parameter bounds Θ, we construct the following optimal control problem for robustness against the unknown parameters.

Problem SD (Swarm Defense). For K defenders and N attackers, determine the defender controls $u_k(t)$ that minimize:

$$J = \int_\theta \left[1 - P_0(t_f, \theta) \right] \phi(\theta) d\theta \tag{27}$$

subject to:

$$\begin{cases} \dot{y}_k(t) = f(y_k(t), u_k(t)), & y_k(0) = y_{k0} \\ \ddot{x}_j(t,\theta) = & \\ F_S(t,\theta) + F_{HVU}(t,\theta) + F_D(t,\theta), & x_j(0,\theta) = x_{j0}(\theta) \\ \dot{Q}_j(t,\theta) = & \\ -Q_j(t,\theta)\sum_{k=1}^{K} P_k(t,\theta) d_y^{j,k}(x_j(t,\theta), y_k(t)), & Q_j(0,\theta) = 1 \\ \dot{P}_k(t) = & \\ -P_k(t,\theta)\sum_{j=1}^{N} Q_j(t,\theta) d_x^{k,j}(y_k(t), x_j(t,\theta)), & P_k(0,\theta) = 1 \end{cases}$$

for swarm attackers $j = 1, \ldots, N$ and controlled defenders $k = 1 \ldots, K$.

We implement Problem **SD** for both swarm models in Section 2.1, for a swarm of $N = 100$ attackers and $K = 10$ defenders.

5.1. Example Model 1: Virtual Body Artificial Potential

The cooperative swarm forces F_S are defined with the Virtual Body Artificial Potential of Section 2.1 with parameters α, d_0 and d_1. In lieu of a potential for the virtual leaders, we assign the HVU tracking function:

$$f_{HVU} = -\frac{K_1(x_i - y_0)}{\|x_i - y_0\|} \tag{28}$$

where $y_0 \in \mathbb{R}^3$ is the position of the HVU. The dissipative force $f_{v_i} = -K_2\dot{x}_i$ is employed to guarantee stability of the swarm system. K_1 and K_2 are positive constants. The swarm's collision avoidance response to the defenders is defined by Equation (4) with parameters α_h, h_0 and h_1. Since there is only a repulsive force between swarm members and defenders, not an attractive force, we set $h_1 = h_0$. For attrition, we use the the damage function defined in Equation (21) of [20]:

$$F_D = \lambda\Phi\left(\frac{F - ar^2}{\sigma}\right), \qquad r = \|x_i - y_j\|_2 \tag{29}$$

where Φ is the cumulative normal distribution and $\|\cdot\|_2$ is the vector 2-norm. This function smoothly penalizes proximity, with the impact decreasing with distance. The parameters λ, F, a, and σ shape the steepness of this function and the decline of damage over distance. For the damage rate of defenders inflicted on attackers, we calibrate by the parameters λ_D, σ_D. For the damage rate of attackers inflicted on defenders, we calibrate by the parameters λ_A, σ_A. In both cases, the parameters F and a in [20] are set to $F = 0$, $a = 1$. Table 1 provides the parameter values that remain fixed in each simulation, and and Table 2 provides the parameters we consider as uncertain.

Table 1. Model 1 Fixed Parameter Values.

Parameter	Value	Meaning
t_f	45	final time
K_1	5	tracking coefficient
K	10	number of defenders
h_1	h_0	interaction parameter
λ_D	2	defender weapon intensity
σ_D	2	defender weapon range
N	100	number of attackers
K_2	5	dissipative force

Table 2. Model 1 Varied Parameter Values.

Parameter	Nominal	Range	Meaning
α	0.5	[0.1, 0.9]	control gain
d_0	1	[0.5, 1.5]	lower range limit
d_1	6	[4, 8]	upper range limit
λ_A	0.05	[0.01, 0.09]	weapon intensity
σ_A	2	[1.5. 2.5]	weapon range
α_h	6	[5, 7]	herding intensity
h_0	3	[2, 4]	herding range

We first use the nominal parameter values provided in Tables 1 and 2 to find a nominal solution defender trajectories that result in the minimum probability of HVU destruction. With the results of these simulations as a reference point, we consider as uncertain each of the parameters that define attacker swarm model and weapon capabilities. In this simulation, these parameters are considered individually. The number of discretization nodes for parameter space was chosen by examination of the Hamiltonian. To illustrate this method and the results obtained in Section 4 we compute Hamiltonians for the Problem **SD** and Model 1 with $\theta = d_0, d_0 \in [0.5, 1.5]$ and $M = [5, 8, 11]$. As M increases the sequence of Hamiltonians should converge to the optimal Hamiltonian for the Problem **SD**. For Problem **SD** that should result in a constant, zero-valued Hamiltonian. Figure 3 shows the respective Hamiltonians for $M = [5, 8, 11]$. The value $M = 11$ was chosen for simulations, based on the approximately zero-valued Hamiltonian it generates.

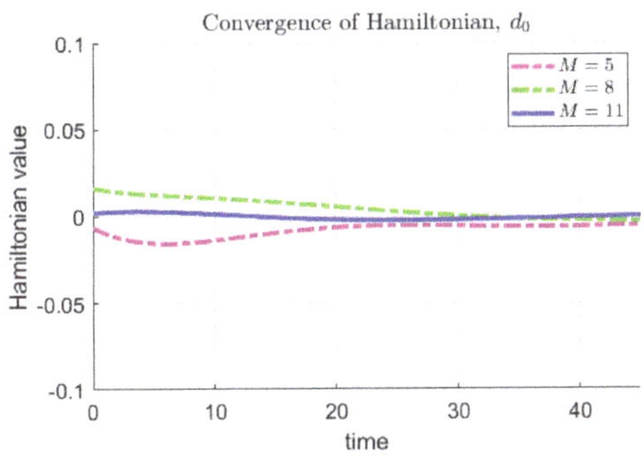

Figure 3. Convergence of Hamiltonion as number of parameter nodes M increases.

We compare the performance of the solution generated using uncertain parameter optimal control Problem **SD** versus a solution obtained with the nominal values. Figure 4 shows the nominal solution trajectories. The comparitive results of the nominal solutions vs the uncertain parameter control solutions are shown in Figure 5, where the performance of each is shown for different parameters values.

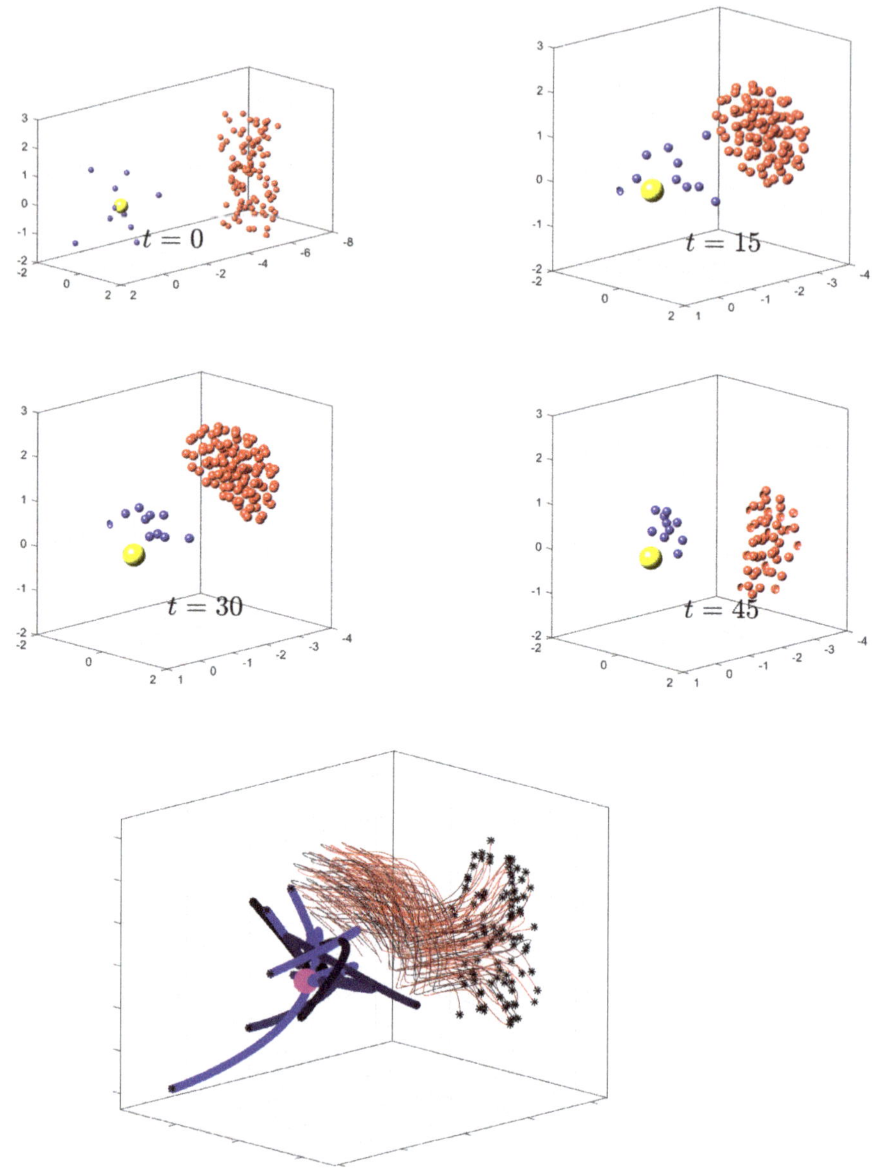

Figure 4. Shown are four snapshots during a simulations at $t = 0$, 15, 30, and 45 (time units are arbitrary). Defenders are represented by blue spheres and attackers by red spheres. The HVU is the yellow sphere. Below these snapshots, we show full trajectories for the entire simulation, which is the result of an optimization protocol using the parameters shown in Table 1.

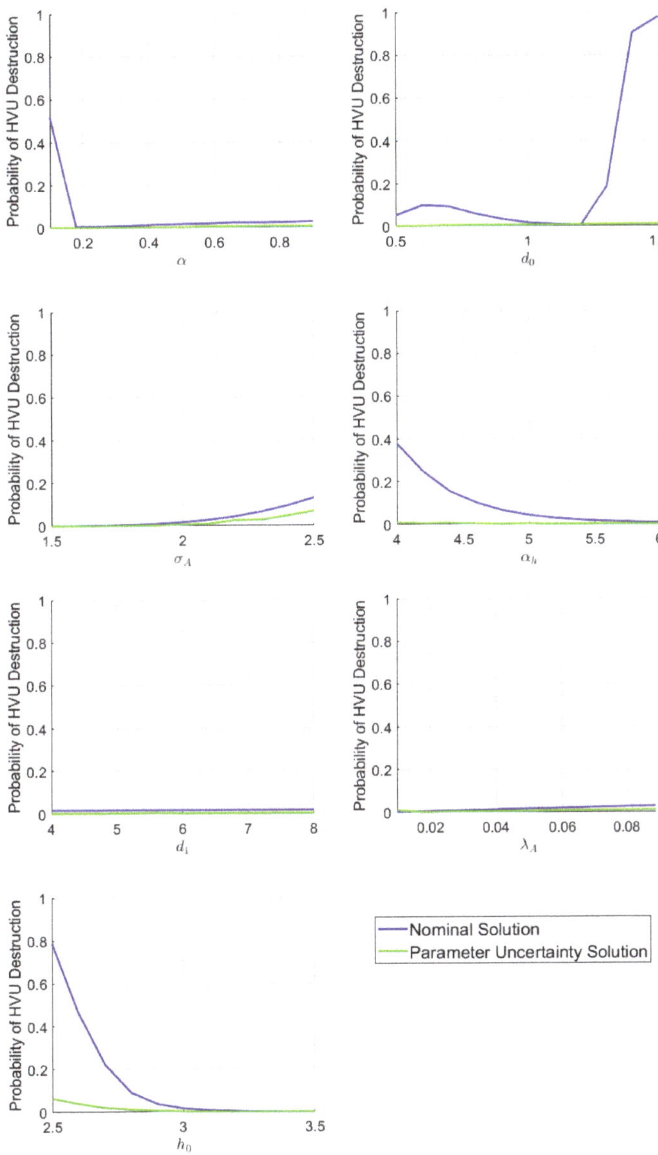

Figure 5. Performance of Solutions of Swarm Model 1 as parameter values are varied. Each panel illustrates a different varied parameter, stated on the x-axis.

As seen in Figure 5 the trajectories generated by optimization using the nominal values perform poorly over a range of α, d_0, σ_A, α_k and h_0. In the case of h_0, for example, this is because the attackers are less repelled by the defenders when h_0 is decreased, and they are more able to destroy the HVU from a longer distance as σ_A is increased. The parameter uncertainty solution, however, demonstrates that using the uncertain parameter optimal control framework a solution can be provided which is robust over a range of parameter values. We contrast these results with the case of uncertain parameters d_1 and λ_A, also shown in Figure 5. It can be seen that robustness improvements are modest to non-existent for these parameters. This suggests an insensitivity of the problem d_1 and λ_A parameters. This kind of analysis can be used to guide inference and observability priorities.

5.2. Example Model 2: Reynolds Boid Model

To demonstrate flexibility of the proposed framework to include diverse swarm models we have applied the same analysis as was done in Section 5.1 to the Reynolds Boid Model introduced in Section 2.1. We apply the same HVU tracking function as Equation (28). The herding force F_D of the defenders repelling attackers is applied as a separation force in the form of Equation (10). The fixed parameter values are the same as those in Table 1; the uncertain parameters and ranges are given in Table 3. The results are shown in Figure 6. Again, we see that the tools developed in this paper can be used to gain an insight into the robustness properties of the nominal versus uncertain parameter solutions. For example, we can see that the uncertain parameter solutions perform much better than the nominal ones for the cases where λ, σ and w_I are uncertain.

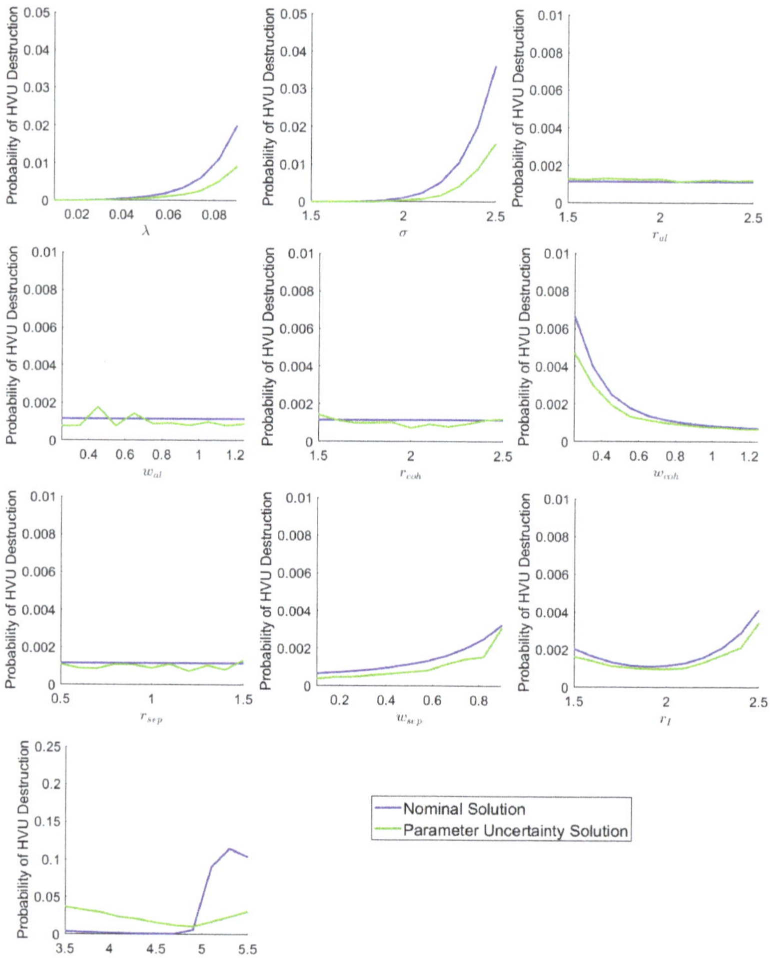

Figure 6. Performance of Solutions of Swarm Model 2 as parameter values are varied.

Table 3. Model 2 Varied Parameter Values.

Parameter	Nominal	Range	Meaning
λ_A	0.05	[0.01, 0.09]	weapon intensity
σ_A	2	[1.5, 2.5]	weapon range
r_{al}	2	[1.5, 2.5]	alignment range
w_{al}	0.75	[0.25, 1.25]	alignment intensity
r_{coh}	2	[1.5, 2.5]	cohesion range
w_{coh}	0.75	[0.25, 1.25]	cohesion intensity
r_{sep}	1	[0.5, 1.5]	separation range
w_{sep}	0.5	[0.1, 0 .9]	separation intensity
r_I	2	[1.5, 2.5]	herding range
w_I	4.5	[3.5, 5.5]	herding intensity

6. Conclusions

In this paper, we have built on our previous work on developing an efficient numerical framework for solving uncertain parameter optimal control problems. Unlike uncertainties introduced into systems due to stochastic "noise", parameter uncertainties do not average or cancel out in regard to their effects. Instead, each possible parameter value creates a specific profile of possibility and risk. The uncertain optimal control framework which has been developed for these problems exploits this inherent structure by producing answers which have been optimized over all parameter profiles. This approach takes into account the possible performance ranges due to uncertainty, while also utilizing what information is known about the uncertain features, such as parameter domains and prior probability distributions over the parameters. Thus, we are able to contain risk, while providing plans which have been optimized for performance under all known conditions. The results reported in this paper include analysis of the consistency of the adjoint variables of the numerical solution. In addition, the paper includes a numerical analysis of a large scale adversarial swarm engagement that clearly demonstrates the benefits of using the proposed framework.

There are many directions for future work for the topics of this paper. The numerical simulations in this paper consider the parameters individually, as one-dimensional parameter spaces. However, Problem **P** allows for multi-dimensional parameter spaces. A more dedicated implementation, taking advantage of the parallelizable form of Equation (16), for example, could certainly manage several simultaneous parameters. Exponential growth as parameter space dimension increases is an issue for both the quadrature format of Equation (15) and handling of the state space size for Equation (16). This can be somewhat mitigated by using sparse grid methods for high-dimensional integration to define the nodes in Equation (15). For large enough sizes, Monte Carlo sampling, rather than quadrature might be more appropriate for designating parameter nodes.

A further direction for future work would be to incorporate these methods into the design of more responsive closed-loop control solutions. The optimization methods in this paper provide open-loop controls. While useful, closed-loop controls would be more ideal for dynamic situations with uncertainty. There are many ways, however, that open-loop solutions can provide stepping stones to developing closed-loop solutions. For instance, Ref. [26] utilizes closed-loop solutions to train a neural network to learn an optimal closed-loop control strategy. Open-loop solutions can also be used to provide initial guesses to discretized closed-loop optimizations, seeding the optimization algorithm.

Another direction for future work is in the greater application of the duality results of Section 4. The numerical results in this paper simply utilize the Hamiltonian consistency. The proof of Theorem 1, however, additionally demonstrates the consistency of the adjoint variables for the problem. As the results demonstrate, parameter sensitivity for these swarm models is highly non-linear. The numerical solutions of Section 5 are able to demonstrate this sensitivity by applying the solution to varied parameter values. However, this is actually a fairly expensive method for a large swarm, as it involves re-evaluation of the swarm ODE for each parameter value. More importantly, it would not be scalable to high-dimensional parameter spaces, as the exponential growth of that approach to sensitivity analysis would be unavoidable. The development of an analytical adjoint sensitivity method for this problem could be of great utility for paring down numerical simulations to only focus on the parameters most relevant to success.

Author Contributions: Formal analysis, T.T.; Writing—original draft, C.W. and I.K.; Writing—review & editing, Q.G. and A.H.C. All authors have read and agreed to the published version of the manuscript.

Funding: This research was funded by the Office of Naval Research under Grant No. N0001421WX01974. Additionally, this work was supported by the Department of the Navy, Office of Naval Research, Consortium for Robotics Unmanned Systems Education and Research at the Naval Postgraduate School.

Institutional Review Board Statement: Not applicable.

Informed Consent Statement: Not applicable.

Data Availability Statement: Not applicable.

Acknowledgments: We would like to thank the University of Texas at San Antonio, the University of California Santa Cruz, and the Naval Postgraduate School for their administrative and research support.

Conflicts of Interest: The authors declare no conflict of interest.

Appendix A. Assumptions and Definitions

We we impose the assumptions in Section 2 of [4]. The definition of accumulation point used in the following proof can be found in Definition 3.2 of [4]. The following assumption is placed on the choice of numerical integration scheme to be utilized in approximating Problem **P**:

Assumption A1. *For each $M \in \mathbb{N}$, there is a set of nodes $\{\theta_i^M\}_{i=1}^M \subset \Theta$ and an associated set of weights $\{\alpha_i^M\}_{i=1}^M \subset \mathbb{R}$, such that for any continuous function $h : \Theta \to \mathbb{R}$,*

$$\int_\Theta h(\theta)d\theta = \lim_{M\to\infty} \sum_{i=1}^{M} h(\theta_i^M)\alpha_i^M.$$

This is the same as Assumption 3.1 of [4]; we include it for reference.

In additions to the assumptions of [4], we also impose the following:

Assumption A2. *The functions f and r are C^1. The set Θ is compact and $x_0 : \Theta \mapsto \mathbb{R}^{n_x}$ is continuous. Moreover, for the compact sets X and U defined in Assumptions 2.3 and 2.4 of [4], and for each $t \in [0, T]$, $\theta \in \Theta$, the Jacobians r_x and f_x are Lipschitz on the set $X \times U$, and the corresponding Lipschitz constants L_r and L_f are uniformly bounded in θ and t. The function F is C^1 on X for all $\theta \in \Theta$; in addition, F and F_x are continuous with respect to θ.*

Appendix B. Main Theorem Proof

The theorem relies on the following lemma:

Lemma A1. *Let $\{u_M\}$ be a sequence of optimal controls for Problem $\mathbf{P^M}$ with an accumulation point u^∞ for the infinite set $V \subset \mathbb{N}$. Let $(x^\infty(t,\theta), \lambda^\infty(t,\theta))$ be the solution to the dynamical system:*

$$\begin{cases} \dot{x}^\infty(t,\theta) = f(x^\infty(t,\theta), u^\infty(t), \theta) \\ \dot{\lambda}^\infty(t,\theta) = -\frac{\partial H(x^\infty(t,\theta),\lambda^\infty(t,\theta),u^\infty(t),t,\theta)}{\partial x} \end{cases} \quad \text{(A1)}$$

$$\begin{cases} x^{\infty}(0,\theta) = x_0(\theta) \\ \lambda^{\infty}(T,\theta) = \frac{\partial F(x^{\infty}(T,\theta),\theta)}{\partial x} \end{cases} \tag{A2}$$

where H is defined as per Equation (19), and let $\{(x_M(t,\theta), \lambda_M(t,\theta))\}$ for $M \in V$ be the sequence of solutions to the dynamical systems:

$$\begin{cases} \dot{x}_M(t,\theta) = f(x_M(t,\theta), u_M(t), \theta) \\ \dot{\lambda}_M(t,\theta) = -\frac{\partial H(x_M(t,\theta),\lambda_M(t,\theta),u_M(t),t,\theta)}{\partial x} \end{cases} \tag{A3}$$

$$\begin{cases} x_M(0,\theta) = x_0(\theta) \\ \lambda_M(T,\theta) = \frac{\partial F(x_M(T,\theta),\theta)}{\partial x} \end{cases} \tag{A4}$$

Then, the sequence $\{(x_M(t,\theta), \lambda_M(t,\theta))\}$ converges pointwise to $(x^{\infty}(t,\theta), \lambda^{\infty}(t,\theta))$ and this convergence is uniform in θ.

Proof. The convergence of $\{x_M(t,\theta)\}$ is given by Lemmas 3.4 and 3.5 of [4]. The convergence of the sequence of solutions $\{\lambda_M(t,\theta)\}$ is guaranteed by the optimality of $\{u_M\}$. The convergence of $\{\lambda_M(t,\theta)\}$ then follows the same arguments given the convergence of $\{x_M(t,\theta)\}$, utilizing the regularity assumptions placed on the derivatives of F, r, and f with respect to x to enable the use of Lipschitz conditions on the costate dynamics and transversality conditions. □

Remark A1. *Note that $\lambda_M(t,\theta)$ is not a costate of Problem $\mathbf{P^{\lambda M}}$, since it is a function of θ. However, when $\theta = \theta_i^M$, then $\lambda_M(t,\theta_i^M) = \tilde{\lambda}_i^M(t)$, where $\tilde{\lambda}_i^M$ is the costate of Problem $\mathbf{P^{\lambda M}}$ generated by the pair of solutions to Problem $\mathbf{P^M}$, $(\tilde{x}_i^M, u_M^*)$. In other words, the function $\lambda_M(t,\theta)$ matches the costate values at all collocation nodes. Since these values satisfy the dynamics equations of Problem $\mathbf{P^{\lambda M}}$, a further implication of this is that the values of $\lambda_M(t,\theta_i^M)$ produce feasible solutions to Problem $\mathbf{P^{\lambda M}}$.*

Remark A2. *Since the functions $\{(x_M(t,\theta), \lambda_M(t,\theta))\}$ obey the respective identities $x_M(t,\theta_i^M) = \tilde{x}_i^M(t)$ and $\lambda_M(t,\theta_i^M) = \tilde{\lambda}_i^M(t)$, their convergence to $(x^{\infty}(t,\theta), \lambda^{\infty}(t,\theta))$ also implies the convergence of the sequence of discretized primals and duals, $\{\tilde{X}_M\}$ and $\{\tilde{\Lambda}_M\}$, to accumulation points given by the relations*

$$\lim_{M\in V} \tilde{x}_i^M(t) = x^{\infty}(t,\theta_i^M), \quad \lim_{M\in V} \tilde{\lambda}_i^M(t) = \lambda^{\infty}(t,\theta_i^M)$$

We now prove Theorem 1. Let $\{(x_M(t,\theta), \lambda_M(t,\theta))\}$ for $M \in V$ be the sequence of solutions defined by Equation (A3) and let $(x^{\infty}(t,\theta), \lambda^{\infty}(t,\theta))$ be the accumulation functions defined by Equation (A1). Incorporating Remarks A1 and A2, we have:

$$\lim_{M\in V} \tilde{H}^M(\tilde{X}_M, \tilde{\Lambda}_M, u_M, t) =$$

$$\lim_{M\in V} \sum_{i=1}^{M} \alpha_i^M \left[\tilde{\lambda}_i^M(t) f(\tilde{x}_i^M(t), u(t), \theta_i^M) + r(\tilde{x}_i^M(t), u(t), t, \theta_i^M)\right]$$

$$= \lim_{M\in V} \sum_{i=1}^{M} \alpha_i^M \left[\lambda_M(t,\theta_i^M) f(x_M(t,\theta_i^M), u(t), \theta_i^M) + r(x_M(t,\theta_i^M), u(t), t, \theta_i^M)\right]$$

Due to the results of Lemma A1, and applying Remark 1 of [4] on the convergence of the quadrature scheme for uniformly convergent sequences of continuous functions, we find that:

$$\lim_{M\in V} \tilde{H}^M(\tilde{X}_M, \tilde{\Lambda}_M, u_M, t) =$$

$$\int_{\Theta}[\lambda^{\infty}(t,\theta)f(x^{\infty}(t,\theta),u^{\infty}(t),\theta)+r(x^{\infty}(t,\theta),u^{\infty}(t),t,\theta)]d\theta=\mathbf{H}(x^{\infty},\lambda^{\infty},u^{\infty},t)$$

thus proving the theorem.

References

1. Walton, C.; Gong, Q.; Kaminer, I.; Royset, J.O. Optimal Motion Planning for Searching for Uncertain Targets. *IFAC Proc. Vol.* **2014**, *47*, 8977–8982. [CrossRef]
2. Gong, Q.; Kang, W.; Walton, C.; Kaminer, I.; Park, H. Partial Observability Analysis of an Adversarial Swarm Model. *J. Guid. Control. Dyn.* **2020**, *43*, 250–261. [CrossRef]
3. Phelps, C.; Gong, Q.; Royset, J.O.; Walton, C.; Kaminer, I. Consistent approximation of a nonlinear optimal control problem with uncertain parameters. *Automatica* **2014**, *50*, 2987–2997. [CrossRef]
4. Walton, C.; Kaminer, I.; Gong, Q. Consistent numerical methods for state and control constrained trajectory optimisation with parameter dependency. *Int. J. Control* **2021**, *94*, 2564–2574. [CrossRef]
5. Weinan, E.; Han, J.; Li, Q. A mean-field optimal control formulation of deep learning. *arXiv* **2018**, arXiv:1807.01083.
6. Walton, C.; Phelps, C.; Gong, Q.; Kaminer, I. A Numerical Algorithm for Optimal Control of Systems with Parameter Uncertainty. *IFAC-PapersOnLine* **2016**, *49*, 468–475. [CrossRef]
7. Chung, S.J.; Paranjape, A.A.; Dames, P.; Shen, S.; Kumar, V. A Survey on Aerial Swarm Robotics. *IEEE Trans. Robot.* **2018**, *34*, 837–855. [CrossRef]
8. Mehmood, U.; Paoletti, N.; Phan, D.; Grosu, R.; Lin, S.; Stoller, S.D.; Tiwari, A.; Yang, J.; Smolka, S.A. Declarative vs rule-based control for flocking dynamics. In Proceedings of the 33rd Annual ACM Symposium on Applied Computing, Pau, France, 9–13 April 2018; pp. 816–823.
9. M. Wahab, S. Nefti-Maziani, A.A. A Comprehensive Review of Swarm Optimization Algorithms. *PLoS ONE* **2015**, *10*, e0122827.
10. Mavrovouniotis, M.; Li, C.; Yang, S. A survey of swarm intelligence for dynamic optimization: Algorithms and applications. *J. Swarm Evol. Comput.* **2017**, *33*, 1–17. [CrossRef]
11. Leonard, N.E.; Fiorelli, E. Virtual leaders, artificial potentials and coordinated control of groups. In Proceedings of the 40th IEEE Conference on Decision and Control (Cat. No. 01CH37228), Orlando, FL, USA, 4–7 December 2001; Volume 3, pp. 2968–2973.
12. Ogren, P.; Fiorelli, E.; Leonard, N.E. Cooperative control of mobile sensor networks: Adaptive gradient climbing in a distributed environment. *IEEE Trans. Autom. Control* **2004**, *49*, 1292–1302. [CrossRef]
13. Reynolds, C.W. *Flocks, Herds and Schools: A Distributed Behavioral Model*; ACM: New York, NY, USA, 1987; Volume 21.
14. Haque, M.; Rahmani, A.; Egerstedt, M. A hybrid, multi-agent model of foraging bottlenose dolphins. *IFAC Proc. Vol.* **2009**, *42*, 262–267. [CrossRef]
15. Strömbom, D.; Mann, R.P.; Wilson, A.M.; Hailes, S.; Morton, A.J.; Sumpter, D.J.; King, A.J. Solving the shepherding problem: heuristics for herding autonomous, interacting agents. *J. R. Soc. Interface* **2014**, *11*, 20140719. [CrossRef] [PubMed]
16. Pierson, A.; Schwager, M. Bio-inspired non-cooperative multi-robot herding. In Proceedings of the 2015 IEEE International Conference on Robotics and Automation (ICRA), Seattle, WA, USA, 26–30 May 2015; pp. 1843–1849.
17. Paranjape, A.A.; Chung, S.J.; Kim, K.; Shim, D.H. Robotic herding of a flock of birds using an unmanned aerial vehicle. *IEEE Trans. Robot.* **2018**, *34*, 901–915. [CrossRef]
18. Kolon, C.; Schwartz, I.B. The Dynamics of Interacting Swarms. *arXiv* **2018**, arXiv:1803.08817.
19. Szwaykowska, K.; Schwartz, I.B.; Romero, L.M.y.T.; Heckman, C.R.; Mox, D.; Hsieh, M.A. Collective motion patterns of swarms with delay coupling: Theory and experiment. *Phys. Rev. E* **2016**, *93*, 032307. [CrossRef]
20. Walton, C.; Lambrianides, P.; Kaminer, I.; Royset, J.; Gong, Q. Optimal motion planning in rapid-fire combat situations with attacker uncertainty. *Nav. Res. Logist.* **2018**, *65*, 101–119. [CrossRef]
21. Ruths, J.; Li, J.S. Optimal Control of Inhomogeneous Ensembles. *Trans. Autom. Control* **2012**, *57*, 2012–2032. [CrossRef]
22. Hager, W.W. Runge-Kutta Methods in optimal control and the transformed adjoint system. *Numer. Math.* **2000**, *87*, 247–282. [CrossRef]
23. Gong, Q.; Ross, I.M.; Kang, W.; Fahroo, F. Connections Between the Covector Mapping Theorem and Convergence of Pseudospectral Methods for Optimal Control. *Comput. Optim. Appl.* **2008**, *41*, 307–335. [CrossRef]
24. Gabasov, R.; Kirillova, F.M. *The Maximum Principle in Optimal Control Theory*; Publishing House Nauka i Tekhnika: Minsk, Belarus, 1974. (In Russian)
25. Hartl, R.F.; Sethi, S.P.; Vickson, R.G. A survey of the maximum principles for optimal control problems with state constraints. *SIAM Rev.* **1995**, *37*, 181–218. [CrossRef]
26. Nakamura-Zimmerer, T.; Gong, Q.; Kang, W. Adaptive Deep Learning for High-Dimensional Hamilton–Jacobi–Bellman Equations. *SIAM J. Sci. Comput.* **2021**, *43*, A1221–A1247. [CrossRef]

sensors

MDPI

Article

Over-Actuated Underwater Robots: Configuration Matrix Design and Perspectives †

Tho Dang [1,*], Lionel Lapierre [1,*], Rene Zapata [1], Benoit Ropars [2] and Pascal Lepinay [1]

1 Laboratory of Informatics, Robotics and MicroElectronics (LIRMM) (UMR 5506 CNRS—UM), Université Montpellier, 161 rue Ada, CEDEX 5, 34392 Montpellier, France; zapata@lirmm.fr (R.Z.); lepinay@lirmm.fr (P.L.)
2 Reeds Company, 199 rue Hélène Boucher, 34170 Castelnau-Le-Lez, France; ropars@lirmm.fr
* Correspondence: danghuu@lirmm.fr (T.D.); lapierre@lirmm.fr (L.L.)
† The manuscript is extended by our publication Dang, HT.; Lapierre, L.; Zapata, R.; Lepinay, P.; Ropars, B. Configuration Matrix Design of Over-Actuated Marine Systems. In Proceedings of the OCEANS 2019, Marseille, France, 17–20 June 2019; pp. 1–10, doi:10.1109/OCEANSE.2019.8867445.

Abstract: In general, for the configuration designs of underwater robots, the positions and directions of actuators (i.e., thrusters) are given and installed in conventional ways (known points, vertically, horizontally). This yields limitations for the capability of robots and does not optimize the robot's resources such as energy, reactivity, and versatility, especially when the robots operate in confined environments. In order to optimize the configuration designs in the underwater robot field focusing on over-actuated systems, in the paper, performance indices (manipulability, energetic, reactive, and robustness indices) are introduced. The multi-objective optimization problem was formulated and analyzed. To deal with different objectives with different units, the goal-attainment method, which can avoid the difficulty of choosing a weighting vector to obtain a good balance among these objectives, was selected to solve the problem. A solution design procedure is proposed and discussed. The efficiency of the proposed method was proven by simulations and experimental results.

Keywords: over-actuated underwater robots; multi-objective optimization; underwater robots; performance indices

Citation: Dang, T.; Lapierre, L.; Zapata, R.; Ropars, B.; Lepinay, P. Over-Actuated Underwater Robots: Configuration Matrix Design and Perspectives. *Sensors* **2021**, *21*, 7729. https://doi.org/10.3390/s21227729

Academic Editors: Reza Ghabcheloo and Antonio M. Pascoal

Received: 15 October 2021
Accepted: 18 November 2021
Published: 20 November 2021

1. Introduction

The *Actuation System* (AS) is an important part of marine robots. The AS groups the different actuators carried by the system. Following the generic Navigation–Guidance–Control (NGC) structure, the AS is in charge of realizing the desired force ($\mathbf{F}_B^d$) provided by the control system (see Figure 1).

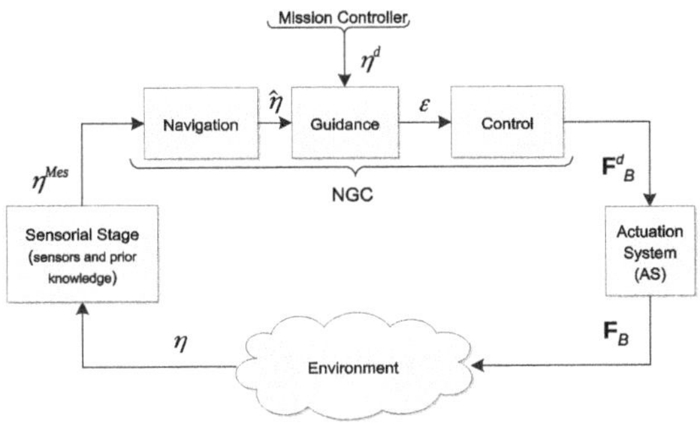

Figure 1. NGC structure augmented with the actuation system and sensorial stage.

Following Figure 1, the sensorial stage using sensors measurement and prior knowledge of the environment provides to the navigation system the necessary information to compute an estimation of the system state ($\hat{\eta}$). Then, the guidance system uses this estimation and the reference system state (η^d) provided by the mission controller to compute the error function (ε). The control system is then in charge of computing the desired force ($\mathbf{F}_B^d$) in order to reduce the error function to zero. Note that classically, this desired force is expressed in the body-frame. Afterwards, the *actuation system* produces in the environment a resulting force ($\mathbf{F}_B$), which should be as close as possible to $\mathbf{F}_B^d$. Note that, in this paper, the desired force ($\mathbf{F}_B^d$) and resulting force ($\mathbf{F}_B$) are (6×1) vectors and include force and torque elements. Inside the AS block, referring to Figure 2, the desired force ($\mathbf{F}_B^d$) is the output of the controller. Then, the dispatcher ($\mathcal{D}$) considers the actuator allocation method (and eventually, redundancy management) to compute the desired actuator force ($\mathbf{F}_m^d$) that each actuator has to produce. The inverse actuator characteristics are then considered in order to compute the actuator inputs ($\mathbf{c}_m$). Once applied, $\mathbf{c}_m$ can produce actuator forces ($\mathbf{F}_m$). The resulting force $\mathbf{F}_B$ is produced with respect to the actuator configuration ($\mathbf{A}$). The properties of the AS are indeed dependent on the actuator configuration (position and attitude of the actuators with respect to the body-frame), actuator dynamics (response characteristics), and dispatcher (control allocation, redundancy management) (see Figure 2) and afford the system different properties. Let us consider in the following that n is the number of Degrees of Freedom (DoFs) of the system and m is the number of actuators. If the system carries less actuators than DoFs, it is said to be *underactuated* (in that case, $\mathbf{A}$ will be an $(n \times m)$ matrix where $n > m$). Long-range Autonomous Underwater Vehicles (AUVs) and, for the terrestrial case, unicycle wheeled vehicles belong to this category [1]. In that case, specific nonlinear guidance strategies have to be used [2]. If the system carries more actuators than DoFs, it is said to be *redundant* ($n < m$). Then, there are different solutions ($\mathbf{c}_m$) to produce an identical resulting force ($\mathbf{F}_B$). Indeed, $\mathcal{D}$ is one of the multiple possible inverses of $\mathbf{A}$, classically $\mathcal{D} = \mathbf{A}^+$, where $\mathbf{A}^+$ is the Moore–Penrose pseudo-inverse. The properties of the AS play a pivotal role in the system performances, in terms of achievable dynamics, maneuverability, robustness, and dependability. The properties of an *overactuated* system have been studied in aerospace control, where critical safety is required [3], and for marine vehicles [4], where the harsh oceanic condition may easily produce actuator failure. Redundancy was also used in [5] in order to compensate different and unknown actuator responses. The domain of robotic manipulators has also extensively studied this question of redundancy, especially with recent works on humanoid robotics, where the task function approach [6] has been used to concurrently achieve equilibriums [7], walking pattern following [8], and multicontact management [9].

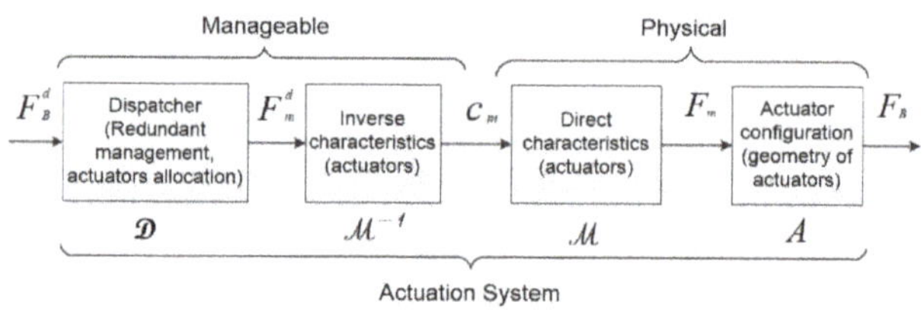

Figure 2. Actuation system scheme.

For a global evaluation of an *actuation system*, we should of course consider many factors, including redundancy management, the control allocation method, the actuator characteristics (inverse and direct), and the actuator configuration. This paper focuses on the study of the actuator configuration; for other problems, the readers can refer to [5] and the references therein.

Different performance criteria related to the actuator configuration design have been proposed. For mobile manipulation, the *manipulability index* [10] measures the manipulation capability of the end-effector. Intuitively, this index regards the set of all end-effector velocities, which is realizable by joint velocities. This set is called the hyper-manipulability ellipsoid. This index is quantified by computing the hyper-manipulability ellipsoid's properties. Based on these properties, there are different ways to quantify the manipulability index, including the volume of the hyper-manipulability ellipsoid, the ratio of the minimum and maximum radii of the hyperellipsoid, and the minimum radius of the hyperellipsoid. The selection depends on the purpose of evaluation. When the uniformity of manipulating ability is important, the ratio of two radii of the hyperellipsoid is chosen (the optimal value will be close to one). Otherwise, the minimum radius of the hyperellipsoid is suited for the case where the minimum manipulating ability might be critical [11]. Another criterion, *attainability* [12–14], was studied using workspace volume estimation.

In the underwater robotics field, the *manipulability index*, *energetic index*, and *force index* were introduced in [15], and the manipulability index was applied in [16]. Specifically, the *manipulability index* is used to measure the system's ability to exert a desired force with a specific actuator configuration. Therefore, the closer to one this index is, the better the robot's isotropy is, i.e., the robot can exert the same forces/torques in any direction. The *energetic index* is a measurement of the variation of system energy when the direction of the desired force changes. This is evaluated by measuring the energy consumption when the direction of a normalized desired force changes over a 3D sphere. The basic idea of the energetic index is to keep the system's energy consumption constant and as low as possible when the direction of action changes. The *force index* is used to measure the ratio between the actual maximum value and the minimum value of realizing forces. However, these studies only considered a given and fixed actuator configuration. Regarding the design of the actuator configuration of an overactuated underwater robot, a general problem is how to achieve an optimal configuration considering different performance indices. This is a challenging issue that raises two specific questions:

1. How do we define general and typical indices to evaluate an actuator configuration of an overactuated underwater robot?
2. How do we solve the complex optimal problem, which is normally nonconvex and has some conflicting objectives?

This paper focuses on the design of the actuator configuration for an overactuated underwater robot with the contributions outlined below:

1. We propose performance indices to evaluate the actuator configuration of underwater robots;
2. We optimize an actuator configuration design of an overactuated underwater robot with respect to different performance indices simultaneously.

This paper focuses on the design of an actuator configuration of an overactuated underwater robot, which optimizes different performance indices. Mathematically, an actuator configuration is a mapping from an actuator force vector to a resulting force vector (*note that these vectors include force and torque elements*). Since we considered an underwater robot equipped with thrusters, the mapping is from a thruster force vector ($\mathbf{F}_m$ space) to a body-frame vector ($\mathbf{F}_B$ space) (see Figure 3). The mapping operator is a matrix, which has different names in the literature such as: control effectiveness matrix [4,17], static transformation matrix [18], geometrical distribution of thrusters [19], configuration matrix [16]. In this paper, the mapping of an actuator configuration is called a *configuration matrix*, denoted as **A**.

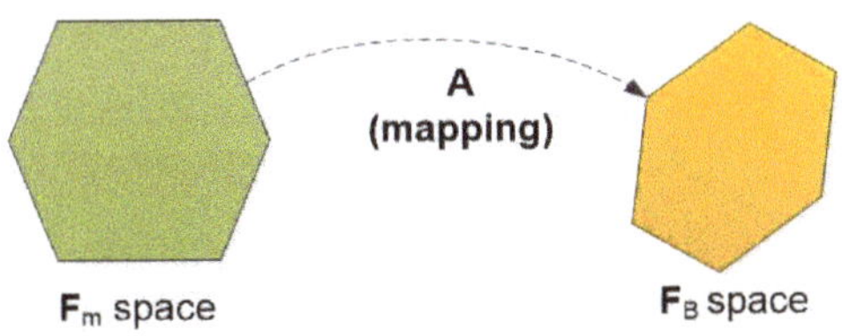

Figure 3. Actuator configuration mapping.

The paper is organized as follows. The notations are given in Section 2. The problem formulation and performance indices are described in Section 3. The problem's solution is displayed in Section 4. Simulation results and analyses are depicted in Section 5. Real experiments are depicted in Section 6. Finally, conclusions and future works are discussed in Section 7.

2. Notation

This section provides most of notations used in the whole paper. However, further notations are introduced when needed. In order to illustrate the notations, a given robot configuration is shown in Figure 4, and detailed explanations are given in Table 1.

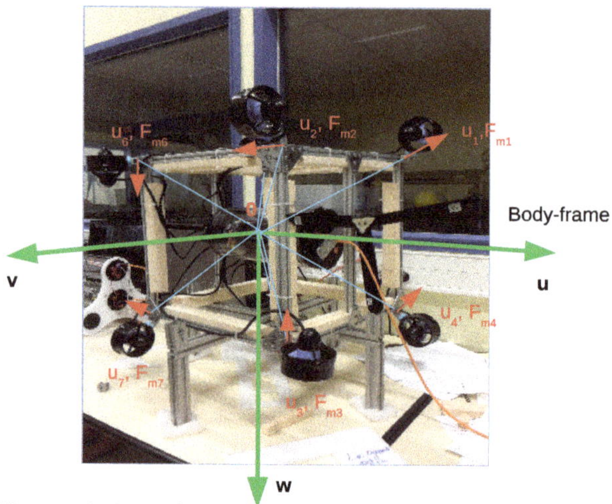

Figure 4. A given robot configuration.

Table 1. Notations.

$\mathbf{A}$	Configuration matrix
$\mathbf{A}^+$	Moore–Penrose pseudo-inverse of $\mathbf{A}$ matrix
$\mathbf{u}_i$	(3×1) unit vector of the direction of the ith thruster
$\mathbf{r}_i$	(3×1) unit vector of the position of the ith thruster
$\mathbf{F}_m$	$(m \times 1)$ force vector of m thrusters
$\mathbf{F}_m^d$	$(m \times 1)$ desired force vector of m thrusters
$F_{m,i}$	Force magnitude of the ith thruster
$\mathbf{F}_B^d$	(6×1) desired force (force and torque elements) w.r.t. the body-frame
$\mathbf{F}_B = \binom{\mathbf{F}}{\boldsymbol{\Gamma}}$	(6×1) resulting force (force and torque elements) w.r.t. the body-frame
$\mathbf{c}_m$	$(m \times 1)$ input vector of thrusters
$\otimes$	Cross product
$\|\cdot\|$	Euclidean norm
$\|\cdot\|_p$	p-norm
m	Number of thrusters
n	Number of Degrees of Freedom (DoFs)
$\mathbf{F}$	(3×1) vector of force elements in the resulting force $\mathbf{F}_B$
$\boldsymbol{\Gamma}$	(3×1) vector of torque elements in the resulting force $\mathbf{F}_B$
$\mathcal{D}$	Dispatcher
d_i	Distance of the ith thruster to the center of the body-frame
$cond(\mathbf{A})$	Condition number of the matrix $\mathbf{A}$
$Vol(.)$	Volume of a space
$rank(.)$	Rank of a matrix

3. Problem Formulation

The relation between the desired force ($\mathbf{F}_B^d$) and resulting force ($\mathbf{F}_B$) depends on different elements (see Figure 2). This paper only focuses on the actuator configuration. Therefore, three assumptions were considered:

1. *The inverse characteristics and direct characteristics of the actuators are perfectly known, i.e.,* $\mathbf{F}_m^d = \mathbf{F}_m$*;*
2. *The dispatcher is the Moore–Penrose pseudo-inverse of the actuators' configuration, i.e., if the actuators' configuration is the* $\mathbf{A}$ *matrix, the dispatcher is* $\mathcal{D} = \mathbf{A}^+$*;*
3. *All actuators have the same characteristics.*

3.1. Model of the Actuator Configuration

This part describes how to model an actuator configuration of an overactuated underwater robot equipped with thrusters. A thruster is modeled by the position and direction of the force produced with respect to the body-frame of the robot. The position of the ith thruster is described by a unit position vector $\mathbf{r}_i$ and the distance d_i to the system's Center of Mass (CM) in the body-frame. The direction of the ith thruster is represented by a unit direction vector $\mathbf{u}_i$ with respect to the body-frame as in Figure 5, and the ith thruster induces a force with the magnitude denoted as $F_{m,i}$. The relation of the thruster force vector and resulting force vector (note that this space includes force elements ($\mathbf{F}$) and torque elements ($\boldsymbol{\Gamma}$)) is described in Equation (1).

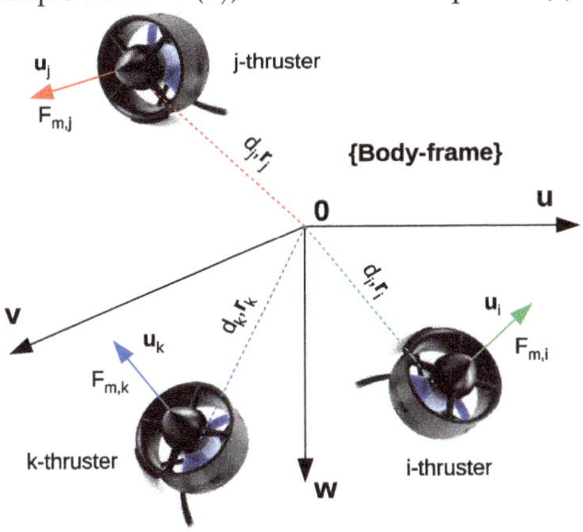

Figure 5. Actuator configuration model.

$$\mathbf{F}_B = \mathbf{A}\mathbf{F}_m = \begin{pmatrix} \mathbf{F} \\ \boldsymbol{\Gamma} \end{pmatrix} \tag{1}$$

where $\mathbf{F}_B = [F_u \quad F_v \quad F_w \quad F_p \quad F_q \quad F_r]^T \in \mathbb{R}^6$, $\mathbf{A} \in \mathbb{R}^{6\times m}$, $\mathbf{F}_m = [F_{m,1} \quad F_{m,2} \ldots F_{m,m}]^T \in \mathbb{R}^m$, and m is the number of thrusters, $m > 6$. The configuration matrix $\mathbf{A}$ is described as:

$$\begin{aligned} \mathbf{A} &= \begin{pmatrix} \mathbf{u}_1 & \mathbf{u}_2 & \cdots & \mathbf{u}_m \\ d_1\mathbf{r}_1 \otimes \mathbf{u}_1 & d_2\mathbf{r}_2 \otimes \mathbf{u}_2 & \cdots & d_m\mathbf{r}_m \otimes \mathbf{u}_m \end{pmatrix} \\ &= \begin{pmatrix} \mathbf{u}_1 & \mathbf{u}_2 & \cdots & \mathbf{u}_m \\ \boldsymbol{\tau}_1 & \boldsymbol{\tau}_2 & \cdots & \boldsymbol{\tau}_m \end{pmatrix} = \begin{pmatrix} \mathbf{A}_1 \\ \mathbf{A}_2 \end{pmatrix} \end{aligned} \tag{2}$$

where $\mathbf{A}_1$ and $\mathbf{A}_2 \in \mathbb{R}^{3\times m}$ are submatrices of $\mathbf{A}$, which correspond to the force and torque elements, respectively. It is obvious to see that $\boldsymbol{\tau}_i^T.\mathbf{u}_i = 0$. This is one constraint of the configuration matrix.

In this paper, we assumed that all distances from the thruster positions to the center of the body-frame are the same, $d_i = d_j = const, i, j = 1, \ldots, m$. Without loss of generality, we can assume that $d_i = 1, i = 1, \ldots, m$.

3.2. Manipulability Index

As mentioned before, the manipulability index was first introduced in [20] for manipulator mechanisms, and there are different ways to quantify the manipulability index. This paper focuses on the isotropic property of a marine robot. Then, the ratio between the maximum and minimum radii of the manipulability ellipsoid was chosen (see Figure 6). Because of the units' consistency, the matrices that relate to the force space, $\mathbf{A}_1$, and torque space, $\mathbf{A}_2$, were investigated separately. However, because of our assumption on d_i, the manipulability index is defined as the condition number of the configuration matrix:

$$I_m = Cond(\mathbf{A}) = \frac{\sigma_{max}}{\sigma_{min}} \tag{3}$$

where σ_{max} and σ_{min} are the maximum and minimum singular values of the configuration matrix, $\mathbf{A}$, respectively.

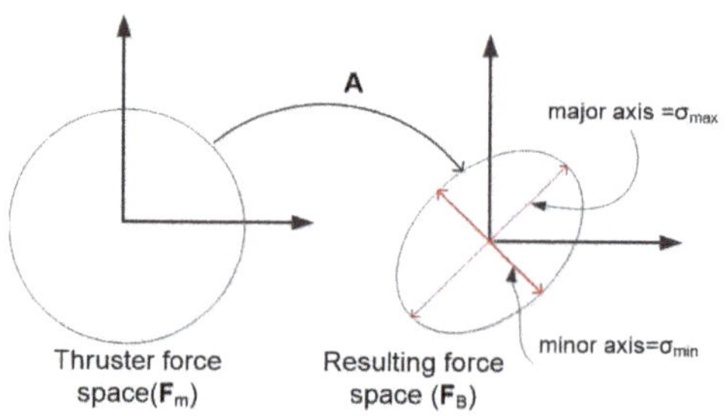

Figure 6. Manipulability ellipsoid with mapping.

Following Figure 6, the manipulability index investigates the resulting force ellipsoid, which is realizable by the thruster forces ($\mathbf{F}_m$) such that $\|\mathbf{F}_m\| \leq 1$ (see Theorem in Appendix A). If $I_m = 1$, the robot is isotropic, or if $I_m = \infty$, the robot cannot act along at least one direction.

3.3. Energetic Index

Energy is very important for marine robots, and the energy consumption of robots depends on many factors such as the mechanical design, the environmental effects, and the specific mission. In order to evaluate the energy performance of an underwater robot, the energetic index was introduced in [15]. Being different from this, in our paper, the two-norm of the thruster force vector, $p_E = \|\mathbf{F}_m\|_2$, was used to quantify the energy that an underwater robot consumes to produce forces and torques, which can be calculated as follows in Equation (4):

$$p_E = \|\mathbf{F}_m\|_2 = \sqrt{\sum_{i=1}^{m} F_{mi}^2} = \|\mathbf{A}^+.\mathbf{F}_B^d\|_2 \tag{4}$$

The energetic index is proposed to measure the variation of the energy consumption of an underwater robot when the direction of the desired force changes. It is quantified by computing the energy consumption when a unit desired force vector, ($\mathbf{F}_B^d$), changes over the unit hypersphere (see Figure 7 for a 3D sphere). Because of the units' consistency, however, the force and torque sphere were computed separately.

For the force sphere case, the unit desired force vector includes a unit vector of force elements and a zero vector of torque elements. For the torque sphere case, the unit desired force vector includes a zero vector of force elements and a unit vector of torque elements. Intuitively, this can be expressed as:

$$\mathbf{F}_B^d = \begin{pmatrix} \mathbf{F} \\ \boldsymbol{\Gamma} \end{pmatrix} = \begin{cases} \begin{pmatrix} \mathbf{u}_s \\ \mathbf{0} \end{pmatrix}, & \text{for the force sphere} \\ \begin{pmatrix} \mathbf{0} \\ \mathbf{u}_s \end{pmatrix}, & \text{for the torque sphere.} \end{cases} \tag{5}$$

where $\mathbf{u}_s = [\cos\alpha\cos\beta \quad \sin\alpha\cos\beta \quad \sin\beta]^T$ is a unit vector in spherical coordinates with $\alpha \in [-\pi, \pi]$ and $\beta \in [-\pi/2, \pi/2]$.

According to these two cases, the norm of the thruster force vector was also divided into two cases as follows:

$$p_E = \begin{cases} p_{Ef} = \|\mathbf{A}^+ \begin{pmatrix} \mathbf{u}_s \\ \mathbf{0} \end{pmatrix}\|, & \text{for the force sphere case} \\ p_{E\Gamma} = \|\mathbf{A}^+ \begin{pmatrix} \mathbf{0} \\ \mathbf{u}_s \end{pmatrix}\|, & \text{for the torque sphere case.} \end{cases} \tag{6}$$

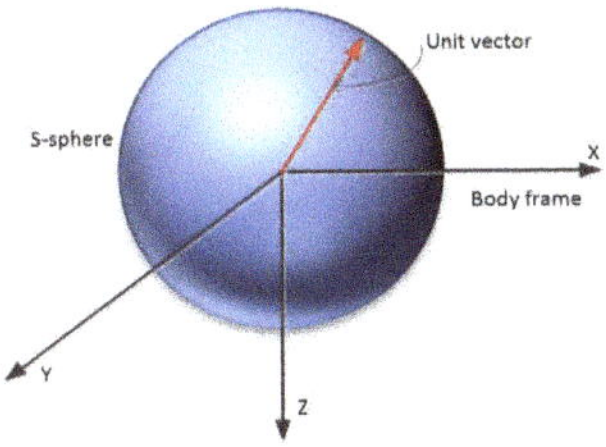

Figure 7. The rotation of the unit desired vector in the 3D sphere.

The energetic index is defined as:

$$I_e = \frac{1}{S}\int_S (w_{ef}p_{Ef} + w_{e\Gamma}p_{E\Gamma})dS \tag{7}$$

where S is the area of the three-dimensional sphere, p_{Ef}, $p_{E\Gamma}$ are the subvectors of p_E corresponding to the force sphere and the torque sphere case, respectively, and w_{ef} and $w_{e\Gamma}$ are the weighting coefficients. Note that the weighting coefficients were chosen to normalize the difference between the force and torque spheres. These choices depends on the robot's characteristics. However, because of the normalization of the spheres, it is normal to assign one to these coefficients.

3.4. Workspace Index

The term workspace volume was first introduced in [13] for manipulator mechanisms. In this paper, the work space index was used to measure the volume of the attainable regions of the resulting force space with respect to (w.r.t.) the body-frame. In general, the characteristics of thrusters always have limitations, namely saturations and dead-zones (in this index, the dead-zone is not considered). This yields the polytope of the thruster force space, the $\mathbf{F}_m$ space, denoted as $\mathbb{M}$. By properly choosing the configuration matrix, $\mathbf{A} = (\mathbf{A}_1\mathbf{A}_2)^T$, the volume of the resulting force space for the force, the $\mathbb{F}_F$ space, and the resulting force space for the torque, the $\mathbb{F}_T$ space, can be maximized (see Figure 8). Note that the resulting spaces for the force and torque were studied separately because of the units' consistency.

In general, the set $\mathbb{M}$ of thruster forces is known (with the given saturations of thrusters), so $\mathbb{M}$ is a polytope and $\mathbb{F}_F$ and $\mathbb{F}_T$ are also polytopes (under the $\mathbf{A_1}$ and $\mathbf{A_2}$ linear transform actions). We define the workspace index as:

$$I_w = \omega_{wf}Vol(\mathbb{F}_F) + \omega_{w\tau}Vol(\mathbb{F}_T) \tag{8}$$

where Vol is the volume of a polytope and ω_{wf} and $\omega_{w\tau}$ are the weighting coefficients.

In control perspectives, the larger the space's volumes are, the less control effort is need. The design objective was to maximize the workspace index, I_w. Normally, the set $\mathbb{M}$ is convex and its vertices are known. It is easy to find the vertices of $\mathbb{F}_F$ and $\mathbb{F}_T$. Of course, $\mathbb{F}_F$ and $\mathbb{F}_T$ are also convex sets (because of the linear transformation). This problem becomes a volume computation of convex polytopes.

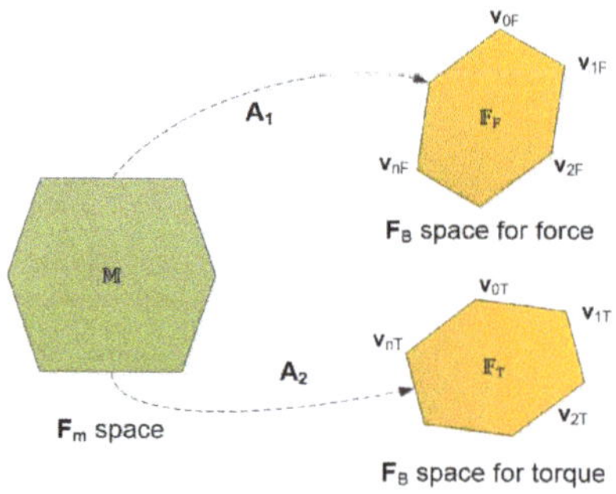

Figure 8. Space mapping ($\mathbf{v}_i$ is denoted as the vertex).

3.5. Reactive Index

The reactive index quantifies how fast the actuation system is able to change the orientation of the resulting force $\mathbf{F}_B$ (ideally, $\mathbf{F}_B^d$). Suppose that the robot is traveling in a direction with a set of thruster forces $\mathbf{F}_{m1}$ induced from desired force vector $\mathbf{F}_{B1}^d$. The robot wants to change to another direction (or the same direction with a different magnitude) with the desired force vector $\mathbf{F}_{B2}^d$, so the thrusters have to produce another set of thruster forces $\mathbf{F}_{m2}$. The two-norm of the deviation of the thruster forces, $\triangle\mathbf{F}_m = \mathbf{F}_{m1} - \mathbf{F}_{m2} = [\triangle F_{m1} \triangle F_{m2} \cdots \triangle F_{mm}]^T$, is considered as the reactive capability of the robot. Referring to the approximation of the characteristics of the thrusters, as Figure 9, the change from F_{m1} to F_{m2} is closer than that from F_{m1} to F_{m3} (in the linear section, the dead-zone of the thruster characteristics is neglected in this paper). Hence, we have:

$$\triangle\mathbf{F}_m = \mathbf{A}^+(\mathbf{F}_{B1}^d - \mathbf{F}_{B2}^d) = \mathbf{A}^+\triangle\mathbf{F}_B^d \tag{9}$$

$$||\triangle\mathbf{F}_m|| = ||\mathbf{A}^+\triangle\mathbf{F}_B^d|| \leq ||\mathbf{A}^+||\,||\triangle\mathbf{F}_B^d|| \tag{10}$$

$$\frac{||\triangle\mathbf{F}_m||}{||\triangle\mathbf{F}_B^d||} \leq ||\mathbf{A}^+|| \tag{11}$$

From Equation (11), the sensitivity of the thruster forces with respect to the desired forces, in other words the variation of the thruster forces w.r.t. the desired forces, is upper-bounded by the norm of the pseudo-inverse of the configuration matrix, $||\mathbf{A}^+||$. We define the reactive index as:

$$I_{re} = ||\mathbf{A}^+|| \tag{12}$$

It is obvious to see that if this index is lower, the robot is more reactive. Then, the objective of the design process is to minimize the reactive index.

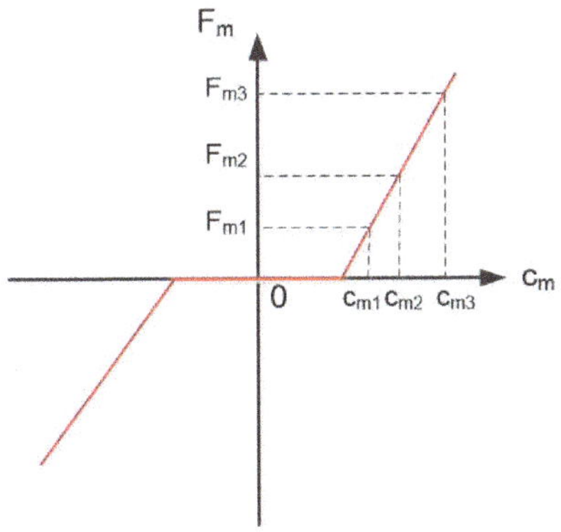

Figure 9. Thruster characteristics' approximation.

3.6. Robustness Index

This criterion measures the robustness level of the AS of an underwater robot. This means that if any thruster of the robot fails, the remaining ones can still perform the robot's mission. In particular, for any $\mathbf{F}_B^d$ vector, there always exists a $\mathbf{F}_m$ vector to satisfy the equation $\mathbf{F}_B = \mathbf{A}\mathbf{F}_m$, and $\mathbf{F}_B$ is as close as possible to $\mathbf{F}_B^d$.

We have:

$$\mathbf{F}_B = \mathbf{A}\mathbf{F}_m = \sum_{i=1}^{m} \mathbf{a}_i F_{m,i} \tag{13}$$

where $\mathbf{a}_i$ is the ith column of the matrix $\mathbf{A}$ and $F_{m,i}$ is the force magnitude of the ith thruster.

When one or more thrusters completely fail, the value of $F_{m,i} = 0$. Note that in the case that the ith thruster has partly failed, the value of $F_{m,i}$ remains small (not addressed in this paper). This is equivalent, as we considered that a corresponding column $\mathbf{a}_i$ of the configuration matrix $\mathbf{A}$ equals the zero vector. Therefore, Equation (13) is equivalent to:

$$\mathbf{F}_B = \mathbf{A}'\mathbf{F}_m \tag{14}$$

where the $\mathbf{A}'$ matrix is the $\mathbf{A}$ matrix with one or more corresponding columns equal to the zero vectors.

We discuss hereafter two questions: What are the conditions of the matrix $\mathbf{A}'$ to guarantee the robustness? What is the maximum number of failed thrusters?

To address these two questions, we supposed that k thrusters fail, and Equation (14) is a linear equation system with six equations (the dimensions of $\mathbf{F}_B$ are 6×1) and $(m-k)$ variables, because the matrix $\mathbf{A}'$ is $6 \times m$, where k columns are zero vectors. It is obvious to see that if $rank(\mathbf{A}') = 6$, for a given $\mathbf{F}_B^d$, there always exits $\mathbf{F}_m$ such that $\mathbf{F}_B = \mathbf{A}'\mathbf{F}_m$ and $\mathbf{F}_B$ is as close as possible to $\mathbf{F}_B^d$. This can be interpreted as $m-k \geq 6$ or $k \leq m-6$. The condition on the configuration matrix and that on the maximum number of failed thruster that guarantee the robustness of an underwater robot are stated as:

1. *The maximum failed thrusters: $m-6$;*
2. *Robustness condition: the rank of the configuration matrix always equals six, i.e., $rank(\mathbf{A}) = 6$, if any columns, from one to a maximum of $(m-6)$, of the $\mathbf{A}$ matrix equal the zero vectors. If $rank(\mathbf{A}) < 6$, the system becomes underactuated, and the guidance and control have to change to guarantee the robot's mission. This problem is not addressed in this paper.*

We define the robustness index as:

$$I_{ro} = rank(\mathbf{A}|_{\leq m-6}) = 6 \tag{15}$$

where $\mathbf{A}|_{\leq m-6}$ is the $\mathbf{A}$ matrix where the maximum number of columns being zero is $(m-6)$. This index is verified in the solving process of the problem.

3.7. Configuration Matrix Design Problem

With all the performance indices discussed above, we propose the following design problem:

$$\begin{aligned} &\min_{\mathbf{A}} \mathbf{V}(\mathbf{A}) = \min_{\mathbf{A}} [I_m \; I_e \; \frac{1}{I_w} \; I_{re}]^T \\ &s.t \quad \mathbf{A} \in \mathbb{A} \end{aligned} \tag{16}$$

where $\mathbf{V}(\mathbf{A})$ is the objective function vector. $\mathbb{A}$ is the feasible set of the configuration matrix $(\mathbf{A})$ including the constraints of the configuration matrix $(\mathbf{A})$ and the robustness index. The reciprocal of the workspace index, $\frac{1}{I_w}$, is in Equation (16), because we wanted to maximize the workspace index.

This is a multi-objective optimization problem, and the unique solution belongs to the convexity of each objective function in the objective vector and the feasible set, $\mathbb{A}$. Note that this optimization problem is with respect to a matrix variable (*matrix optimization*), not a vector variable. However, the optimization techniques for vector variables (*vector optimization*) can be applied here because we do not lose the physical meaning when converting a matrix variable to a vector variable in this optimization problem (because of the independency of each column in the matrix derived from the independent positions and orientations of the thrusters).

Specifically, Equation (16) can be rewritten as:

$$\begin{aligned} &\min_{\mathbf{A}} \mathbf{V}(\mathbf{A}) = \min_{\mathbf{A}} [I_m \; I_e \; \frac{1}{I_w} \; I_{re}]^T \\ &s.t \quad \|\mathbf{u}_i\| = 1, i = 1,2,\ldots,m \\ &\qquad \|\boldsymbol{\tau}_i\| \leq 1, i = 1,2,\ldots,m \\ &\qquad \boldsymbol{\tau}_i^T \mathbf{u}_i = 0, i = 1,2,\ldots,m \\ &\qquad I_{ro} = rank(\mathbf{A}|_{\leq m-6}) = 6 \end{aligned} \tag{17}$$

The problem (17) is to minimize an objective vector $\mathbf{V}(\mathbf{A})$, including the manipulability index, the energetic index, the reciprocal of the workspace index, and the reactive index, with respect to configuration matrix, $\mathbf{A}$, and to satisfy the constraints of the matrix structure itself and the robustness index. It is clear that this is a nonconvex and multi-objective optimization problem, which normally has many solutions. In the following sections, we propose a mathematical analysis and a method for solving the multi-objective optimization problem.

4. Problem Solution

Our final objective was to find a distribution (position and orientation) of the thrusters of an underwater robot. This means obtaining the $\mathbf{u}_i$ and $\mathbf{r}_i$ vectors for $i = 1,2,\ldots,m$. These vectors can be extracted from configuration matrix $\mathbf{A}$, which is the solution of the problem (17). Recall that our problem (17) is the multi-objective optimization problem with nonconvexity, and theoretically, this problem has infinitely many Pareto-optimal solutions. Our objective was to find one Pareto-optimal solution to build the robot. Analyzing the underlying mathematical properties of the problem helped us simplify the solving process. Thus, the mathematical analysis of the problem is shown in the next section.

4.1. Mathematical Analysis

The configuration matrix $\mathbf{A}$ has the form:

$$\mathbf{A} = \begin{pmatrix} \mathbf{u}_1 & \mathbf{u}_2 & \cdots & \mathbf{u}_m \\ \boldsymbol{\tau}_1 & \boldsymbol{\tau}_2 & \cdots & \boldsymbol{\tau}_m \end{pmatrix} \tag{18}$$

We have:

$$\mathbf{B} = \mathbf{A}^T\mathbf{A} = \begin{pmatrix} \mathbf{u}_1 & \mathbf{u}_2 & \cdots & \mathbf{u}_m \\ \boldsymbol{\tau}_1 & \boldsymbol{\tau}_2 & \cdots & \boldsymbol{\tau}_m \end{pmatrix}^T \begin{pmatrix} \mathbf{u}_1 & \mathbf{u}_2 & \cdots & \mathbf{u}_m \\ \boldsymbol{\tau}_1 & \boldsymbol{\tau}_2 & \cdots & \boldsymbol{\tau}_m \end{pmatrix} \tag{19}$$

$\mathbf{B}$ is an $m \times m$ symmetric matrix where each element is denoted as b_{ij}. We have:

$$\begin{aligned} Tr(\mathbf{B}) &= \sum_{i=1}^{m} b_{ii} \\ &= \sum_{i=1}^{m} \lambda_i \end{aligned} \tag{20}$$

where λ_i is the ith eigenvalue of matrix $\mathbf{B}$.

From Equations (19) and (20), we have:

$$\begin{aligned} \sum_{i=1}^{m} \lambda_i &= \sum_{i=1}^{m} \mathbf{u}_i^T\mathbf{u}_i + \boldsymbol{\tau}_i^T\boldsymbol{\tau}_i \\ &= \sum_{i=1}^{m} \|\mathbf{u}_i\|^2 + \|\boldsymbol{\tau}_i\|^2 \\ \sum_{i=1}^{m} \lambda_i &= \sum_{i=1}^{m} (1 + \|\boldsymbol{\tau}_i\|^2) \end{aligned} \tag{21}$$

In the case of manipulability index optimization, the condition of configuration matrix $\mathbf{A}$ is one, $cond(\mathbf{A}) = 1$. This means that the maximum singular value equals the minimum singular value, $\sigma_{max} = \sigma_{min}$. Note that the matrix $\mathbf{A}$ is the $n \times m$ matrix with $n < m$. The matrix $\mathbf{A}$ has n nonzero singular values (we have to guarantee that $rank(\mathbf{A}) = n$), then the matrix $\mathbf{B}$ has n nonzero eigenvalues and $m - n$ zero eigenvalues.

In the optimization case of the manipulability index, $cond(\mathbf{A}) = 1 \Rightarrow \sigma_{max} = \sigma_{min}$, we have $\lambda_i = \lambda_{max} = \lambda_{min} = \lambda$ ($\sigma = \sqrt{\lambda}$). Equation (21) is rewritten as:

$$\begin{aligned} n\lambda &= m + \sum_{i=1}^{m} \|\boldsymbol{\tau}_i\|^2 \\ \lambda &= \frac{m}{n} + \frac{1}{n}\sum_{i=1}^{m} \|\boldsymbol{\tau}_i\|^2 \end{aligned} \tag{22}$$

Considering the fact that $\|\boldsymbol{\tau}_i\|^2 \leq 1$, we have:

$$\lambda \leq 2.\frac{m}{n} \tag{23}$$

Therefore, we have $\lambda_{max} = 2\frac{m}{n}$ when $\|\boldsymbol{\tau}_i\|^2 = 1$.

In the singular-value decomposition of a matrix, when $cond(\mathbf{A}) = 1$, the matrix $\mathbf{A}$ can be written as:

$$\mathbf{A} = \mathbf{U}\mathbf{S}\mathbf{V}^T = \mathbf{U}[\sigma]_{n\times m}\mathbf{V}^T \tag{24}$$

where $\mathbf{U} \in \mathbb{R}^{n\times n}$, $\mathbf{V} \in \mathbb{R}^{m\times m}$ are orthogonal matrices, $\mathbf{S} = [\sigma]_{n\times m} = \begin{pmatrix} \sigma & 0 & \cdots & 0 \\ \vdots & \sigma & \cdots & 0 \\ 0 & \cdots & \sigma & 0 \end{pmatrix} \in \mathbb{R}^{n\times m}$

The pseudo-inverse of matrix $\mathbf{A}$ is $\mathbf{A}^+$ and can be written as:

$$\mathbf{A}^+ = \mathbf{V}\mathbf{S}^+\mathbf{U}^T = \mathbf{V}[\frac{1}{\sigma}]_{m\times n}\mathbf{U}^T \tag{25}$$

where $\mathbf{S}^{+} = [\frac{1}{\sigma}]_{m \times n} = \begin{pmatrix} \frac{1}{\sigma} & \cdots & 0 \\ \vdots & \frac{1}{\sigma} & 0 \\ 0 & 0 & \frac{1}{\sigma} \\ 0 & \cdots & 0 \end{pmatrix} \in \mathbb{R}^{m \times n}$

Our objective for the reactive index was to minimize $||\mathbf{A}^{+}||$. From Equation (25), the reactive index $I_{re} = ||\mathbf{A}^{+}|| = \frac{1}{\sigma}$, and the minimum value of the reactive index is equivalent to the maximum value of σ. This leads to the equality of Equation (23).

In order to minimize the reactive index and the manipulability index, the configuration matrix $\mathbf{A}$ has the following structure:

$$\begin{aligned} \mathbf{A} &= \mathbf{USV}^T \\ &= \mathbf{U} \begin{pmatrix} \sigma & 0 & \cdots & 0 & 0 & 0 \\ 0 & \sigma & 0 & \cdots & 0 & 0 \\ 0 & 0 & \sigma & 0 & \cdots & 0 \\ \vdots & \vdots & \vdots & \vdots & \vdots & \vdots \\ 0 & 0 & 0 & \sigma & 0 & 0 \end{pmatrix} \mathbf{V}^T \end{aligned} \tag{26}$$

where $\mathbf{S}(n \times m)$ is like-diagonal and $\sigma = \sqrt{\lambda} = \sqrt{2\frac{m}{n}}$; $\mathbf{U}(n \times n)$ and $\mathbf{V}(m \times m)$ are orthogonal matrices ($\mathbf{UU}^T = \mathbf{I}, \mathbf{VV}^T = \mathbf{I}$). This result can be used as the initial value of the numerical optimization process and is useful for solving the problem.

We continue to discuss the energetic index. First, we introduce a proposition as follows:

Proposition 1. *Let* $\mathbf{M}$ *be a* $p \times q$ *matrix* $(p \geq q)$, $\mathbf{M} \in \mathbb{R}^{p \times q}$. *For all* $\mathbf{x} \in \mathbb{R}^{q}$, *if* $\mathbf{M} = \mathbf{P\Sigma Q}^T$, *where* $\mathbf{P} \in \mathbb{R}^{p \times p}, \mathbf{Q} \in \mathbb{R}^{q \times q}$ *are orthogonal matrices,* $\mathbf{\Sigma} = \begin{pmatrix} \mu & 0 & \cdots & 0 \\ 0 & \mu & \cdots & 0 \\ 0 & \cdots & \mu & 0 \\ 0 & \cdots & 0 & \mu \\ \vdots & \vdots & \vdots & \vdots \\ 0 & 0 & 0 & 0 \end{pmatrix} \in \mathbb{R}^{p \times q}$, *then* $||\mathbf{Mx}|| = ||\mathbf{M}|| \, ||\mathbf{x}||$.

Proof. We have:

$$||\mathbf{Mx}||^2 = (\mathbf{Mx})^T(\mathbf{Mx}) = \mathbf{x}^T\mathbf{M}^T\mathbf{Mx} \tag{27}$$

With $\mathbf{M} = \mathbf{P\Sigma Q}^T$:

$$\begin{aligned} ||\mathbf{Mx}||^2 &= \mathbf{x}^T(\mathbf{P\Sigma Q}^T)^T(\mathbf{P\Sigma Q}^T)\mathbf{x} \\ &= \mathbf{x}^T\mathbf{Q\Sigma}^T\mathbf{P}^T\mathbf{P\Sigma Q}^T\mathbf{x} \\ &= \mathbf{x}^T\mathbf{Q\Sigma}^T\mathbf{\Sigma Q}^T\mathbf{x} \end{aligned} \tag{28}$$

We have:

$$\boldsymbol{\Sigma}^T\boldsymbol{\Sigma} = \begin{pmatrix} \mu & 0 & \cdots & 0 \\ 0 & \mu & \cdots & 0 \\ 0 & \cdots & \mu & 0 \\ 0 & \cdots & 0 & \mu \\ \vdots & \vdots & \vdots & \vdots \\ 0 & 0 & 0 & 0 \end{pmatrix}^T \begin{pmatrix} \mu & 0 & \cdots & 0 \\ 0 & \mu & \cdots & 0 \\ 0 & \cdots & \mu & 0 \\ 0 & \cdots & 0 & \mu \\ \vdots & \vdots & \vdots & \vdots \\ 0 & 0 & 0 & 0 \end{pmatrix}$$
$$= \begin{pmatrix} \mu^2 & 0 & \cdots & 0 \\ 0 & \mu^2 & \cdots & 0 \\ \vdots & \vdots & \vdots & \vdots \\ 0 & \cdots & 0 & \mu^2 \end{pmatrix} = \mu^2\mathbf{I} \tag{29}$$

where $\mathbf{I}$ is a $q \times q$ identity matrix.

Replacing Equation (29) to (28), we have:

$$\begin{aligned} ||\mathbf{Mx}||^2 &= \mathbf{x}^T\mathbf{V}\mu^2\mathbf{I}\mathbf{V}^T\mathbf{x} \\ &= \mu^2\mathbf{x}^T\mathbf{x} = ||\mathbf{M}||^2||\mathbf{x}||^2 \end{aligned} \tag{30}$$

Therefore, $||\mathbf{Mx}|| = ||\mathbf{M}||\,||\mathbf{x}||$. □

The energetic index is stated as:

$$I_e = \frac{1}{S}\int_S (w_{ef}||\mathbf{A}^+(\mathbf{F}_B^d(f)|| + w_{e\Gamma}||\mathbf{A}^+\mathbf{F}_B^d(\Gamma)||)dS \tag{31}$$

Choosing $w_{ef} = w_{e\Gamma} = 1$ (because the desired force vectors, $\mathbf{F}_B^d(f), \mathbf{F}_B^d(\tau)$, are units), we have:

$$I_e = \frac{1}{S}\int_S (||\mathbf{A}^+\mathbf{F}_B^d(f)|| + ||\mathbf{A}^+\mathbf{F}_B^d(\Gamma)||)dS \tag{32}$$

In the case in which the reactive index and the manipulability index are minimum, the configuration matrix $\mathbf{A}(n \times m)$ has the form of Equation (26); therefore, the pseudo-inverse matrix $\mathbf{A}^+(m \times n, \quad m > n)$ has the following structure:

$$\mathbf{A}^+ = \mathbf{V}\mathbf{S}^+\mathbf{U}^T = \mathbf{V}\begin{pmatrix} \frac{1}{\sigma} & 0 & \cdots & 0 \\ 0 & \frac{1}{\sigma} & \cdots & 0 \\ 0 & \cdots & \frac{1}{\sigma} & 0 \\ 0 & \cdots & 0 & \frac{1}{\sigma} \\ \vdots & \vdots & \vdots & \vdots \\ 0 & 0 & 0 & 0 \end{pmatrix}\mathbf{U}^T \tag{33}$$

where $\mathbf{V}, \mathbf{U}$ are orthogonal matrices.

It is clear that matrix $\mathbf{A}^+$ satisfies the condition of Proposition 1. Applying this proposition, we have: $||\mathbf{A}^+\mathbf{F}_B^d(f)|| = ||\mathbf{A}^+||\,||\mathbf{F}_B^d(f)||$ and $||\mathbf{A}^+\mathbf{F}_B^d(\Gamma)|| = ||\mathbf{A}^+||\,||\mathbf{F}_B^d(\Gamma)||$. Therefore, Equation (32) becomes:

$$\begin{aligned} I_e &= \frac{1}{S}\int_S (||\mathbf{A}^+||\,||\mathbf{F}_B^d(f)|| + ||\mathbf{A}^+||\,||\mathbf{F}_B^d(\Gamma)||)dS \\ &= \frac{1}{S}||\mathbf{A}^+||\int_S (||\mathbf{F}_B^d(f)|| + ||\mathbf{F}_B^d(\Gamma)||)dS \\ &= 2||\mathbf{A}^+|| \end{aligned} \tag{34}$$

From the aforementioned mathematical analysis of the energetic index, we can see that the energetic index belongs to the norm of the pseudo-inverse of the configuration matrix, $I_{re} = 2\|\mathbf{A}^+\|$, when the configuration matrix $\mathbf{A}$ has the form of (26).

We then discuss the upper-bound of the workspace index. For the units' consistency, the workspace index for the force space and that for the torque space were investigated separately, denoted as I_{wf} and $I_{w\tau}$, respectively. Recall that the objective of the workspace index is to maximize the volume of the resulting force space ($\mathbf{F}_B$ space), including the resulting space for the force and the resulting space for the torque with given thrusters' force (the $\mathbf{F}_m$ space).

The fact that for all vectors $\mathbf{F}_m \in \mathbb{R}^m$, $\|\mathbf{A}\mathbf{F}_m\| \leq \|\mathbf{A}\|\|\mathbf{F}_m\|$. The volume of the resulting force space is maximum when the equality holds.

Following Figure 10, the volumes of the resulting force spaces ($\mathbf{F}_B$) (the force and torque spaces) are always less than the volumes of the exterior hypersphere of $\mathbf{F}_B$ spaces of the force and torque (this may be the circumscribed spheres or not). This means that:

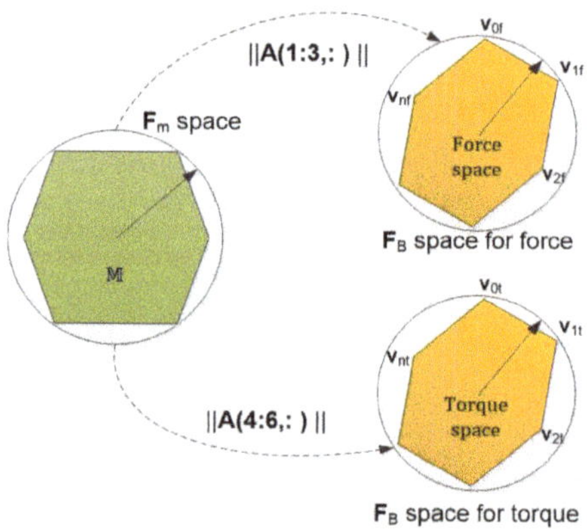

Figure 10. Upper-bound of the resulting force space.

$$\begin{aligned} I_{wF} &\leq Vol(\mathbf{B}(R1)) \\ I_{wT} &\leq Vol(\mathbf{B}(R2)) \end{aligned} \tag{35}$$

where $\mathbf{B}(R1)$ and $\mathbf{B}(R2)$ are Euclidean balls of radius $R1 = \|\mathbf{A}(\mathbf{1}:\mathbf{3},:)\|\|\mathbf{F}_m\| = \|\mathbf{A}_1\|\|\mathbf{F}_m\|$ and of radius $R2 = \|\mathbf{A}(\mathbf{4}:\mathbf{6},:)\|\|\mathbf{F}_m\| = \|\mathbf{A}_2\|\|\mathbf{F}_m\|$, respectively; $\mathbf{A}(\mathbf{1}:\mathbf{3},:)$ is composed of the first three rows of $\mathbf{A}$, and $\mathbf{A}(\mathbf{4}:\mathbf{6},:)$ is composed of the last three rows of $\mathbf{A}$.

The volume of a Euclidean ball of radius R in n-dimensional Euclidean space is [21]:

$$V_n(R) = \begin{cases} \frac{\pi^k}{k!}R^{2k}, & \text{if} n = 2k \\ \frac{2^{k+1}\pi^k}{(2k+1)!!}R^{2k+1}, & \text{if} n = 2k+1. \end{cases} \tag{36}$$

where $(2k+1)!! = 1.3.5, \ldots, (2k-1).(2k+1)$.

Proposition 2. *If the configuration matrix* $\mathbf{A}$ *has the form of* (26), *then* $cond(\mathbf{A}_1) = cond(\mathbf{A}_2) = 1$ *and* $\|\mathbf{A}_1\| = \|\mathbf{A}_2\| = \sigma$.

Proof. We have:

$$\begin{aligned} \mathbf{A}\mathbf{A}^T &= (\mathbf{U}\mathbf{S}\mathbf{V}^T)(\mathbf{U}\mathbf{S}\mathbf{V}^T)^T = \mathbf{U}\mathbf{S}\mathbf{V}^T\mathbf{V}\mathbf{S}^T\mathbf{U}^T \\ &= \mathbf{U}\mathbf{S}\mathbf{S}^T\mathbf{U}^T = \sigma^2\mathbf{I} \end{aligned} \tag{37}$$

On the other hand:

$$\mathbf{A}\mathbf{A}^T = \begin{pmatrix}\mathbf{A}_1\\ \mathbf{A}_2\end{pmatrix}\begin{pmatrix}\mathbf{A}_1\\ \mathbf{A}_2\end{pmatrix}^T = \begin{pmatrix}\mathbf{A}_1\\ \mathbf{A}_2\end{pmatrix}(\mathbf{A}_1^T\mathbf{A}_2^T)$$
$$= \begin{pmatrix}\mathbf{A}_1\mathbf{A}_1^T & \mathbf{0}\\ \mathbf{0} & \mathbf{A}_2\mathbf{A}_2^T\end{pmatrix} \quad (38)$$

From (37) and (38), we have:

$$\mathbf{A}_1\mathbf{A}_1^T = \sigma^2\mathbf{I}_1$$
$$\mathbf{A}_2\mathbf{A}_2^T = \sigma^2\mathbf{I}_2 \quad (39)$$

where $\mathbf{I}_1$ and $\mathbf{I}_2$ are partitioned matrices of matrix $\mathbf{I}$.

From (39) and the uniqueness of singular-value decomposition [22], it is obvious to see that the structures of $\mathbf{A}_1$ and $\mathbf{A}_2$ are the same as (26) with different dimensions. Therefore, $cond(\mathbf{A}_1) = cond(\mathbf{A}_1) = 1$ and $\|\mathbf{A}_1\| = \|\mathbf{A}_2\| = \sigma$. □

From (35) and (36) and Proposition 2, it is obvious to obtain the upper-bound of the resulting spaces of the force and torque of the system and then the upper-bound of the workspace index. Normally, the weighting coefficients in the workspace index are chosen as one because of our assumption of d_i.

4.2. Problem Solution

Based on the above mathematical analysis, the goal-attainment method was chosen to solve the problem with the given desired values. The idea of this method is to minimize the deviation of the desired values and the obtained values. One advantage of the goal-attainment method is that the problem does not need to be normalized to a dimensionless problem. The solution of this method has been proven to be Pareto-optimal. This method is also suitable when the feasible objective set is nonconvex [23]. All Pareto-optimal solutions may be found by changing the attainment vector; however, this depends on the properties of the problem.

Our problem using the goal-attainment method becomes:

$$\min_{\mathbf{A},\gamma} \gamma$$
$$s.t \quad \mathbf{A} \in \bar{\mathbb{A}}$$
$$\mathbf{V}(\mathbf{A}) - \mathbf{w}\gamma \le \mathbf{V}_{goal} \quad (40)$$

where $\bar{\mathbb{A}} = \mathbb{A} \setminus I_{ro}$, i.e., the feasible set of configuration matrices, $\mathbf{A}$, without robustness index I_{ro}, γ is a slack vector variable, and $\mathbf{V}_{goal} = [I_m^d \quad I_e^d \quad \frac{1}{I_w^d} \quad I_{re}^d]$ is the desired objective vector, $\mathbf{w}$ is an attainment vector, which can be chosen. The goal-attainment method with two objective functions is illustrated in Figure 11. By altering $\mathbf{w}$ vector, we searched for the Pareto-optimal solutions.

Therefore, our solving process included two phases:

1. Phase 1: Find one Pareto-optimal solution of the configuration matrix with the goal-attainment method;
2. Phase 2: Check the robustness index of the chosen solution in Phase 1.

The optimization toolbox in the MATLAB environment was used to solve our problem. Note that our problem was formulated as a multi-objective optimization problem. One objective has one desired value excluding the robustness index, which is as a constraint, and therefore, the desired vector is set up. The goal-attainment method was used to solve the problem. An attainment vector was chosen as a trade-off between the underachievement and overachievement of the objective functions. In multi-objective optimization, an optimal solution depends on a decision maker. Theoretically, there is no method for this choice. In our work, this vector was selected by trial and error. In particular, for the manipulability, reactivity, and energetic indices, we know exactly the desired values, so the corresponding values in the attainment vector were chosen as zero, which means that these are hard constraints. For the workspace index, we only know the upper-bound of the desired value; therefore, a positive value was chosen for underachievement.

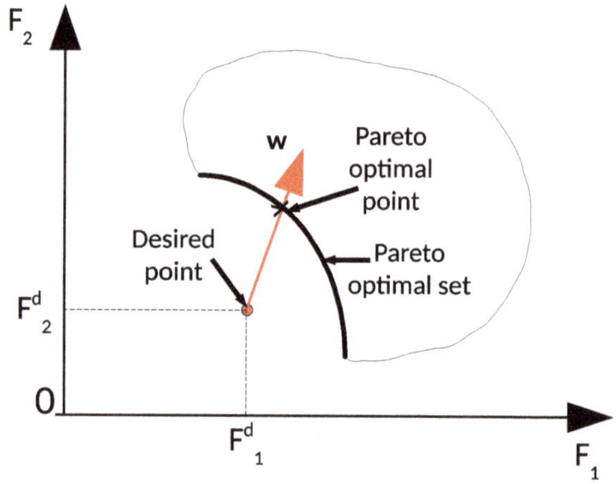

Figure 11. Goal-attainment method with two objective functions.

5. Simulation Results

We designed an overactuated underwater robot with $m = 8$ thrusters and $n = 6$ degrees of freedom. Two cases were simulated: the general case and the given position case. In the general case, we have to identify both the positions and orientations of eight thrusters optimizing the performance indices. In the given position case, the thrusters are installed at the corners of a cube, and we only have to determine the directions of the thrusters. In this simulation, the thruster characteristics were chosen as in [5], then the maximum and minimum values of the thrusters forces were $F_{imax} = 1.1N$ and $F_{imin} = -0.4N$, respectively. The desired values of the performance indices were subsequently $I_m^d = 1$, $I_e^d = 1.2248, I_{wF}^d = 597.7, I_{wT}^d = 597.7, I_{re}^d = 0.6124$ ($\sigma^{max} = \sqrt{2\frac{m}{n}} = 1.6330$; see Table 2 for more details).

Table 2. Desired values of the indices.

Index	Optimal Formula and Condition	Desired Value
I_m^d	$\sigma_{max} = \sigma_{min}$	1
I_e^d	$2 \|\|\mathbf{A}^+\|\|$	1.2248
$\frac{1}{I_w^d}$	see Equations (35) and (36) and $\frac{1}{I_w^d} = \frac{1}{I_{wF}^d} + \frac{1}{I_{wT}^d}$	0.0033
I_{re}^d	$\frac{1}{\sigma^{max}}$	0.6124

5.1. General Case

In this case, the robot is called a ball robot, and the positions and orientations of the thrusters are not known. The problem (40) is solved as follows.

5.1.1. Phase 1

The optimization toolbox was used to solve the problem (40) with the desired goal vector, and the constraints were $\mathbf{V}_{goal} = [I_m^d \quad I_e^d \quad \frac{1}{I_w^d} \quad I_{re}^d] = [1 \quad 1.2248 \quad 0.0033 \quad 0.6124]^T$, the constraint set $\bar{\mathbb{A}} = \{\mathbf{A} \in \mathbb{R}^{6\times 8} / \|\mathbf{u}_i\| = 1, \|\boldsymbol{\tau}_i\| \leq 1, \boldsymbol{\tau}_i^T \mathbf{u_i} = 0\}$, and the attainment vector $\mathbf{w} = [0 \quad 0 \quad 0 \quad 0.0036]^T$. The attainment vector allows setting the overachievement or underachievement of the individual goals. At the moment, there is no general method to choose this attainment vector. It was chosen by trial and error. By our approach, we found that some values of this vector can be assigned zero (this imposes hard constraints), except the workspace index (because of the upper-bound value).

The simulation results are shown in Figures 12 and 13a,b. The configuration matrix **A** and optimal values are shown in Table 3. Specifically, in Figure 12, the positions of the thrusters are at the top of the blue line, and the orientations of the thrusters are shown as the red arrow. Furthermore, we can see that the isotropic property of the robot is guaranteed (see Figure 13a,b) with the sphere shapes of the attainable spaces of the forces and torques. From Table 3, the obtained values of the manipulability index, the energetic index, and the reactive index were almost the same as the desired values. However, the obtained workspace index was smaller than the desired one.

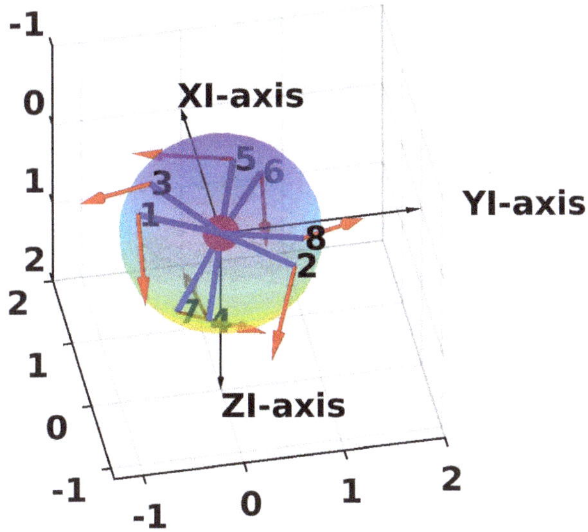

Figure 12. Positions and directions of the thrusters (general case) (XI-axis = **u**-axis; YI-axis = **v**-axis; ZI-axis=**w**-axis).

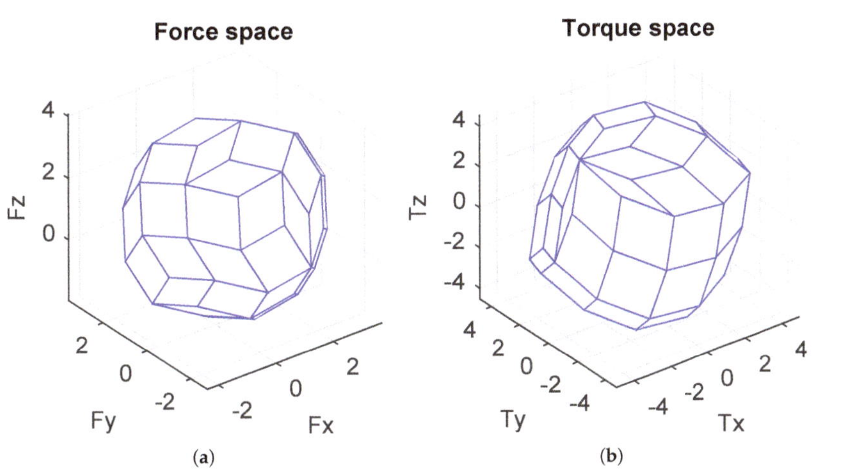

Figure 13. Attainable spaces in the general case (Fx-axis = **u**-axis; Fy-axis = **v**-axis; Fz-axis = **w**-axis). (**a**) Force space. (**b**) Torque space.

Table 3. Configuration matrix in the general case.

Configuration Matrix	Optimal Value	Attainment Factor
$\mathbf{A} = \begin{pmatrix} -0.8891 & -0.3645 & 0.5438 & 0.9879 & 0.3134 & 0.0148 & 0.0495 & 0.6090 \\ -0.0985 & -0.3036 & -0.5911 & -0.0608 & -0.9493 & 0.0515 & 0.8919 & 0.7158 \\ 0.4471 & 0.8803 & 0.5957 & 0.1429 & 0.0260 & 0.9986 & 0.4495 & 0.3417 \\ -0.4308 & 0.4701 & -0.8386 & 0.0379 & -0.1336 & 0.5628 & -0.9972 & 0.4758 \\ 0.5107 & 0.7561 & -0.4103 & 0.9868 & -0.0712 & -0.8259 & 0.0690 & 0.0149 \\ -0.7441 & 0.4554 & 0.3583 & 0.1577 & -0.9885 & 0.0342 & -0.0272 & -0.8794 \end{pmatrix}$	$Fval = \begin{pmatrix} 1.0000 \\ 1.2200 \\ 0.0050 \\ 0.6124 \end{pmatrix}$	0.3896

5.1.2. Phase 2

In this phase, the robustness index was checked. The optimal configuration matrix **A** in Table 3 satisfies the robustness constraint. Specifically, the maximum number of thrusters that are acceptable (for failures) is two.

5.2. Given Position Case

In this case, the robot is called the cube robot, and the positions of thrusters are at the corners of the cube. We only had to find their orientations. The number of variables in the problem (40) was reduced. The desired objective vector and attainment vector were the same as in general case. The results are presented in the sequel.

5.2.1. Phase 1

The optimization toolbox was used to solve our problem, and the simulation results are shown in Figures 14 and 15a,b and Table 4. The directions of the thrusters are depicted as red arrows in Figure 14. Being similar to the general case, the isotropic property is also guaranteed in this case (see Figure 15a,b). One Pareto-optimal configuration matrix is shown in Table 4. We can see that the obtained objective values in Table 4 are the same as the general case.

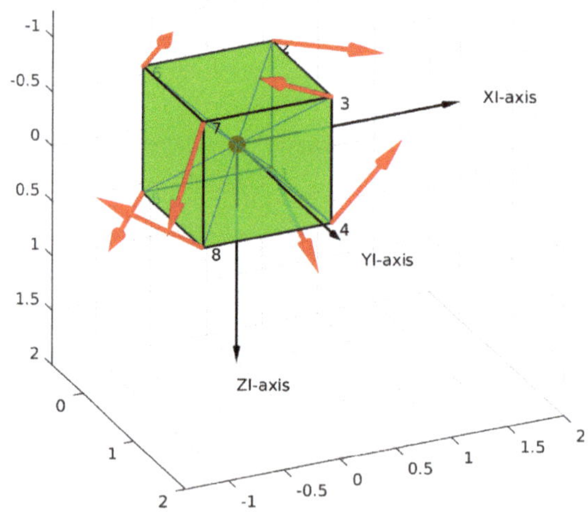

Figure 14. Robot design with the directions of the thrusters (given position case) (XI-axis = **u**-axis; YI-axis = **v**-axis; ZI-axis = **w**-axis).

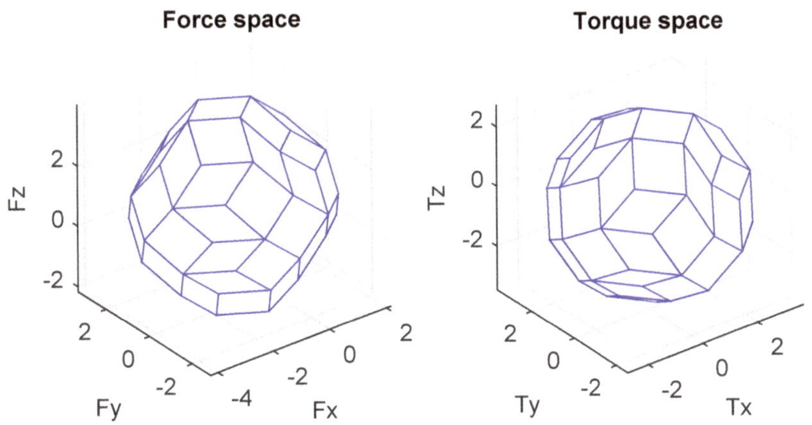

(a) (b)

Figure 15. Attainable spaces in the given position case (Fx-axis = **u**-axis; Fy-axis = **v**-axis; Fz-axis = **w**-axis). (**a**) Force space. (**b**) Torque space.

Table 4. Configuration matrix in the given position case.

Configuration Matrix	Optimal Value	Attainment Factor
$\mathbf{A} = \begin{pmatrix} 0.0836 & 0.6616 & -0.8122 & 0.4785 & -0.6616 & -0.0836 & -0.4785 & -0.8122 \\ 0.7452 & 0.7452 & 0.3337 & 0.3337 & 0.7452 & 0.7452 & 0.3337 & -0.3337 \\ 0.6616 & -0.0836 & -0.4785 & -0.8122 & 0.0836 & -0.6616 & 0.8122 & -0.4785 \\ -0.8122 & 0.4785 & -0.0836 & -0.6616 & -0.4785 & 0.8122 & 0.6616 & -0.0836 \\ -0.3337 & -0.3337 & 0.7452 & 0.7452 & -0.3337 & -0.3337 & 0.7452 & -0.7452 \\ 0.4785 & 0.8122 & 0.6616 & -0.0836 & -0.8122 & -0.4785 & 0.0836 & 0.6616 \end{pmatrix}$	$Fval = \begin{pmatrix} 1.0000 \\ 1.2200 \\ 0.0050 \\ 0.6124 \end{pmatrix}$	0.3868

5.2.2. Phase 2

The optimal configuration matrix **A** in Table 4 satisfies the conditions of the robustness index. Similarly, the maximum number of thrusters that can fail is two.

5.3. A Comparison of Two Configurations

In this section, a comparison of two configurations is illustrated. The choice of the configurations corresponds to a real robot (cube robot), which is used in the experiments in the next section. The first one is a normal configuration (denoted as $\mathbf{C}^1$) in which the thrusters are distributed vertically or horizontally (in practice, this configuration is easier to install, as Figure 21 shows). The configuration matrix of the $\mathbf{C}^1$ configuration, denoted as $\mathbf{A}_1$, is shown in Equation (41).

$$\mathbf{A}_1 = \begin{pmatrix} 0 & 1 & 0 & 0 & 0 & 0 & -1 & 0 \\ 1 & 0 & 0 & -1 & 1 & 0 & 0 & 0 \\ 0 & 0 & -1 & 0 & 0 & 1 & 0 & -1 \\ 0.27 & 0 & -0.27 & 0.27 & 0.27 & 0.27 & 0 & 0.27 \\ 0 & -0.27 & 0.27 & 0 & 0 & 0.27 & -0.27 & -0.27 \\ 0.27 & -0.27 & 0 & 0.27 & 0.27 & 0 & 0.27 & 0 \end{pmatrix} \quad (41)$$

The second one (denoted as $\mathbf{C}^2$) is an optimal configuration, denoted as $\mathbf{A}_2$, which is a solution of the optimization problem (given position case) thanks to the thruster characteristics of BlueRobotics (Figure 16), and the optimal configuration matrix is shown in Equation (42).

$$\mathbf{A}_2 = \begin{pmatrix} 0.6616 & -0.8122 & 0.4785 & 0.0836 & -0.0836 & -0.4785 & -0.8122 & -0.6616 \\ 0.7452 & 0.3337 & 0.3337 & 0.7452 & 0.7452 & 0.3337 & -0.3337 & 0.7452 \\ -0.0836 & -0.4785 & -0.8122 & 0.6616 & -0.6616 & 0.8122 & -0.4785 & 0.0836 \\ 0.1608 & 0.0111 & -0.2459 & -0.3708 & 0.3642 & 0.2015 & 0.0011 & -0.1658 \\ -0.0989 & 0.3556 & 0.3633 & -0.0989 & -0.1056 & 0.3508 & -0.3456 & -0.1056 \\ 0.3906 & 0.2292 & 0.0044 & 0.1583 & -0.1649 & -0.0254 & 0.2392 & -0.3708 \end{pmatrix} \quad (42)$$

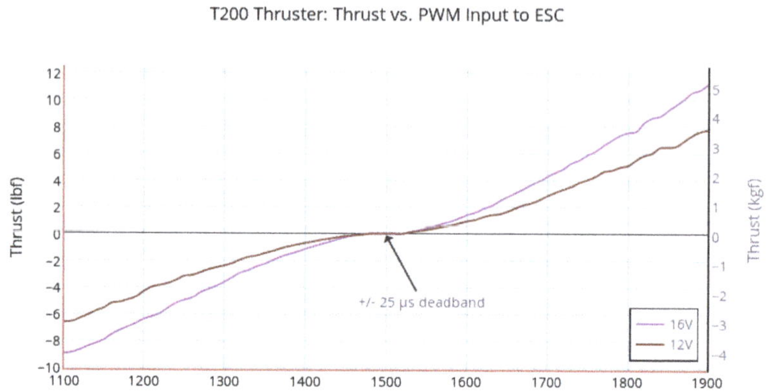

Figure 16. Thruster characteristics (BlueRobotics) [24].

Note that the configuration matrices $\mathbf{A}_1$ and $\mathbf{A}_2$ were calibrated with the corresponding geometrical properties of the real cube robot at the LIRMM Institute, Montpellier University. The attainable force space and torque space corresponding to the two configurations $\mathbf{C}^1$ and $\mathbf{C}^2$ are illustrated in Figure 17a,b. It is obvious that the $\mathbf{C}^2$ configuration is more isotropic than the $\mathbf{C}^1$ configuration. However, for some specific points of the attainable force and torque spaces, the $\mathbf{C}^1$ configuration is better than the $\mathbf{C}^2$ configuration.

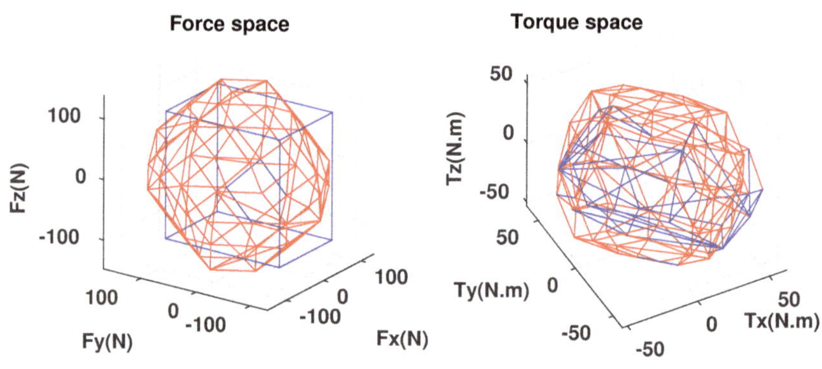

(**a**) (**b**)

Figure 17. Attainable spaces for different configurations (X-axis = **u**-axis; Y-axis = **v**-axis; Z-axis = **w**-axis). (**a**) $\mathbf{C}^1$(blue), $\mathbf{C}^2$(red). (**b**) $\mathbf{C}^1$(blue), $\mathbf{C}^2$(red).

Thanks to the properties of matrices $\mathbf{A}_1$ and $\mathbf{A}_2$ (Equations (41) and (42)) and the thruster characteristics (Figure 16), Table 5 shows the values of the performance indices for both configurations. The performances of the $\mathbf{C}^2$ configuration are better than $\mathbf{C}^1$. Because of the calibration (the distance d_i is different between the motors), the manipulability index (I_m) is larger than one (Note that, theoretically, the distances of all thrusters with respect to the center of mass were assumed the same (without loss of generality, they were assigned one). However, in practice, for our cube robot, these distances were not completely the same, and we had to calibrate the configuration matrix.).

Table 5. Comparison between the two configurations (I_{ro} shows the maximum number of thrusters that can fail to make sure that $rank(\mathbf{A} = 6)$).

No.	Indices	C^1	C^2
1	I_m	7.12	2.559
2	I_e	3.32	2.09
3	I_w	6,511,536.45	10,919,428.13
4	I_{re}	4.05	1.56
5	I_{ro}	0	2

In order to verify the attainability of the two configurations (workspace index), incremental torques were applied about the **u**-, **v**-, and **w**-axis, respectively (Figures 18a, 19a and 20a), and the corresponding Pulse-Width Modulation (PWM) inputs ($\mathbf{c}_m$) of the eight thrusters were computed. The results are shown in Figures 18b,c, 19b,c and 20b,c, in which the two PWM saturation values of the thrusters (upper saturation value: 1900, lower saturation value: 1100) are plotted with two bold lines. We can see that the performances of the robot with the two configurations are almost the same for the rotation about the **u**- and **v**-axis. However, the $\mathbf{C}^2$ configuration showed better performance for the rotation about the **w**-axis. In fact, the thrusters with the $\mathbf{C}^1$ configuration reached saturations very earlier in comparison with the thrusters with the $\mathbf{C}^2$ configuration (Figure 20b,c).

In order to validate the robustness of the optimal configuration ($\mathbf{C}^2$) in comparison with the normal configuration ($\mathbf{C}^1$), the rank of matrices $\mathbf{A}_1$ and $\mathbf{A}_2$ was checked when one or two arbitrary columns have been nullified. When the resulting matrices are rank deficient, this means that the robustness is not guaranteed because one direction is not actuated. Therefore, we cannot control all 6 DoFs independently. The robustness index in Table 5 shows the checking results. In particular, when the fifth thruster of the $\mathbf{C}^1$ configuration fails, the robustness is not guaranteed.

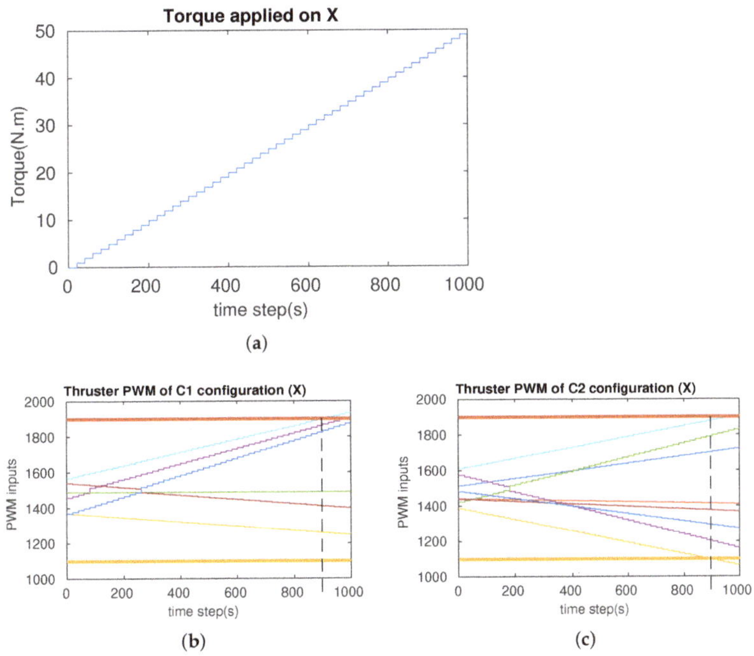

Figure 18. The simulation of the cube rotation about the **u**-axis for $\mathbf{C}^1$ and $\mathbf{C}^2$ (X-axis = **u**-axis). (**a**) Applied torque about the **u**-axis. (**b**) PWM inputs of $\mathbf{C}^1$. (**c**) PWM inputs of $\mathbf{C}^2$.

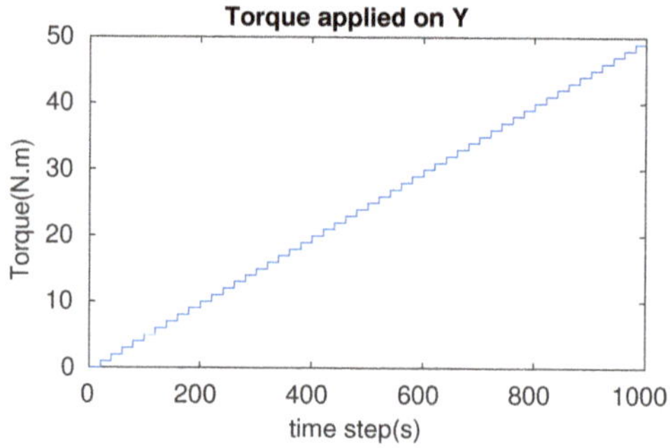

(a)

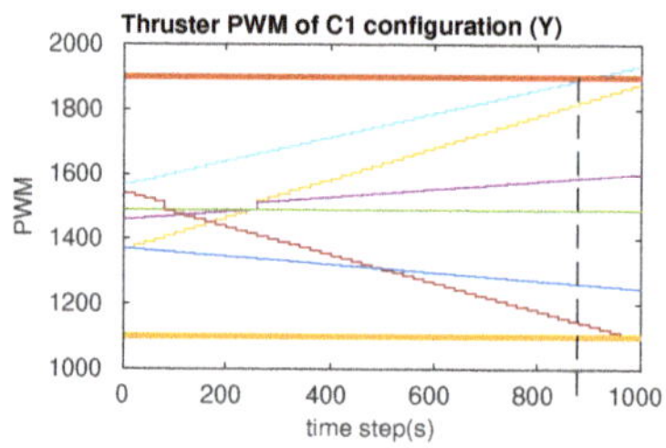

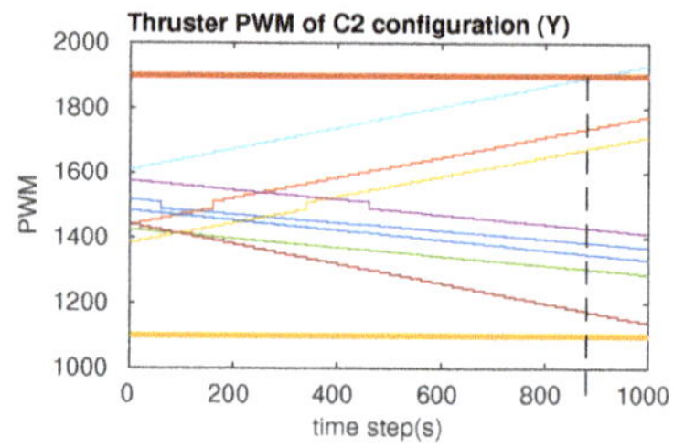

(b) (c)

Figure 19. The simulation of the cube rotation about the **v**-axis for $\mathbf{C}^1$ and $\mathbf{C}^2$ (Y-axis = **v**-axis). (**a**) Applied torque about the **v**-axis. (**b**) PWM inputs of $\mathbf{C}^1$. (**c**) PWM inputs of $\mathbf{C}^2$.

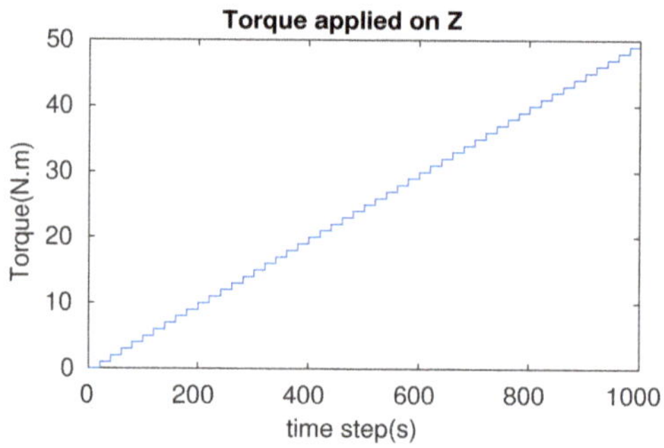

(a)

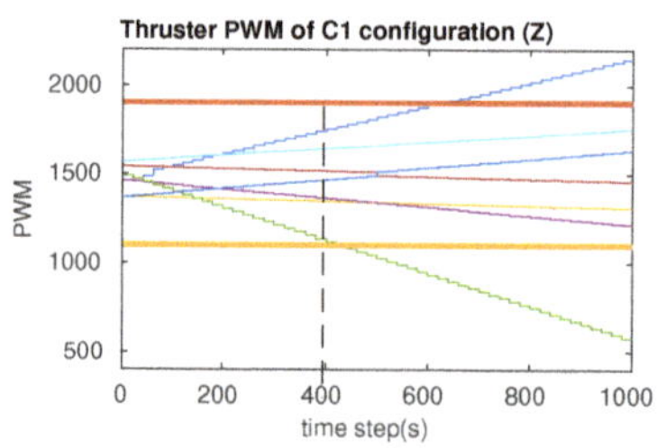

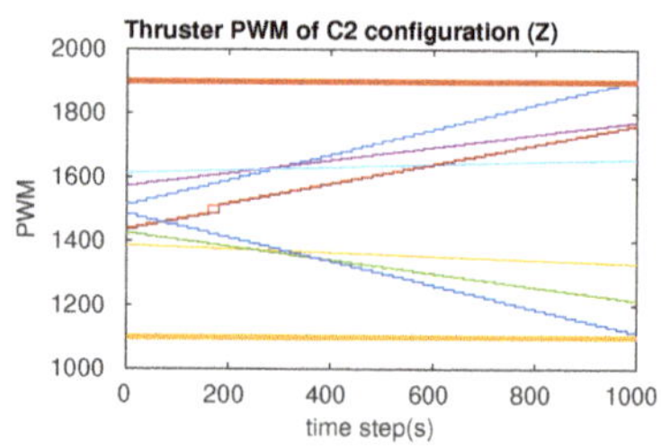

(b) (c)

Figure 20. The simulation of the cube rotation about the Z-axis for $\mathbf{C}^1$ and $\mathbf{C}^2$ (Z-axis = **w**-axis). (**a**) Applied torque about the **w**-axis. (**b**) PWM inputs of $\mathbf{C}^1$. (**c**) PWM inputs of $\mathbf{C}^2$.

6. Experimental Results

Experiments were carried out on the cube robot to compare between the two configurations, $\mathbf{C}^1$ (see Figure 21) and $\mathbf{C}^2$ (see Figure 22), in the swimming pool at Montpellier University. The cube in the water and a video link for the cube's operations can be seen in Figure 23.

Figure 21. $\mathbf{C}^1$ of the cube robot.

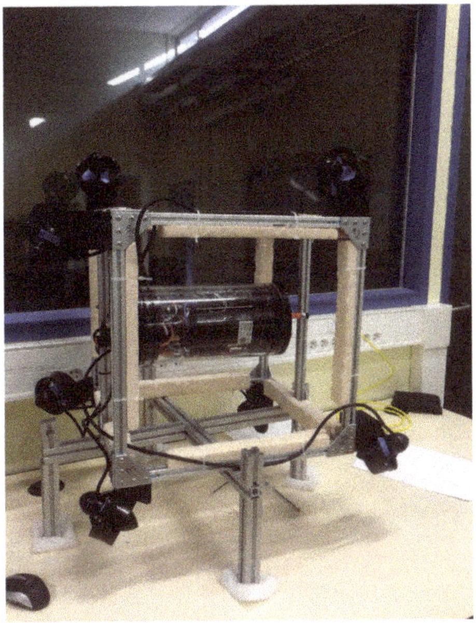

Figure 22. $\mathbf{C}^2$ of the cube robot.

Figure 23. Cube robot in the water https://www.youtube.com/watch?v=RKiWUOxDKdw (accessed on 18 October 2019)

6.1. Attainability Validation

The incremental torques about the **u**-axis, **v**-axis, and **w**-axis were applied to cube robot respectively, and angular velocities and PWM input values were stored to evaluate these two configurations. For safety, the experiments were stopped when one thruster reached the saturation values. The experimental results are shown in Figures 24–26. For rotating about the **u**-axis (Figure 24), the attainability of configurations $\mathbf{C}^1$ and $\mathbf{C}^2$ was almost the same: all thrusters operated in a feasible region. Otherwise, for rotating about the **v**-axis or **w**-axis, the attainability of configuration $\mathbf{C}^2$ was better than that of $\mathbf{C}^1$. In particular, with the **v**-axis experiment (Figure 25), the cube robot with $\mathbf{C}^1$ stopped the mission earlier than with $\mathbf{C}^2$ (at Time Step 771) because one thruster reached saturation. The same thing happened with the **w**-axis experiment (at Time Step 451) (see Figure 26).

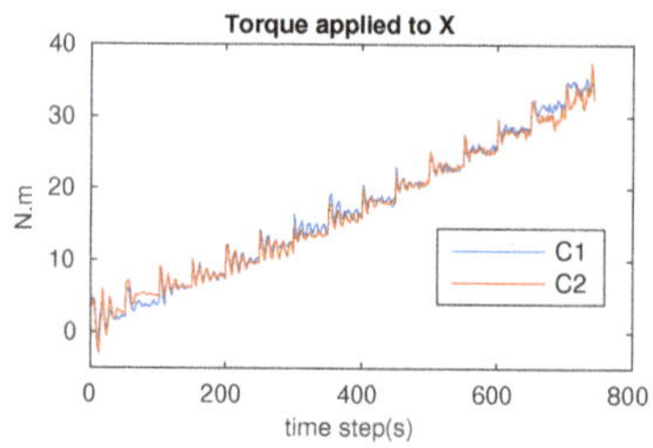

(**a**) Applied torque about the **u**-axis

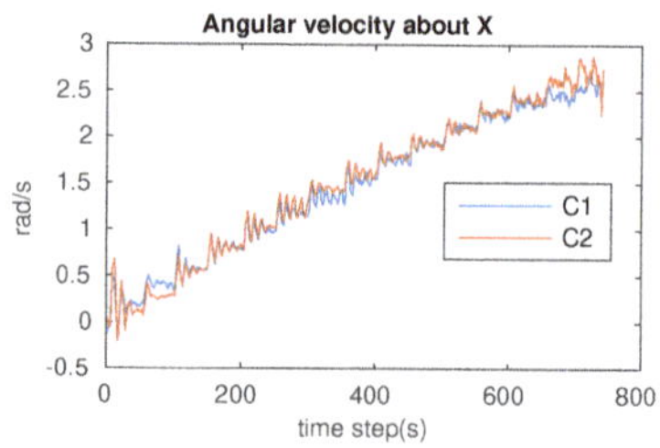

(**b**) Angular velocities

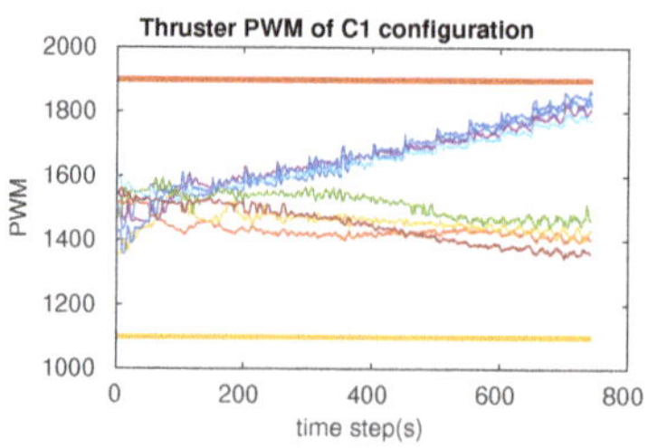

(**c**) PWM inputs of $\mathbf{C}^1$

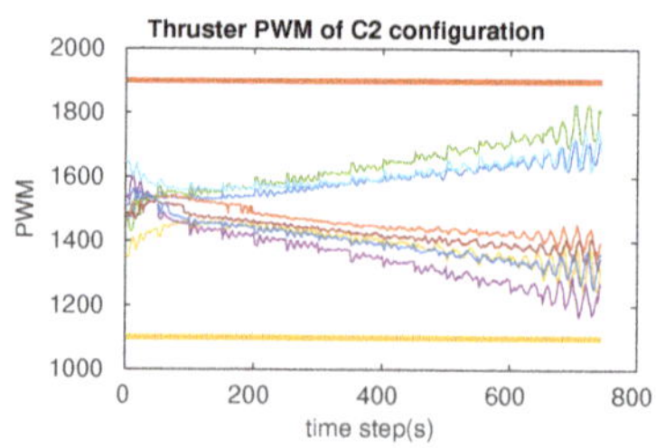

(**d**) PWM inputs of $\mathbf{C}^2$

Figure 24. The cube rotates about the **u**-axis for $\mathbf{C}^1$ and $\mathbf{C}^2$ (X-axis = **u**-axis).

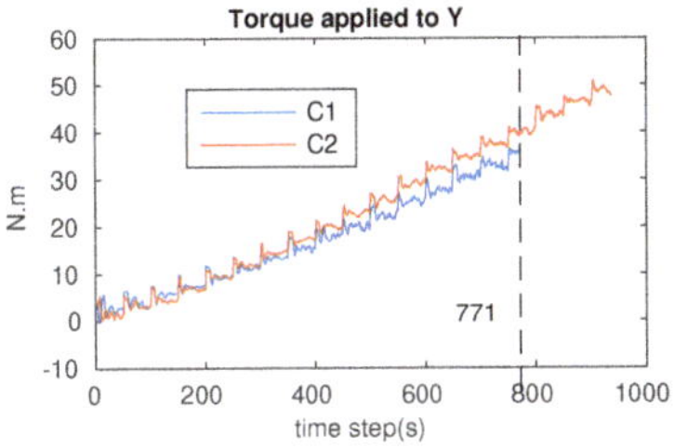

(**a**) Applied torque about the **v**-axis

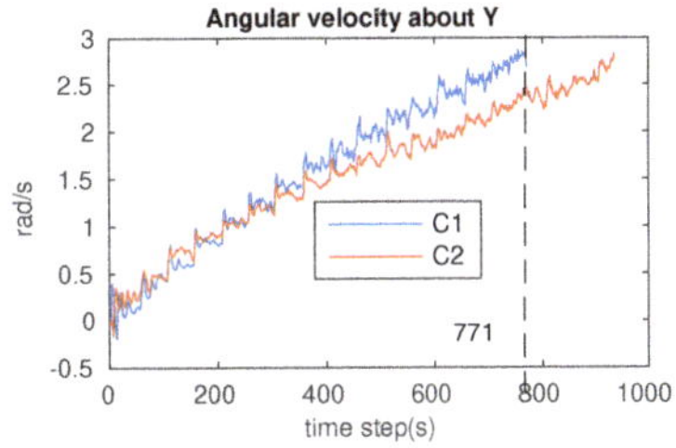

(**b**) Angular velocities

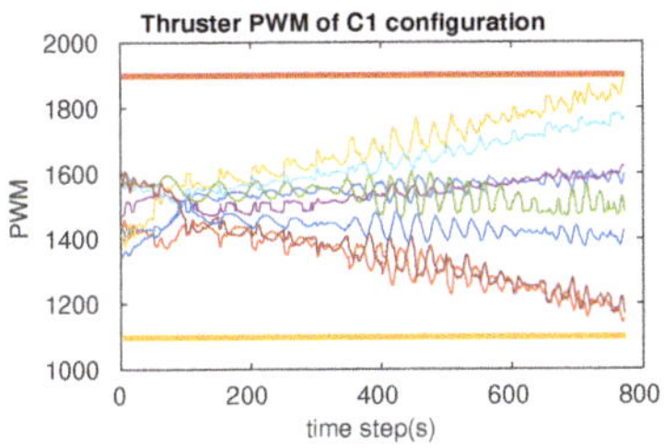

(**c**) PWM inputs of $\mathbf{C}^1$

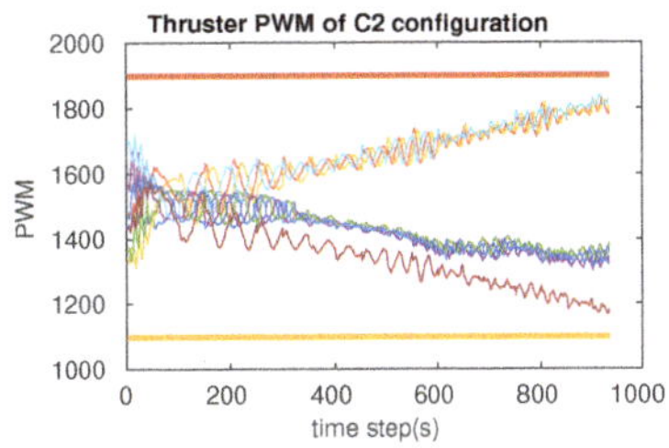

(**d**) PWM inputs of $\mathbf{C}^2$

Figure 25. The cube rotates about the **v**-axis for $\mathbf{C}^1$ and $\mathbf{C}^2$ (Y-axis = **v**-axis).

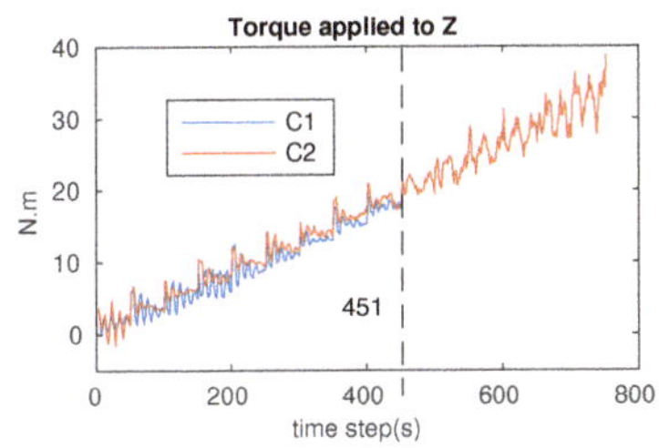

(**a**) Applied torque about the **w**-axis

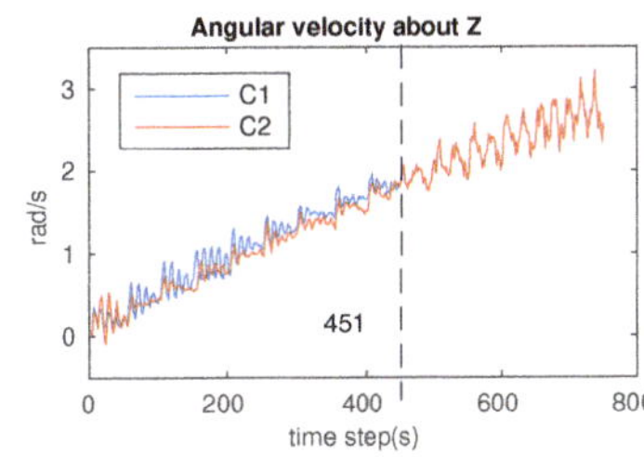

(**b**) Angular velocities

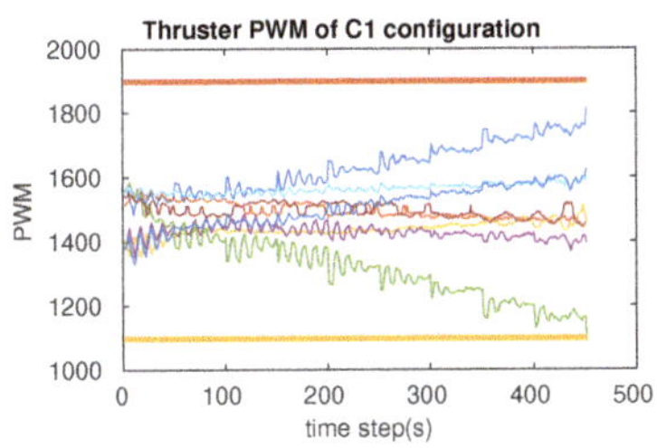

(**c**) PWM inputs of $\mathbf{C}^1$

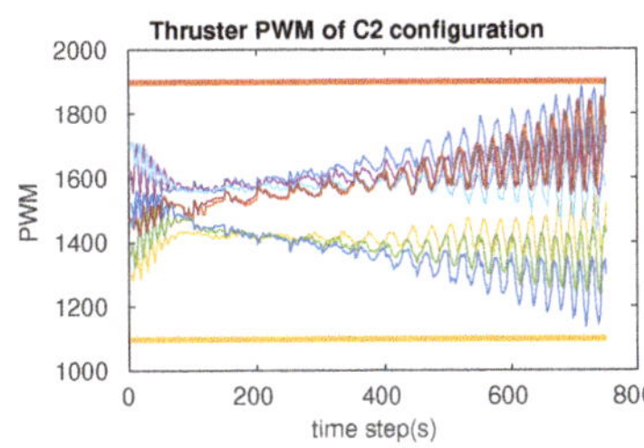

(**d**) PWM inputs of $\mathbf{C}^2$

Figure 26. The cube rotates about the Z–axis for $\mathbf{C}^1$ and $\mathbf{C}^2$ (Z-axis = **w**-axis).

6.2. Energetic Validation

In this section, we verify the energy consumption during these experiments for the two configurations. An energy-like criterion is proposed:

$$\mathbf{E} = \sum_{i=1}^{m} \int_{t=0}^{T} |PWM^i(t) - 1500| dt \quad (43)$$

where m is the number of thrusters, T is the time of the experiment, and $PWM^i(t)$ is the PWM inputs of the ith thruster.

Table 6 shows the energy consumption of the robot during the three rotation experiments. For **u**-axis rotation, the attainability of the two configurations was the same, but the energy consumption of $\mathbf{C}^2$ was lower than that of $\mathbf{C}^1$. For **v**-axis and **w**-axis rotations, the duration of the experiments of $\mathbf{C}^2$ was longer than that of $\mathbf{C}^1$, and the energy consumption, therefore, was higher.

Table 6. Energy consumption of the two configurations.

No.	Rotation	E_{C^1}	E_{C^2}
1	**u**	7.2303×10^4	6.9603×10^4
2	**v**	7.5480×10^4	1.0590×10^5
3	**w**	3.1637×10^4	7.4350×10^4

Table 7 shows the comparison of the energy consumption of the two configurations with the same time duration. For the **v**-axis rotation, the energy value of $\mathbf{C}^2$ was lower than that of $\mathbf{C}^1$. However, for the **w**-axis, the energy value of $\mathbf{C}^2$ was higher. This happened because the robot dived deeper in $\mathbf{C}^2$ in the experiment of the **w**-axis rotation, and the robot had to deliver more power to maintain a greater constant depth.

Table 7. Energy consumption of the two configurations with the same time duration.

No.	Rotation	E_{C^1}	E_{C^2}
1	**v**	7.5480×10^4	7.2715×10^4
2	**w**	3.1637×10^4	3.3312×10^4

6.3. Robustness and Reactive Validation

This section validates the robustness and reactivity of the optimal configuration ($\mathbf{C}^2$) in comparison to the normal one ($\mathbf{C}^1$). For robustness, the robot performed a mission, and one or two thrusters were turned off. For the normal configuration $\mathbf{C}^1$, the mission would fail, and for the optimal configuration $\mathbf{C}^2$, the mission would be guaranteed. Specifically, for the robustness index, we carried out the following experiments:

1. The cube robot dives to a predefined depth with all motors being in the normal operating conditions;
2. The cube robot dives to the same predefined depth with one vertical motor being stopped;
3. The cube robot dives to the same predefined depth with two vertical motors being stopped;
4. The cube robot dives to the same predefined depth with three motors being stopped (two vertical motors and one arbitrary motor);
5. The cube robot simultaneously dives to the same predefined depth and rotates about the **w**-axis with three motors being stopped (two vertical motors and one horizontal motor)

For the reactive index, we measured how fast the robot changed missions. The following experiments were carried out:

1. The cube robot goes down to the predefined depth and goes up to another predefined depth, then goes down again to the former predefined depth;
2. In the sequel, the cube robot goes down to the predefined depth, rotates about the **u**-axis, and after that, rotates about the **v**-axis. The rotation time of each axis should be 60 s or longer;
3. Next, the cube robot goes down to the predefined depth, rotates about the **u**-axis, and after that, rotates about the diagonal-axis (diagonal of the cube robot). The rotation time of each axis should be 60 s or longer.

The experimental results for the robustness validation of $\mathbf{C}^1$ and $\mathbf{C}^2$ are shown in Figures 27–29. In the case of one or two motors stopped, the depth control performances of $\mathbf{C}^1$ and $\mathbf{C}^2$ were almost the same (see Figure 27). The differences are clear in the case of three thrusters stopped (Figure 29): the performance of $\mathbf{C}^1$ was not guaranteed (Figure 28) and violations of the PWM values occurred (see Figure 29a).

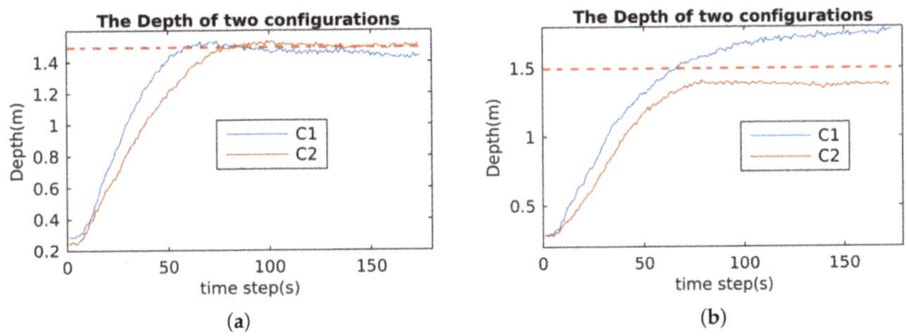

Figure 27. Depth control for $\mathbf{C}^1$ and $\mathbf{C}^2$ with one and two motors stopped. (**a**) Depth control of two configurations with one motor stopped. (**b**) Depth control of two configurations with two motors stopped.

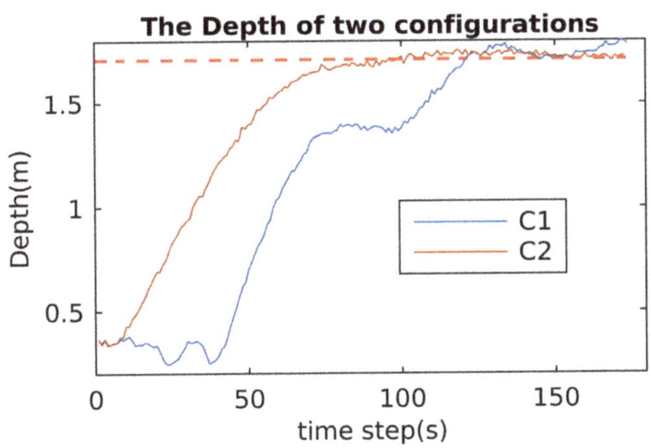

Figure 28. Depth control for $\mathbf{C}^1$ and $\mathbf{C}^2$ with three motors stopped.

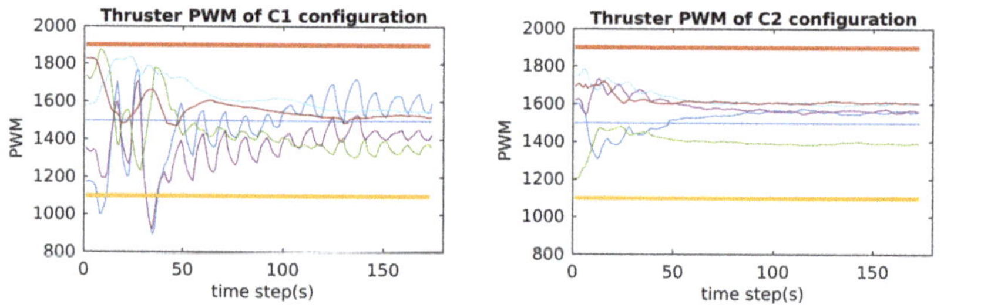

(**a**) (**b**)

Figure 29. PWM evaluation for $\mathbf{C}^1$ and $\mathbf{C}^2$ with 3 motors stopped. (**a**) PWM of $\mathbf{C}^1$. (**b**) PWM of $\mathbf{C}^2$.

The results for the reactive validation are shown in Figures 30–32. We measured the reactive time of the angular velocities when the directions of the cube's actions changed. It is clear that the reactive time of $\mathbf{C}^2$ was faster than that of $\mathbf{C}^1$. Specifically, the reactive time is the region formed by the vertical dashed lines in Figures 30–32. It is obvious that the reactive time of $\mathbf{C}^2$ was smaller than that of $\mathbf{C}^2$ (see Figures 31 and 32).

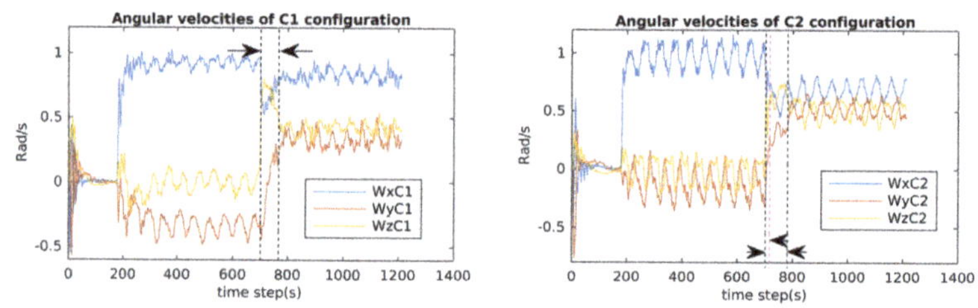

(**a**) (**b**)

Figure 30. Angular velocity evaluation for $\mathbf{C}^1$ and $\mathbf{C}^2$: diving, rotating about the **u**-axis, and rotating about the diagonal-axis (**Wx** = **p**; **Wy** = **q**; **Wz** = **r**). (**a**) Angular velocities of $\mathbf{C}^1$. (**b**) Angular velocities of $\mathbf{C}^2$.

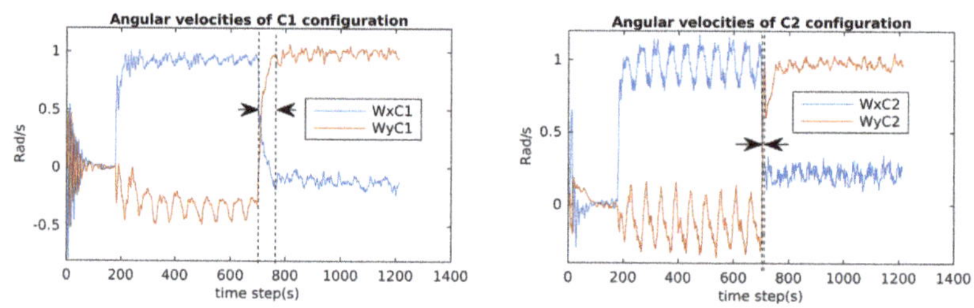

(**a**) (**b**)

Figure 31. Angular velocity evaluation for $\mathbf{C}^1$ and $\mathbf{C}^2$: diving, rotating about the **u**-axis, and rotating about the **v**-axis (**Wx** = **p**; **Wy** = **q**). (**a**) Angular velocities of $\mathbf{C}^1$. (**b**) Angular velocities of $\mathbf{C}^2$.

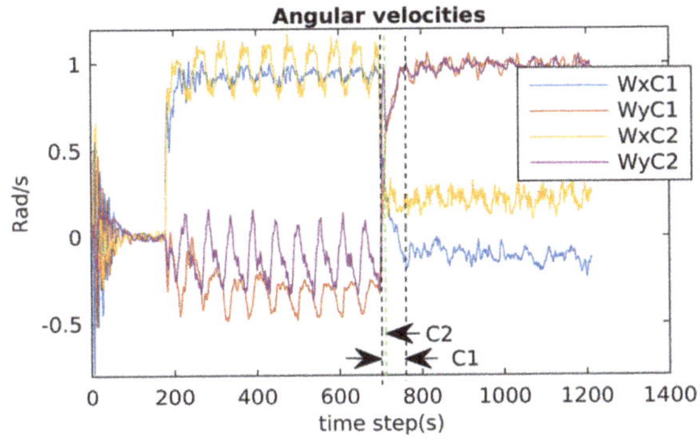

Figure 32. Angular velocity evaluation for $\mathbf{C}^1$ and $\mathbf{C}^2$: diving, rotating about the **u**-axis, and rotating about the **v**-axis ($\mathbf{Wx} = \mathbf{p}$; $\mathbf{Wy} = \mathbf{q}$).

7. Conclusions and Future Work

In this paper, an approach for designing an optimal configuration matrix (which depends on the positions and directions of the thrusters) of overactuated underwater robots was presented. The performance indices (related to manipulability, energy, workspace, reactivity, and robustness) were proposed and analyzed. Specifically, the manipulability index shows the isotropic properties of a robot; the energetic index minimizes the energy consumption under some assumptions; the workspace index is related to the attainable spaces (i.e., the force and torque spaces) of the robot; the reactive index presents how fast the robot changes the direction of the resulting actuation force; finally, the robustness index is related to the capacity of the robot to maintain its performance in the case of failures (i.e., some thrusters are completely stopped). It was formulated as a multi-objective optimization problem. Because the different indices exhibit different magnitudes and physical meanings, the goal-attainment method was chosen to find one Pareto-optimal solution. Simulation and experimental results showed that the performances of the optimal configuration were better than a "normal" configuration, which is often used (thrusters are installed vertically or horizontally). Because of the nonconvexity of the problem, finding all Pareto-optimal solutions, the Pareto front, remains a challenging problem and will be future work. Moreover, a design problem relaxing the assumptions (i.e., perfectly known characteristics of the actuators, pseudo-inverse dispatcher) is also an interesting direction for future research.

Author Contributions: Conceptualization, T.D., L.L., and R.Z.; methodology, T.D., L.L., and R.Z.; software, T.D., B.R., and P.L.; validation, T.D., L.L., R.Z., B.R., and P.L.; writing—original draft preparation, T.D.; writing—review and editing, L.L.; supervision, L.L. and R.Z.; project administration, L.L.; funding acquisition, L.L. All authors have read and agreed to the published version of the manuscript.

Funding: This project was supported by the LabEx NUMEV (ANR-10-LABX- 0020) within the I-SITE MUSE (ANR-16-IDEX-0006) and the Region Occitanie (french FEDER funds).

Institutional Review Board Statement: Not applicable.

Informed Consent Statement: Not applicable.

Data Availability Statement: Not applicable.

Acknowledgments: The authors would like to thank Numev Labex, MUSE, Montpellier University; Region Occitanie; and FEDER for supporting this research.

Conflicts of Interest: The authors declare no conflict of interest.

Abbreviations

The following abbreviations are used in this manuscript:

AS	Actuation System
NGC	Navigation–Guidance–Control
DoFs	Degrees of Freedom
AUVs	Autonomous Underwater Vehicles
CM	Center of Mass
PWM	Pulse-Width Modulation

Appendix A

Theorem A1. *The image of the unit hypersphere under any $n \times m$ matrix is a hyperellipsoid.*

Proof. Let $\mathbf{A}$ be an $n \times m$ matrix with rank r. Let $\mathbf{A} = \mathbf{USV}^T$ be a singular-value decomposition of $\mathbf{A}$. The left and right singular vectors of $\mathbf{A}$ are denoted as $\mathbf{u}_1, \mathbf{u}_2, \ldots, \mathbf{u}_n$ and $\mathbf{v}_1, \mathbf{v}_2, \ldots, \mathbf{v}_m$, respectively. Since $rank(A) = r$, the singular values of $\mathbf{A}$ have the properties: $\sigma_1 \geq \sigma_2 \geq, \ldots, \geq \sigma_r > 0$ and $\sigma_{r+1} = \sigma_{r+2} =, \ldots, = \sigma_m = 0$.

Let $\mathbf{x} = \begin{pmatrix} x_1 \\ \vdots \\ x_m \end{pmatrix}$ be an unit vector in $\mathbb{R}^m$. Because $\mathbf{V}$ is an orthogonal matrix and $\mathbf{V}^T$ is also, we have that $\mathbf{V}^T\mathbf{x}$ is an unit vector (it is easy to see that $\|\mathbf{V}^T\mathbf{x}\| = \|\mathbf{x}\|$). Therefore, $(\mathbf{v}_1^T\mathbf{x})^2 + (\mathbf{v}_2^T\mathbf{x})^2+, \ldots, +(\mathbf{v}_m^T\mathbf{x})^2 = 1$.

On the other hand, we have $\mathbf{A} = \sigma_1\mathbf{u}_1\mathbf{v}_1^T + \sigma_2\mathbf{u}_2\mathbf{v}_2^T+, \ldots, +\sigma_r\mathbf{u}_r\mathbf{v}_r^T$. Therefore:

$$\begin{aligned} \mathbf{Ax} &= \sigma_1\mathbf{u}_1\mathbf{v}_1^T\mathbf{x} + \sigma_2\mathbf{u}_2\mathbf{v}_2^T\mathbf{x}+, \ldots, +\sigma_r\mathbf{u}_r\mathbf{v}_r^T\mathbf{x} \\ &= (\sigma_1\mathbf{v}_1^T\mathbf{x})\mathbf{u}_1 + (\sigma_2\mathbf{v}_2^T\mathbf{x})\mathbf{u}_2+, \ldots, +(\sigma_r\mathbf{v}_r^T\mathbf{x})\mathbf{u}_r \\ &= \mathbf{y}_1\mathbf{u}_1 + \mathbf{y}_2\mathbf{u}_2+, \ldots, +\mathbf{y}_r\mathbf{u}_r \\ &= \mathbf{Uy} \end{aligned} \tag{A1}$$

where $\mathbf{y}_i$ denotes the $\sigma_i\mathbf{v}_i^T\mathbf{x}$ and $\mathbf{y} = \begin{pmatrix} y_1 \\ \vdots \\ y_r \end{pmatrix}$.

From (A1), we have: $\|\mathbf{Ax}\| = \|\mathbf{Uy}\| = \|\mathbf{y}\|$ (since $\mathbf{U}$ is an orthogonal matrix). Moreover, $\mathbf{y}$ has the following property:

$$\begin{aligned} &(\frac{y_1}{\sigma_1})^2 + (\frac{y_2}{\sigma_2})^2+, \ldots, +(\frac{y_r}{\sigma_r})^2 = \\ &= (\mathbf{v}_1^T\mathbf{x})^2 + (\mathbf{v}_2^T\mathbf{x})^2+, \ldots, +(\mathbf{v}_r^T\mathbf{x})^2 \leq 1 \end{aligned} \tag{A2}$$

Specifically:

1. If $r = m$ (of course, we must have $m \leq n$), the equality in Equation (A2) holds, and the image of the unit hypersphere forms the surface of a hyperellipsoid;
2. If $r < m$, the image of the unit hypersphere corresponds to a solid hyperellipsoid.

This completes the proof. □

References

1. Lapierre, L. Robust diving control of an AUV. *Ocean Eng.* **2009**, *36*, 92–104. [CrossRef]
2. Lapierre, L.; Jouvencel, B. Robust nonlinear path-following control of an AUV. *IEEE J. Ocean. Eng.* **2008**, *33*, 89–102. [CrossRef]
3. Levine, W.S. *The Control Systems Handbook: Control System Applications*; CRC Press: Boca Raton, FL, USA, 2010.
4. Johansen, T.A.; Fossen, T.I. Control allocation—A survey. *Automatica* **2013**, *49*, 1087–1103. [CrossRef]
5. Ropars, B.; Lapierre, L.; Lasbouygues, A.; Andreu, D.; Zapata, R. Redundant actuation system of an underwater vehicle. *Ocean Eng.* **2018**, *151*, 276–289. [CrossRef]

6. Nakamura, Y.; Hanafusa, H.; Yoshikawa, T. Task-priority based redundancy control of robot manipulators. *Int. J. Robot. Res.* **1987**, *6*, 3–15. [CrossRef]
7. Adorno, B.V.; Fraisse, P.; Druon, S. Dual position control strategies using the cooperative dual task-space framework. In Proceedings of the IROS'10: International Conference on Intelligent Robots and Systems, Taipei, Taiwan, 18–22 October 2010; pp. 3955–3960.
8. Agravante, D.J.; Sherikov, A.; Wieber, P.B.; Cherubini, A.; Kheddar, A. Walking pattern generators designed for physical collaboration. In Proceedings of the ICRA 2016: 2016 IEEE International Conference on Robotics and Automation (ICRA), Stockholm, Sweden, 16–21 May 2016; pp. 1573–1578.
9. Pham, T.H.; Caron, S.; Kheddar, A. Multicontact Interaction Force Sensing From Whole-Body Motion Capture. *IEEE Trans. Ind. Inform.* **2018**, *14*, 2343–2352. [CrossRef]
10. Yoshikawa, T. Dynamic manipulability of robot manipulators. In Proceedings of the 1985 IEEE International Conference on Robotics and Automation, St. Louis, MO, USA, 25–28 March 1985; Volume 2, pp. 1033–1038. [CrossRef]
11. Yoshikawa, T. *Foundations of Robotics: Analysis and Control*; MIT Press: Cambridge, MA, USA, 1990.
12. Kumar, A.; Waldron, K. The workspaces of a mechanical manipulator. *J. Mech. Des.* **1981**, *103*, 665–672. [CrossRef]
13. Paden, B.; Sastry, S. Optimal kinematic design of 6R manipulators. *Int. J. Robot. Res.* **1988**, *7*, 43–61. [CrossRef]
14. Park, F.C.; Brockett, R.W. Kinematic dexterity of robotic mechanisms. *Int. J. Robot. Res.* **1994**, *13*, 1–15. [CrossRef]
15. Pierrot, F.; Benoit, M.; Dauchez, P. Optimal thruster configuration for omni-directional underwater vehicles. SamoS: A Pythagorean solution. In Proceedings of the OCEANS '98 Conference Proceedings, Nice, France, 28 September–1 October 1998; Volume 2, pp. 655–659. [CrossRef]
16. Kharrat, H. Optimization of Thruster Configuration for Swimming Robots. Master's Thesis, Rice University, Houston, TX, USA, 2015.
17. Stephan, J.; Fichter, W. Fast Exact Redistributed Pseudoinverse Method for Linear Actuation Systems. *IEEE Trans. Control Syst. Technol.* **2017**, *27*, 451–458. [CrossRef]
18. Grechi, S.; Caiti, A. Comparison between Optimal Control Allocation with Mixed Quadratic & Linear Programming Techniques. In Proceedings of the 10th IFAC Conference on Control Applications in Marine SystemsCAMS, Trondheim, Norway, 13–16 September 2016; Volume 49, pp. 147–152.doi: 10.1016/j.ifacol.2016.10.335. [CrossRef]
19. Ropars, B.; Lasbouygues, A.; Lapierre, L.; Andreu, D. Thruster's dead-zones compensation for the actuation system of an underwater vehicle. In Proceedings of the Control Conference (ECC), Linz, Austria, 15–17 July 2015; pp. 741–746.
20. Yoshikawa, T. Manipulability of robotic mechanisms. *Int. J. Robot. Res.* **1985**, *4*, 3–9. [CrossRef]
21. Olver, F.W.; Lozier, D.W.; Boisvert, R.F.; Clark, C.W. *NIST Handbook of Mathematical Functions Hardback and CD-ROM*; Cambridge University Press: Cambridge, UK, 2010.
22. Trefethen, L.N.; Bau, D., III. *Numerical Linear Algebra*; Siam: Philadelphia, PA, USA, 1997; Volume 50.
23. Gembicki, F.; Haimes, Y. Approach to performance and sensitivity multiobjective optimization: The goal-attainment method. *IEEE Trans. Autom. Control* **1975**, *20*, 769–771. [CrossRef]
24. BlueRobotics. Available online: https://bluerobotics.com/ (accessed on 20 December 2018).

sensors

MDPI

Article

Robust Nonlinear Tracking Control with Exponential Convergence Using Contraction Metrics and Disturbance Estimation

Pan Zhao *, Ziyao Guo and Naira Hovakimyan

Department of Mechanical Science and Engineering, University of Illinois at Urbana-Champaign, Urbana, IL 61801, USA; ziyaog2@illinois.edu (Z.G.); nhovakim@illinois.edu (N.H.)
* Correspondence: panzhao2@illinois.edu

Abstract: This paper presents a tracking controller for nonlinear systems with matched uncertainties based on contraction metrics and disturbance estimation that provides exponential convergence guarantees. Within the proposed approach, a disturbance estimator is proposed to estimate the pointwise value of the uncertainties, with a pre-computable estimation error bounds (EEB). The estimated disturbance and the EEB are then incorporated in a robust Riemannian energy condition to compute the control law that guarantees exponential convergence of actual state trajectories to desired ones. Simulation results on aircraft and planar quadrotor systems demonstrate the efficacy of the proposed controller, which yields better tracking performance than existing controllers for both systems.

Keywords: robust control; nonlinear control; uncertain systems; disturbance estimation; robot safety

Citation: Zhao, P.; Guo, Z.; Hovakimyan, N. Robust Nonlinear Tracking Control with Exponential Convergence Using Contraction Metrics and Disturbance Estimation. *Sensors* **2022**, *22*, 4743. https://doi.org/10.3390/s22134743

Academic Editor: Baochang Zhang

Received: 9 May 2022
Accepted: 20 June 2022
Published: 23 June 2022

1. Introduction

Robotic systems generally have nonlinear dynamics and are subject to model uncertainties and disturbances. Moreover, many robotic systems are underactuated, i.e., having fewer independent control inputs than degrees of freedom, including fixed-wing aircraft, quadrotors and dynamic walking robots. The design of tracking controllers for underactuated robotic systems is a much more challenging problem compared to that for fully-actuated systems. Recently, the concept of a control contraction metric (CCM) was introduced in [1] to synthesize trajectory tracking controllers for general nonlinear systems, including underactuated ones. The CCM extends contraction theory [2] from analysis to constructive control design, while contraction theory is focused on analyzing nonlinear systems in a differential framework by studying the convergence between pairs of state trajectories toward each other. It was shown in [3] that CCM reduces to conventional sliding and energy-based designs for fully-actuated systems. On the other hand, for underactuated systems, compared to prior approaches based on local linearization [4], the CCM approach leads to a convex optimization problem for controller synthesis and generates controllers that stabilize *every* feasible trajectory in a region, instead of just a single target trajectory that must be known a priori [3].

On the other hand, control design methods to deal with dynamic uncertainties in the deterministic setting can be roughly classified into adaptive and robust approaches. Robust approaches, such as H_∞ control [5], μ synthesis [6] and robust/tube model predictive control (MPC) [7,8], usually consider parametric uncertainties or bounded disturbances and aim to find controllers with performance guarantees for the worst case of such uncertainties. The consideration of worst-case scenarios associated with robust approaches often leads to conservative nominal performance. Disturbance–observer (DOB) based control and related methods such as active disturbance rejection control (ADRC) [9] lump all uncertainties that may include parametric uncertainties, unmodeled dynamics and external disturbances, together as a "total disturbance", estimate it via an observer and then compute control actions to compensate for the estimated disturbance [10] to recover the nominal performance. However, for state-dependent uncertainties, DOB-based control methods usually ignore the dependence of "disturbance" on system states and rely on assumptions on the derivative of the "disturbance" that are difficult to verify for theoretical guarantees [10,11]. Alternatively, adaptive control methods such as model reference adaptive control (MRAC) [12] usually need a parametric structure for the uncertainties, rely on online estimation of the parameters for control law construction and provide asymptotic performance guarantees in most cases. One of the exceptions is $\mathcal{L}_1$ adaptive control [13] that does not need a parameterization of the uncertainties (similar to DOB-based control) and focuses on transient performance guarantees in terms of uniformly bounded error between the ideal and uncertain systems.

Both robust and adaptive control approaches have been explored in the context of CCM-based control in the presence of uncertainties and disturbances. In particular, adaptive control was combined with CCM to handle nonlinear control-affine systems with both parametric [14] and non-parametric uncertainties [15]. The case of bounded disturbances in CCM-based control was addressed by leveraging input-to-state stability analysis [16] or robust CCM [17,18]. CCM for stochastic systems was developed in [19] to minimize the mean squared tracking error in the presence of stochastic disturbances. Closely relevant to this paper, [15] designed an $\mathcal{L}_1$ adaptive controller to augment a baseline CCM-based controller to compensate for matched nonlinear non-parametric uncertainties that can depend on both time and states. The authors of [15] proved that transient tracking performance was guaranteed in the sense that the actual state trajectory exponentially converges to a neighborhood or a tube around the desired one. Compared to [15], our approach relies on a disturbance observer that yields an estimation error bound and robust Riemannian energy condition and ensures that the actual state trajectory exponentially converges to the nominal one.

Statement of Contributions: We present a tracking controller for nonlinear systems subject to matched uncertainties that can depend on both time and states based on contraction metrics and disturbance estimation. Our controller leverages a disturbance estimator to estimate the pointwise value of the uncertainties, with a pre-computable estimation error bound. The estimated disturbance and the estimation error bound are then incorporated into a robust Riemannian energy condition to compute the control law that guarantees exponential convergence of actual state trajectories to nominal ones. We validate the efficacy of our controller on two simulation examples and demonstrate its advantages over existing controllers.

The idea presented in this paper is leveraged in [20] for safe learning of uncertain dynamics using deep neural networks. Compared to [20], this paper is not relevant to learning and allows the uncertainty to be dependent on both time and states, as opposed to the dependence on states only in [20]. Additionally, this paper includes an additional aircraft example for performance illustration and conducts extensive comparisons with existing adaptive approaches in simulations that are not available in [20].

Notations: Let $\mathbb{R}^n$, $\mathbb{R}^+$ and $\mathbb{R}^{m\times n}$ denote the n-dimensional real vector space, the set of non-negative real numbers and the set of real m by n matrices, respectively. I and 0 denote an identity matrix, and a zero matrix of compatible dimensions, respectively; $\|\cdot\|$ denotes the 2-norm of a vector or a matrix. For a vector y, y_i denotes its ith element. For a matrix-valued function $M : \mathbb{R}^n \to \mathbb{R}^{n\times n}$ and a vector $y \in \mathbb{R}^n$, $\partial_y M(x) \triangleq \sum_{i=1}^n \frac{\partial M(x)}{\partial x_i} y_i$ denotes the directional derivative of $M(x)$ along y. For symmetric matrices P and Q, $P > Q$ ($P \geq Q$) means $P - Q$ is positive definite (semidefinite). $\langle X \rangle$ is the shorthand notation of $X + X^\top$. Finally, $\ominus$ denotes the Minkowski set difference.

2. Problem Statement and Preliminaries

Consider a nonlinear control-affine system with uncertainties

$$\dot{x}(t) = f(x(t)) + B(x(t))(u(t) + d(t, x(t))), \quad (1)$$

where $x(t) \in \mathcal{X} \subset \mathbb{R}^n$ is the state vector, $u(t) \in \mathcal{U} \subset \mathbb{R}^m$ is the control input vector, $f : \mathbb{R}^n \to \mathbb{R}^n$ and $B : \mathbb{R}^n \to \mathbb{R}^m$ are known and locally Lipschitz continuous functions, and $d(t,x)$ represents the *matched* model uncertainty that can depend on both time and states. We assume that $B(x)$ has full column rank for any $x \in \mathcal{X}$. Suppose $\mathcal{X}$ is a compact set that contains the origin, and the control constraint set $\mathcal{U}$ is defined as $\mathcal{U} \triangleq \{u \in \mathbb{R}^m : \underline{u} \le u \le \bar{u}\}$, where $\underline{u}, \bar{u} \in \mathbb{R}^m$ denote the lower and upper bounds of all control channels, respectively. Furthermore, we make the following assumptions on $B(x)$ and $d(t,x)$.

Assumption 1. *There exist known positive constants L_B, L_d, l_d and b_d such that for any $x, y \in \mathcal{X}$ and $t, \tau \ge 0$, the following inequalities hold:*

$$\|B(x) - B(y)\| \le L_B \|x - y\|, \quad (2)$$

$$\|d(t,x) - d(\tau,y)\| \le L_d \|x - y\| + l_d |t - \tau|, \quad (3)$$

$$\|d(t,x)\| \le b_d. \quad (4)$$

Remark 1. *Assumption 1 indicates that the uncertain function $d(t,x)$ is locally Lipschitz in both t and x with known Lipschitz constants and is uniformly bounded by a known constant in the compact set $\mathcal{X}$.*

In fact, given the local Lipschitz constants L_d and l_d, a uniform bound on $d(t,x)$ in $\mathcal{X}$ can always be derived by using Lipschitz continuity properties if the bound on $d(t,x^*)$ for an arbitrary x^* in $\mathcal{X}$ and any $t \ge 0$ is known. For instance, assuming $\|d(t,0)\| \le b_d^0$, from (3), we have $\|d(t,x)\| \le b_d^0 + L_d \max_{x \in \mathcal{X}} \|x\|$ for any $x \in \mathcal{X}$ and $g \ge 0$. In practice, some prior knowledge about the actual system and the uncertainty may be leveraged to obtain a tighter bound than the one based on the Lipschitz continuity explained earlier, which is why we directly make an assumption on the uniform bound. With Assumption 1, we will show (in Section 3.3) that the pointwise value of $d(t,x(t))$ at any time t can be estimated with a pre-computable estimation error bound.

For the system in (1), assume we have a nominal state and input trajectory, $x^\star(\cdot)$ and $u^\star(\cdot)$, which satisfy the nominal, i.e., uncertainty-free, dynamics:

$$\dot{x}^\star = f(x^\star) + B(x^\star)u^\star. \quad (5)$$

We would like to design a state-feedback controller in the form of

$$u(t) = k(t, x(t), x^\star(t)) + u^\star(t), \quad (6)$$

so that the actual state trajectory $x(\cdot)$ exponentially converges to the nominal one $x^\star(\cdot)$. Our solution is based on CCM and disturbance estimation. Next, we briefly review CCM for uncertainty-free systems.

Control Contraction Metrics (CCMs)

We first introduce some notations related to Riemannian geometry, most of which are from [1]. A Riemannian metric on $\mathbb{R}^n$ is a symmetric positive-definite matrix function $M(x)$, smooth in x, which defines a "local Euclidean" structure for any two tangent vectors δ_1 and δ_2 through the inner product $\langle \delta_1, \delta_2 \rangle_x \triangleq \delta_1^\top M(x) \delta_2$ and the norm $\sqrt{\langle \delta_1, \delta_2 \rangle_x}$. A metric is called *uniformly bounded* if $a_1 I \le M(x) \le a_2 I$ holds $\forall x$ and for some scalars $a_2 \ge a_1 > 0$. Let $\Gamma(a,b)$ be the set of smooth paths connecting two points a and b in $\mathbb{R}^n$, where each $c \in \Gamma(a,b)$ is a piecewise smooth mapping, $c : [0,1] \to \mathbb{R}^n$, satisfying $c(0) = a, c(1) = b$. We use the notation $c(s)$, $s \in [0,1]$, and $c_s(s) \triangleq \frac{\partial c}{\partial s}$. Given a metric $M(x)$ and a curve $c(s)$, we define the Riemannian energy of $c(s)$ as $E(c) \triangleq \int_0^1 c_s^\top M(c(s))c_s(s)ds$. The Riemannian energy between a and b is defined as $E(a,b) \triangleq \inf_{c\in\Gamma(a,b)} E(c)$.

Contraction theory [2] draws conclusions on the convergence between pairs of state trajectories toward each other by studying the evolution of the distance between any two infinitesimally close neighbouring trajectories. CCM generalizes contraction analysis to the controlled dynamics setting in which the analysis jointly searches for a controller and a metric that describes the contraction properties of the resulting closed-loop system. Following [1,14], we now briefly review CCMs by considering the nominal, i.e., uncertainty-free, system:

$$\dot{x} = f(x) + B(x)u, \tag{7}$$

where $x(t) \in \mathbb{R}^n$ and $u(t) \in \mathbb{R}^m$. The differential form of (7) is given by $\dot{\delta}_x = A(x,u)\delta_x + B(x)\delta_u$, where $A(x,u) \triangleq \frac{\partial f}{\partial x} + \sum\limits_{i=1}^{m} \frac{\partial b_i}{\partial x} u_i$ with $b_i(x)$ denoting the ith column of $B(x)$. Consider a function $V(x,\delta_x) = \delta_x^\top M(x)\delta_x$ for some positive definite metric $M(x)$, which can be viewed as the Riemannian squared differential length at point x. Differentiating and imposing that the squared length decreases exponentially with rate 2λ, one obtains

$$\dot{V}(x,\delta_x) = \delta_x^\top (\langle MA \rangle + \dot{M})\delta_x + 2\delta_x^\top MB\delta_u \leq -2\lambda\delta_x^\top M\delta_x, \tag{8}$$

where $\dot{M} = \partial_{f+Bu}M = \partial_f M + \sum_{i=1}^m \partial_{b_i} M u_i$. We first recall some basic results related to CCM.

Definition 1 ([1]). *The system (7) is said to be universally exponentially stabilizable if, for any feasible desired trajectory $x^\star(t)$ and $u^\star(t)$, a feedback controller can be constructed that for any initial condition $x(0)$, a unique solution to (7) exists and satisfies $\|x(t) - x^\star(t)\| \leq R\|x(0) - x^\star(0)\| e^{-\lambda t}$, where λ and R are the convergence rate and overshoot, respectively, independent of the initial conditions.*

Lemma 1 ([1]). *If there exists a uniformly bounded metric $M(x)$, i.e., $\alpha_1 I \leq M(x) \leq \alpha_2 I$ for some positive constants α_1 and α_2, such that for all x and $\delta_x \neq 0$ satisfying $\delta_x^\top MB = 0$,*

$$\delta_x^\top \left(\left\langle M\frac{\partial f}{\partial x} \right\rangle + \partial_f M + 2\lambda M \right) \delta_x \leq 0, \tag{9a}$$

$$\delta_x^\top \left(\left\langle M\frac{\partial b_i}{\partial x} \right\rangle + \partial_{b_i} M \right) \delta_x = 0,\ i = 1,\dots,m \tag{9b}$$

hold, then the system (7) is universally exponentially stabilizable in the sense of Definition 1 via continuous feedback defined almost everywhere, and everywhere in the neighborhood of the target trajectory with the convergence rate λ and overshoot $R = \sqrt{\frac{\alpha_2}{\alpha_1}}$.

The condition (9) ensures that the dynamics orthogonal to the input are contracting, i.e., (8) holds in the presence of $\delta_x^\top MB = 0$ and is often termed as the strong CCM condition [1]. In particular, the condition (9b) can be satisfied by enforcing that each column of $B(x)$ forms a killing vector field for the metric $M(x)$, i.e., $\left\langle M\frac{\partial b_i}{\partial x} \right\rangle + \partial_{b_i} M = 0$ for all $i = 1,\dots,m$. The CCM condition (9) can be transformed into a convex constructive condition for the metric $M(x)$ by a change of variables. Let $W(x) = M^{-1}(x)$ (commonly referred to as the *dual metric*), and $B_\perp(x)$ be a matrix whose columns span the null space of the input matrix B (i.e., $B_\perp^\top B = 0$). Then, condition (9) can be cast as convex constructive conditions for $W(x)$:

$$B_\perp^\top\left(\left\langle\frac{\partial f}{\partial x}W\right\rangle - \partial_f W + 2\lambda W\right)B_\perp \le 0 \tag{10a}$$

$$\left\langle\frac{\partial b_i}{\partial x}W\right\rangle - \partial_{b_i} W = 0, \text{ for } i = 1, \cdots, m. \tag{10b}$$

The existence of a contraction metric $M(x)$ is sufficient for stabilizability via Lemma 1. What remains is constructing a feedback controller that achieves the universal exponential stabilizability (UES). As mentioned in [1,16], one way to derive the controller is to interprete the Riemann energy, $E(x^\star(t), x(t))$, as an incremental control Lyapunov function and use it to construct a min-norm controller that renders for any time t

$$\dot{E}(x^\star(t), x(t)) \le -2\lambda E(x^\star(t), x(t)). \tag{11}$$

Specifically, at any time $t > 0$, given the metric $M(x)$ and a desired/actual state pair $(x^\star(t), x(t))$, a minimum-energy path, i.e., a geodesic, $\gamma(\cdot, t)$ connecting these two states (i.e., $\gamma(0,t) = x^\star(t)$ and $\gamma(1,t) = x(t)$), can be computed (e.g., using the pseudospectral method in [21] to solve a nonlinear programming problem). Consequently, the Riemannian energy of the geodesic is defined as $E(x^\star(t), x(t)) = \int_0^1 \gamma_s(s,t)^\top M(\gamma(s,t))\gamma_s(s,t)ds$, where $\gamma_s(s) \triangleq \frac{\partial \gamma}{\partial s}$, can be calculated. As noted in [16], from the formula for the first variation of energy [22], $\dot{E}(x^\star(t), x(t)) = 2\gamma_s^\top(1,t)M(x(t))\dot{x}(t) - 2\gamma_s^\top(0,t)M(x^\star(t))\dot{x}^\star(t)$. Therefore, (11) can be rewritten as

$$\gamma_s^\top(1,t)M(x(t))\dot{x}(t) - \gamma_s^\top(0,t)M(x^\star(t))\dot{x}^\star(t) \le -\lambda E(x^\star(t), x(t)), \tag{12}$$

where $\dot{x}(t) = f(x(t)) + B(x(t))u(t)$ and $\dot{x}^\star(t) = f(x^\star(t)) + B(x^\star(t))u^\star(t)$. Therefore, the control signal with a minimum norm for $u(t) - u^\star(t)$ can then be obtained by solving the following quadratic programming (QP) problem:

$$u(t) = \underset{k\in\mathbb{R}^m}{\operatorname{argmin}} \left\| k - u^\star(t) \right\|^2 \text{ subject to (12)} \tag{13}$$

at each time t, which is guaranteed to be feasible under condition (9) [1]. The minimization problem (13) is often termed as the *pointwise min-norm control* problem and has an analytic solution [23]. The above discussions can be summarized in the following theorem. The proof is trivial by following Lemma 1 and the subsequent discussions and is thus omitted.

Theorem 1 ([1]). *Given a nominal system (7), assume that there exists a uniformly bounded metric $W(x)$ that satisfies (10) for all $x \in \mathbb{R}^n$. Then, the control law constructed by solving (13) with $M(x) = W^{-1}(x)$, universally exponentially stabilizes the system (7) in the sense of Definition 1, where $R = \sqrt{\frac{\alpha_2}{\alpha_1}}$ with α_1 and α_2 being two positive constants satisfying $\alpha_1 I \le M(x) \le \alpha_2 I$.*

Remark 2. *According to Definition 1 and Theorem 1, under the conditions of Theorem 1, given any feasible trajectory $(x^\star((\cdot), u^\star(\cdot))$ of (7), a controller can always be constructed to ensure that the actual state trajectory $x(\cdot)$ exponentially converges to $x^\star(\cdot)$.*

3. Robust Trajectory Tracking Using CCM and Disturbance Estimation

In Section 2, we have shown that existence of a CCM for a *nominal* (i.e., uncertainty-free) system can be used to construct a feedback control law to guarantee the universal exponential stabilizability (UES) of the system. In this section, we present a controller based on CCM and disturbance estimation to ensure the UES of the uncertain system (1), whose architecture is depicted in Figure 1.

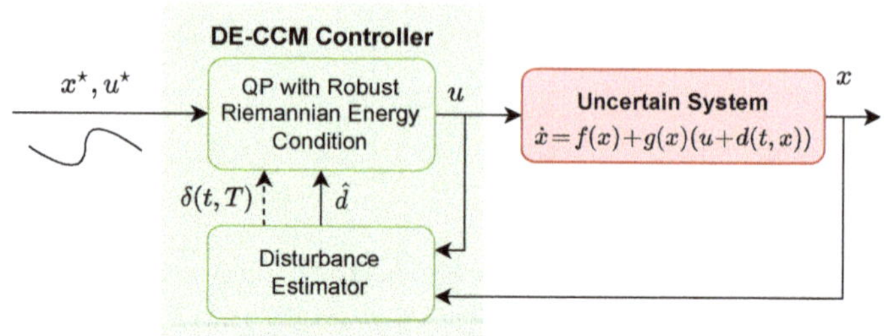

Figure 1. Block diagram of the closed-loop system with the proposed DE-CCM controller.

3.1. CCMs for the Actual System

To apply the contraction method to design a controller to guarantee the UES of the uncertain system (1), we need to first search a valid CCM for it. Following Section 2, we can derive the counterparts of the strong CCM condition (9) or (10). Due to the particular structure with (1) attributed to the matched uncertainty assumption, we have the following lemma. A similar observation has been made in [14] for the case of matched parametric uncertainties. The proof is straightforward and thus omitted. One can refer to [14] for more details.

Lemma 2. *The strong (dual) CCM condition for the uncertain system* (1) *is the same as the strong (dual) CCM condition, i.e.,* (9) *and* (10)*, for the nominal system.*

Remark 3. *As a result of Lemma 2, a metric $M(x)$ (dual metric $W(x)$) satisfying the condition* (9) *and* (10) *for the nominal system* (7) *is always a CCM (dual CCM) for the true system* (1).

Define $\mathcal{D} = \{y \in \mathbb{R}^m : \|y\| \leq b_d\}$, where b_d is introduced in Assumption 1. Assumption 1 indicates $d(t, x) \in \mathcal{D}$ for any $t \geq 0$ and $x \in \mathcal{X}$. As mentioned in Section 2, given a CCM and a desired trajectory $x^\star(t)$ and $u^\star(t)$ for a nominal system, a control law can be constructed to ensure exponential convergence of the actual state trajectory $x(t)$ to the desired state trajectory $x^\star(t)$. In practice, we have access to only the nominal dynamics (5) instead of the true dynamics to plan a trajectory $x^\star(t)$ and $u^\star(t)$. The following lemma gives the condition when $x^\star(t)$, planned using the nominal dynamics (5), is also a feasible state trajectory for the true system.

Lemma 3. *Given a desired trajectory $x^\star(t)$ and $u^\star(t)$ satisfying the nominal dynamics* (5) *with $x^\star(t) \in \mathcal{X}$, if*

$$u^\star(t) \in \mathcal{U} \ominus \mathcal{D}, \quad \forall t \geq 0, \tag{14}$$

then $x^\star(t)$ is also a feasible state trajectory for the true system (1).

Proof. Define $\bar{u}^\star(t) \triangleq u^\star(t) - d(t, x^\star(t))$. Since $u^\star(t) \in \mathcal{U} \ominus \mathcal{D}$ and $-d(t, x^\star(t)) \in \mathcal{D}$, which is due to $x^\star(t) \in \mathcal{X}$ and Assumption 1, we have $\bar{u}^\star(t) \in \mathcal{U}$. By comparing the dynamics in (1) and (5), we conclude that $x^\star(t)$ and $\bar{u}^\star(t)$ satisfy the true dynamics (1) and thus are a feasible state and input trajectory for the true system. □

Lemma 3 provides a way to verify whether a trajectory planned using the nominal dynamics is a feasible trajectory for the true system in the presence of actuator limits. In the absence of such limits, any feasible trajectory for the learned dynamics is also a feasible trajectory for the true dynamics due to the particular structure of (1) associated with the matched uncertainty assumption.

3.2. Robust Riemannian Energy Condition

Section 2 shows that, given a nominal system and a CCM for such a system, a control law can be constructed via solving a QP problem (13) with a condition to constrain the decreasing rate of the Riemannian energy, i.e., condition (12). When considering the uncertain dynamics in (1), the condition (12) becomes

$$\gamma_s^\top(1,t)M(x(t))\dot{x}(t) - \gamma_s^\top(0,t)M(x^\star(t))\dot{x}^\star(t) \leq -\lambda E(x^\star(t), x(t)), \tag{15}$$

where $\dot{x}(t) = f(x(t)) + B(x(t))(u(t) + d(x(t)))$ represents the true dynamics evaluated at $x(t)$, and $\dot{x}^\star(t) = f(x^\star) + B(x^\star)u^\star$ as defined in (5). Several observations follow immediately. First, it is clear that (15) is *not implementable* due to its dependence on the true uncertainty $d(x(t))$ through $\dot{x}(t)$. Second, if we could have access to the *pointwise value* of $d(x(t))$ at each time t, (15) will become implementable even when we do not know the exact functional representation of $d(x)$. Third, if we could estimate the pointwise value of $d(x(t))$ at each time t with a bound to quantify the estimation error, then we could derive a robust condition for (15). Specifically, assume $d(x(t))$ is estimated as $\hat{d}(t)$ at each time t with a uniform estimation error bound (EEB) δ, i.e., $\left\|\hat{d}(t) - d(x(t))\right\| \leq \delta$, $\forall t \geq 0$. Then, we could immediately get the following sufficient condition for (15):

$$\gamma_s^\top(1,t)M(x)\dot{\hat{x}}(t) - \gamma_s^\top(0,t)M(x^\star)\dot{x}^\star + \left\|\gamma_s^\top(1,t)M(x)B(x)\right\|\delta \leq -\lambda E(x^\star, x), \tag{16}$$

where

$$\dot{\hat{x}}(t) \triangleq f(x) + B(x)(u(t) + \hat{d}(t)). \tag{17}$$

Moreover, since $M(x)$ satisfies the CCM condition (9), $u(t)$ that satisfies (16) is guaranteed to exist for any $t \geq 0$, regardless of the size of δ, if the input constraint set $\mathcal{U}$ is sufficiently large. We term condition (16) the *robust Riemannian energy* (RRE) condition.

3.3. Disturbance Estimation with a Computable EEB

We now introduce a disturbance estimation scheme to estimate the pointwise value of the uncertainty $d(x)$ with a pre-computable EEB, which can be systematically improved by tuning a parameter in the estimation law. The estimation scheme is based on the piecewise-constant estimation (PWCE) law in [24], which was originally from [25]. The PWCE law consists of two elements, namely a state predictor and a piecewise-constant update law. The state predictor is defined as:

$$\dot{\hat{x}}(t) = f(x(t)) + B(x(t))u(t) + \hat{\sigma}(t) - a\tilde{x}(t), \ \hat{x}(0) = x(0), \tag{18}$$

where $\tilde{x}(t) \triangleq \hat{x}(t) - x(t)$ is the prediction error, and a is an arbitrary positive constant. The estimation, $\hat{\sigma}(t)$, is updated in a piecewise-constant way:

$$\begin{cases} \hat{\sigma}(t) = \hat{\sigma}(iT), & t \in [iT, (i+1)T), \\ \hat{\sigma}(iT) = -\dfrac{a}{e^{aT} - 1}\tilde{x}(iT), \end{cases} \tag{19}$$

where T is the estimation sampling time, and $i = 0, 1, 2, \cdots$. Finally, the pointwise value of $d(x(t))$ at time t is estimated as

$$\hat{d}(t) = B^\dagger(x(t))\hat{\sigma}(t), \tag{20}$$

where $B^\dagger(x(t))$ is the pseudoinverse of $B(x(t))$. The following lemma establishes the EEB associated with the estimation scheme in (18) and (19). The proof is similar to that in [24]. For completeness, it is given in Appendix A.

Lemma 4. *Given the dynamics (1) subject to Assumption 1, and the estimation law in (18) and (19), if $x \in \mathcal{X}$ and $u \in \mathcal{U}$ for any $t \geq 0$, the estimation error can be bounded as*

$$\left\|\hat{d}(t)-d(t,x(t))\right\| \leq \delta(t,T) \triangleq \begin{cases} b_d, \ \forall\, 0 \leq t < T, \\ \alpha(T)\max_{x\in\mathcal{X}} B^{\dagger}(x), \ \forall\, t \geq T, \end{cases} \tag{21}$$

where

$$\alpha(T) \triangleq \left(2\sqrt{n}T(L_d\phi + l_d) + (1-e^{-aT})\sqrt{n}b_d\right)\max_{x\in\mathcal{X}}\|B(x)\| + 2\sqrt{n}TL_Bb_d, \tag{22}$$

$$\phi \triangleq \max_{x\in\mathcal{X},u\in\mathcal{U}}\|f(x) + B(x)u\| + b_d\max_{x\in\mathcal{X}}\|B(x)\|, \tag{23}$$

with constants L_B, L_d and b_d from Assumption 1, and ϕ defined in (23). Moreover, $\lim_{T\to 0}\delta(t,T) = 0$, for any $t \geq T$.

Proof. See Appendix A. □

Remark 4. *Lemma 4 implies that theoretically, for $t \geq T$, the disturbance estimation after a single sampling interval can be made arbitrarily accurate by reducing T, which further indicates that the conservatism with the RRE condition can be arbitrarily reduced after a sampling interval.*

In practice, the value of T is subject to the limitations related to computational hardware and sensor noise. Additionally, using a very small T tends to introduce high frequency components in the control loop, potentially harming the robustness of the closed-loop system, e.g., against time delay. This is similar to the use of a high adaptation rate in model reference adaptive control schemes as discussed in [13]. Therefore, one should avoid the use of a very small T for the sake of robustness unless a low-pass filter is used to filter the estimated disturbance before fed into (16), as suggested by the $\mathcal{L}_1$ adaptive control theory [13].

Remark 5. *The estimation in $[0,T)$ cannot be arbitrarily accurate. This is because the estimation in $[0,T)$ depends on $\tilde{x}(0)$ according to (19). Considering that $\tilde{x}(0)$ is purely determined by the initial state of the system, $x(0)$, and the initial state of the predictor, $\hat{x}(0)$, it does not contain any information of the uncertainty. Since T is usually very small in practice, lack of a tight estimation error bound for the interval $[0,T)$ will not cause an issue from a practical point of view. Additionally, the estimation of ϕ defined in (23) could be quite conservative. Further considering the frequent use of Lipschitz continuity and inequalities related to matrix/vector norms in deriving the constant $\alpha(T)$, $\alpha(T)$ can be overly conservative. Therefore, for practical implementation, one should leverage some empirical study, e.g., performing simulations under a few user-selected functions of $d(t,x)$ and determining a bound for $\delta(t,T)$. In our experiments, we found the theoretical bound $\delta(t,T)$ computed according to (21) was usually at least 10 and could be 10^4 times more conservative.*

3.4. Exponentially Convergent Trajectory Tracking

Based on the review of contraction control in Section 2 and the discussions in Sections 3.2 and 3.3, the control law can be obtained by solving the following QP problem at each time t:

$$u(t) = \underset{k\in\mathbb{R}^m}{\operatorname{argmin}}\|k - u^\star(t)\|^2 \tag{24}$$

subject to

$$\gamma_s^\top(1,t)M(x)\dot{\hat{x}} - \gamma_s^\top(0,t)M(x^\star)\dot{x}^\star + \left\|\gamma_s^\top(1,t)M(x)B(x)\right\|\delta(t,T) \leq -\lambda E(x^\star,x), \tag{25}$$

where $\dot{\hat{x}}(t) = f(x) + B(x)(k + \hat{d}(t))$, according to (17), depends on $\hat{d}(t)$, which is from the disturbance estimation law defined by (18) to (20), $\delta(t,T)$ as defined in (21), and $\dot{x}^\star(t) = f(x^\star) + B(x^\star)u^\star$ as defined in (5). Similar to (13), problem (24) is a pointwise min-norm control problem and has an analytic solution [23]. Specifically, denoting $\phi_0(t,x^\star,x) \triangleq \gamma_s^\top(1,t)M(x)(f(x) + B(x)(u^\star(t) + \hat{d}(t)) + \left\|\gamma_s^\top(1,t)M(x)B(x)\right\|\delta(t,T) -$

$\gamma_s^\top(0,t)M(x^\star)\dot{x}^\star + \lambda E(x^\star, x)$ and $\phi_1(x^\star, x) \triangleq B^\top(x)M(x)\gamma_s(1,t)$, (25) can be written as $\phi_0(t, x^\star, x) + \phi_1^\top(x^\star, x)(k - u^\star(t)) \leq 0$, and the solution for (24) is given by

$$u(t) = k^\star = \begin{cases} u^\star(t) & \text{if } \phi_0(t,x^\star,x) \leq 0, \\ u^\star(t) - \dfrac{\phi_0(t,x^\star,x)\phi_1(x^\star,x)}{\|\phi_1(x^\star,x)\|^2} & \text{if } \phi_0(t,x^\star,x) > 0. \end{cases} \tag{26}$$

To move forward with analysis, we need to verify that when $x(t), x^\star(t) \in \mathcal{X}$, the control signal $u(t)$ resulting from solving the QP problem (24) satisfies $u(t) \in \mathcal{U}$. Deriving verifiable conditions to ensure this set bound is outside the scope of this paper and will be addressed as future work. We are now ready to state the main result of the paper.

Theorem 2. *Given an uncertain system represented by (1) satisfying Assumption 1, assume that there exists a metric $W(x)$ such that for all $x \in \mathcal{X}$, (10) holds and $\alpha_1 I \leq M(x) = W^{-1}(x) \leq \alpha_2 I$ holds for positive constants α_1 and α_2. Furthermore, suppose that a nominal trajectory $(x^\star(\cdot), u^\star(\cdot))$ planned using the nominal dynamics (5) and the initial actual states $x(0)$ satisfy (14) and*

$$\Omega(t) \triangleq \left\{ y \in \mathbb{R}^n : y \leq \|x^\star(t)\| + \sqrt{\frac{\alpha_2}{\alpha_1}}\|x(0) - x^\star(0)\| e^{-\lambda t} \right\} \subset \mathcal{X}, \tag{27}$$

for any $t \geq 0$. Then, if $u(t)$ from solving (24) satisfies $u(t) \in \mathcal{U}$ for any $t \geq 0$, the control law constructed by solving (24) ensures $x(t) \in \mathcal{X}$ for any $t \geq 0$, and furthermore, universally exponentially stabilizes the uncertain system (1) in the sense of Definition 1 with $R = \sqrt{\frac{\alpha_2}{\alpha_1}}$, i.e.,

$$\|x(t) - x^\star(t)\| \leq \sqrt{\frac{\alpha_2}{\alpha_1}}\|x(0) - x^\star(0)\| e^{-\lambda t}, \quad \forall t \geq 0. \tag{28}$$

Proof. We use contradiction to show $x(t) \in \mathcal{X}$ for all $t \geq 0$. Assume this is not true. According to (27), $x(0) \in \mathcal{X}$. Since $x(t)$ is continuous, there must exist a time τ such that

$$x(t) \in \mathcal{X}, \ \forall t \in [0, \tau^-] \text{ and } x(\tau) \notin \mathcal{X}. \tag{29}$$

Now let us consider the system evolution in $[0, \tau^-]$. Since $u(t) \in \mathcal{U}$ by assumption and $x(t) \in \mathcal{X}$ for any t in $[0, \tau^-]$, the EEB in (21) holds in $[0, \tau^-]$. As a result, the control law obtained from solving (24) ensures satisfaction of the RRE condition (16) and thus satisfaction of the Riemannian energy condition (15) for the uncertain system (1), and thereby universally exponentially stabilizes the uncertain system (1) in $[0, \tau^-]$, in the sense of Definition 1 with $R = \sqrt{\frac{\alpha_2}{\alpha_1}}$, according to Theorem 1. On the other hand, satisfaction of (14) implies that $x^\star(t)$ is a feasible state trajectory for the uncertain system (1) according to Lemma 3. Further considering Theorem 1, we have $\|x(t)\| \leq \|x^\star(t)\| + \sqrt{\frac{\alpha_2}{\alpha_1}}\|x(0) - x^\star(0)\| e^{-\lambda t}$ for any t in $[0, \tau^-]$. Due to (27), the preceding inequality indicates that $x(t)$ remains in the interior of $\mathcal{X}$ for t in $[0, \tau^-]$. This, together with the continuity of $x(t)$, immediately implies $x(\tau) \in \mathcal{X}$, which contradicts (29). Therefore, we conclude that $x(t) \in \mathcal{X}$ for all $t \geq 0$. From the development of the proof, it is clear that with the control law given by the solution of (24), the UES of the closed-loop system in the sense of Definition 1 with $R = \sqrt{\frac{\alpha_2}{\alpha_1}}$ for all $t \geq 0$ is achieved, which is mathematically represented by (28). The proof is complete. □

3.5. Discussion

Theorem 2 essentially states that under certain assumptions, the proposed controller guarantees exponential convergence of the actual state trajectory $x(t)$ to a desired one $x^\star(t)$. With the exponential guarantee, if the actual trajectory meets the desired trajectory at certain time τ, then these two trajectories will stay together afterward. While the exponential convergence guarantee is stronger than the performance guarantees provided by existing adaptive CCM-based approaches [14,15] that deal with similar settings (i.e., matched uncertainties), the proposed method requires the knowledge of the Lipschitz bound of the uncertainty $d(t,x)$ and the input matrix function $B(x)$ to be in a compact set known a priori (see Assumption 1), and the actual control inputs to stay in a compact set known a priori, which cannot be verified at this moment due to the lack of a bound on the control inputs. These requirements are not needed in [14,15].

The approach here is related to the robust control Lyapunov-based approaches [23] which provide robust stabilization around an equilibrium point (as opposed to a trajectory considered in this paper) in the presence of uncertainties.

Remark 6. *The exponential convergence guarantee stated in Theorem 2 is based on a continuous-time implementation of the controller. In practice, a controller is normally implemented on a digital processor or controller with a fixed sampling time. As a result, the property of exponential convergence may be slightly violated.*

Computational cost: As can be seen from Sections Section 2 and 3.2–3.4, computation of the control signal at each time t includes three steps: (i) updating the estimated disturbance $\hat{d}(t)$ via (18) to (20), (ii) computing the geodesic $\gamma(\cdot,t)$ connecting the actual and nominal states (see the discussion below (11)), and (iii) computing the control signal $u(t)$ via (26). The computation costs of steps (i) and (iii) are quite low as they only involve integration and algebraic calculation. In comparison, step (ii) has a relatively high computational cost as it necessitates solving a nonlinear programming (NLP) problem. However, since the NLP problem does not involve dynamic constraints, it is much easier to solve than a nonlinear model predictive control (MPC) problem [21]. Following [21], such a problem can be efficiently solved by applying a pseudospectral method.

4. Simulation Results

In this section, we illustrate the performance of our proposed tracking controller based on the RRE condition and disturbance estimation, denoted as DE-CCM, using aircraft and planar quadrotor examples. For both examples, we perform comparisons of DE-CCM with standard CCM controllers that ignore the uncertainties and adaptive CCM (Ad-CCM) controllers considering parametric uncertainties designed using the approach in [14]. All the computations and simulations were performed in Matlab R2021b.

4.1. Longitudinal Dynamics of an Aircraft

We first implement our method on the simplified pitch dynamics of an aircraft borrowed from [26]:

$$\dot{x} \triangleq \begin{bmatrix} \dot{\theta} \\ \dot{\alpha} \\ \dot{q} \end{bmatrix} = \begin{bmatrix} q \\ q - \bar{L}(\alpha) \\ -k_q q + \bar{M}(\alpha) \end{bmatrix} + \begin{bmatrix} 0 \\ 0 \\ 1 \end{bmatrix} u, \tag{30}$$

where θ, α and q are the pitch angle (in rad), angle of attack (in rad), and pitch rate (in rad/s). Here $\bar{L}(\alpha)$ and $\bar{M}(\alpha)$ are aerodynamic lift and moment, respectively. Using the flat plat theory, these two aerodynamic terms are approximated by $\bar{L}(\alpha) = 0.8\sin(2\alpha)$ and $\bar{M}(\alpha) = -l_\alpha \bar{L}(\alpha)$ with unknown parameters $k_q \in [0.1 \quad 0.8]$ and $l_\alpha \in [-3 \quad 1]$. For all the simulations, the true values are chosen to be $k_q = 0.8$, $l_\alpha = -3$, while the nominal values of these parameters used in designing all the tested controllers are limited to $k_q^{nom} = 0.2$, $l_\alpha^{nom} = -1$. As a result, the dynamics can be recast in the form of (30) with $f(x) = [q, q - \bar{L}(\alpha), -k_q^{nom} q - l_\alpha^{nom}\bar{L}(\alpha)]^\top$, $B(x) = [0,0,1]^\top$ and $d(t,x) = -(k_q - k_q^{nom})q - (l_\alpha - l_\alpha^{nom})\bar{L}(\alpha)$. The control objective is to drive the system from nominal initial states $[0,0,0]^T$ to terminal states $[180^\circ,0,0]^T$. For CCM search and trajectory planning, the following constraints are enforced: $x \in \mathcal{X} = [-10^\circ, 180^\circ] \times [-5^\circ, 40^\circ] \times [-10, 50]^\circ/\mathrm{s}$, $u \in \mathcal{U} = [-15, 15]^\circ/\mathrm{s}^2$.

We set the convergence rate λ to 1. By gridding the set of α and evaluating the constraints (9) in those grid points, we found a CCM $W(x)$ as a quadratic function of α with the SPOT toolbox [27] (to formulate the convex optimization problem) and Mosek solver [28].

Additionally, the constants α_1 and α_2 in (28) such that $\alpha_1 I \leq M(x) = W^{-1}(x) \leq \alpha_2 I$ for all $x \in \mathcal{X}$ were found to be $\alpha_1 = 0.1$ and $\alpha_2 = 396.5$. We planned a nominal trajectory ($x^\star(\cdot)$, $u^\star(\cdot)$) using OptimTraj [29,30], to drive the system from the initial states $[0, 0, 0]^T$ to the terminal states $[180°, 0, 0]^T$, while minimizing the task completion time (T_a) and energy consumption characterized by the cost function $J = \int_0^{T_a} u(t)^2 dt + 5T_a$. For simulation, OPTI [31] and Matlab `fmincon` solvers were used to solve the geodesic optimization problem (see Section 2). The initial states of the actual system were chosen to be $[5°, 5°, 0]^T$, slightly deviated from that planned ones to better illustrate the tracking performance. We implemented our proposed DE-CCM, Ad-CCM from [14] and a standard CCM which neglects all the uncertainty. For Ad-CCM design, the adaptation gain was chosen to be $\text{diag}([10^3, 10^3])$ to achieve a relatively good tracking result, while further increasing it did not help much with the tracking performance. The design procedure for Ad-CCM in [14] requires a parametric structure for the uncertainty, which is given by

$$d(t, x) = \underbrace{\begin{bmatrix} q & \bar{L}(\alpha) \end{bmatrix}}_{\Delta(t,x)} \underbrace{\begin{bmatrix} k_q^{nom} - k_q \\ l_\alpha^{nom} - l_\alpha \end{bmatrix}}_{\theta} = \begin{bmatrix} q & 0.8\sin(2\alpha) \end{bmatrix} \begin{bmatrix} -0.6 \\ 2 \end{bmatrix}, \quad (31)$$

where $\Delta(t, x)$ is the known base function and θ is the unknown parameter vector to be estimated by the adaptive law proposed in [14]. The control signals under all three controllers were updated at 200 Hz.

It is easy to notice from Assumption 1 that $L_B = 0$ and $l_d = 0$ since the input matrix is constant, and the uncertainty is time-invariant. We can also verify that the disturbance is bounded by $b_d = 2.12$ and has a Lipschitz constant $L_d = 3.80$. By gridding the space $\mathcal{X}$ and making use of the control input bound, the system derivative can also be bounded by a constant $\phi = 2.11$. According to (21), if we want to achieve an EEB $\delta(t, T) = 0.05$ for all $t \geq T$, the maximum value for the estimation sample time T is 7.76×10^{-4} s. However, as mentioned in Remark 5, the way to compute the EEB is quite conservative. In the simulations we found that $T_s = 0.005$ was more than enough to ensure the EEB and therefore used $T_s = 0.005$ for implementing the DE-CCM controller.

As shown in Figures 2 and 3, due to ignoring the uncertainties, CCM yielded a large tracking error between 2 and 6 s. The state trajectories under Ad-CCM had some oscillations, which lasted roughly up to 8 s. All three states yielded by DE-CCM achieve good tracking performance without large deviations from the planned trajectories, unlike the performance yielded by Ad-CCM and CCM. From Figure 3 we notice that the tracking error represented by $x - x^\star$ under DE-CCM monotonically decreases and achieves the smallest steady-state error. The small non-zero tracking error at the end under DE-CCM, which is inconsistent with the performance guarantee in (25), is due to the limited control update frequency, while the performance guarantee in Lemma 4 holds under continuous update of the control signal, i.e., corresponding to an infinitely high update frequency. Table 1 shows the mean squared error (MSE) for state trajectory tracking defined by

$$\text{MSE} = \frac{1}{N} \sum_{i=1}^{N} \|x(t_i) - x^\star(t_i)\|^2, \quad (32)$$

where N is the number of data points, under DE-CCM, Ad-CCM and CCM. We observe that DE-CCM outperforms CCM and Ad-CCM in terms of MSE by 54% and 2%, respectively.

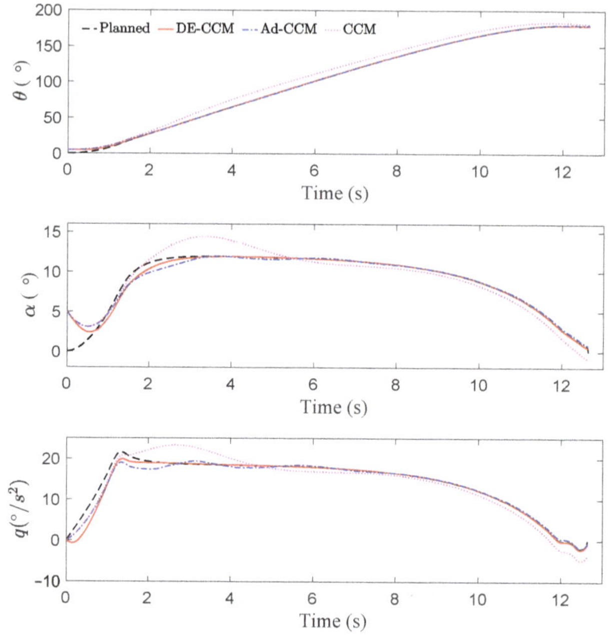

Figure 2. Trajectory tracking performance of different controllers.

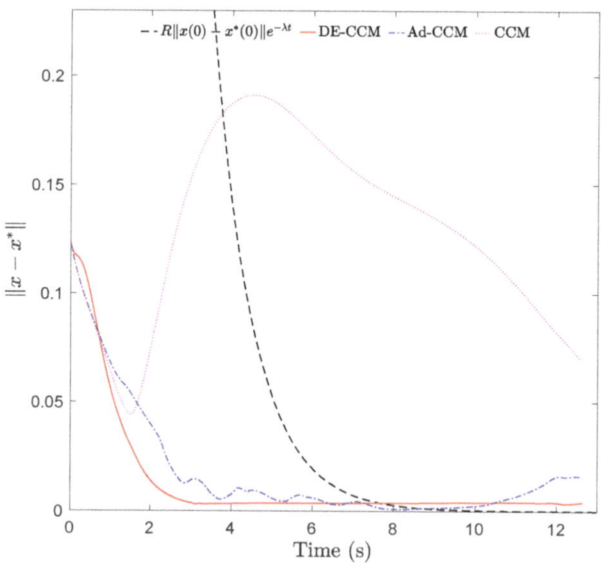

Figure 3. Tracking error under different controllers.

Table 1. MSE for state trajectory tracking for the aircraft example.

	CCM	Ad-CCM	DE-CCM
MSE	2.994×10^{-3}	1.397×10^{-3}	1.369×10^{-3}

From Figure 4, we observe that the input of DE-CCM is smoother than Ad-CCM. The small oscillations between 2 s to 8 s in DE-CCM input are due to the finite tolerance in the geodesic optimization. Decreasing the tolerance and the sample time will reduce the oscillations but request more iterations (and thus more time) to compute the control signal at each time step.

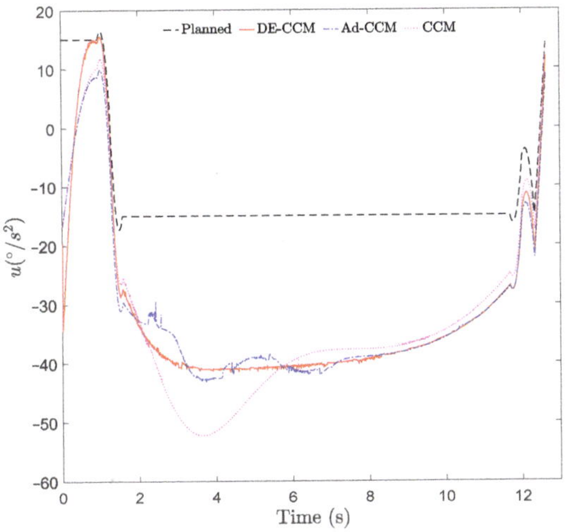

Figure 4. Control inputs yielded by different controllers.

4.2. Planar Quadrotor

A planar quadrotor system is borrowed from [16]. The state vector is defined as $x = [p_x, p_z, \phi, v_x, v_z, \dot{\phi}]^\top$, where p_x and p_z are the positions in x and z directions, respectively, v_x and v_z are the slip velocity (lateral) and the velocity along the thrust axis in the body frame of the vehicle, ϕ is the angle between the x direction of the body frame and the x direction of the inertia frame. The input vector $u = [u_1, u_2]$ contains the thrust force produced by each of the two propellers. The dynamics of the vehicle are given by

$$\dot{x} = \begin{bmatrix} \dot{p}_x \\ \dot{p}_z \\ \dot{\phi} \\ \dot{v}_x \\ \dot{v}_z \\ \ddot{\phi} \end{bmatrix} = \begin{bmatrix} v_x \cos(\phi) - v_z \sin(\phi) \\ v_x \sin(\phi) + v_z \cos(\phi) \\ \dot{\phi} \\ v_z \dot{\phi} - g \sin(\phi) \\ -v_x \dot{\phi} - g \cos(\phi) \\ 0 \end{bmatrix} + \begin{bmatrix} 0 & 0 \\ 0 & 0 \\ 0 & 0 \\ 0 & 0 \\ \frac{1}{m} & \frac{1}{m} \\ \frac{l}{J} & -\frac{l}{J} \end{bmatrix} (u + d(t,x)),$$

where m and J denote the mass and moment of inertia about the out-of-plane axis, and l is the distance between each of the propellers and the vehicle center, and $d(t,x)$ denotes the unknown disturbances exerted on the propellers. The parameters were set as $m = 0.486$ kg, $J = 0.00383$ Kg m^2, and $l = 0.25$ m. The uncertainty $d(t,x)$ was set to be $d(t,x) = 0.15(v_x^2 + v_z^2)[-1 + 0.3\sin(2t); -1 + 0.3\cos(2t)]$. We imposed the following constraints: $x \in \mathcal{X} \triangleq [0,10] \times [0,10] \times [-\frac{\pi}{3}, \frac{\pi}{3}] \times [-2,2] \times [-1,-1] \times [-\frac{\pi}{3}, \frac{\pi}{3}]$, $u \in \mathcal{U} \triangleq [0, \frac{3}{2}mg] \times [0, \frac{3}{2}mg]$.

When searching for CCM, we parameterized the CCM W by ϕ and v_x and imposed the constraint $W \geq 0.01I$. The convergence rate λ was chosen to be 0.8. More details about synthesizing the CCM can be found in [17]. For estimating the disturbance using (18) to (20), we set $a = 10$. It is easy to verify that $L_d = 1.23$, $l_d = 0.64$, $b_d = 1.38$, and $L_B = 0$ (due to the fact that B is constant) satisfy (2). By gridding the space $\mathcal{X}$, the constant ϕ in (23) can be determined as $\phi = 584.6$. According to (21), if we want to achieve an EEB $\delta(t,T) = 0.1$ for all $t \geq T$, then the estimation sampling time needs to satisfy $T \leq 6.42 \times 10^{-7}$ s. However, as noted in Remark 5, the EEB computed according to (21) could be overly conservative.

In the simulations, we found that the estimation sampling time of 0.002 s was more than enough to ensure the desired EEB and therefore simply set $T = 0.002$ s.

We consider the task of navigation from $(2, 0)$ to $(8, 8)$ while avoiding three obstacles depicted by black circles in Figure 5. A nominal trajectory $(x^\star(\cdot), u^\star(\cdot))$ was generated using OptimTraj [29] to minimize the cost $J = \int_0^{T_a} \|u(t)\|^2 dt + 5T_a$, where T_a is the arrival time. OPTI [31] and Matlab `fmincon` solvers were used to solve the geodesic optimization problem (see Section 2). The actual start point was set to be $(0, 0)$, which was different from the planned start point, to reveal the trajectory convergence pattern.

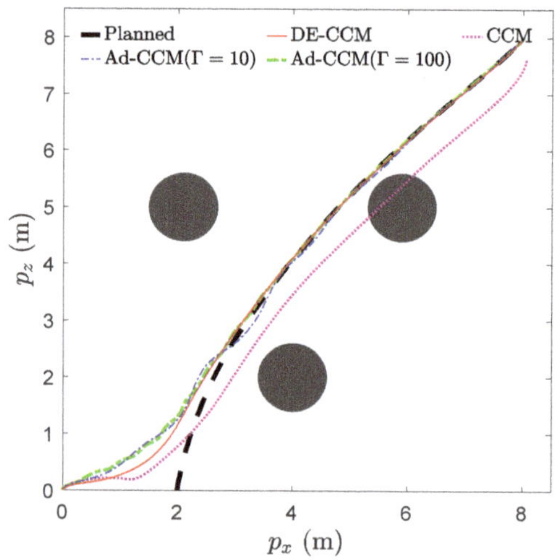

Figure 5. Trajectory tracking performance of different controllers.

For comparison, we also designed a standard CCM controller by completely ignoring the uncertainty and two adaptive CCM (Ad-CCM) controllers following the approach in [14]. To apply the approach in [14] which can only handle parametric uncertainties, we parameterized the uncertainty as

$$d(t, x) = \underbrace{\begin{bmatrix} v_x^2(-1 + 0.3\sin(2t)) & v_z^2(-1 + 0.3\sin(2t)) \\ v_x^2(-1 + 0.3\cos(2t)) & v_z^2(-1 + 0.3\cos(2t)) \end{bmatrix}}_{\Delta(t,x)} \underbrace{\begin{bmatrix} 0.15 \\ 0.15 \end{bmatrix}}_{\theta}, \tag{33}$$

where $\Delta(t, x)$ is the basis function that is assumed to be known, and θ is th vector of unknown parameters. With the parametric structure (33), we designed two adaptive CCM controllers using $\Gamma = 10$ and $\Gamma = 100$, respectively, where Γ denotes the adaptive gain. Figure 5 shows the planned and actual trajectories under the CCM, Ad-CCM, and our proposed controller based on the RRE condition and disturbance estimation, denoted as DE-CCM, while Figures 6 and 7 show the control inputs and Riemannian energy. One can see that the actual trajectories yielded by the CCM controller deviated quite a lot from the planned ones and collided with one obstacle. On the other hand, the actual trajectories yielded by the DE-CCM controller converged to the desired trajectory as expected and almost overlapped with it afterward. In fact, the slight deviations of actual trajectories from the desired ones under the DE-CCM controller were due to the finite step size associated with the ODE solver used for the simulations (see Remark 6). Table 2 shows the MSE for state trajectory tracking defined in (32). We observe that DE-CCM outperforms CCM and Ad-CCM in terms of MSE by 46% and 14%, respectively.

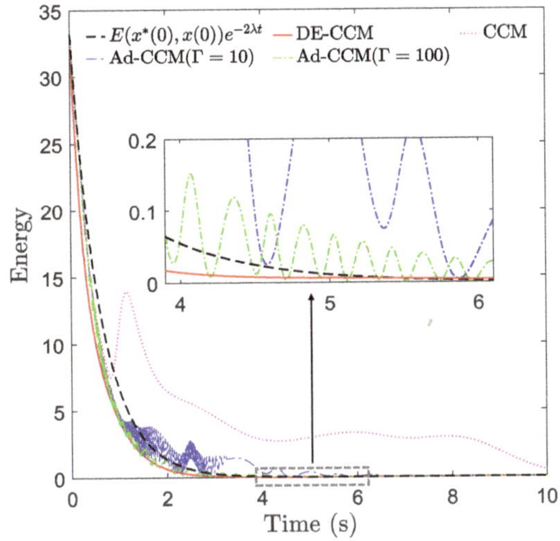

Figure 6. Riemannian energy under different controllers. $E_0 \triangleq E(x^\star(0), x(0))$.

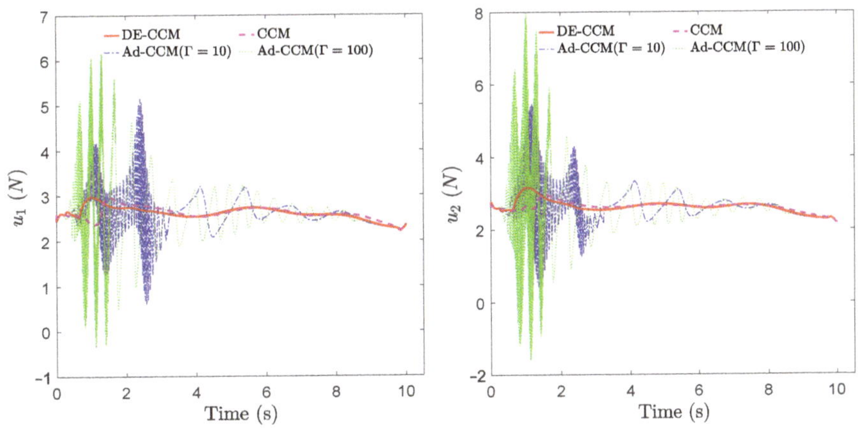

Figure 7. Control inputs yielded by different controllers.

Table 2. MSE for state trajectory tracking for the planar quadrotor example. For Ad-CCM, only the result for $\Gamma = 100$, which corresponds to better tracking performance compared to $\Gamma = 10$, is included.

	CCM	Ad-CCM ($\Gamma = 100$)	DE-CCM
MSE	1.175	0.738	0.634

From Figure 6, one can see that the magnitude of $E(x^\star, x)$ under the RD-CCM controller decreased exponentially, and the magnitude was bounded by the curve $E(x^\star(0), x(0))e^{-2\lambda t}$ from above except at the very end when the energy is close to zero. In comparison, Ad-CCM with $\Gamma = 100$ yielded similar tracking performance to DE-CCM, while the tracking performance of Ad-CCM with $\Gamma = 10$ was relatively worse and characterized by larger oscillations. Additionally, from Figure 7, one can see that the control inputs generated by both of the Ad-CCM controllers have high-frequency oscillations before 3 s, which is undesired for practical deployment. Finally, Figure 8 shows the actual and estimated disturbances as well as the estimation error. One can see that the estimated disturbance is quite close to the actual one for both channels, and the EEB of 0.1 is respected throughout the simulation.

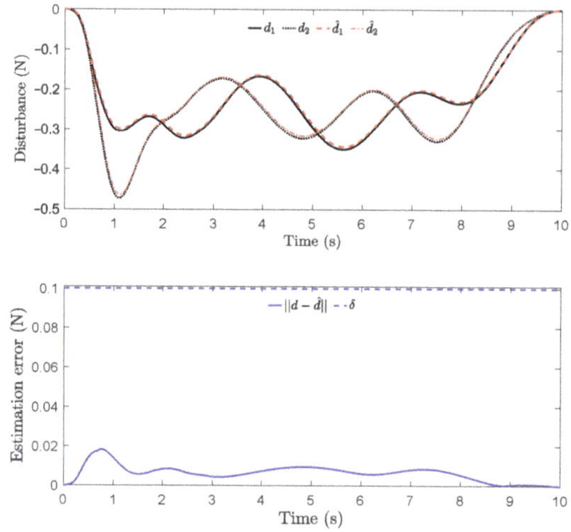

Figure 8. Actual and estimated disturbances (top) and the estimation error (bottom). Note that d_i and $\hat{d}_i$ $(i = 1, 2)$ represent the ith element of d (actual disturbance) and $\hat{d}$ (estimated disturbance), respectively. The blue dashed line in the bottom plot denotes the EEB used in computing the control inputs.

5. Conclusions

This paper presents a robust trajectory tracking controller with exponential convergence for uncertain nonlinear systems based on control contraction metrics (CCM) and disturbance estimation. The controller uses a disturbance estimator to estimate the pointwise value of the uncertainty with a pre-computable estimation error bound (EEB). The estimated disturbance and the EEB are then incorporated into a robust Riemannian energy condition, which guarantees exponential convergence of actual trajectories to desired ones. The efficacy of the proposed controller is validated in simulations. In particular, the proposed controller outperforms an existing adaptive CCM controller in terms of tracking performance by 2% for the aircraft example and 14% for the planar quadrotor example, while not needing to know the basis functions to parameterize the uncertainties that are needed by the adaptive CCM controller.

This paper considers only matched uncertainties, which are added to the system through the same channels as control inputs. In the future, we would like to address unmatched uncertainties that widely exist in practical systems. Additionally, we would like to experimentally validate the proposed controller on real hardware.

Author Contributions: The individual contributions of the authors are as follows. Conceptualization, methodology and formal analysis, P.Z.; software and investigation, Z.G. and P.Z.; validation and visualization, Z.G.; supervision, N.H.; writing—original draft preparation, P.Z.; writing—review and editing, Z.G. and N.H.; project administration, P.Z.; funding acquisition, N.H. All authors have read and agreed to the published version of the manuscript.

Funding: This research was funded in part by AFOSR, in part by NASA, and in part by NSF under the RI grant #2133656 and NRI grant #1830639.

Institutional Review Board Statement: Not available.

Informed Consent Statement: Not applicable.

Data Availability Statement: Not available.

Conflicts of Interest: The authors declare no conflict of interest.

Appendix A

Proof of Lemma 4: Hereafter, we use the notations $\mathbb{Z}_i$ and $\mathbb{Z}_1^n$ to denote the integer sets $\{i, i+1, i+2, \cdots\}$ and $\{1, 2, \cdots, n\}$, respectively. Additionally, for notation brevity, we define $\sigma(t) \triangleq B(x(t))d(x(t))$. From (1) and (18), the prediction error dynamics are obtained as

$$\dot{\tilde{x}}(t) = -a\tilde{x}(t) + \hat{\sigma}(t) - \sigma(t), \quad \tilde{x}(0) = 0. \tag{A1}$$

Note that $\hat{\sigma}(t) = 0$ (and thus $\hat{d}(t) = 0$ due to (20)) for any $t \in [0, T)$ according to (19). Further considering the bound on $d(t, x)$ in (4), we have

$$\left\|\hat{d}(t) - d(t, x(t))\right\| \leq b_d, \quad \forall t \in [0, T). \tag{A2}$$

We next derive the bound on $\|\hat{\sigma}(t) - \sigma(t)\|$ for $t \geq T$. For any $t \in [iT, (i+1)T)$ ($i \in \mathbb{Z}_0$), we have

$$\tilde{x}(t) = e^{-a(t-iT)}\tilde{x}(iT) + \int_{iT}^{t} e^{-a(t-\tau)}(\hat{\sigma}(\tau) - \sigma(\tau))d\tau.$$

Since $\tilde{x}(t)$ is continuous, the preceding equation implies

$$\begin{aligned}
\tilde{x}((i+1)T) &= e^{-aT}\tilde{x}(iT) + \int_{iT}^{(i+1)T} e^{-a((i+1)T-\tau)}d\tau\hat{\sigma}(iT) - \int_{iT}^{(i+1)T} e^{-a((i+1)T-\tau)}\sigma(\tau)d\tau \\
&= e^{-aT}\tilde{x}(iT) + \frac{1-e^{-aT}}{a}\hat{\sigma}(iT) - \int_{iT}^{(i+1)T} e^{-a((i+1)T-\tau)}\sigma(\tau)d\tau \\
&= -\int_{iT}^{(i+1)T} e^{-a((i+1)T-\tau)}\sigma(\tau)d\tau,
\end{aligned} \tag{A3}$$

where the first and last equalities are due to the estimation law (19).

Since $x(t)$ is continuous, $\sigma(t)$ $(= B(x)d(x))$ is also continuous given Assumption 1. Furthermore, considering that $e^{-a((i+1)T-\tau)}$ is always positive, we can apply the first mean value theorem in an element-wise manner (Note that the mean value theorem for definite integrals only holds for scalar valued functions) to (A3), which leads to

$$\tilde{x}((i+1)T) = -\int_{iT}^{(i+1)T} e^{-a((i+1)T-\tau)}d\tau[\sigma_j(\tau_j^*)] = -\frac{1}{a}(1-e^{-aT})[\sigma_j(\tau_j^*)], \tag{A4}$$

for some $\tau_j^* \in (iT, (i+1)T)$ with $j \in \mathbb{Z}_1^n$ and $i \in \mathbb{Z}_0$, where $\sigma_j(t)$ is the j-th element of $\sigma(t)$, and

$$[\sigma_j(\tau_j^*)] \triangleq [\sigma_1(\tau_1^*), \cdots, \sigma_n(\tau_n^*)]^\top.$$

The estimation law (19) indicates that for any t in $[(i+1)T, (i+2)T)$, we have $\hat{\sigma}(t) = -\frac{a}{e^{aT}-1}\tilde{x}((i+1)T)$. The preceding equality and (A4) imply that for any t in $[(i+1)T, (i+2)T)$ with $i \in \mathbb{Z}_0$, there exist $\tau_j^* \in (iT, (i+1)T)$ ($j \in \mathbb{Z}_1^n$) such that

$$\hat{\sigma}(t) = e^{-aT}[\sigma_j(\tau_j^*)]. \tag{A5}$$

Note that

$$\left\|\sigma(t) - [\sigma_j(\tau_j^*)]\right\| \leq \sqrt{n}\left\|\sigma(t) - [\sigma_j(\tau_j^*)]\right\|_\infty = \sqrt{n}\left|\sigma_{\bar{j}_t}(t) - \sigma_{\bar{j}_t}(\tau_{\bar{j}_t}^*)\right| \leq \sqrt{n}\left\|\sigma(t) - \sigma(\tau_{\bar{j}_t}^*)\right\|, \tag{A6}$$

where $\bar{j}_t = \arg\max_{j \in \mathbb{Z}_1^n} \left|\sigma_j(t) - \sigma_j(\tau_j^*)\right|$. Similarly,

$$\left\|[\sigma_j(\tau_j^*)]\right\| \leq \sqrt{n}\left\|[\sigma_j(\tau_j^*)]\right\|_\infty = \sqrt{n}\left|\sigma_{\hat{j}}(\tau_{\hat{j}}^*)\right| \leq \sqrt{n}\left\|\sigma(\tau_{\hat{j}}^*)\right\| \leq \sqrt{n} b_d \max_{x \in \mathcal{X}} \|B(x)\|, \tag{A7}$$

where $\hat{j} = \arg\max_{j \in \mathbb{Z}_1^n} \left|\sigma_j(\tau_j^*)\right|$, and the last inequality is due to the fact $\|B(x)d(t,x)\| \leq \|B(x)\|\|d(t,x)\|$ and (4). Therefore, for any $t \in [(i+1)T, (i+2)T)$ $(i \in \mathbb{Z}_0)$, we have

$$\begin{aligned}\|\sigma(t) - \hat{\sigma}(t)\| = \left\|\sigma(t) - e^{-aT}[\sigma_j(\tau_j^*)]\right\| &\leq \left\|\sigma(t) - [\sigma_j(\tau_j^*)]\right\| + (1 - e^{-aT})\left\|[\sigma_j(\tau_j^*)]\right\| \\ &\leq \sqrt{n}\left\|\sigma(t) - \sigma(\tau_{\bar{j}_t}^*)\right\| + (1 - e^{-aT})\sqrt{n} b_d \max_{x \in \mathcal{X}} \|B(x)\|,\end{aligned} \tag{A8}$$

for some $\tau_{\bar{j}_t}^* \in (iT, (i+1)T)$, where the equality is due to (A5), and the last inequality is due to (A6) and (A7). The dynamics in (1) indicates that

$$\|\dot{x}\| \leq \|f(x) + B(x)u\| + \|B(x)\|\|d(t,x)\| \leq \phi, \tag{A9}$$

where ϕ is defined in (23). As a result, the inequality (A9) implies that

$$\left\|x(t) - x(\tau_{\bar{j}_t}^*)\right\| \leq \int_{\tau_{\bar{j}_t}^*}^t \|\dot{x}(\tau)\| \, d\tau \leq \int_{\tau_{\bar{j}_t}^*}^t \phi d\tau = \phi(t - \tau_{\bar{j}_t}^*) \leq 2\phi T, \tag{A10}$$

where the last inequality is due to the fact that

$$t \in [(i+1)T, (i+2)T) \text{ and } \tau_{\bar{j}_t}^* \in (iT, (i+1)T) \tag{A11}$$

Therefore, we have

$$\begin{aligned}&\left\|\sigma(t) - \sigma(\tau_{\bar{j}_t}^*)\right\| = \left\|B(x(t))\left(d(t, x(t)) - d(\tau_{\bar{j}_t}^*, x(\tau_{\bar{j}_t}^*))\right) + \left(B(x(t)) - B(x(\tau_{\bar{j}_t}^*))\right) d(\tau_{\bar{j}_t}^*, x(\tau_{\bar{j}_t}^*))\right\| \\ &\leq \|B(x(t))\|\left\|d(t, x(t)) - d(\tau_{\bar{j}_t}^*, x(\tau_{\bar{j}_t}^*))\right\| + \left\|B(x(t)) - B(x(\tau_{\bar{j}_t}^*))\right\|\left\|d(\tau_{\bar{j}_t}^*, x(\tau_{\bar{j}_t}^*))\right\| \\ &\leq \left(L_d\left\|x(t) - x(\tau_{\bar{j}_t}^*)\right\| + l_d(t - \tau_{\bar{j}_t}^*)\right)\max_{x \in \mathcal{X}}\|B(x)\| + L_B\left\|x(t) - x(\tau_{\bar{j}_t}^*)\right\| b_d \\ &\leq 2T\left((L_d\phi + l_d)\max_{x \in \mathcal{X}}\|B(x)\| + L_B b_d\right),\end{aligned} \tag{A12}$$

where the second inequality is due to Assumption 1 and the last inequality is due to (A10) and (A11). Finally, plugging (A12) into (A8) leads to

$$\begin{aligned}\|\sigma(t) - \hat{\sigma}(t)\| &\leq 2\sqrt{n}T\left((L_d\phi + l_d)\max_{x \in \mathcal{X}}\|B(x)\| + L_B b_d\right) + (1 - e^{-aT})\sqrt{n} b_d \max_{x \in \mathcal{X}}\|B(x)\| \\ &= \left(2\sqrt{n}T(L_d\phi + l_d) + (1 - e^{-aT})\sqrt{n} b_d\right)\max_{x \in \mathcal{X}}\|B(x)\| + 2\sqrt{n}T L_B b_d \qquad \text{(A13)} \\ &= \alpha(T), \qquad \text{(A14)}\end{aligned}$$

for any $t \geq T$. From (A2) and (A14) and the relation between $\hat{\sigma}(t)$ and $\hat{d}(t)$ in (20), we arrive at (21). Considering Assumption 1 and the assumption that $\mathcal{X}$ and $\mathcal{U}$ are compact, the constants involved in the definition of $\alpha(T)$ in (22) are all finite. As a result, we have $\lim_{T \to 0} \alpha(T) = 0$, which further indicates that $\lim_{T \to 0} \delta(t, T) = 0$, for any $t \geq T$. The proof is complete. □

References

1. Manchester, I.R.; Slotine, J.J.E. Control contraction metrics: Convex and intrinsic criteria for nonlinear feedback design. *IEEE Trans. Autom. Control* **2017**, *62*, 3046–3053. [CrossRef]
2. Lohmiller, W.; Slotine, J.J.E. On contraction analysis for non-linear systems. *Automatica* **1998**, *34*, 683–696. [CrossRef]
3. Manchester, I.R.; Tang, J.Z.; Slotine, J.J.E. Unifying robot trajectory tracking with control contraction metrics. In *Robotics Research*; Springer: Cham, Switzerland, 2018; pp. 403–418.
4. Tedrake, R.; Manchester, I.R.; Tobenkin, M.; Roberts, J.W. LQR-trees: Feedback motion planning via sums-of-squares verification. *Int. J. Robot. Res.* **2010**, *29*, 1038–1052. [CrossRef]
5. Doyle, J.; Glover, K.; Khargonekar, P.; Francis, B. State-space solutions to standard H_2 and H_∞ control problems. *IEEE Trans. Autom. Control* **1989**, *34*, 831–847. [CrossRef]
6. Packard, A.; Doyle, J. The complex structured singular value. *Automatica* **1993**, *29*, 71–109. [CrossRef]
7. Mayne, D.Q.; Seron, M.M.; Raković, S. Robust model predictive control of constrained linear systems with bounded disturbances. *Automatica* **2005**, *41*, 219–224. [CrossRef]
8. Mayne, D.Q. Model predictive control: Recent developments and future promise. *Automatica* **2014**, *50*, 2967–2986. [CrossRef]
9. Han, J. From PID to active disturbance rejection control. *IEEE Trans. Ind. Electron.* **2009**, *56*, 900–906. [CrossRef]
10. Chen, W.H.; Yang, J.; Guo, L.; Li, S. Disturbance-observer-based control and related methods—An overview. *IEEE Trans. Ind. Electron.* **2015**, *63*, 1083–1095. [CrossRef]
11. Li, S.; Yang, J.; Chen, W.; Chen, X. Generalized extended state observer based control for systems With mismatched uncertainties. *IEEE Trans. Ind. Electron.* **2012**, *59*, 4792–4802. [CrossRef]
12. Ioannou, P.A.; Sun, J. *Robust Adaptive Control*; Dover Publications, Inc.: Mineola, NY, USA, 2012.
13. Hovakimyan, N.; Cao, C. *$\mathcal{L}_1$ Adaptive Control Theory: Guaranteed Robustness with Fast Adaptation*; Society for Industrial and Applied Mathematics: Philadelphia, PA, USA, 2010.
14. Lopez, B.T.; Slotine, J.J.E. Adaptive nonlinear control with contraction metrics. *IEEE Control Syst. Lett.* **2020**, *5*, 205–210. [CrossRef]
15. Lakshmanan, A.; Gahlawat, A.; Hovakimyan, N. Safe feedback motion planning: A contraction theory and $\mathcal{L}_1$-adaptive control based approach. In Proceedings of the 59th IEEE Conference on Decision and Control (CDC), Jeju Island, Korea, 14–18 December 2020; pp. 1578–1583.
16. Singh, S.; Landry, B.; Majumdar, A.; Slotine, J.J.; Pavone, M. Robust feedback motion planning via contraction theory. *Int. J. Robot. Res.* 2019, *under review*.
17. Zhao, P.; Lakshmanan, A.; Ackerman, K.; Gahlawat, A.; Pavone, M.; Hovakimyan, N. Tube-certified trajectory tracking for nonlinear systems with robust control contraction metrics. *IEEE Robot. Autom. Lett.* **2022**, *7*, 5528–5535. [CrossRef]
18. Manchester, I.R.; Slotine, J.J.E. Robust control contraction metrics: A convex approach to nonlinear state-feedback H_∞ control. *IEEE Control Syst. Lett.* **2018**, *2*, 333–338. [CrossRef]
19. Tsukamoto, H.; Chung, S.J. Robust controller design for stochastic nonlinear systems via convex optimization. *IEEE Trans. Autom. Control* **2020**, *66*, 4731–4746. [CrossRef]
20. Zhao, P.; Guo, Z.; Cheng, Y.; Gahlawat, A.; Kang, H.; Hovakimyan, N. Guaranteed nonlinear tracking control in the presence of DNN-learned dynamics with contraction metrics and disturbance estimation. *IEEE Conf. Decis. Control.* 2022, *under review*.
21. Leung, K.; Manchester, I.R. Nonlinear stabilization via control contraction metrics: A pseudospectral approach for computing geodesics. In Proceedings of the American Control Conference, Seattle, WA, USA, 24–26 May 2017; pp. 1284–1289.
22. Do Carmo, M.P.; Flaherty Francis, J. *Riemannian Geometry*; Springer: Boston, MA, USA, 1992.
23. Freeman, R.; Kokotovic, P.V. *Robust Nonlinear Control Design: State-Space and Lyapunov Techniques*; Springer Science & Business Media: Berlin, Germany, 2008.
24. Zhao, P.; Mao, Y.; Tao, C.; Hovakimyan, N.; Wang, X. Adaptive robust quadratic programs using control Lyapunov and barrier functions. In Proceedings of the 59th IEEE Conference on Decision and Control, Jeju Island, Korea, 14–18 December 2020; pp. 3353–3358.
25. Cao, C.; Hovakimyan, N. $\mathcal{L}_1$ adaptive output feedback controller for non strictly positive real reference systems with applications to aerospace examples. In Proceedings of the AIAA Guidance, Navigation and Control Conference and Exhibit, Honolulu, HI, USA, 18–21 August 2008; p. 7288.
26. Lopez, B.T.; Slotine, J.J.E.; How, J.P. Robust Adaptive Control Barrier Functions: An Adaptive and Data-Driven Approach to Safety. *IEEE Control Syst. Lett.* **2021**, *5*, 1031–1036. [CrossRef]
27. Megretski, A. Systems Polynomial Optimization Tools (SPOT). 2010. Available online: https://github.com/spot-toolbox/spotless (accessed on 1 September 2021).
28. Andersen, E.D.; Andersen, K.D. The MOSEK interior point optimizer for linear programming: An implementation of the homogeneous algorithm. In *High Performance Optimization*; Springer: New York, NY, USA, 2000; pp. 197–232.
29. Kelly, M. An introduction to trajectory optimization: How to do your own direct collocation. *SIAM Rev.* **2017**, *59*, 849–904. [CrossRef]
30. Kelly, M.P. OptimTraj User's Guide, Version 1.5. 2016 . Available online: https://github.com/MatthewPeterKelly/OptimTraj (accessed on 1 September 2021).
31. Currie, J.; Wilson, D.I. OPTI: Lowering the barrier between open source optimizers and the industrial MATLAB user. In Proceedings of the Foundations of Computer-Aided Process Operations, Savannah, GA, USA, 8–11 January 2012.

MDPI
St. Alban-Anlage 66
4052 Basel
Switzerland
Tel. +41 61 683 77 34
Fax +41 61 302 89 18
www.mdpi.com

Sensors Editorial Office
E-mail: sensors@mdpi.com
www.mdpi.com/journal/sensors

www.ingramcontent.com/pod-product-compliance
Lightning Source LLC
Chambersburg PA
CBHW061607220926
51850CB00005B/440
* 9 7 8 3 0 3 6 5 6 3 2 8 2 *